Springer Series in
MATERIALS PROCESSING

Springer
Berlin
Heidelberg
New York
Barcelona
Hong Kong
London
Milan
Paris
Tokyo

Physics and Astronomy ONLINE LIBRARY
http://www.springer.de/phys/

Springer Series in

MATERIALS PROCESSING

Series Editors: H. Warlimont E. Weber W. Michaeli

This series is focused on the science and application of materials processing as an essential part of progress in the materials field. It addresses researchers and process engineers alike. The scope of the series includes all classes of materials – metals, inorganic non-metallic materials, polymers and composites – in the form of bulk materials, thin films and layered structures as well as micro- and nanostructured forms. All aspects of materials processing from fundamental understanding to innovative strategies and methods of practical process implementation and control are covered. It is the aim of the series to provide comprehensive information on the science and application of leading-edge and well-established processing technologies.

Low-Pressure Synthetic Diamond. Manufacturing and Applications
Editors: B. Dischler and C. Wild

Supercritical Fluids Molecular Interactions, Physical Properties, and New Applications
Editors: Y. Arai, T. Sako, and Y. Takebayashi

Purification Process and Characterization of Ultra High Purity Metals. Application of Basic Science to Metallurgical Processing
Editors: Y. Waseda and M. Isshiki

Series homepage – http://www.springer.de/phys/books/ssmp/

Y. Arai   T. Sako   Y. Takebayashi (Eds.)

Supercritical Fluids

Molecular Interactions, Physical Properties, and New Applications

With 301 Figures and 37 Tables

Springer

Professor Yasuhiko Arai
Kyushu University, Graduate School of Engineering
Hakozaki 6-10-1, Higashi-ku, 812-0053 Fukuoka, Japan
arai@chem-eng.kyushu-u.ac.jp

Professor Takeshi Sako
Shizuoka University, Faculty of Engineering
Johoku 3-5-1, Hamamatsu-shi, 432-8561 Shizuoka, Japan
ttsako@ipc.shizuoka.ac.jp

Dr. Yoshihiro Takebayashi
National Institute of Advanced Industrial Science and Technology
Higashi 1-1, Tsukuba-shi, 305-8565 Ibaraki, Japan
chikulin@ni.aist.go.jp

Series Editors:

Professor Dr. H. Warlimont
Institut für Festkörper- und Werkstofforschung e.V.,
Helmholtzstrasse 20, 01069 Dresden, Germany

Professor Dr. E. Weber
Materials Science and Mineral Engineering, University of California
587 Evans Hall, Berkeley, CA 94720-1760, USA

Professor Dr.-Ing. Walter Michaeli
Institut für Kunststoffverarbeitung, RWTH Aachen
Pontstraße 49, 52056 Aachen, Germany

ISSN 1434-9795
ISBN 3-540-41248-4 Springer-Verlag Berlin Heidelberg New York

Library of Congress Cataloging -in-Publication-Data
Supercritical fluid science and technology / Y. Arai, T. Sako, Y. Takebayashi (eds.). p. cm. – (Springer series in materials processing, ISSN 1434-9795) Includes bibliographical references and index.
ISBN 3540412484 (alk. paper) 1. Supercritical fluids. I. Arai, y. (Yasuhiko) II. Sako, T. (Takeshi) III. Takebayashi, Y. (Yoshihiro) IV. Series. TP156.E8.S837 2002 660'.28423–dc21 00-049717

Springer-Verlag Berlin Heidelberg New York
a member of BertelsmannSpringer Science+Business Media GmbH

http://www.springer.de

Printed in Germany

Typesetting by the editors
Cover concept: eStudio Calamar Steinen
Cover production: *design & production* GmbH, Heidelberg

Printed on acid-free paper SPIN: 10784347 57/3141/mf – 5 4 3 2 1 0

Preface

Scientists have been aware of the novel characteristics of a dense gas or supercritical fluid as a solvent for more than a century. More recently, supercritical fluids or substances in a dense gas state have been recognized to have many favorable characteristics for use in the chemical industry. Research into supercritical fluid science and technology has lead to the development of industrial processes, including coffee and tea decaffeination, hops and flavor extraction, and pharmaceutical separations. During the past decade, there has been renewed interest in supercritical fluids especially for water, carbon dioxide, and alcohols, since these solvents can be expected to provide new technologies for green or sustainable chemistries.

The physical properties of supercritical fluids and their mixtures continue to be of importance in these developments and there is increased activity regarding reactions in supercritical fluids. In Japan, one of the earliest reports on anomalous behavior of reactions at supercritical conditions was by Toriumi and Hara (1941), who noted a pronounced increase in the oxidation of ammonia and sulfur dioxide in the critical region. Since then, there have been many developments in fractionation, drying, materials processing, chemical recycling, and waste treatment. Since 1992, the fundamentals and applications of supercritical fluids have been studied as a priority research area, supported by the Japanese government and also through various other research initiatives. This book summarizes some of the latest research into supercritical fluid science and technology in terminology that is suitable for industrialists, scientists, and engineers.

Supercritical fluids can lead to promising technologies in many fields such as separation, chemical reaction, material processing, waste treatment, plastic recycling and others. They can provide environmentally friendly and advanced industrial processes with less energy consumption and high efficiency. Supercritical carbon dioxide, water and alcohol are excellent candidates for industrial use because of their unique and suitable solvent properties, safety, small impact on the Earth's environment, and their economic advantages. Supercritical carbon dioxide with a critical temperature near room temperature can be used as an alternative for toxic organic solvents in the fields of separation, reaction and material processing. Supercritical water with a high critical temperature and pressure can decompose refractory harmful organics easily and depolymerize polymers to monomers and oligomers rapidly. This fluid has been applied to waste treatment, recycling of plastics, inorganic material processing, hydrogen production from heavy hydrocarbons, and so on. Supercritical methanol has a critical temperature between those of carbon dioxide and water. It is employed in selective methylation with or without a catalyst and in chemical recycling of waste plastics.

To date several books on supercritical fluid science and its applications have been published in specific areas. This text, on the other hand, attempts to provide six well-balanced chapters so that the reader can obtain an overview of some of the main future areas in supercritical fluid science and its potential for supporting the development of new technologies. The first section (Chaps. 1-3) provides a description of supercritical fluids and their interactions which is based on our knowledge after many years of research in the spectroscopic, molecular simulation, and thermodynamic fields. The second section (Chaps. 4-6) covers applications in extractions, separations, materials processing, and reactions. It is our hope that the text will contribute to our understanding of fluids in their supercritical state and help to inspire both young and established researchers to explore and develop new chemical and related processes which are friendly to our environment. We would like to take this opportunity to thank all the authors for their participation. We would also like to thank Springer for their guidance and encouragement during the preparation of the manuscript.

Fukuoka *Yasuhiko Arai*
Hamamatsu *Takeshi Sako*
Tsukuba *Yoshihiro Takebayashi*

October 2001

Contents

List of Contributors

Tadafumi Adshiri
Sects. 5.4, 6.1. 6.3
Graduate School of Engineering
Tohoku University
Aza-aoba 07, Aramaki, Aoba-ku,
Sendai, 980-8579 Miyagi, Japan
ajiri@arai.che.tohoku.ac.jp

Seiji Akimoto
Sect. 1.2
Graduate School of Engineering
Hokkaido University
Nishi 8-chome, Kita 13-jo,
Kita-ku, 060-8628 Sapporo,
Hokkaido, Japan
akimoto@eng.hokudai.ac.jp

Kunio Arai
Sects. 5.4, 6.1. 6.3
Graduate School of Engineering
Tohoku University
Aza-aoba 07, Aramaki, Aoba-ku,
Sendai, 980-8579 Miyagi, Japan
karai@arai.che.tohoku.ac.jp

Yasuhiko Arai
Sect. 2.4
Graduate School of Engineering
Kyushu University
Hakozaki 6-10-1, Higashi-ku,
812-0053 Fukuoka, Japan
arai@chem-eng.kyushu-u.ac.jp

Dmitry Bulgarevich
Sect. 6.2
Research Institute
of Green Technology
National Institute of Advanced
Industrial Science and Technology
Higashi 1-1-5, Tsukuba,
305-8565 Ibaraki, Japan
dmitry.bulgarevich@aist.go.jp

Ichiro Fujihara
Sect. 1.4
College of General Education
Osaka Sangyo University
Nakagaichi 3-1-1, Daito,
574-8530 Osaka, Japan
ichiro@las.osaka-sandai.ac.jp

Toshitaka Funazukuri
Sect. 3.3
Faculty of Science and Engineering
Chuo University
Kasuga 1-13-27, Bunkyo-ku,
112-8551 Tokyo, Japan
funazo@apchem.chem.chuo-u.ac.jp

Kaoru Fujimoto
Sect. 6.6
Faculty of Environmental
Engineering
The University of Kitakyushu
Hibikino 1-1, Wakamatsu-ku,
Kitakyushu, 808-0135 Fukuoka,
Japan
fujimoto@env.kitakyu-u.ac.jp

Motonobu Goto
Sect. 4.3
Department of Applied Chemistry and Biochemistry
Kumamoto University
Kurokami 2-39-1, Kumamoto
860-8555, Japan
mgoto@kumamoto-u.ac.jp

Yukio Hirata
Sect. 4.4
School of Materials Science
Toyohashi University of Technology
Hibarigaoka 1-1, Tempaku-cho,
Toyohashi, 441-8580 Aichi, Japan
hirata@chrom.tutms.tut.ac.jp

Masaru Hongo
Sect. 2.3
College of Industrial Technology
Nihon University
Izumi-cho 1-2-1, Narashino,
275-8575 Chiba, Japan

Takao Ikariya
Sect. 6.4
Graduate School of Science and Engineering
Tokyo Institute of Technology
Ookayama 2-12-1, Meguro-ku,
152-8552 Tokyo, Japan
tikariya@o.cc.titech.ac.jp

Yutaka Ikushima
Sect. 6.5
Supercritical Fluid Research Center
National Institute of Advanced Industrial Science and Technology
Nigatake 4-2-1, Miyagino-ku,
Sendai, 983-8551 Miyagi, Japan
y-ikushima@aist.go.jp

Hiroshi Inomata
Sect. 1.5
Research Center of Supercritical Fluid Technology
Tohoku University
Aza-aoba 07, Aramaki, Aoba-ku,
Sendai, 980-8579 Miyagi, Japan
inomata@scf.che.tohoku.ac.jp

Yoshio Iwai
Sect. 1.4
Graduate School of Engineering
Kyushu University
Hakozaki 6-10-1, Higashi-ku,
812-0053 Fukuoka, Japan
iwai@chem-eng.kyushu-u.ac.jp

Okitsugu Kajimoto
Sect. 1.2
Graduate School of Science,
Kyoto University
Oiwake-cho, Kitashirakawa,
Sakyo-ku, 606-8502 Kyoto, Japan
kajimoto@kuchem.kyoto-u.ac.jp

Yosuke Kataoka
Sects. 1.3, 1.5
Faculty of Engineering
Hosei University
Kajino-cho 3-7-2, Koganei,
184-8584 Tokyo, Japan
kataoka@mailer.k.hosei.ac.jp

Masahiro Kato
Sect. 2.2
College of Engineering
Nihon University
Nakagawara 1, Aza-tokusada,
Tamura-machi, Kooriyama,
963-8642 Fukushima, Japan
mkatoh@chem.ce.nihon-u.ac.jp

Yoshihito Kayaki
Sect. 6.4
Graduate School of Science
and Engineering
Tokyo Institute
of Technology & CREST
Ookayama 2-12-1, Meguro-ku,
152-8552 Tokyo, Japan

Mitsutaka Kitamura
Sect. 5.2
Graduate School of Engineering
Hiroshima University
Kagamiyama 1-4-1,
Higashi-Hiroshima,
739-0046 Hiroshima, Japan
mkitamu@ipc.hiroshima-u.ac.jp

Osamu Kitao
Sect. 1.3
Photoreaction Control
Research Center
National Institute of Advanced
Industrial Science and Technology
Higashi 1-1-5, Tsukuba,
305-8565 Ibaraki, Japan
osamu-kitao@aist.go.jp

Seiichiro Koda
Sect. 6.1
Graduate School of Engineering
The University of Tokyo
Hongo 7-3-1, Bunkyo-ku,
113-8656 Tokyo, Japan
koda@chemsys.t.u-tokyo.ac.jp

Hirokatsu Masuoka
Sect. 5.1
Graduate School of Engineering
Hiroshima University
Kagamiyama 1-4-1,
Higashi-Hiroshima,
739-0046 Hiroshima, Japan
masuoka@ipc.hiroshima-u.ac.jp

Kunio Nagahama
Sects. 4.1, 4.2, 5.2
Faculty of Technology
Tokyo Metropolitan University
Minami-Ohsawa, Hachioji,
192-0397 Tokyo, Japan
nagahama-kunio@c.metro-u.ac.jp

Koichiro Nakanishi
Sect. 1.1
College of Science
and Industrial Technology
Kurashiki University of Science
and the Arts
Nishinoura 2640, Tsurajima-cho,
Kurashiki, 712-8505 Okayama,
Japan
nakanisi@chem.kusa.ac.jp

Hideo Nishiumi
Sect. 3.3
Faculty of Engineering
Hosei University
Kajino-cho 3-7-2, Koganei,
184-8584 Tokyo, Japan
nishi@k.hosei.ac.jp

Tomoshige Nitta
Sect. 1.4
Graduate School
of Engineering Science
Osaka University
Machikaneyama-cho 1-3,
Toyonaka, Osaka 560-8531, Japan
nitta@chemeng.cheng.es.osaka-u.ac.jp

Susumu Okazaki
Sect. 1.3
Interdisciplinary Graduate School
of Science and Engineering
Tokyo Institute of Technology
Nagatsuda-cho, 4259, Midori-ku,
Yokohama, 226-8502 Kanagawa,
Japan
okazaki@echem.titech.ac.jp

Yoshito Oshima
Sect. 6.3
Enironmental Center
The University of Tokyo
Hongo 7-3-1, Bunkyo-ku,
113-0033 Tokyo, Japan
oshima@esc.u-tokyo.ac.jp

Takeshi Sako
Sect. 6.2
Faculty of Engineering
Shizuoka University
Johoku 3-5-1, Hamamatsu,
432-8561 Shizuoka, Japan
ttsako@ipc.shizuoka.ac.jp

Yoshiyuki Sato
Sect. 5.3
Graduate School of Engineering
Hiroshima University
Kagamiyama 1-4-1,
Higashi-Hiroshima,
739-0046 Hiroshima, Japan
ysato@ipc.hiroshima-u.ac.jp

Richard L. Smith
Sect. 1.5
Research Center of Supercritical
Fluid Technology
Tohoku University
Aza-aoba 07, Aramaki, Aoba-ku,
Sendai, 980-8579 Miyagi, Japan
smith@scf.che.tohoku.ac.jp

Shigeki Takishima
Sect. 5.1
Graduate School of Engineering
Hiroshima University
Kagamiyama 1-4-1,
Higashi-Hiroshima,
739-0046 Hiroshima, Japan
r736735@ipc.hiroshima-u.ac.jp

Hiroyuki Tanaka
Sect. 2.2
College of Engineering
Nihon University
Nakagawara 1, Aza-tokusada,
Tamura-machi, Kooriyama,
963-8642 Fukushima, Japan
htanaka@chem.ce.nihon-u.ac.jp

Yoshiyuki Tanaka
Sect. 3.2
Faculty of Engineering
Kobe University
Rokkodai-cho 1-1, Nada-ku, Kobe,
Hyougo 657-8501, Japan
ytanaka@kobe-u.ac.jp

Katsumi Tochigi
Sect. 2.5
College of Science and Technology
Nihon University
Surugadai 1-8, Kanda,
Chiyoda-ku,
101-8308 Tokyo, Japan
tochigi@chem.cst.nihon-u.ac.jp

Noritatsu Tsubaki
Sect. 6.6
Faculty of Engineering
Toyama University
Gofuku 3190, 930-8555 Toyama,
Japan
tsubaki@eng.toyama-u.ac.jp

Tomoya Tsuji
Sect. 2.3
College of Industrial Technology
Nihon University
Izumi-cho 1-2-1, Narashino,
275-8575 Chiba, Japan
tsuji@dic.cit.nihon-u.ac.jp

Hirohisa Uchida
Sect. 2.4
Faculty of Technology
Tokyo University of Agriculture
and Technology
Naka-machi 2-24-16, Koganei,
184-8588 Tokyo, Japan
huchida@cc.tuat.ac.jp

Masahiko Uematsu
Sect. 2.1
Faculty of Science and Technology
Keio University
Hiyoshi 3-14-1, Yokohama,
223-8522 Kanagawa, Japan
uematsu@mech.keio.ac.jp

Chiaki Yokoyama
Sect. 3.1
Laboratory of Chemical Reaction Engineering
Tohoku University
Katahira 2-1-1, Aoba-ku, Sendai, 980-8577 Miyagi, Japan
chiaki@tagen.tohoku.ac.jp

Noriyuki Yoshii
Sect. 1.3
Interdisciplinary Graduate School of Science and Engineering
Tokyo Institute of Technology
Nagatsuda-cho, 4259, Midori-ku, Yokohama, 226-8502 Kanagawa, Japan

Abbreviations of Symbols Frequently Used

Latin Alphabets

A	area
A	Helmholtz free energy
a	thermal diffusivity
a , b	parameters in van der Waals equation of state
C_P	specific heat at constant pressure
C_V	specific heat at constant volume
D	diffusion coefficient
E	total internal energy
E_T	polarity parameter
f	fugacity
G	Gibbs free energy
$g(r)$	radial distribution function
H	enthalpy
i , j	component of fluid
i	molecule index
J	flux
K	equilibrium constant
K	kinetic energy
k	rate constant
k_B	Boltzmann constant
k_{ij} , l_{ij}	binary interaction parameters
K_w	ion product
M	molecular weight
m	molecular mass
N_A	Avogadro number
n	number of molecules
n	refractive index
P	pressure
R	gas constant
r	position
T	temperature
t	time
U	internal energy
V	volume

v	specific volume
v	molecular velocity
w	speed of sound
x	position
x_i, y_i	mole fraction of component i
Z	compressibility factor ($= PV / nRT$)
C.P.	critical point
$scCO_2$	supercritical carbon dioxide
SCF	supercritical fluid
SCW	supercritical water
T.P.	triple point

Greek Alphabets

α_P	thermal expansion coefficient
γ	activity coefficient
γ	Euler constant
δ	solubility parameter
ε	dielectric constant
ε	energy parameter in Lennard-Jones potential function
η	viscosity
κ_T	isothermal compressibility
λ	thermal conductivity
μ	chemical potential
μ	dipole moment
ξ	correlation length
π	the ratio of the circumference of a circle to its diameter
π^*	Kamlet-Taft parameter
ρ	density
σ	length parameter in Lennard-Jones potential function
ϕ	fugacity coefficient
$\phi(r)$	potential function
ω	Pitzer's acentric factor

Subscripts and Superscripts

X_c	critical property
X^E	excess property
X^*	reduced property ($= X / X_c$)
ΔX	residual property

Definitions of other symbols are presented on the spots.

1 Solution Structure in Supercritical Fluids

1.1 Introduction

1.1.1 What is a Supercritical Fluid?

The position of the supercritical fluid (SCF) region can be shown on the conventional *PVT* phase diagram. If the projection onto the $P-T$ plane and that onto the $P-V$ plane are made, we obtain two diagrams as in Fig. 1.1.1 (a) and (b) with the critical point (C.P.) at each center. The SCF is in the region above C.P. on the $P-T$ and $P-V$ diagrams.

Large fluctuations are seen in the density and energy at C.P. and in its close vicinity, where some physical properties exhibit a so-called critical anomaly. How to approach these anomalies has long been of interest in statistical mechanics. Such a large fluctuation rapidly disappears as the temperature T increases from T_C, which allows us to use a classical treatment. A characteristic feature of a SCF is the fact that in the SCF region the slope of the $P-V$ isotherm shown in Fig. 1.1.1 (b) is exceptionally small. This means that a small change in pressure leads to a large volume change. In other words, the isothermal compressibility of a SCF is very large. This is to be compared with the case of a fluid at higher pressure, where the compressibility is quite small.

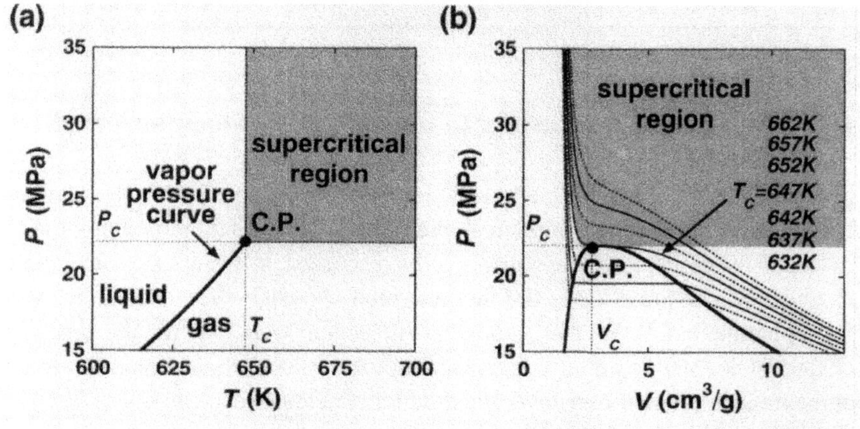

Fig. 1.1.1 Projections of phase diagram of water (a) in the *P-T* plane and (b) in the *P-V* plane: C.P. = critical point

The critical constants of some typical molecules used as SCF solvents are shown in Table 1.1.1. For these relatively simple molecules, the critical compressibility factor $Z_C = P_C V_C / RT_C$ takes similar values between 0.235 (H_2O) and 0.296 (Ar). The SCF region usually used for carbon dioxide is within relatively low P and T (50–100 bar, 304–350 K), while that for water has rather high P and T (220 bar and 650 K) or more. The range considered to be the SCF region may depend on the problem to be solved, but generally it will be $1.01\,T_C < T < 1.2\,T_C$.

Table 1.1.1 Critical constants of typical molecules used as SCF solvents

molecule	critical temperature T_C (K)	critical pressure P_C (atm)	critical compressibility factor $Z_C = P_C V_C / RT_C$
Ar	151	49	0.296
CH_4	191	45	0.287
Xe	290	58	0.287
CF_3H	299	47	0.259
CO_2	304	73	0.274
C_2H_6	305	48	0.285
C_3H_8	370	42	0.281
NH_3	406	111	0.244
H_2O	647	218	0.235

1.1.2 Studies of Supercritical Fluids at the Molecular Level

Studies of SCFs are performed mostly at the macroscopic level. The features of typical SCF systems summarized above are described quantitatively by the macroscopic properties, which will be fully discussed in Chaps. 2 and 3. In this chapter, we call attention to the fact that many physico-chemical studies should be supplemented by investigations "at molecular level". Thus, the rest of this chapter is devoted to an introduction to molecular level studies on SCFs. Studies at the molecular level may be divided into three categories, namely,

(i) theoretical studies based on statistical thermodynamics;
(ii) molecular simulation studies by molecular dynamics (MD) and Monte Carlo (MC) methods;
(iii) experimental studies especially by spectroscopic measurements.

Since SCFs may be classified as high-density condensed matter, the range that the traditional approach based on statistical mechanics can be applied to is quite limited. We will not discuss approach (i).

In order to supplement this lack, molecular simulation will play an important role. Methods of molecular simulation can be divided into two different approaches: MD and MC calculations. In this chapter, after a brief explanation of the intermolecular interaction potential which is indispensable in molecular simulation (Sect. 1.3.1), we will present some details of the principles and techniques of the molecular simulation (Sect. 1.3.2). Those readers who have less interest in the details of the computation may skip this part. Then, examples of contemporary studies on static and dynamic properties are shown in Sects 1.3.3–1.5. Their descriptions are not comprehensive but specific and topical.

The third approach, such as spectroscopic and diffraction measurements, leads directly to microscopic properties. Successful results of spectroscopic studies are given in detail in Sect. 1.2. In contrast to molecular simulation, spectroscopic methods cannot always be applied to any kind of solution phenomena, but they are still useful in many cases. We note that an MD calculation has been done by the same authors to supplement extensive experimental information from spectroscopy. This may suggest that, in future, any experimental studies in SCFs should be combined more closely with the appropriate molecular simulation approach.

1.2 Solvation Structure from Spectroscopy

1.2.1 Introduction

Spectroscopy provides a variety of information not only about the molecular structure of isolated molecules but also about the molecular interaction between a chromophore molecule and its environment. Chromophore molecules in supercritical fluids (SCFs) are believed to be solvated by fluid molecules in various ways depending on the temperature and pressure of the fluid [1]. The best way to elucidate the static and dynamic aspects of solvation in SCFs is to measure the spectra of chromophore molecules in the fluids. Electronic spectra, i.e., ultraviolet (UV) and visible spectra, are usually sensitive to the bulk environment around a chromophore molecule as a whole and are strongly affected by the local dielectric constant of the immediate vicinity. On the other hand, vibrational spectra, i.e., infrared (IR) and Raman spectra, reflect the local anisotropic interaction between the chromophore and the fluid molecules, and their sensitivity to the interaction depends on the vibrational mode to be observed. In this section, the experimental and theoretical aspects of the electronic and vibrational spectroscopy of molecules solvated in SCFs are summarized in relation to the solvation phenomena.

Spectroscopy in SCFs was first studied back in 1895 by Villard [2]. He attempted to observe the change in the color of I_2 when the solvent fluid CO_2 passed through the critical point. Later, in the 1980s, Hyatt tried to estimate the polarity of liquid and supercritical CO_2 by using the polarity parameter E_T [3]. The IR and Raman spectra of carbonyl compounds were also observed in SCFs to demonstrate the local environmental effects on the vibrational frequency and spectral width [4]. In both the electronic and vibrational spectra, the spectral shift from the gas-phase frequency was used to estimate the magnitude of the interaction with the environment. Theoretical studies of the environmental effects on the spectrum have been conducted since the 1960s, mainly for molecules in the liquid phase or under high pressure [5, 6, 7]. The theoretical concepts required to explain liquid phase spectra can also be applied to the spectra of the molecules in SCFs. The clustering effect, however, has to be taken into account in order to fully explain the spectral shift observed in SCFs which show a quite large compressibility near the critical temperature. In turn, the spectra of molecules in SCFs provide essential information on the solvation structure around chromophore molecules [1].

In the following, the spectral shifts in electronic spectra and vibrational spectroscopy are discussed in both experimental observation and theoretical interpretation. This section does not present the entire collection of spectral studies in SCFs, and therefore readers are recommended to read other reviews appearing in the ACS monograph series [8] and *Chemical Reviews* [9].

1.2.2 Electronic Transitions

(a) Experimental Observations

Absorption

UV or visible absorption usually corresponds to the $S_0 \rightarrow S_1$ electronic transition and the absorption energy is the energy difference between these two electronic states. When a chromophore molecule in the ground (S_0) state possesses a dipole moment, its S_0 state is stabilized by an attractive electrostatic interaction with a polar solvent. The excited (S_1) state which usually has larger dipole moment and polarizability is much more stabilized than the S_0 state. The combination of these two effects makes the $S_0 \rightarrow S_1$ transition energy in a polar solvent lower than that in the isolated state. In turn, the electronic transition of a specific chromophore molecule could be a measure of the solvent polarity. The E_T value expresses the solvent polarity by using the $S_0 \rightarrow S_1$ electronic transition energy of betaine. Since the ground and excited states of betaine are ionic and neutral, respectively, the polar environment stabilizes the ground state more than the excited state and consequently gives a blue shift in the absorption spectrum. Therefore, the larger the E_T value, the greater the polarity of the solvent.

Hyatt [3] first determined the E_T value of liquid and supercritical CO_2 and compared the value (34 kcal/mol) with other liquid solvents: 30.9 kcal/mol for n-hexane and 46.0 kcal/mol for acetonitrile. Sigman et al. [10] then examined the density dependence of the solvent parameter π^* for supercritical CO_2. They observed a large variation of π^* with varying density, though the π^* value for liquid CO_2 is similar to that of supercritical CO_2 at the same density. Yonker et al. [11, 12] also determined the π^* parameter for supercritical CO_2, N_2O, CCl_3F, NH_3, Xe, SF_6 and CO_2 + CH_4 mixtures. Yonker and Smith [13, 14] further discussed the relation between the π^* value and the local solvent density in terms of a dielectric continuum model.

Kim and Johnston [15] tried to use the bathochromic shift of absorption spectra, E_T, as a measure of the local solvent density. They measured the density dependence of the E_T value of phenol blue in supercritical ethylene, CF_3H, and CF_3Cl, and found that the E_T value at low density was smaller than that predicted by the perturbation theory of transition energies. That is, the excited state of phenol blue was more stabilized than expected. This observation suggests that the local dielectric constant around a solute molecule is larger than that of bulk fluid, indicating the occurrence of local density enhancement, i.e., clustering. They estimated the local density around the solute molecule as equal to the density whose bulk dielectric constant reproduces the observed E_T value according to the perturbation equation, as is shown in the following.

Kajimoto et al. [16, 17] measured the spectral shift in the absorption spectra of 4-(*N*,*N*-dimethylamino)benzonitrile in supercritical CF_3H as a function of fluid density (Fig. 1.2.1). They found that the spectral shift was much larger around and below the critical density than that predicted from the Lippert–Mataga equation based on Onsager's reaction field theory. This is due to the excess local dielectric constant caused by the enhanced local concentration of solvent molecules. They further tried to evaluate the number of molecules clustering around the solute molecule by using a simple model based on the Langmuir-type adsorp-

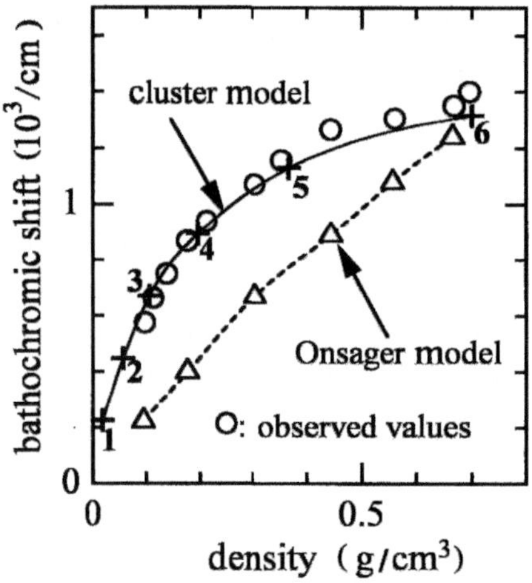

Fig. 1.2.1 The shift of the absorption maximum of DMABN in supercritical CF_3H is plotted as a function of fluid density (○). Comparing with the prediction from Onsager's theory (△), the observed shifts grows more rapidly in the low-density region. This trend is well reproduced by the clustering model based on the simple Langmuir-type approximation [17]. The numbers given in the figure are the clustering number predicted by the clustering model

tion equilibrium and the solute−solvent interaction potential. The spectral shifts predicted by the simple theory well reproduced the observed trend.

Fluorescence

A molecule consisting of electron-donating and electron-accepting parts forms, on photoexcitation, a charge-transfer (CT) state where the electron moves from the donor part to the acceptor part. This reaction is called the intramolecular CT reaction and is known to be quite sensitive to the solvent polarity because of the large dipole moment induced by the charge separation. The fluorescence emitted from the CT state is largely red-shifted from the excitation wavelength because of the large stabilization of the CT state by the polar solvent. Using this spectral shift, called a Stokes shift, one can easily evaluate the polar nature of the solvent including SCFs. Kajimoto et al. [16] first applied the Stokes shift to demonstrate solvent clustering around the CT molecule in a polar SCF, CF_3H. They showed that the Stokes shift of *N,N*-dimethylaminobenzonitrile (DMABN) in low-density CF_3H is much larger than the shift expected from simple Onsager's reaction field theory and attributed this to the local enhancement of solvent density.

Sun et al. [18] observed both the absorption and fluorescence spectra of DMABN and ethyl *p*-(*N,N*-dimethylamino)benzoate (DMAEB) in supercritical CF_3H, CO_2 and ethane in a wide density range. They found that DMAEB could form the CT state even in nonpolar supercritical CO_2 because the excited state potential surface of DMAEB directly leads to the CT state without curve crossing. The trend in the bathochromic shifts in the absorption spectrum was found to be quite similar in both DMABN and DMAEB. In the bathochromic shifts in the CT fluorescence spectra of DMAEB measured in CF_3H, they observed the same trend as that reported previously for DMABN.

(b) Theoretical Interpretation

The observed spectral shifts have been interpreted mainly within two types of theoretical framework, a continuum model and a molecular model. In the continuum model, the electrostatic stabilization of a solute molecule in a medium having a dielectric constant ε is formulated. The solvent is considered to be a continuum medium where the electrostatic interaction is operative. The molecular model treats the solvent as a vast number of actual molecules and the interactions between the solute molecule and the individual solvent molecule are taken into account. To estimate the distribution of molecules around the solute, one has to perform molecular dynamics calculations or evaluate a radial distribution function based on the solute–solvent and solvent–solvent intermolecular potentials.

Onsager's Reaction Field Theory in a Continuum Medium

The quantitative treatment of the electrostatic solvent effects in continuum media is mostly based on Onsager's reaction field theory. The reaction field theory [5, 19] provides a simple formalism to estimate the stabilization energy of a dipolar molecule in a polar environment. The model consists of a solute molecule of dipole moment μ in a cavity of radius a formed in a continuum medium whose dielectric constant and refractive index are ε and n, respectively. The dipole of the solute molecule generates the electric field which affects the continuum outside the cavity. The polarization of the continuum medium is increased by this electric field and exerts an additional field on the central solute molecule. This additional field is called the "reaction field" and becomes the origin of the stabilization of the solute molecule. According to his theory, the stabilization energy is given by,

$$\Delta E = \frac{\mu^2}{a^3}\frac{2(\varepsilon-1)}{2\varepsilon+1}. \tag{1.2.1}$$

Using this equation, one can derive the formula known as *the Lippert–Mataga–Ooshika equation*, expressing the absorption spectral shift $\Delta\nu_{\rm abs}$ of a molecule whose dipole moments of the S_0 and S_1 states are $\mu_{\rm g}$ and $\mu_{\rm e}$, respectively, as [20, 21]

$$\Delta\nu_{\rm abs} = \frac{\mu_{\rm g}^2 - \mu_{\rm e}^2}{a^3}\frac{n^2-1}{2n^2+1} + \frac{2\mu_{\rm g}(\mu_{\rm g}-\mu_{\rm e})}{a^3}\left(\frac{\varepsilon-1}{\varepsilon+2} - \frac{n^2-1}{n^2+2}\right). \tag{1.2.2}$$

This equation has been successfully applied to the solvent spectral shifts in liquid solutions. To apply this formalism to SCF solutions, one has to take into account the large compressibility of the medium or the clustering of solvent molecules around a solute molecule.

Compressible Continuum Model

In SCFs, the dielectric constant just around the solute molecule and that of the bulk SCF are significantly different because of clustering. Therefore the local dielectric constant should be used for the estimation of the electrostatic interaction between the solute molecule and the continuum medium of the immediate vicinity. In addition, both the simple electrostatic stabilization and the compression of the medium occur simultaneously in dynamic processes such as photoexcitation and ionic reactions. Recently, Luo and Tucker [22] proposed a new method, a compressible continuum model. Their idea is that the charge distribution in the solute or reacting molecules produces an electrostatic field $E(r)$ which induces the solvent compression or enhances the solvent density $\rho(E)$ as a function of the field strength. However, since the dielectric constant $\varepsilon(\rho)$ is a function of the solvent density and the electric field $E(\varepsilon)$ is a function of ε, this compression again changes the electrostatic field. In order to reach the self-consistent solution of E and ρ, Luo and Tucker solved Poisson's equation iteratively with the finite-difference grid algorithm.

The above formalism can be used for the calculation of spectral shifts of a strongly polar molecule or an ionic species, and Luo and Tucker applied this method to estimate the local solvent density of water in the course of the anisole hydrolysis reaction [23],

$$PhOCH_3 + OH_2 \rightarrow PhO^- + CH_3OH_2^+ . \tag{1.2.3}$$

They used the Hartree–Fock 6–31G** calculations to obtain the charge density distribution within the reacting system. They found that, even at the low bulk density of 0.25 g/cm^3, the local density of solvent water within 2 Å from the reactant molecule was almost the same as that at 0.5 g/cm^3. In the ordinary continuum model, the free energy of solvation is calculated as the sum of the work for carrying the charges in vacuum together in the medium with the dielectric constant ε. In the compressible continuum model, the charges are carried together through the medium of *varying* dielectric constant and electronic field $E(r)$. The free energy of solvation was found to be 11.6 kcal/mol lower than the value expected from the ordinary continuum (incompressible) model. They [24] also applied the model to a typical S_N2 reaction,

$$Cl^- + CH_3Cl \rightarrow CH_3Cl + Cl^-, \tag{1.2.4}$$

and compared it with the MD calculations [25, 26]. As compared with the incompressible model, the compressible continuum model gave much better agreement with the MD calculation in the solvation energy along the reaction path.

Clustering Model and Radial Distribution Function

When the dipole moment, quadrupole moment and polarizability of a solute molecule as well as those of a solvent molecule in vacuum are known, one can calculate the molecular interaction between the solute molecule and the individual solvent molecule. Under the assumption that the molecular interaction between the solute molecule and the solvent molecules of the immediate vicinity is additive, one can estimate the total interaction from the clustering number of solvent molecules and the interaction between the 1:1 solute–solvent pair.

The clustering number is obtained by integration of the radial distribution function within the range of the first solvation shell. Alternatively, one can use the Langmuir-type approximation for the evaluation of the clustering number as a function of the bulk fluid density [1]. It is important to note that the total interaction may not be proportional to the number of solvent molecules and one should take into account the contribution of the solvent–solvent interaction. Nevertheless, this scheme can well reproduce the observed trend in both the UV and IR spectra as shown in the next subsection.

Molecular Dynamics Calculations

With the rapid progress in fast-computer technology, classical molecular dynamics (MD) calculations of a large system consisting of more than 1000 molecules become possible. The calculations dealing with pure supercritical fluid have been performed for a variety of fluids such as Ar, Xe, CO_2, H_2O, and CF_3H [27, 28]. The binary systems consisting of a nonreactive solute molecule and supercritical fluid molecules have been studied with MD calculations to demonstrate the existence of clusters and to characterize them in various fluid conditions [29–31]. When the solute–solvent interaction potential is known, the MD calculation can provide the time-averaged number of solvent molecules in the first solvation shell. Using this number one can evaluate the spectral shift applying the method of the preceding section. The MD calculation is utilized to evaluate the solvation number in place of using the radial distribution function.

More sophisticated results would be obtained by the combination of MD and molecular-orbital (MO) calculations. Using the self-consistent-field MO (SCF-MO) method, one can calculate the S_0 and S_1 electronic energies of the total system including the solute molecule and the solvent molecules at a specific spatial distribution which is derived by the MD calculation. Monte Carlo (MC) calculations can also be utilized for this purpose. The transition energy of the solute molecule averaged over various distributions of the solvent molecules then gives the absorption frequency of the solute molecule in SCFs of a given density and temperature. A variety of methods have been used for incorporating the effect of solvent molecules into the electronic energy of the solute molecule. With the charge distribution of the solute molecule being fixed, one can calculate the Coulombic interaction energies between the charge distribution of the solute molecule and an ensemble of solvent molecules. One can also explicitly include the effect of the solute–solvent interaction into the Hamiltonian of the reacting molecules in SCF-MO calculations. Furthermore, the Hamiltonian of the total system consisting of all the solvent molecules and the reacting molecules could be solved directly by an ab initio SCF-MO method. Although the last method formally gives the most precise results, the calculations require enormous time on a high-speed computer.

1.2.3 Vibrational Spectra

The vibrational frequency of a molecule in solvents is generally shifted from the gas-phase value because of the intermolecular interactions with solvent molecules. This shift provides information about the solvation structure around a solute molecule. In terms of the intermolecular interactions, the spectral shift is controlled by the balance between the solute–solvent attractive and repulsive interactions. In SCFs, the fluid molecules cluster around a solute molecule and the attractive interaction should give the dominant contribution to the spectral shift. The observation of the spectral shifts in a wide density range of SCFs provides us with detailed information on the changes of clustering structure in the gas–liquid transition region.

To analyze the effects of the solute–solvent interaction on the spectral shifts, two types of approaches have been adopted. One approach treats the solvent fluid as a continuum with a dielectric constant ε and a refractive index n, and the other treats the solvent as being composed of real molecules which have effective diameters, polarizabilities and dipole moments. In the following sections, we first report the observed spectral shifts and then compare them with the predictions based on the above two treatments. The radial distribution functions of the solvent fluid molecules are evaluated with the Percus–Yevick (PY) or hypernetted chain (HNC) procedure and used for predicting the density dependence of the shifts.

(a) Raman Spectra

Dependence of Raman Shift on Fluids and Their Densities

The spectral shifts in the Raman bands of acetone and acetonitrile, in particular the C=O stretching and the C≡N stretching vibrations, respectively, have been observed in a wide density range of supercritical Ar, CO_2, CF_3Cl and CF_3H [4]. These vibrations are known to be significantly influenced by the solute–solvent interaction [32].

Figure 1.2.2 shows the change of the C=O stretching Raman band of acetone in supercritical CF_3H as a function of fluid density. Although the observed Raman band is actually composed of several peaks, the shift in the main peak is used for the following discussion. The density dependence of the spectral shift in supercritical Ar, CO_2, CF_3Cl and CF_3H is illustrated in Fig. 1.2.3. With increasing solvent density, the frequency monotonically shifts to lower values. The magnitude of the shifts strongly depends on the solvent fluid, reflecting the strength of the interaction between the C=O bond and the solvent molecules. CF_3H gives by far the largest shift compared with other solvents. This may be related to the large dipole moment of CF_3H. Argon shows the least shift, probably due to its small polarizability. In spite of the zero dipole moment and the relatively small polarizability, CO_2 gives much larger shift than CF_3Cl. This unexpected trend will be discussed later.

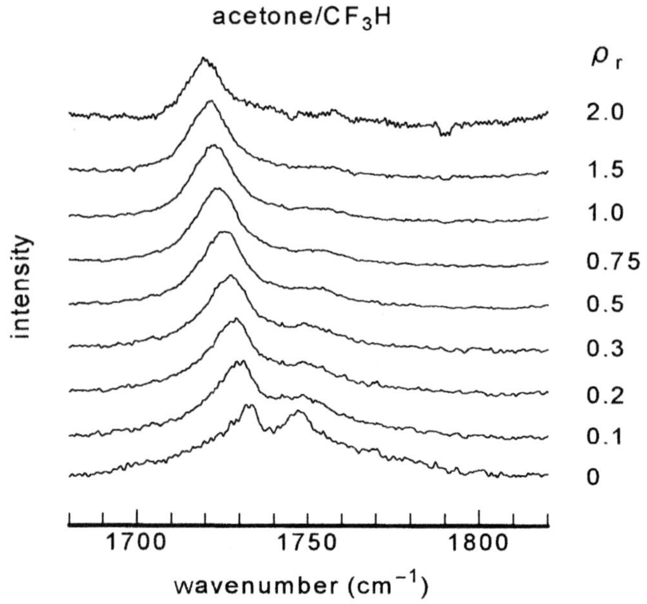

Fig. 1.2.2 The change of Raman band of the C=O stretching vibration of acetone as a function of fluid density in supercritical CF_3H

Theoretical Interpretations of Spectral Shifts

Approach Using the Continuum Model. The simplest theoretical approach to interpret the solvent-induced spectral shifts is the Kirkwood–Bauer–Magat treatment [33], which is essentially similar to the Lippert–Mataga–Ooshika treatment [20, 21] for the spectral shifts in electronic transitions. According to Onsager's reaction field theory [5], a medium of large dielectric constant stabilizes the vibrationally excited state more than the ground state because of its larger dipole moment and hence causes a red shift. The lines in Fig. 1.2.3 indicate the values of the "polarity parameter" for each fluid as a function of the reduced density. The theory expects the proportionality of the spectral shifts of this parameter to be,

$$\frac{\Delta \nu}{\nu} = C\frac{\varepsilon - 1}{2\varepsilon + 1}. \tag{1.2.5}$$

The observed trends of the shift in the C=O stretching vibration of acetone are in fair accordance with the continuum model except for CO_2. However, a

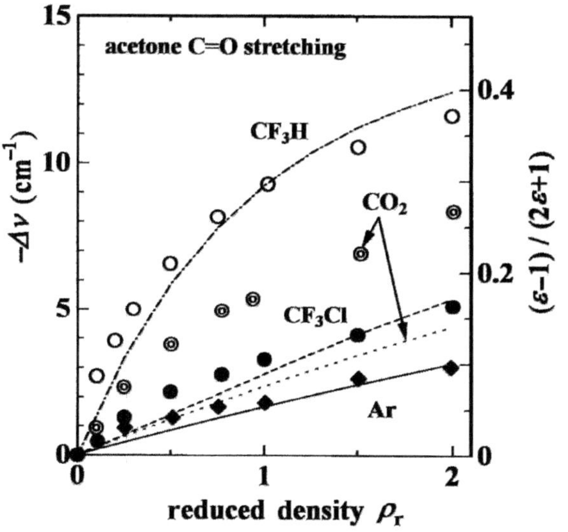

Fig. 1.2.3 Density dependence of the C=O stretching vibrational frequency of acetone in various SCFs. The lines indicate the parallelism between the observed shifts and the polarity parameters derived from the continuum model

closer look at the plots reveals that the observed shifts are larger than those expected from the polarity parameter of homogeneous SCFs at a lower density region in all of the fluid solvents. This discrepancy indicates the enhancement of the local concentration by the formation of clusters just around the solute molecule. The large discrepancy in CO_2 may reflect the anisotropy in the polarizability of a linear CO_2 molecule. The orientation of CO_2 should be taken into account for the correct prediction.

Electrostatic Approach Using Buckingham's Treatment. In order to obtain a detailed molecular-level picture, one can use Buckingham's approach [6] with the electrostatic interaction. According to this theory, the spectral shift of a diatomic molecule can be expressed as,

$$\frac{\Delta\nu}{\nu} = \frac{1}{f}\left[-\left(\frac{3g}{2f}\right)\frac{\partial V(r,x)}{\partial x} + \frac{1}{2}\frac{\partial^2 V(r,x)}{\partial x^2}\right], \tag{1.2.6}$$

where $V(r,x)$ is the intermolecular potential energy as a function of both the intramolecular vibrational coordinate x and the intermolecular distance r (between the center of gravity of each molecule); f and g are the harmonic and anharmonic force constants of the stretching vibrational potential of a diatomic molecule in the gas phase,

$$U_{\text{vib}} = \frac{1}{2} f(x - x_{\text{e}})^2 + \frac{1}{2} g(x - x_{\text{e}})^3 + \cdots . \qquad (1.2.7)$$

The shifts are calculated using (1.2.6), on the assumption that the vibrations are localized in the C=O bond. The attractive interaction in terms of the three contributions are taken into account: the dipole–dipole (V_{dip}), dispersive (V_{dis}), and dipole–induced-dipole (V_{ind}) interactions are considered.

The spectral shift for a 1:1 complex between the solute and the solvent molecules is calculated according to (1.2.6) at the equilibrium distance and fixed orientation. Averaging of this shift over the angular distribution, as well as the Boltzmann distribution, gives the final prediction of the spectral shifts. Since the calculated shift for CO_2 was found to be much smaller than the observed shift, the molecular shape of CO_2 was explicitly included through the angular dependent interaction potential and the dipole–quadrupole interaction, V_{dq}. The inclusion of the molecular shape much improved the agreement between prediction and observation. The calculated spectral shifts are listed in Table 1.2.1 together with the contribution from the various types of interactions.

Table 1.2.1 Partial contributions to the spectral shifts of the C=O stretching vibration of acetone in Ar, CO_2, CF_3Cl, and CF_3H. The shifts originated from the dipole–dipole ($\Delta \nu_{\text{dip}}$), dispersive ($\Delta \nu_{\text{dis}}$), dipole–induced-dipole ($\Delta \nu_{\text{ind}}$) and dipole–quadrupole ($\Delta \nu_{\text{dq}}$) interactions, and the sum ($\Delta \nu_{\text{tot}}$) of these values is listed. The observed spectral shifts ($\Delta \nu_{\text{obs}}$) at the solvent reduced density of 0.5 are also listed. The normalized shifts are given in units of 10^{-4}.

	$\Delta\nu_{\text{dip}}/\nu$	$\Delta\nu_{\text{dis}}/\nu$	$\Delta\nu_{\text{ind}}/\nu$	$\Delta\nu_{\text{dq}}/\nu$	$\Delta\nu_{\text{tot}}/\nu$	$\Delta\nu_{\text{obs}}/\nu$ /5
Ar	0	-0.989	-1.33	0	-2.32	-1.5
CO_2	0	-1.53	-2.18	-1.70	-5.42	-4.4
CF_3Cl	-0.534	-0.899	-1.301	-0.296	-3.03	-2.5
CF_3H	-6.12	-0.611	-0.897	-0.597	-8.23	-7.6

Theoretical Prediction of Density Dependence of Spectral Shifts. Although the predicted solvent dependence of the shifts is roughly in accordance with those observed, the shift in Ar is overestimated. This is probably due to the smaller clustering efficiency of Ar at 333.2 K. Since the cluster efficiency is most clearly expressed by the radial distribution function for the solvent molecules around a solute molecule, the density dependence of the shift is calculated by using the solute–solvent radial distribution functions.

The frequency shift at a certain density can be obtained by multiplying the shift for a 1:1 complex by the number of nearest neighbors at the given density:

$$\frac{\Delta \nu}{\nu} = N\left(\frac{\Delta \nu}{\nu}\right)_{1:1}, \tag{1.2.8}$$

where N is the number of nearest neighbors and $(\Delta \nu / \nu)_{1:1}$ is the shift predicted for a 1:1 complex between the solute and solvent molecules. In order to estimate N, the solute–solvent radial distribution function, $g(r)$, is integrated over the equilibrium solute–solvent distance as,

$$N = 4\pi\rho \int_{r_e-a}^{r_e+a} g(r) r^2 \, \mathrm{d}r, \tag{1.2.9}$$

where a defines the range of the nearest neighbors, ρ denotes the number density of the solvent, and r_e is taken to be the peak position of $g(r)$. The radial distribution functions at several densities are calculated by using the standard integral equations derived from the PY or HNC theories.

The result is shown in Fig. 1.2.4. The large shift observed for the C=O stretching vibration in low-density CF_3H was attributed to the enhancement of the shift by cluster formation. However, the shift calculated using radial distribution functions was still lower than the observed shift. This is most probably due to the presence of the specific interaction between acetone and CF_3H. The first solvent molecule must be located near the oxygen atom of acetone to form the hydrogen bonding, and this CF_3H molecule causes a larger solvent shift than the other solvent molecules which are distant from the oxygen atom.

Density Dependence of Shifts in C–C and C–H Stretching Vibrations

The density dependence of the other vibrational modes has also been observed in various SCFs. As shown in Fig. 1.2.5, the features are strongly dependent on the vibrational mode. The spectral shifts in both the C–C and C–H symmetric stretching vibrations are in the opposite direction to that of C=O stretching. In order to interpret these trends, the *indirect through-bond* interaction with the solvent molecules should be taken into account, in addition to the *direct* interaction dealt with in the preceding sections. That is, the C–C and

C–H bonds of acetone are affected by the change of charge distribution in the molecule caused by the strong interaction between the terminal C=O bond and the solvent molecules. In order to take this effect into account, we performed the MO calculation including solvent effects by means of image approximation formulated by Sato. [34].

The interaction of Ar with the solute molecule is not strong enough to cause the *indirect through-bond* effect whereas all kinds of *direct* interactions and the *through-bond* interaction could be operative in the case of CF_3H. The difference between the observed shifts of CF_3H and Ar, therefore, represents the pure *indirect* contribution which corresponds to the shift predicted by the MO calculation. The differences obtained in this way for the C–C symmetric and C–H symmetric stretching vibrations of acetone are listed in Table 1.2.2. They actually show qualitative agreement with the shifts obtained from the MO calculation combined with the *image* approximation for both vibrational modes.

Table 1.2.2 Calculated shifts in a continuum of dielectric constant 10, and observed shifts in Ar and CF_3H, for the C–C symmetric and C–H symmetric stretching vibrations of acetone (* cavity size representing a solute molecule, ** shifts obtained at the reduced density of 2.0, *** $\Delta\Delta\nu_{CF_3H} = \Delta\nu_{CF_3H} - \Delta\nu_{Ar}$).

calculated /cm^{-1}			observed** /cm^{-1}			
gas	$\Delta\nu_{sol.}$(4D)*	$\Delta\nu_{sol.}$(6D)*	$\Delta\nu_{Ar}$	$\Delta\nu_{CF_3H}$	$\Delta\Delta\nu_{CF_3H}$***	assignment
847.3	+2.5	+0.2	+0.5	+6.2	+5.7	C-C symm. stretching
3189.7	+7.0	+0.8	-0.5	+2.2	+2.7	C-H symm. stretching

(b) Infrared Spectroscopy

In the 1960s, several research groups observed the infrared and near-infrared spectra of water in the supercritical region to elucidate the structure of water at high temperatures [35, 36]. They measured the spectral shift and the change of the band shape in the O–D and O–H stretching vibration and tried to estimate the amount of free water against the hydrogen-bonded water molecules. Later, Zerda et al. [37] measured the spectral shift of the C–H stretching vibration of naphthalene in supercritical CO_2 with the FTIR technique in the course of determining the solubility of anthracene in CO_2. They found that the C–H stretching frequency shifted blue as the density of CO_2 increased and attributed this to the repulsive interaction between CO_2 and the C–H bond in the anthracene plane.

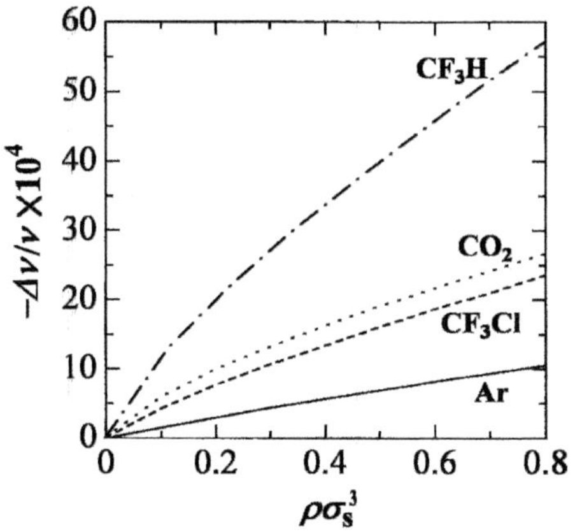

Fig. 1.2.4 The density dependence of the frequency shift in the C=O stretching vibration of acetone in various SCFs. The curves represent the prediction by Buckingham's model combined with the clustering number. The cluster number as a function of density was estimated by integration of the radial distribution function over the assumed range of the first solvation shell. The abscissa is the density multiplied by the volume of the solvent molecule σ_S^3, where σ_S is the collision diameter of the solvent molecule

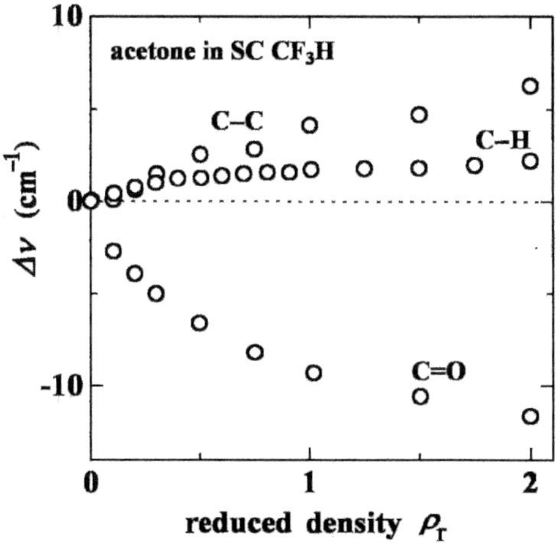

Fig. 1.2.5 The density dependence of the spectral shifts in various vibrational modes of acetone in supercritical CF_3H

Just like the explanation of the spectral shifts in UV absorption spectra, Onsager's reaction field theory was utilized by Blitz et al. [38] to interpret the spectral shift in the $\nu_2 + \nu_3$ band of water in supercritical CO_2. They found that the red shift increases with increasing density of supercritical CO_2 and noticed the rapid increase of the shift at lower density regions as observed in the UV absorption spectra of DMABN. They ascribed the observed trend to the clustering of CO_2 molecules around a polar water molecule.

Recently, Wada et al. [39] measured the IR spectra of substituted benzene derivatives in supercritical CO_2 and proposed a method to estimate the local excess density using the measured molar absorption coefficients of the substituent deformation stretching and the C−C ring stretching vibrations. They derived the equation for estimating the hypothetical molar absorption coefficients as a function of the polarity parameter ($[\varepsilon(\rho_B)-1]/[2\varepsilon(\rho_B)+1]$, where $\varepsilon(\rho_B)$ is the dielectric constant of the fluid whose bulk density is ρ_B. By comparing the observed molar coefficient with that obtained from the equation, they estimate the local density around the vibrating molecule. The underlying concept is the electrostatic interaction between the dipole moment of the vibrating bond and the medium with dielectric constant ε. The same concepts are adopted in the cases of the spectral shits in UV absorption spectra and Raman frequencies as given in the preceding sections.

1.3 Solvation Structure from Molecular Simulations

1.3.1 Molecular Interactions in Supercritical Fluids

Knowing the molecular interactions is important to understand the solution structure in supercritical fluids (SCF). Moreover, molecular simulation studies of SCFs require a set of analytical pair potential functions for the molecular interaction. This section explains how to construct the analytical function for carbon dioxide (CO_2), which is the most frequently used molecule in SCF technology.

(a) Molecular Interaction Based on ab initio Molecular Orbital Calculations

One example of an analytical pair potential function is

$$V = \sum_{i=1}^{3}\sum_{j=1}^{3} f(r_{ij}), \tag{1.3.1}$$

$$f(r_{ij}) = \frac{a_{ij}}{r_{ij}^{12}} - \frac{b_{ij}}{r_{ij}^{6}} + \frac{q_i q_j}{r_{ij}}, \tag{1.3.2}$$

where i and j are the mean interaction sites on the carbon and oxygen atoms. Each function $f(r_{ij})$ in V consists of three terms. The first and second terms, the so-called Lennard–Jones (LJ) potentials, describe the van der Waals interaction, and the third term describes the Coulomb interaction. This function requires a set of three parameters: a_{ij}, b_{ij} and q_i.

There are two methods commonly used to construct this set of parameters. The first method involves adjusting the parameters to reproduce specific experimental data, such as the compressibility of a gas, the heat of evaporation of a liquid, and the heat of sublimation of a crystal. Murthy et al. [1] proposed an effective model for CO_2 with empirical adjustable parameters: LJ with an electrostatic part consisting of a quadruple moment. Their model excellently reproduced several properties by using a different set of parameters for each property [2–5]. The second method involves doing the ab initio molecular orbital (MO) calculations for several configurations of the CO_2 dimer, and then adjusting the set of parameters to reproduce the calculated potential energy surface. Due to recent advances in computer speed, many researchers can use this computationally intensive method. We therefore describe this second method here.

The input data for the ab initio MO calculations are the geometries of the dimer configuration and the quality of the calculations. Four coordinates (i.e., three angles and the distance between the molecules) determine the potential energy surface of the dimer. As described by examples later in this section [6–10], the CO_2 dimer requires 40–200 unique configurations to set up the potential energy surface. Because the molecular interaction of CO_2 is extremely weak, the quality of the MO calculations requires careful choice of the basis set and electron correlation. Although a detailed description of ab initio MO calculations is given in textbooks [11], we stress two requirements for setting up the calculation [12].

The first requirement is that we must use a large basis set for the calculation; for example, recent studies use 6–311+G(2df) [6] and 8s6p4d1f [10]. In addition, we must monitor the basis set superposition error (BSSE) due to the incompleteness of the basis set. This error is not negligible; for example, it is about 15% of the interaction energy for 8s6p4d1f in the minimum energy configuration [10]. The counterpoise calculation [13] removes the BSSE in following manner. The following equation overestimates the interaction energy:

$$V = W_{AB}(AB) - W_A(A) - W_B(B). \tag{1.3.3}$$

Table 1.3.1 Molecular interaction of CO_2 dimer (kcal/mol) [6]

method	HF	MP2	MP3	MP4
energy	-0.29	-0.78	-0.67	-0.78

Here, the notation $W_X(Y)$ means that molecule Y uses a basis set on X. This overestimation is due to an inadequate description for each monomer in (1.3.3). The alternative is to calculate the interaction energy by the equation,

$$V = W_{AB}(AB) - W_{AB}(A) - W_{AB}(B). \tag{1.3.4}$$

Here, $W_{AB}(A)$ means that molecule A uses a basis set on both A and B.

The second requirement is that we must include the electron correlation in the description of the van der Waals contribution. Because the CO_2 molecule has no dipole moment, the leading interaction is a quadruple moment. Therefore, the dominant term is the Coulomb interaction, which the naive Hartree–Fock (HF) calculation can describe. As expected from this interaction, the experimental data confirmed that the stable structure of the dimer is slipped parallel with the C_{2h} symmetry or T-shape [14, 15]. However, the dispersion energy component is not small as summarized in Table 1.3.1 [6], and the appropriate description requires a large basis set and the inclusion of electron correlation. The second-, third-, and fourth-order Moller–Plesset perturbation theories (MP2, MP3, and MP4, respectively) add the electron correlation to the HF energy. Tsuzuki et al. [6] calculated these intermolecular interaction energies with the 6–311G(2d) basis set at the T-shape and with 4.2 Å for the C–C distance.

Using any software package for ab initio MO calculations, all researchers obtain the same result for the potential energy surface if they use the same calculation level. However, simulation results depend on how they construct the analytical function. To illustrate this dependence, we next compare two recent simulation results.

(b) Examples of Constructing Analytic Pair Function for CO_2 Molecules

Tsuzuki et al. [6] and Steinebrunner et al. [10] recently reported the intermolecular potential energy surface based on ab initio MO calculation of high quality. The two groups also studied the molecular dynamics of a supercritical CO_2 fluid. Although in their calculations, both groups used ab initio data of a similar level, they used different approaches in constructing the analytical pair potential function.

Tsuzuki et al. [6] calculated the intermolecular interaction energies of 40 geometrical configurations of dimers by MP2/6−311+G(2df) level with the BSSE correction. Using the calculated energies, they optimized the Buckingham type equation:

$$E_{ij} = \sum_{i=1}^{3}\sum_{j=1}^{3} E(r_{ij}), \tag{1.3.5}$$

$$E(r_{ij}) = B_{ij}\exp(-C_{ij}r_{ij}) - \frac{A_{ij}}{r_{ij}^{6}} + \frac{q_i q_j}{r_{ij}}. \tag{1.3.6}$$

After locating the interaction sites at each van der Waals interaction center, Tsuzuki et al. then incorporated the effect of anisotropy by relocating the interaction sites so that they are close to the connected carbon atoms from the original oxygen atomic position. The best relocation of the oxygen interaction site was 9% of the bond distance. By fitting the electrostatic potentials, Tsuzuki et al. determined the point charges on each atom [16]: $0.68e$ for carbon and $-0.34e$ for oxygen.

Table 1.3.2 summarizes the optimized value [6]. Because the C–C interaction is still attractive at the C−C distance of 1 Å, this value for the C−C interaction loses physical meaning as the van der Waals interaction. However, the sum of the values for the C−C, C−O and O−O interactions well reproduced the ab initio data within 0.14 kcal/mol. Tsuzuki et al. confirmed that their model potential reproduces experimental lattice constants and lattice vibration frequencies of the molecular crystal, and molar volumes and heat of evaporation of the liquid.

Using this model, Tsuzuki et al. [7] simulated the molecular dynamics of supercritical CO_2. They used 216 rigid CO_2 molecules in cubic unit cells with a periodic boundary condition and *NVT* ensembles with a time step of 2 fs. After 20 ps for the initial equilibration, they used the next 50 ps for estimating the pressure, and then the next 100 ps for estimating the self-diffusion coeffi-

Table 1.3.2 Parameters used in the model potential of CO_2 [6]

	A (kcal Å^6/mol)	*B* (kcal/mol)	*C* (Å^{-1})
C–C	356	217	2.27
C–O	372	8760	3.16
O–O	389	354000	4.40

cient. They did the simulation for five different values of density ρ (0.004–0.020 mol/cm^3) and four values of temperature T (320–400 K). The calculated pressure agreed well with the experimental value: the largest deviation was 18.4%. Similarly, the calculated self-diffusion coefficients agreed well with the experimental values: the deviation ranged from 3 to 20%. They concluded that the overall performance of their model is satisfactory for simulating an SCF.

In contrast to Tsuzuki et al. [7], Steinebrunner et al. [10] calculated over 200 energy points with a 8s6p4d1f basis set. They estimated the energy by MP2 with BSSE correction. They fitted the calculated results by the following analytical function, which includes two additional interaction cites B at the center of each C–O bond:

$$E_{ij} = \sum_{i=1}^{5}\sum_{j=1}^{5} E(r_{ij}), \tag{1.3.7}$$

$$E(r_{ij}) = a_{ij}^{(1)} \exp\left(-a_{ij}^{(2)} r_{ij}\right) + \left[\frac{a_{ij}^{(3)}}{r_{ij}^{12}} + \frac{a_{ij}^{(4)}}{r_{ij}^{10}} + \frac{a_{ij}^{(5)}}{r_{ij}^{8}} + \frac{a_{ij}^{(6)}}{r_{ij}^{6}} + \frac{q_i q_j}{r_{ij}}\right] D(r_{ij}), \tag{1.3.8}$$

$$D(r_{ij}) = \frac{1}{\left(1+\exp\left\{-2\left(\frac{r_{ij}}{a_0}-2.1\right)\right\}\right)^8}. \tag{1.3.9}$$

The damping function D provides a smooth transition at 4.2 a_0 (a_0= 52.9177 pm) and a strong decay at shorter distances. They set the charges q_i to reproduce the calculated quadruple moment of the monomer. The result was $-0.3125e$ and $0.625e$ for oxygen and carbon, respectively. Table 1.3.3 lists the fitted parameters [10]. Steinebrunner et al. also accounted for the intramolecular interaction by using a harmonic potential.

Using this model, Steinebrunner et al. [10] carried out the molecular dynamics calculation of supercritical CO_2. They used 343 CO_2 molecules in cubic unit cells with a periodic boundary condition and *NVE* ensembles with a time step of 0.25 fs. After 2.5 ps for the initial equilibration, they used the next 2.5 ps for the simulation. They compared the simulation results from this model with those from a model based on a small basis set, 5s4p2d [9].

As a result, they confirmed that the large basis set improved the calculated fluid properties, such as second virial coefficient, pressure, and internal energy. In the supercritical region, 380 K and ρ = 0.020455 mol/cm^3, the

Table 1.3.3 Parameters used in the model potential of CO_2 (atomic unit) [10]

	$a^{(1)}$	$a^{(2)}$	$a^{(3)}$	$a^{(4)}$	$a^{(5)}$	$a^{(6)}$
O–O	26.709	1.7744	$+1.4502\times10^5$	-4.7153×10^4	+7027.0	−405.47
C–O	10.436	1.6944	-4.1617×10^2	-1.0802×10^5	+18684.0	−1105.40
C–C	26.095	1.6943	$+9.1846\times10^5$	-3.0615×10^5	+33817.0	−981.45
B–B	0.0	0.0	$+8.8873\times10^5$	-3.7482×10^5	+48251.0	−1949.80
B–O	0.0	0.0	-3.6825×10^5	$+1.4878\times10^5$	−18543.0	+931.70
B–C	0.0	0.0	-1.3251×10^6	$+4.5281\times10^5$	−51294.0	+1844.00

simulation well reproduced the pair distribution function after 3 Å. However, the left side of the first peak corresponding to the O–O distance was too low compared with the experimental data. Steinebrunner et al. attributed this deviation to an inaccurate relative well-depth of the potential for the slipped parallel and T-shaped dimers and to experimental problems.

We make two concluding remarks for the model of a CO_2 dimer based on these two examples:

(i) Successful description of the weak interaction of a CO_2 molecule requires large basis sets, such as 6–311+G(2df) and 8s6p4d1f, and requires electron correlation at the MP2 level.

(ii) Analytical pair functions based on the above ab initio MO calculations can well reproduce certain experimental data, such as those confirmed by Tsuzuki et al., but there is not complete agreement as confirmed by Steinebrunner et al.

1.3.2 Introduction to Molecular Simulation

Computer simulation methods, Monte Carlo (MC) and molecular dynamics (MD) simulations, provide many kinds of detailed molecular information on the liquid and solution [17]. Liquid-state theory also gives molecular level structure by the pair correlation function [18]. These methods show how the macroscopic properties depend on the microscopic structures on the basis of the assumed molecular interaction in the system [17, 18].

The MD simulation supplies the molecular trajectories by solving the equation of motion of the molecular system, whereas the MC simulation generates the so-called canonical ensemble of the liquid and solution by the random number [17]. Many applications of these methods will be found in the following sections. Before doing so, in this section, we present an elementary introduction to the methodology of molecular simulation, focusing on the MD method.

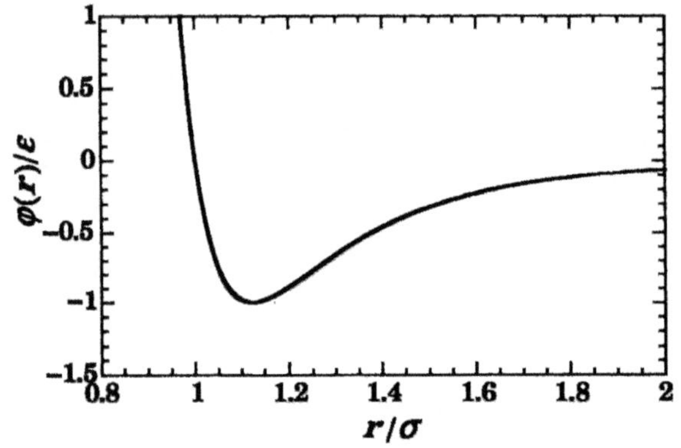

Fig. 1.3.1 Lennard-Jones function vs. atomic distance r

(a) Molecular Interaction

Let us assume a very simple system with N Ar-like particles as an example. The molecular interaction is expressed by the Lennard–Jones function, which has repulsion and attraction part

$$\phi(r) = 4\varepsilon\left\{\left(\frac{\sigma}{r}\right)^{12} - \left(\frac{\sigma}{r}\right)^{6}\right\}, \tag{1.3.10}$$

where r is the atomic distance. This is shown in Fig. 1.3.1. This function has the minimum value $-\varepsilon$ at $r = 2^{1/6}\sigma$. The length and energy will be measured by σ and ε, respectively.

(b) Classical Mechanics

MD simulation gives the molecular trajectories by numerical classical mechanics with the following practical conditions.

Periodic Boundary Condition

The periodic boundary condition (Fig. 1.3.2) is used to avoid the large effects of the wall when a physical box is used. A cubic cell is usually assumed, because liquid is isotropic.

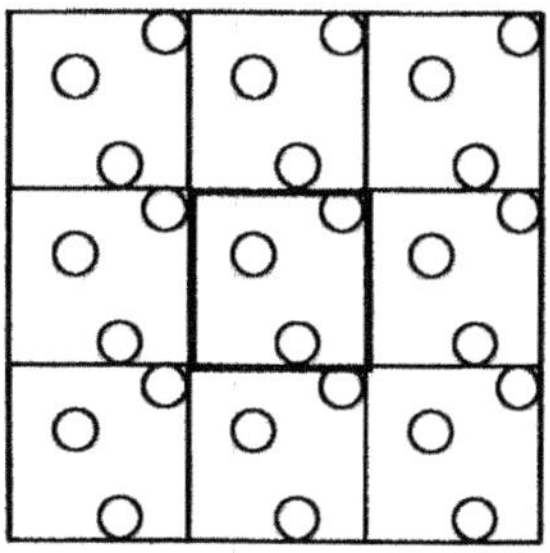

Fig. 1.3.2 A two-dimensional version of the periodic boundary condition

Initial Configuration

When a simulation is started, some initial configuration must be given. Usually a distorted face-centered-cubic (FCC) lattice is used for a dense liquid as the normal liquid state, because it is so dense that the random configuration generated only by random numbers has too high a potential energy. The initial velocity distribution is prepared as a Maxwell distribution corresponding to the desired temperature of the system. The velocity of the center of mass is set as zero, because the motion of the whole cell has no meaning in the properties of the liquid state.

Numerical Solution of the Equation of Motion

As already mentioned, a monatomic liquid is treated in this introduction for simplicity. A molecular liquid can be studied in a similar way, where the motion of the each atom is solved. The motion of the position vector $\boldsymbol{r}_i = (x_i, y_i, z_i)$ of the i-th atom with mass m is written as follows by means of the force $\boldsymbol{f}_i = (f_{ix}, f_{iy}, f_{iz})$ which the atom i feels (only the x-component of a vector will be given explicitly):

$$m\frac{\mathrm{d}^2 x_i}{\mathrm{d}t^2} = f_{ix} = -\sum_j \frac{\partial \phi(r_{ij})}{\partial x_i}, \tag{1.3.11}$$

where the function $\phi(r_{ij})$ is the potential energy between atoms i and j.

The initial time is set as $t = 0$. After a small time interval Δt, the position $r(0 + \Delta t)$ is written as follows:

$$x(0+\Delta t) = x(0) + \frac{\mathrm{d}x}{\mathrm{d}t}\Delta t + \frac{\mathrm{d}^2 x}{\mathrm{d}t^2}(\Delta t)^2 . \tag{1.3.12}$$

Now we assume the positions of the atoms are known up to time t. Then, we have the following Taylor expansion of r with respect to Δt:

$$x(t \pm \Delta t) = x(t) \pm \frac{\mathrm{d}x}{\mathrm{d}t}\Delta t + \frac{\mathrm{d}^2 x}{\mathrm{d}t^2}(\Delta t)^2 , \tag{1.3.13}$$

where the higher order terms are neglected because Δt is small enough. The sum and difference of the (1.3.13) are

$$x(t+\Delta t) + x(t-\Delta t) = 2x(t) + 2\frac{\mathrm{d}^2 x}{\mathrm{d}t^2}(\Delta t)^2 , \tag{1.3.14}$$

$$x(t+\Delta t) - x(t-\Delta t) = 2\frac{\mathrm{d}x}{\mathrm{d}t}\Delta t . \tag{1.3.15}$$

Equation (1.3.14) shows that the position at $t+\Delta t$ is calculated from the positions at $t-\Delta t$ and t and the acceleration $\mathrm{d}^2 x/\mathrm{d}t^2$, which can be obtained from (1.3.11), as:

$$x(t+\Delta t) = 2x(t) - x(t-\Delta t) + 2\frac{\mathrm{d}^2 x}{\mathrm{d}t^2}(\Delta t)^2 . \tag{1.3.16}$$

The velocity $\boldsymbol{v} = (v_{x,} v_{y,} v_{z,})$ at time t is estimated as

$$v_x(t) = \{x(t+\Delta t) - x(t-\Delta t)\}/(2\Delta t) . \tag{1.3.17}$$

The value of the time interval Δt is chosen so that there are only slight departures from the conservation of the total energy of the system (the sum of the kinetic energy and the potential energy).

The above mentioned method is called the Verlet method and is very popular, and a similar method (the leap-frog method) is also used frequently. There is another method, the predictor-corrector method which gives more precise solutions if the time interval Δt is chosen carefully [17].

Dimensionless Equation of Motion

The equation of motion should be made dimensionless, when it is solved numerically by computer. The Lennard–Jones system is again used as an example. The unit of mass is the mass of the atom m. The unit of length is the size parameter σ in the potential function (1.3.10). The energy unit is the energy parameter ε in the function. The unit of time τ is chosen so that the equation of motion can be written as simply as possible. The reduced variables x^r, ϕ^r and t^r are related to the original variables as follows:

$$x = x^r \sigma \ , \ \phi = \phi^r \varepsilon \, , \ t = t^r \varepsilon \tag{1.3.18}$$

When these are put into the equation of motion, we have

$$\frac{m\sigma}{\tau^2} \frac{\mathrm{d}^2 x^r}{\mathrm{d}(t^r)^2} = -\frac{\varepsilon}{\sigma} \frac{\mathrm{d}\phi^r}{\mathrm{d}x^r} \, . \tag{1.3.19}$$

Here the unit of time τ is defined by the equation

$$\frac{m\sigma}{\tau^2} = \frac{\varepsilon}{\sigma} \, . \tag{1.3.20}$$

Then the equation of motion has the simple form:

$$\frac{\mathrm{d}^2 x^r}{\mathrm{d}(t^r)^2} = -\frac{\mathrm{d}\phi^r}{\mathrm{d}x^r} \, . \tag{1.3.21}$$

In this unit system, all physical quantities are expressed in a useful way in the system where the Lennard–Jones potential can be assumed.

Example

An example of a system (the number of particle is $N = 256$, the volume per particle is $V/N = 3.5\,\sigma^3$) is shown in Fig. 1.3.3. The x-coordinate $x(t)$ and the x-component of the particle velocity $v_x(t)$ are plotted as a function of time t. The assumed density is low like in the critical region. For this reason, the frequency of the molecular collision is low. It is seen that the coordinate jumps at certain times, because of the periodic boundary condition.

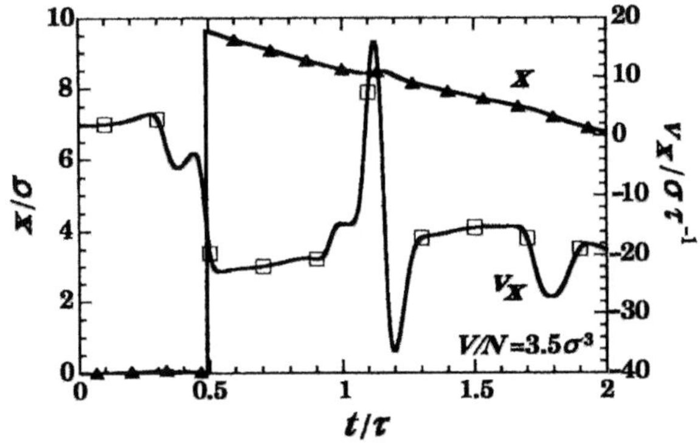

Fig. 1.3.3 An example of trajectories

(c) Statistical Averages

In the MD simulation, the thermal average $\langle A \rangle$ of a physical quantity A is calculated by the time average:

$$\langle A \rangle = \lim_{t \to \infty} \frac{1}{t} \int_0^t A(t')\,\mathrm{d}t' . \tag{1.3.22}$$

This average is valid only when the molecular configurations simulated by that MD run cover almost all important representative configurations in the equilibrium state. Whether this criterion is satisfied is usually decided by the time evolution of the potential energy. The plot of the potential energy vs. time is shown in Fig. 1.3.4. The simulated system is believed to be in the equilibrium state after the relaxation stage.

The average temperature T is obtained by the equipartition theorem:

$$(1/2) f k_{\mathrm{B}} T = \langle E_k \rangle , \tag{1.3.23}$$

where f is the total number of degrees of freedom of the system, k_{B} is the Boltzmann constant and E_k is the total kinetic energy of the system.

The above mentioned MD average corresponds to that in the microcanonical (*NVE*) ensemble in statistical mechanics. Other MD methods are also available, e.g., *NVT*, *NPH* and *NPT* [17]. In the *NPT* ensemble, the volume of the system fluctuates as shown in Fig. 1.3.5, where the state is close to the critical point of the Lennard–Jones fluid [19].

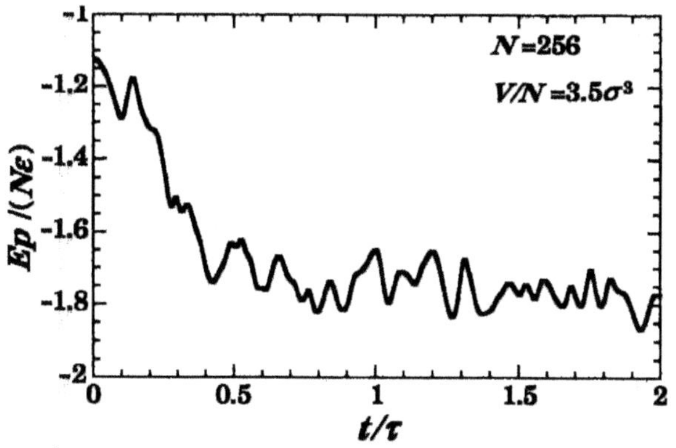

Fig. 1.3.4 Potential energy of the system vs. time plot

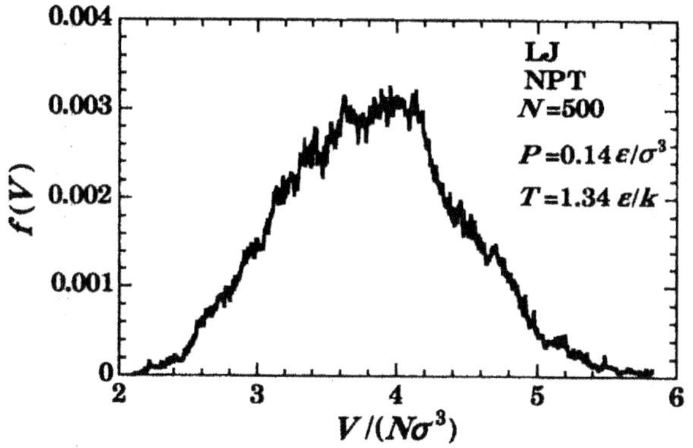

Fig. 1.3.5 Distribution of the volume of the system in the *NPT* ensemble

(d) Dynamical Properties

By using MD simulations, one can save the trajectories of all atoms in the system to obtain the many kinds of dynamical properties of the molecular system. One of the most popular quantities is the self-diffusion coefficient, which is calculated by the plot of the mean square displacement of the center of mass of a molecule as a function of time. A linear part is usually found in this plot. The tangent of this part corresponds to $6D$, where D is the self-diffusion coefficient. Time integration of the velocity autocorrelation function also gives the self-diffusion coefficient D [17]. The Fourier transform of the time correlation function contains other information on the motion of the molecules.

(e) Liquid and Solution Structures

All the coordinates of the system are known in a MD run. One can analyze these to get the required kinds of structures in liquid or solution. The most important quantity is the radial distribution function $g(r)$. This is calculated by the histogram of the atomic distance in the MD sample. The histogram is normalized to that in a complete random distribution at the same density. Some examples will be given in the following sections.

1.3.3 Solvation Structure in Supercritical Fluids

Molecular simulations for supercritical LJ fluids have been reported from as early as the 1960s [20–24], although they were mostly concerned with *PVT* properties. Based upon current interest in supercritical fluids (SCFs), molecular simulations have actively been done from the end of the 1980s. In particular, two pioneering works excited the interest of theoretical chemists. One is a molecular dynamics calculation for LJ mixtures by Debenedetti's group [25]. The other is a neutron diffraction measurement of supercritical water (SCW) by Neilson's group [26, 27]. In the former calculation [25], clustering of solvent molecules around a solute which shows a strongly attractive interaction with the solvent was found. This successfully explains the largely negative partial molar volume of solvation observed by Eckert et al. [28]. The calculation also showed a lean solvation of solute which weakly interacts with the solvent. The calculation was later extended to aqueous systems by some groups [29–34], presenting an interesting picture for the mixtures of SCFs. The latter experiment [26, 27] was a great surprise to researchers on SCW. It is reported that the hydrogen bond has almost completely disappeared in the SCW, denying the results of molecular dynamics (MD) calculations up to that time. This gave rise to much controversy among physical chemistry of SCW. The hydrogen bond in SCW has thus actively been investigated, based upon computer simulations as well as various kinds of experiments.

In this subsection, a number of structural features of SCW are presented in addition to the hydrogen bond. The solvation structure of the solute in SCW is also discussed [29–34], representing the common behavior of SCF mixtures. Here, we skip the short-range orientational correlations between molecules in SCFs of dipolar and quadrupolar molecules such as freons and carbon dioxide, respectively [35–37].

(a) Supercritical Water

Most of the empirical potential functions for water were determined so that they successfully reproduce several physical properties of water at ambient condition, i.e. 298 K and 1 atm. Charges on interaction sites of this kind of model are fixed at certain values. Molecular dynamics calculations for SCW were first done using these fixed charge potential models such as CC [38], TIP4P [39–41], SPC and SPC/E [42–44], and modified flexible models [45, 46]. With respect to the hydrogen bond, Mountain [39] reported that 70% of the hydrogen bonds were still found in supercritical TIP4P water compared with that at ambient conditions. The first peak of $g_{OH}(r)$ at about $r = 2$ Å was certainly observed, although it is rather small. However, on the contrary, $g_{OH}(r)$ obtained by the isotopic substitution method of neutron diffraction measurements [26, 27] showed no peak there for SCW. That is, no hydrogen bond was found. The neutron diffraction measurement claimed that the contradiction was caused by neglect of many-body effect in the intermolecular interaction. A number of molecular simulation studies [40–52] for SCW as well as other experimental works [53–57] started entering into the controversy. Most of them showed that the hydrogen bondings in SCW persist. In response to these investigations, Soper et al. [58, 59] reanalyzed their experiment and reached the conclusion that there exist hydrogen bonds. However,

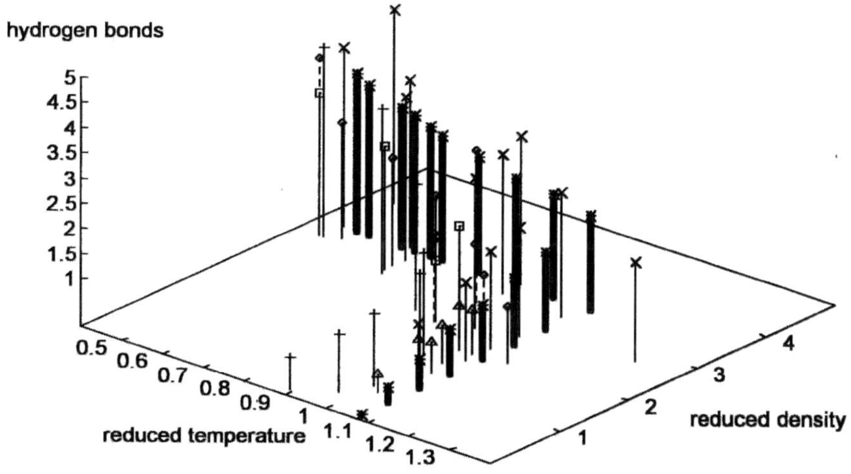

Fig. 1.3.6 Density and temperature dependence of the average number of hydrogen bonds in supercritical water: △, NMR [55]; ◇, X-ray [61]; □, neutron diffraction [59]; ◇ with dotted line, neutron diffraction [58]; +, MD (SPC/E) [43]; ×, MD (SPC/E) [44] and ∗, MD (RPOL) [51]

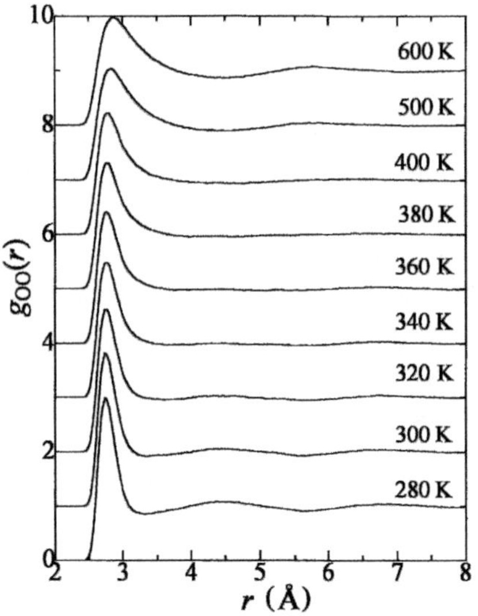

Fig. 1.3.7 Temperature dependence of the oxygen–oxygen pair correlation function for water along an isochore at 1 g/cm^3 [51]

the experimental number of hydrogen bondings is small compared with that estimated from molecular simulations based upon fixed charge models as shown in Fig. 1.3.6. Considering the fact that the dipole moment of a water molecule in ice is 2.6 D but decreases to 1.85 D in the gas phase, the dipole moment 2.2–2.4 D of the fixed charge models must overestimate the interaction in SCW. Although ab initio MD [47, 48] might be the best way to describe the many-body effect of the hydrogen bonds, the computational cost of this method is very high. Thus, Okazaki et al. adopted [51, 52] the RPOL model [60] for water which is economical with respect to the computational cost. Since atomic polarizabilities are explicitly taken into account in the model, RPOL water can describe a state-dependent many-body potential. This may be essential to reproduce hydrogen bonds in high-temperature and low-density SCW. Their result of the number of hydrogen bonds for RPOL water at sub- and supercritical states is also presented in Fig. 1.3.6. [43, 44, 51, 55, 58, 59, 61] It shows a good correspondence with the experimental results by NMR, neutron, and X-ray diffraction. With increasing temperature and decreasing density, the number of hydrogen bonds decreases.

The structure of the fluid has been investigated in detail for RPOL water at sub- and supercritical states. Figure 1.3.7 presents the oxygen–oxygen pair correlation function $g_{OO}(r)$ for water along an isochoric line at 1.0 g/cm^3 [51]. It is interesting to find that the second peak considerably changes its position.

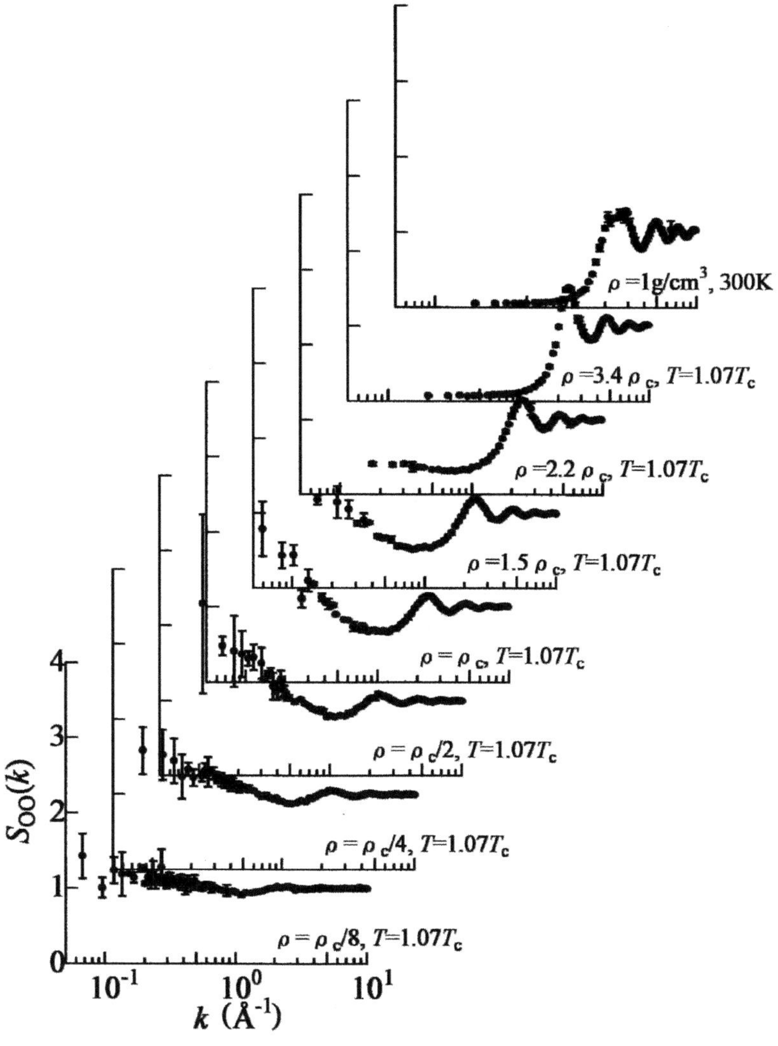

Fig. 1.3.8 Oxygen–oxygen static structure factor $S_{OO}(k)$ for supercritical water at various densities along an isotherm at $T = 1.07\,T_C$ [52]

At $T = 280$ K, the ratio of the r value of the second peak to that of the first one is 1.6, which represents ice-like tetrahedral coordination of water molecules around a water molecule. However, it is about 2.0 for the fluid at $T = 600$ K indicating that the fluid structure is similar to that of simple liquids such as argon and xenon. The Arrhenius plot of the self-diffusion coefficient clearly shows different activation energies reflecting this structural change, 13 kJ/mol in the low-temperature region and 6.8 kJ/mol in the high-

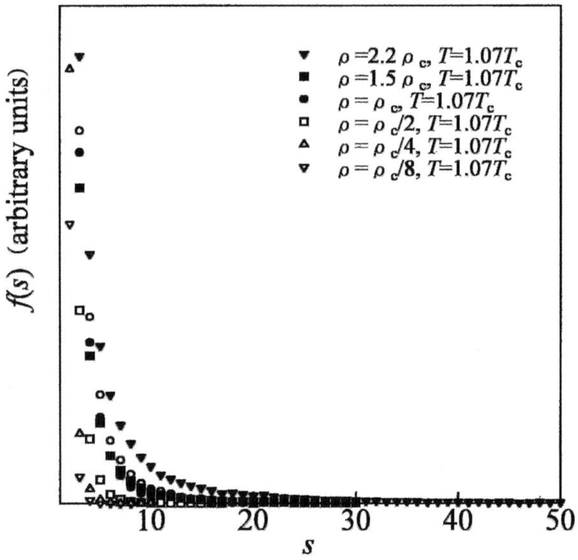

Fig. 1.3.9 Probability $f(s)$ that a hydrogen-bonded cluster composed of s molecules is found in supercritical water at various densities along an isotherm at $T = 1.07\,T_C$ [52]

temperature region. The time constant of the total dipole moment relaxation also decreases according to the structural change to the simple liquid type where the hydrogen bonding network is not strong, although the system is still percolated by the hydrogen bonds.

Along an isotherm at $1.07\,T_C$, the oxygen–oxygen structure factor $S_{OO}(k)$ presented in Fig. 1.3.8 [52] shows strong small-angle scattering near the critical density ρ_C. The calculation successfully reproduces the long-range strong density fluctuation. The percolation threshold for the hydrogen bonds was found at a density a few times higher than ρ_C. The fluid at densities lower than this threshold may be regarded as an assembly of clusters. The clusters formed in SCW have been analyzed in detail. The size of the cluster was defined by the number of molecules which are connected to each other by the hydrogen bonds. The distribution of cluster size $f(s)$ is given in Fig. 1.3.9. [52] The higher the density, the more often the large clusters are observed. The size distribution was found to follow Fisher's droplet model, $f(s) \propto s^{-\tau} \exp(-as^{2\sigma})$, where $\tau = 2.1$ and $\sigma = 1/2$. A stereograph of a cluster of $s = 104$ is presented in Fig. 1.3.10. Clusters are bulky, having a lot of open space in them. A log–log plot of the radius of gyration clearly shows a linear line with a slope of 2.2, showing that the clusters are fractal. Furthermore, it was found that the larger the cluster, the longer its lifetime. Together

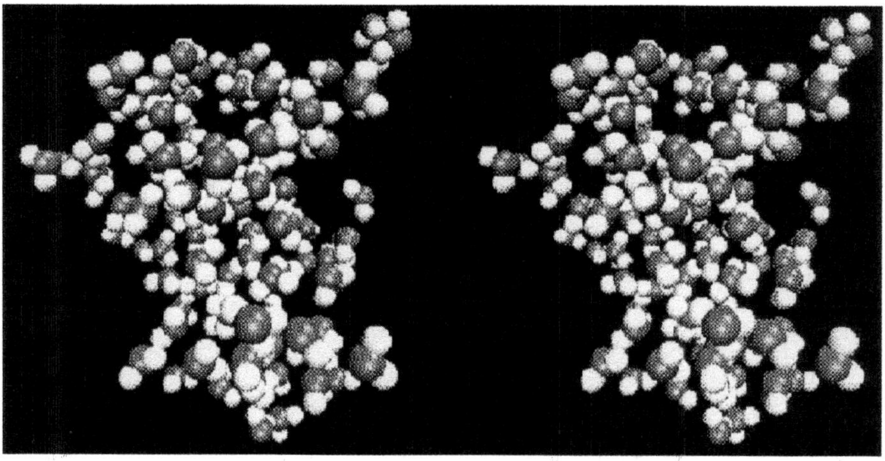

Fig. 1.3.10 Stereograph of a cluster of size $s = 104$ formed in supercritical water

with the density dependence of the size distribution, this shows the possibility of finding a kind of critical slowing down with respect to the properties related to the hydrogen bonding network.

(b) Supercritical Aqueous Solutions

The solvation structure of various solutes in SCW has been investigated since the end of the 1980s [29–34]. Cummings et al. [29–31] performed a series of molecular dynamics calculations for supercritical aqueous solution of salt, rare gas, and organic molecules. Figure 1.3.11 [31] presents the pair correlation function $g(r)$ between the solute and water. For NaCl solution, the first peak shows a very strong correlation between Na^+ and water, representing a rigid solvation shell. The function is greater than unity over a long range of r. These clearly show that clustering occurs around the attractive solute in the fluid. On the contrary, the solvent density is low around nonionic solutes such as benzene, benzonitrile, and toluene compared with the bulk density. Similar behavior is found around argon, too. Thus, these solutes result in a reduction of the solvent distribution around them, that is, an apparently repulsive correlation with water molecules. These present a molecular based picture for the large negative partial molar volume of the attractive solute and the largely positive one for the repulsive or weakly interacting solute.

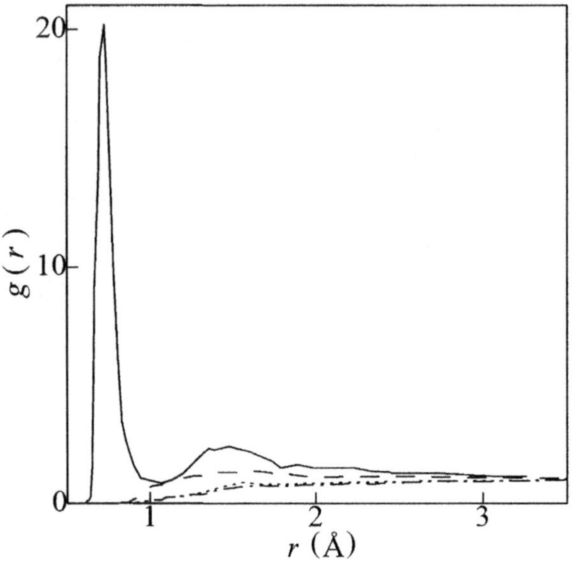

Fig. 1.3.11 Solute–water pair correlation function for supercritical aqueous solutions. Na^+–water (——), benzonitrile–water (— —), benzene–water (- - - -), and toluene–water (— · —) [31]

1 Solution Structure in Supercritical Fluids

1.4 Static Properties of Supercritical Fluids

If the intermolecular interaction $\phi(r)$ is obtained, various thermodynamic properties, such as the internal energy U or pressure P at a given point of volume V and temperature T, can be calculated by molecular simulations. It is not so difficult to calculate these properties above the critical temperature, where the system can be assumed to be homogeneous. Below the critical temperature, the calculated *PVT* relation shows the so-called van der Waals loops. However, it was recently proved that these loops vanish by employing a larger number of particles in the calculations [1, 2].

Computer simulation has several merits in these studies. It can be applied even to fluids at high temperature and high pressure where experimental studies are not easy. Besides, molecular simulation can extract general relations between the microscopic potential and the macroscopic properties. The macroscopic properties of fluids are closely related to their microscopic structures. In this section, we will describe several applications of Monte Carlo (MC) molecular simulation to the thermodynamic properties (excess functions, solubility and adsorption) of supercritical fluids.

1.4.1 Excess Thermodynamic Quantities

(a) *PVT* Relation

Since Nicolas et al. [3] reported the *PVT* relations for the LJ particles calculated by the MC method, attempts have been continuously made to arrange data by an empirical equation of state. For example, Johnson et al. [4] and Kolafa et al. [5] have reported optimized values for the 32 adjustable parameters in the modified Benedict–Webb–Rubin (BWR) equations of state. Now, we can estimate the *PVT* relation of substances precisely by using the BWR equation of state if suitable ε and σ values are available. Some equations of state, including the BWR type, are explained in Sect. 2.4.

(b) Definition and Features of Excess Functions

Recently, the physical and chemical properties of mixtures near or in the supercritical region has become of great interest, as well as those of pure components, especially in the field of chemical engineering. The excess thermodynamic quantity, X^{E}, for a binary system is defined as follows,

$$X^{\mathrm{E}} = X_{\mathrm{m}} - \left\{x_1 X_1^0 - x_2 X_2^0\right\}, \tag{1.4.1}$$

where X_{m} and X_i^0 are the thermodynamic quantity of the mixture and that of the pure component, respectively. x_i is the mole fraction of component i. For example, the molar excess volume, V_P^{E}, the molar excess enthalpy, H_P^{E}, and the molar excess heat capacity, c_P^{E}, are often used.

The excess quantities can be measured quite precisely under ordinary laboratory condition, and provide us with fundamental information on the intermolecular interactions in a mixture. Abundant experimental data of H_P^{E} and V_P^{E} have been accumulated for normal mixtures. It has been revealed that the behavior of H_P^{E} and V_P^{E} is not too complex, unless the system contains a strong and specific intermolecular interaction like the dipole–dipole interaction. H_P^{E} or V_P^{E} for mixtures of paraffin hydrocarbons are nearly symmetrical with respect to the mole fraction or the volume fraction, and the absolute values of them are small. The mixtures of simple liquids exhibit thermodynamic properties which are fairly close to those of the ideal mixture, and the behavior of excess thermodynamic functions for such mixtures are well described by the regular solution theories proposed by Guggenheim [6] and Hildebrand-Scatchard [7].

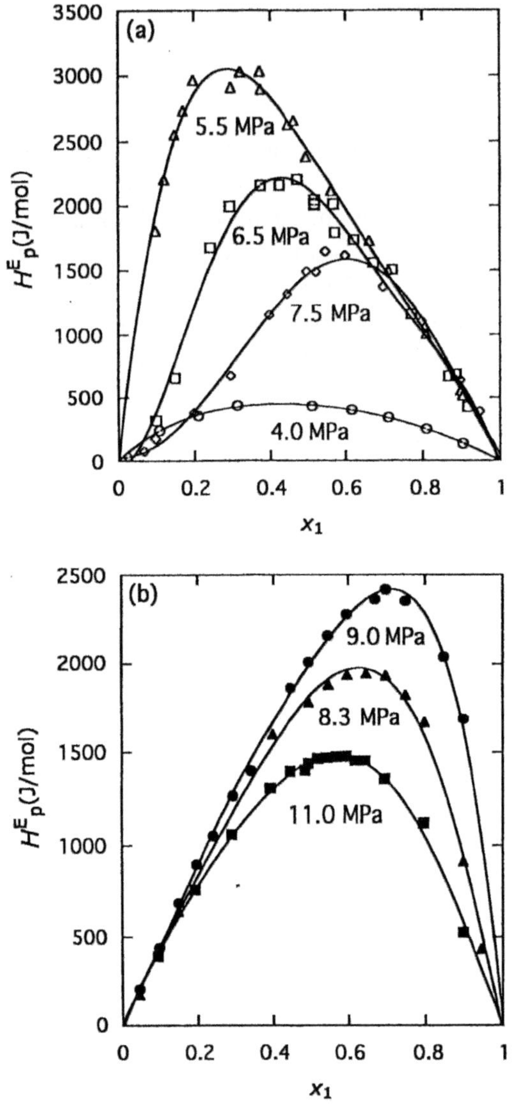

Fig. 1.4.1 a, b Concentration dependence of H_P^E for a carbon dioxide + ethane system at 308.4 K [8]. Lines are drawn for visual aid

Under high-temperature and high-pressure conditions, including the supercritical region, on the other hand, it is extremely difficult to perform direct thermodynamic measurements. The measurement of H_P^E and V_P^E for binary solutions near to or in the supercritical region were carried out by Wormald et

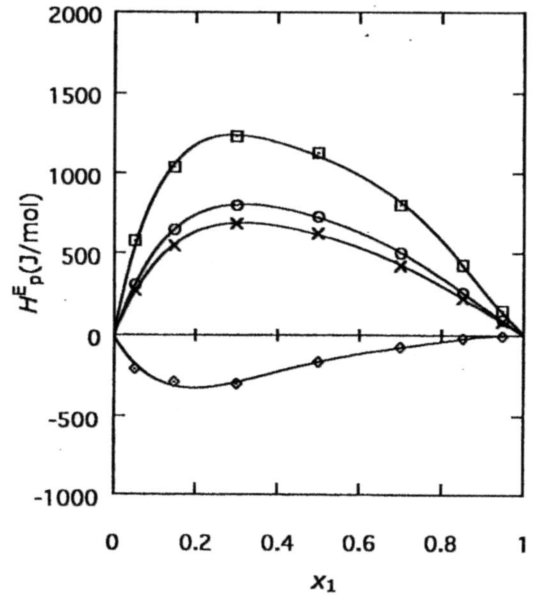

Fig. 1.4.2 Concentration dependence of H_P^E for X-2-1 systems at 250 K and 120 atm. □, X = A; ○, X = LB; ×, X = ID; ◇, X = S [11]. Lines are drawn for visual aid

al. [8] and Ott et al. [9]. According to their work, the thermodynamic behavior of a mixture in the supercritical region is quite different from that at low temperature and low pressure. Figure 1.4.1 shows the concentration dependence of H_P^E for a carbon dioxide + ethane system, at different pressures just above the critical temperatures for both components, measured by Wormald et al. [8]. H_P^E for the mixtures in the supercritical region has characteristic features, i.e., extremely large absolute values and very complex dependencies on concentration and pressure.

(c) Calculation of Excess Functions

Since McDonald [10] developed the MC method in the NPT ensemble, it became easier to calculate excess thermodynamic functions such as H_P^E and V_P^E for mixtures by molecular simulation than before. Fujihara et al. [11] carried out MC calculations for supercritical Lennard–Jones fluids and their mixtures with different kinds of combining rules. Figure 1.4.2 shows the concentration dependence of H_P^E for so-called X-2-1 (X = A, S, LB and ID) systems [11]. The second and third numerals indicate the ratio $\varepsilon_{22}/\varepsilon_{11}$ and σ_{22}/σ_{11}, respectively. Energy and size parameters for the pure components are

Table 1.4.1 Intermolecular potential parameters for pure components

	ε/k (K)	σ (nm)
component 1	84.71	0.3405
component 2	169.42	0.3405

Table 1.4.2 Mixing rules for unlike interactions

type of mixture X	unlike interaction
LB (Lorenz–Berthelot)	$\varepsilon_{12}=(\varepsilon_{11}*\varepsilon_{22})^{1/2}$
A (association)	$\varepsilon_{12}=\varepsilon_{11}$
S (solvation)	$\varepsilon_{12}=\varepsilon_{22}$
ID (quasi-ideal)	$\varepsilon_{12}=(\varepsilon_{11}+\varepsilon_{22})/2$
for all X	$\sigma_{12}=(\sigma_{11}+\sigma_{22})/2$

summarized in Table 1.4.1 and the four combining rules for unlike interactions are shown in Table 1.4.2. [11] The calculated H_P^{E} for the model systems in the supercritical region represents the characteristics of the real supercritical fluid mixtures mentioned above.

In order to investigate the origin of this unique thermodynamic behavior, two different types of mixing process were considered which eventually lead to the same mixture state. One is the usual mixing at constant pressure. We mix the two components 1 and 2 at constant temperature and pressure of T_0 and P_0. The resultant mixture has a volume of V_{m} and an internal energy of $U_{\mathrm{m}}(x_1, V_{\mathrm{m}})$ and the excess internal energy in this isobaric process is given by

$$U_P^{\mathrm{E}} = U_{\mathrm{m}}(x_1, V_{\mathrm{m}}) - \{x_1 U_1(V_1) + x_2 U_2(V_2)\}\,. \tag{1.4.2}$$

Alternately, we can prepare the same mixture as follows. In the first step, we compress or expand component 1 and 2 isothermally at T_0 to have the same molar volume of V_{m}. In the second step, we mix these two components isothermally at T_0 and at constant volume V_{m}. Since the temperature and volume of this mixture are the same as those of the mixture made under isobaric conditions and the internal energy is a state function, the pressure and the internal energy of this mixture are also P_0 and $U_{\mathrm{m}}(x_1, V_{\mathrm{m}})$. The excess internal energy during this isochore process, which is the second step of the two-step mixing process, is given by

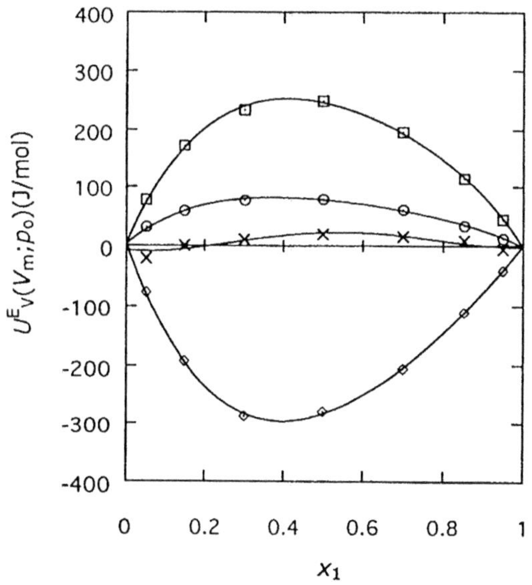

Fig. 1.4.3 Concentration dependence of $U_V^E(V_m;P_0)$ for X-2-1 systems at 250 K and 120 atm. □, X= A; ○, X= LB; ×, X= ID; ◇, X= S [11]. Lines are drawn for visual aid

$$U_V^E(V_m;P_0) = U_m(x_1,V_m) - \{x_1U_1(V_m) + x_2U_2(V_m)\} \tag{1.4.3}$$

The difference between $U_V^E(V_m;P_0)$ and U_P^E becomes

$$U_V^E(V_m;P_0) = U_P^E - \left\{ x_1 \int_{V_1}^{V_m} \left(\frac{\partial U_1}{\partial V}\right)_T \mathrm{d}V + x_2 \int_{V_2}^{V_m} \left(\frac{\partial U_2}{\partial V}\right)_T \mathrm{d}V \right\},$$

$$\left(\frac{\partial U}{\partial V}\right)_T = T\left(\frac{\partial P}{\partial T}\right)_V - P = T\left(\frac{\alpha_P}{\kappa_T}\right) - P, \tag{1.4.4}$$

where α_P and κ_T are the thermal expansion coefficient and the compressibility, respectively. If we have data for H_P^E, V_P^E and the equation-of-state for the pure components, then we can evaluate $U_V^E(V_m;P_0)$. Figure 1.4.3 shows the concentration dependence of $U_V^E(V_m;P_0)$ for the corresponding X-2-1 systems [11].

In contrast to H_P^E and U_P^E, $U_V^E(V_m;P_0)$ shows quite simple behavior, that is, the magnitude is small and the concentration dependence is symmetrical as the regular solution theory predicts. $U_V^E(V_m;P_0)$ for the carbon dioxide + ethane system was also estimated by applying this data conversion method to H_P^E results measured by Wormald et al. As in the case of model mixtures, $U_V^{\mathrm{E}}(V_{\mathrm{m}};P_0)$ of this real system became small and quite symmetrical with concentration. Considering that the difference between U_P^{E} and $U_V^{\mathrm{E}}(V_{\mathrm{m}};P_0)$ is due to the change of internal energy during the compression and expansion process of pure components, the volumetric behavior of the system plays a predominant role in the unique thermodynamic behavior in the supercritical region.

1.4.2 Solubility

(a) Fundamental Equations

In this subsection, we consider the solubility of a solid compound (2) in a supercritical fluid (1). The residual chemical potential of the solid compound in the supercritical fluid, μ_2^{r}, can be related to Henry's constant, H_2, at infinite dilution,

$$H_2 = \rho_1 kT \exp\left(\frac{\mu_2^{\mathrm{r}}}{kT}\right), \tag{1.4.5}$$

where ρ_1 is the number density of the supercritical fluid. The solubility of the solid compound in the supercritical fluid, y_2, is usually very low. Therefore, y_2 can be calculated by

$$H_2 = \frac{f_2^{\mathrm{G}}}{y_2}, \tag{1.4.6}$$

where f_2^{G} is the fugacity in the supercritical phase. Since f_2^{G} is equal to the fugacity of compound 2 in the solid phase (pure solid), f_2^{S}, at the equilibrium condition, the following equation is obtained:

$$f_2^{\mathrm{G}} = f_2^{\mathrm{S}} = P_2^{\mathrm{sat}} \exp\left\{\frac{v_2^{\mathrm{S}}\left(P - P_2^{\mathrm{sat}}\right)}{RT}\right\}, \tag{1.4.7}$$

where P_2^{sat} is the saturation pressure and v_2^{S} is the solid-state molar volume. From the above equations, y_2 is given as follows:

$$y_2 = \frac{P_2^{\mathrm{sat}}}{H_2} \exp\left\{\frac{v_2^{\mathrm{S}}\left(P - P_2^{\mathrm{sat}}\right)}{RT}\right\}, \tag{1.4.8}$$

where P_2^{sat} and v_2^{S} are the properties of the pure compound, and can be obtained from experimental data or correlation methods. Only the residual chemical potential μ_2^{r} should be evaluated by computer simulation, where the test particle method proposed by Widom [12] and the isothermal-isobaric Kirkwood method [13] are applied.

(b) Supercritical Carbon Dioxide (1) + Naphthalene (2) System

The solubilities of naphthalene in supercritical carbon dioxide have been measured by many researchers as a standard system. Shing and Chung [14] calculated the solubilities of naphthalene in supercritical carbon dioxide using molecular simulation and showed the applicability of the MC method. However, the potential parameters were not optimized and the calculated results were not in good agreement with the experimental data. Eya et al. [15] used the Lennard–Jones potential and applied the test particle method proposed by Widom to calculate the residual chemical potential of naphthalene in supercritical carbon dioxide. They introduced the interaction parameter k_{ij} for unlike pairs,

$$\varepsilon_{ij} = \left(1 - k_{ij}\right)\sqrt{\varepsilon_{ii}\varepsilon_{jj}} \ . \tag{1.4.9}$$

Figure 1.4.4 shows the calculated results of solubilities [15]. The calculated results are in good agreement with the experimental data with $k_{12} = 0.05$.

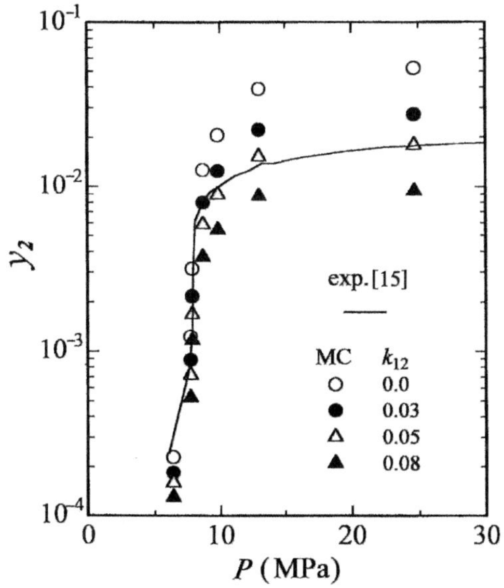

Fig. 1.4.4 Solubilities of naphthalene in supercritical carbon dioxide at 308.15 K [15]

(c) Supercritical Carbon Dioxide (1) + Aromatic Compound (2) Systems

Iwai et al. [17, 18] applied the MC method to calculate the solubilities of phenol, xylenol isomers, naphthalene, naphthol isomers, dimethylnaphthalene isomers, phenanthrene and anthracene in supercritical carbon dioxide. Carbon dioxide was treated as a single-site molecule, and aromatic compounds were treated as multisite molecules. For aromatic compounds, a benzene ring, a methyl group and a hydroxyl group were adopted as sites (groups). The Lennard–Jones potential was used as the site–site potential, and the Lorentz–Berthelot combining rules were adopted for unlike pair sites. A modified test particle method was adopted to calculate the residual chemical potentials of aromatic compounds in supercritical carbon dioxide based on the canonical ensemble. The structures of aromatic compounds and the position of sites are shown in Fig. 1.4.5 [17]. The common potential parameters are listed in Table 1.4.3 [17]. As shown in Fig. 1.4.6 [17], the calculated results are in good

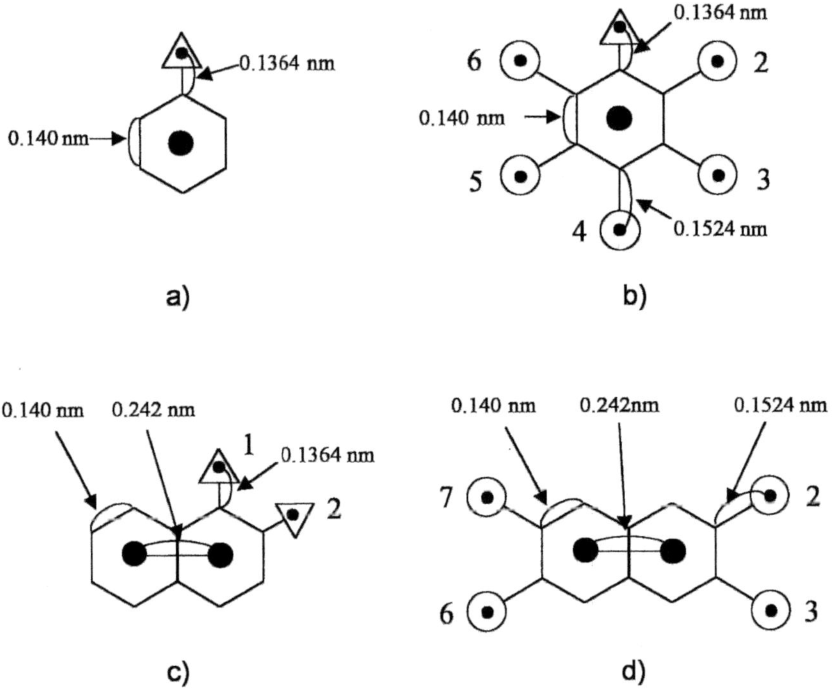

Fig. 1.4.5 Structures of aromatic compounds: (**a**) phenol; (**b**) xylenol; (**c**) naphthol; (**d**) dimethylnaphthalene: ●, site of benzene ring; △, site of hydroxyl group; O, site of methyl group [17]

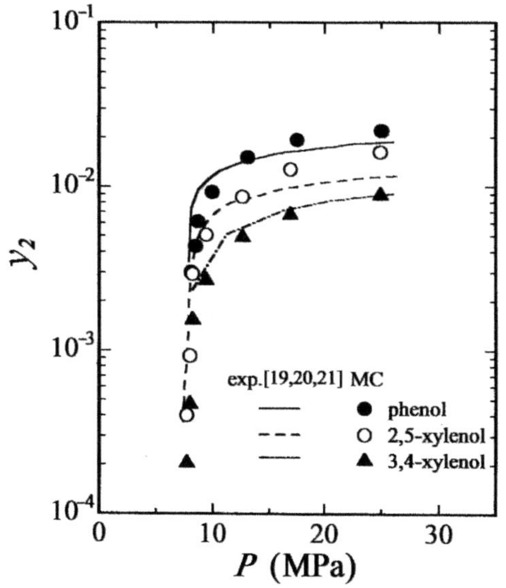

Fig. 1.4.6 Solubilities of phenol and 2,5- and 3,4-xylenol in supercritical carbon dioxide at 308.15 K [17]

agreement with the experimental data [19–21] in a wide range of pressures. The solubilities of isomers can be quantitatively distinguished by the group contribution site model. Figure 1.4.7 shows the radial distribution functions of supercritical carbon dioxide around the hydroxyl group of xylenol isomers [17]. This figure shows that the radial distribution functions are different in spite of using the common potential parameters for isomers. This means that the screening effect of substituents can be evaluated easily by the site model.

Table 1.4.3 Lennard-Jones potential parameters [17]

site	σ (nm)	ε/k (K)
CO_2	0.3912	225.3
benzene ring (one ring)	0.5501	360.2
benzene ring (one of two rings)	0.5501	251.0
hydroxyl group	0.3070	308.8
methyl group	0.3775	60.19

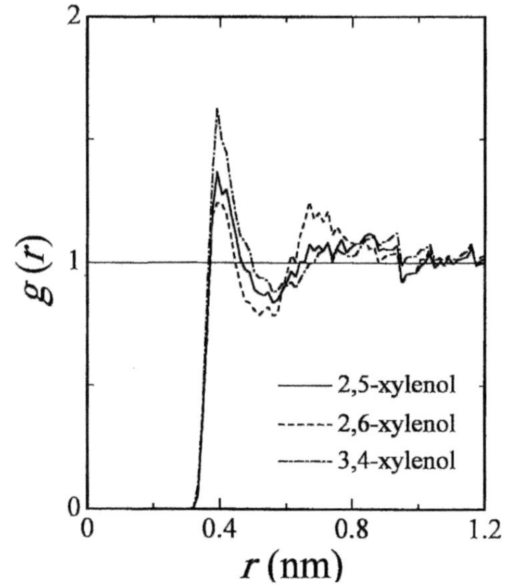

Fig. 1.4.7 Monte Carlo calculation results for radial distribution functions, $g(r)$, of supercritical carbon dioxide around OH groups of xylenol isomers at 308.15 K and 13 MPa [17].

(d) Supercritical Fluid (1) + Chain Molecule (2) Systems

MC simulation was applied to calculate the static properties of paraffins, higher alcohols and fatty acids in supercritical ethane and supercritical carbon dioxide [21–24]. Ethane and carbon dioxide were treated as single-site molecules, while chain molecules were approximated as many-site molecules. The Lennard–Jones potential was used as the site–site potential. The solubilities of chain molecules can be calculated quantitatively by introducing only one intermolecular parameter between unlike sites. Figure 1.4.8 shows the reduced mean-square end-to-end separations of paraffins [23], $\langle r^2 \rangle / \sum L^2$, in supercritical ethane and in supercritical carbon dioxide, where L is the bond length between sites. The calculated results of the mean-square end to end separations of paraffins increase with pressure both in supercritical ethane and carbon dioxide. The results in supercritical ethane are slightly longer than those in supercritical carbon dioxide because of the more emphatic interaction between ethane and paraffins.

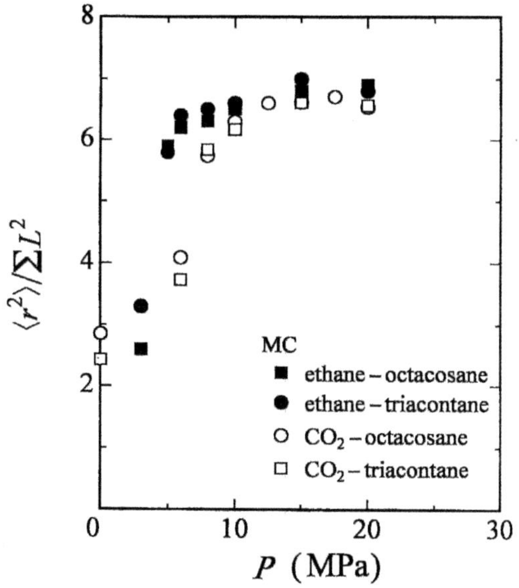

Fig. 1.4.8 Monte Carlo calculation results for reduced mean-square end-to-end separations of paraffins at 308.2 K [23]. Reprinted from: Y. Iwai, Y. Koga, Y. Arai: Monte Carlo simulation of *n*–paraffins and higher alcohols in supercritical carbon dioxide: *Fluid Phase Equilibria* (Elsevier Science) **116,** 267 (1996).

1.4.3 Adsorption

The enrichment of substances onto a solid surface from a fluid is called *adsorption*. The surface potential that attracts molecules is due to dispersion, induction and electrostatic interactions between solid atoms and molecules. Since molecular concentrations obey the Boltzmann distribution law in a classical equilibrium system, substances that attractively interact with a surface are preferentially concentrated in the first monolayer on the surface where the potential is the deepest. This explains the reason why substances are enriched on the surface from gases diluted in an inert gas; however, the understanding of adsorption characteristics of substances diluted in a supercritical fluid (SCF) requires two more factors: the competitive adsorption of solvent molecules in the adsorbed phase and the enhanced solubility in the SCF phase. It is essentially important to recognize that the adsorption in SCF is characterized by mixed-gas adsorption.

Adsorption equilibrium in an SCF is determined from the thermodynamic criterion of equilibrium: the equality of the chemical potentials of each substance in two phases, i.e. the adsorbed and fluid phases. Therefore, the surface concentration of an adsorbate is strongly influenced by the surface concentration of the SCF component as well as the change in the chemical potential (or solubility) of the adsorbate in the SCF phase. The former influence is called the competitive or coadsorption effect and the latter the enhanced solubility effect.

In molecular simulations, the standard method for calculating the density (and composition) in a pore in equilibrium with a fluid (mixture) is the grand canonical (μVT) ensemble MC method, in which the chemical potentials of each component (μ) composing a fluid are specified as well as the pore volume (V) and the temperature (T) [25, 26]. Nitta and Yoneya [27] applied the μVT ensemble MC method to investigate the adsorption characteristics of a solute diluted in SCF by choosing a system CO_2 (1) + benzene (2) / carbonaceous slitlike pore. Their calculation procedures are as follows:

(i) Carry out the NVT ensemble MC computation for pure CO_2 at a specified temperature (T) and density (ρ) with insertions of ghost molecules of CO_2 and benzene in order to calculate the residual chemical potentials of each component (μ_1^{r}, μ_2^{r}).

(ii) Calculate the chemical potentials (μ_1, μ_2) by specifying a small composition of benzene in the fluid (for example, $y_1 = 0.999$ and $y_2 = 0.001$),

$$\mu_i = k_{\mathrm{B}} T \ln(y_i \rho) + \mu_i^{\mathrm{r}} . \tag{1.4.10}$$

(iii) Calculate the pore density (ρ_{p}) and the mole fractions (x_1, x_2) in a slitlike pore by performing the μVT ensemble MC simulation at a specified set of μ_1, μ_2, V and T.

The structureless wall model composed of graphite basal planes, suggested by Steele [28], was used for the slitlike pore. The Lennard–Jones (LJ) potential function was used to describe the intermolecular interactions between molecules–molecules and molecules–carbon atoms. The LJ parameters are listed in

Table 1.4.4 Lennard-Jones parameters used in simulations [28]

	ε/k [K]	σ[nm]
CO_2 (1)	225.3	0.3910
benzene (2)	468.5	0.5506
carbon (s)	28.0	0.340

k_{12}=0.1232, k_{1s}=0.170, k_{2s}=0.0

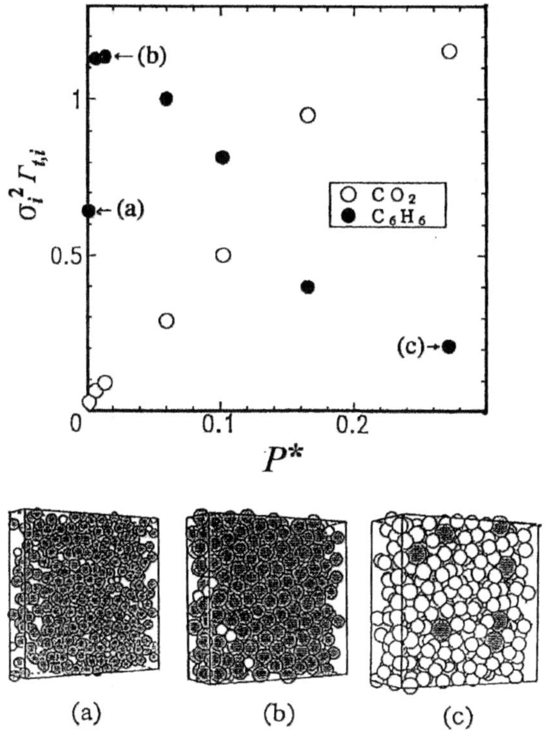

Fig. 1.4.9 The true adsorptions of CO_2 and benzene in a slit pore of $H = 2$ nm in equilibrium with a fluid mixture of $y_2 = 0.001$ at 313.2 K [28]

Table 1.4.4 [28], where k_{ij} is the binary energy parameter representing the deviation from the geometric mean of the pair potential wells, ε_i and ε_j.

Figure 1.4.9 shows the true adsorptions of CO_2 and benzene ($\sigma_{ii}^2 \Gamma_{t,i}$) in a slit pore of width 2 nm against the pressure ($P^* = P\sigma^3/\varepsilon$) in the case where $y_2 = 0.001$ and $T = 313.2$ K which is 9 K higher than the critical temperature of CO_2 [28]. The adsorbed amount of benzene shows a maximum at a pressure ($P^* = 0.007$) far below the critical pressure of CO_2 ($P_C^* = 0.142$) and then it decreases with increasing pressure while the adsorbed amount of CO_2 increases monotonically. It is noteworthy that the simulation curve for benzene resembles the experimental curve for adsorption of benzene on a commercial activated carbon in supercritical CO_2 as reported by Shojibara et al. [29]. The lower part of Fig. 1.4.9 shows the snapshots of the three states, a, b and c. At low pressures, benzene molecules almost fill the pore; however, as the pressure increases, they seem to be washed out by CO_2 molecules. These observations are understandable by referring to the chemical potential curves

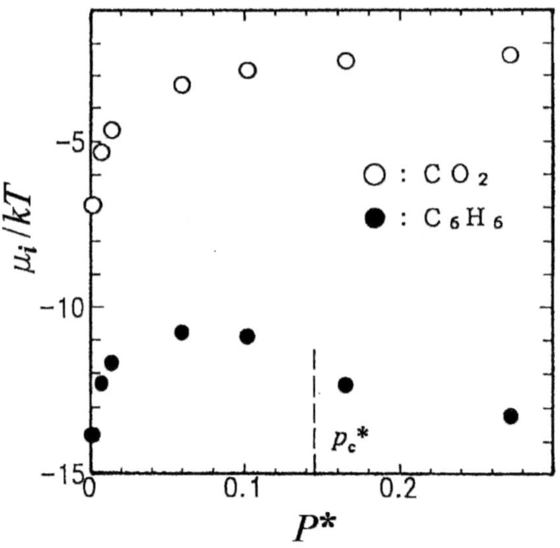

Fig. 1.4.10 The chemical potentials of CO_2 and benzene for a fluid mixture of $y_2 = 0.001$ at 313.2 K [28]

of each component as shown in Fig. 1.4.10 [28]. The chemical potential of CO_2 increases monotonically with increasing pressure while that of benzene shows a maximum at $P^* = 0.06$, which is about four tenths of P_C^* of CO_2, and then it decreases. The decrease in the chemical potential of a solute in a fluid is equivalent to the increase in solubility of the solute while the increase in the chemical potential of CO_2 means the increase in adsorption intensity of the CO_2. Since the pressure corresponding to the maximum adsorption ($P^* = 0.007$) is much lower than the pressure corresponding to the maximum chemical potential ($P^* = 0.06$), one can conclude that the decrease in adsorption of benzene in the SCF is ascribed to two factors: the increase in solubility in the SCF and the competitive adsorption of the SCF molecules; both factors work to suppress the adsorption of benzene.

Molecular simulations have the advantage that they independently evaluate the influence of any microscopic properties, such as the LJ parameters or the pore width, as well as state variables on the macroscopic properties of a system since similar calculations are easily performed by changing only the target parameter or variable. Following this strategy, some simulation results were reported to elucidate the effects of the temperature (T), the binary parameter between a solute molecule and solid carbons (k_{2s}), the slit width (H), and the solute composition in the fluid (y_2) on the adsorption isotherm of a solute diluted in an SCF [30, 31]. Figure 1.4.11 is one of their results to

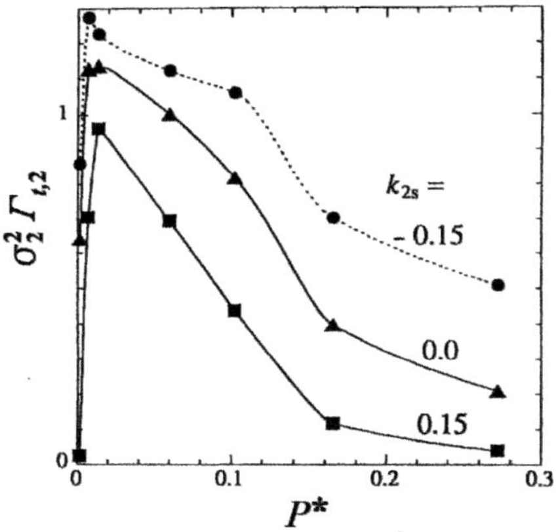

Fig. 1.4.11 Effect of k_{2s} on the true adsorption of a benzene-like solute in a slit pore of $H = 2$ nm from a fluid mixture of $y_2 = 0.001$ at 313.2 K [30]

show the effect of the binary parameter k_{2s} on the true adsorption of a benzene-like solute [30]. The adsorbed amount of solute increases as k_{2s} decreases toward negative values, which corresponds to more attractive interactions between the solute and the surface. The partition behavior of a solute between a slitlike pore and an SCF has been studied by use of MC simulations by Shigeta et al. [31–33].

1.5 Dynamic Properties of Supercritical Fluids

The molecular dynamics (MD) simulation method is a powerful tool for studying the dynamic properties in supercritical fluids at the molecular level. In this section, we will present the usefulness of the simulation method in two examples: one is on the time evolution of the cluster structure, and the other is on transport properties such as the diffusion coefficient and viscosity.

1.5.1 Dynamics of Clusters in Supercritical Fluids

As described in Sect. 1.2, the clustering phenomenon or local density enhancement is regarded as a fundamental feature in supercritical fluids (SCFs) and their mixtures. Cluster formation generally influences the solution structure and affects transport properties such as mass transfer coefficients and is thought to be important for effectively using SCFs as separation, material production and reaction media.

The cluster in SCF solutions is different from that in solids and liquids. The member molecules are bound to each other with relatively weak intermolecular forces such as van der Waals forces and the lifetime of an average cluster is much shorter than that in solids and liquids. In other words, it should be noted that the characteristic feature of clusters in SCFs is their dynamic and softly bound structure. Recent research has suggested that the solvating dynamics have a strong correlation with the electron and atomic nucleus configuration during chemical reactions, which indicates that studies on the dynamics of clusters in SCF solutions are indispensable for understanding reactions in supercritical solutions [1–3].

Since the time scale of the cluster formation in SCF solutions is very small (~ ps), possible methods to follow the dynamics of the cluster are time-resolved spectroscopic experiments and molecular simulations such as molecular dynamics (MD). MD simulation provides us with the static and dynamic structure of clustering phenomenon at the microscopic level.

(a) Definition of Clusters

For discussing the dynamics of clusters in SCFs, it is necessary to specify whether a molecule belongs to a cluster in the solutions. A number of definitions for clusters in SCFs have been proposed [4–7]. Within these proposals, two common formalizations have emerged according to MD simulations: geometric definitions and energetic definitions.

Geometric Definition

In the geometric definition, the cluster (i.e. solvation) region consists of a sphere of radius, R, located at a center solute molecule. Particles within a given distance R from a central particle (mostly solute) are considered to be members of the same cluster. This type of approach has been used by Petsche and Debenedetti [6], Cummings et al. [8] and Inomata et al. [13]

Energetic Definition

The energetic definition was originally proposed by Hill [9]. In Hill's definition, two particles are considered to be bound to each other only when their relative kinetic energy EK_{12} is less than or equal to their pair potential energy U_{12}:

$$EK_{12}(\boldsymbol{v}_{12}) \le U_{12}(\boldsymbol{r}_{12}), \tag{1.5.1}$$

where

$$EK_{12} = \frac{1}{2} \cdot \frac{m_1 m_2}{m_1 + m_2} \boldsymbol{v}_{12}^2 . \tag{1.5.2}$$

(b) MD Simulation of Clustering Phenomena in Supercritical Fluids

Hill's definition was used to discuss clustering in MD simulations together with the geometric definition by Petsche and Debenedetti [6]. A connectivity matrix was used to determine directly and indirectly bound particles. In that study, the first dynamic pictures of SCF clusters were made for Lennard–Jones (LJ) mixtures of Xe + Ne. The time evolution of the geometrically and energetically defined cluster of pure (Ne in Ne), attractive mixtures (Xe in Ne) and repulsive mixtures (Ne in Xe) in the near critical region was followed. Most of the results were analyzed by the energetic definition, although the geometric definition was used in discussing experimental results such as spectroscopic data. This is because the energetic definition can define a cluster without any empirical parameters. According to their MD simulation results, the mean cluster size was 2 for repulsive mixtures whereas that for attractive mixtures was 59, indicating a strong dependence of cluster size on the intermolecular interaction. Figure 1.5.1 shows that there is a tendency for energetically defined clusters to persist in the system [6]. The values were evaluated by counting the number of particles belonging to a particular cluster at time t that also belonged to the same cluster at time zero. The figure indicates that the lifetime of the average cluster was 0.54 ps for the pure Ne system and 1.88 ps for the attractive Xe in Ne system. Although the noise of this simulation was large, the work demonstrated that MD simulation could be a powerful technique for the study of dynamic structures in SCF mixtures.

From the viewpoint of computer simulation, it should be pointed out that Hill's definition has some practical difficulties in determining the time correlation function of the cluster member because of the necessity to compute the connectivity matrix of directly and indirectly bound particles of the entire system. The computation of the cluster connectivity matrix requires a $N(N-1)/2$ loop evaluation, resulting in excessive computation time. In addition, Hill's definition may result in the formation of a supercluster due to the periodic boundary conditions that are commonly used [10, 11].

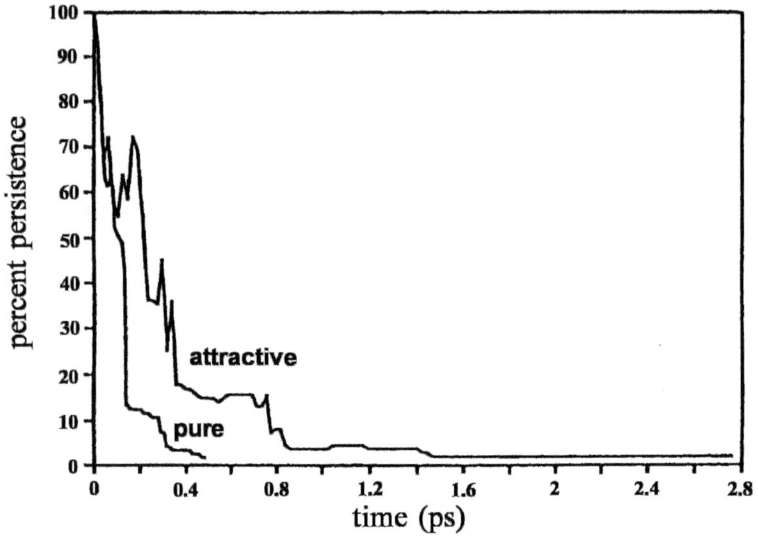

Fig. 1.5.1 Number of solvent molecules in an energetically defined cluster [6]

Most of the practical difficulties of Hill's definition were solved by Liew et al., who developed a method that used the relative radial kinetic energy of particles at a given instant. Suppose that the relative motion of two particles is as shown in Fig. 1.5.2 [12]. The relative velocity of the particle can be divided into two components, a relative radial velocity, v_r, and a circular (or angular) velocity, v_c. The pairwise interaction, which is generally used in molecular simulation, affects mostly the radial relative motion. Therefore, in a mixture, the solvent molecules around a solute can be defined as cluster member when the radial component of their relative kinetic energy $EK_{12}(v_r)$ is less than their

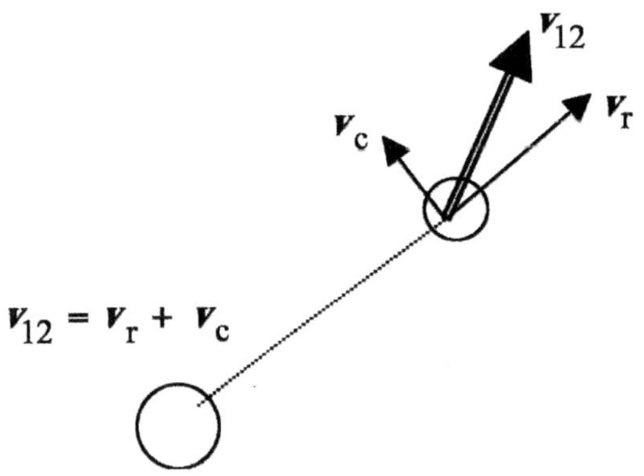

Fig. 1.5.2 Relative velocity between two particles

Table 1.5.1 Lennard-Jones parameters used in simulations [28]

solute	σ/σ_{CO_2}	$\varepsilon/\varepsilon_{CO_2}$	m/m_{CO_2}
CO_2	1.00	1.00	1.00
pyrene	1.88	2.94	4.59
model A	1.88	1.00	1.00
model B	1.00	2.94	1.00
model C	1.00	1.00	4.59

pair potential energy U_{12}. Furthermore Liew et al. [12] consider cluster members to be only those particles that are directly bound to each other:

$$EK_{12}(v_r) \leq U_{12}(r_{12}), \tag{1.5.3}$$

where

$$EK_{12}(v_r) = \frac{1}{2} \cdot \frac{m_1 m_2}{m_1 + m_2} v_r^2 . \tag{1.5.4}$$

The primary difference between (1.5.3) and (1.5.4) and those proposed by Hill, (1.5.1) and (1.5.2), is the use of the relative radial velocity. Liew et al. [12] performed MD simulations for infinitely dilute solution of LJ solutes in supercritical CO_2 as listed in Table 1.5.1. From their analyses, the instantaneous number of CO_2 molecules as cluster members around a pyrene molecule fluctuates around the average as shown in Fig. 1.5.3 [12]. The dotted line shows the number of initial member molecules that remain as members around

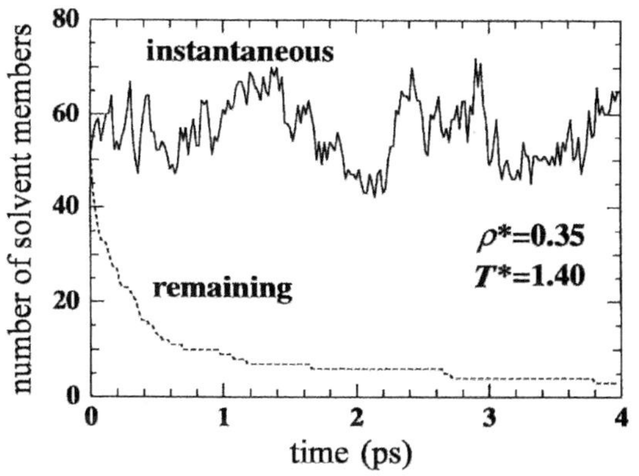

Fig. 1.5.3 Number of CO_2 molecules in a cluster around a pyrene molecule [12] Reprinted from: C.C. Liew, H. Inomata, S. Saito: Molecular dynamics study on solvent clustering in supercritical fluid solutions based on particle radial kinetic energy, *Fluid Phase Equilibria* (Elsevier Science) **104**, 317 (1995)

the pyrene since time zero ($t = 0$). The number of CO_2 molecules remaining in the cluster decays rapidly, which was also observed by Petsche and Debenedetti for the Xe + Ne system.

Some snapshots of the clusters with time are useful for understanding the dynamic structure of the clusters. These are shown in Fig. 1.5.4 for the CO_2 + pyrene system [12]. The number of solvent molecules is coincident with the instantaneous values in Fig. 1.5.3. Figure 1.5.4 (a) shows the dynamical structure of the cluster and Fig. 1.5.4 (b) shows the time evolution of the initial cluster members which remain in the same cluster. It can be seen that the cluster structure or local density enhancement around a solute does not change drastically whereas solvent molecules in the cluster are exchanged rapidly with bulk solvent molecules. In other words, attractive clusters are dynamically stable, but not static entities, i.e., solvent particles continuously diffuse in and out of the sphere of influence of the strongly attractive solute without significantly affecting the enhanced solvent density in the cluster or its structure. From the figures, we can suppose that cluster members in the nearest neighbor of the solute remain even after 2 ps.

For discussing the stability or lifetime of the cluster member, the decay of the number of remaining cluster members can be plotted against time in semi-log coordinates as shown in Fig. 1.5.5 [12]. The number of remaining members initially exhibits a large decay and approaches a constant exponential decay after about 1 ps in all systems. According to the snapshots in Fig. 1.5.4, the late stage of the decay indicates a contribution of strongly bound members in the first solvation shell. Namely, the slope of the linear exponential decay represents the stability of the strongly bound molecules around a solute. Comparing the stability of several LJ models listed in Table 1.5.1, the authors found that the energy parameter affected the stability of a cluster around the solute while the size parameter greatly affected the cluster size.

Inomata et al. [13] performed MD simulations for CO_2 + benzene systems with multisite LJ models and investigated the solvation dynamics using a geometrical definition. The time evolution of the number of tagged solvating molecules at time zero exhibited a similar dependence to that found by Liew et al. [12]. Figure 1.5.6 shows the density and temperature dependence of the decay in the number of initially tagged solvating molecules [13]. Except for the early stage, the decay can be expressed by a single exponential function. The characteristic time, which corresponds to stability, increases with density and tends to decrease with increasing temperature.

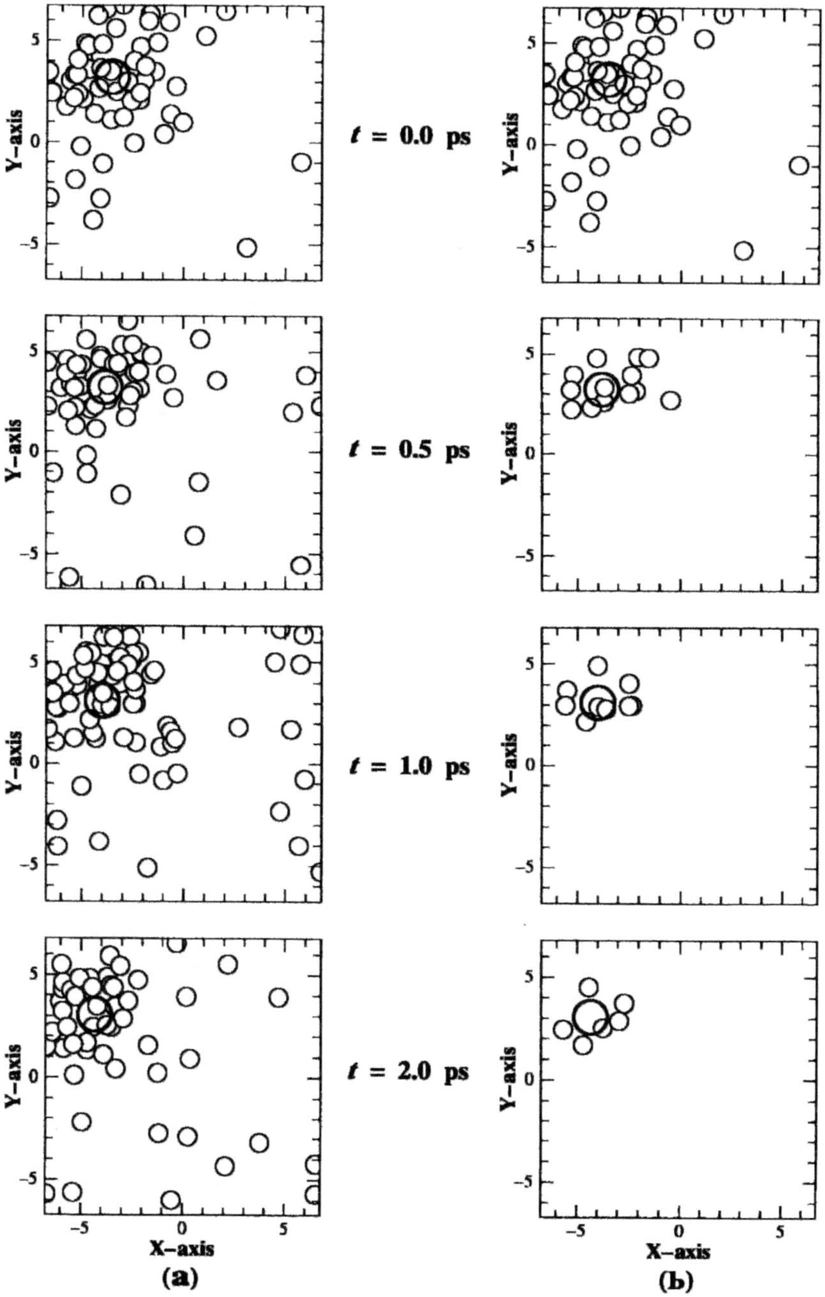

Fig. 1.5.4 Snapshots of instantaneous (**a**) and remaining (**b**) members around a pyrene molecule with time [12]

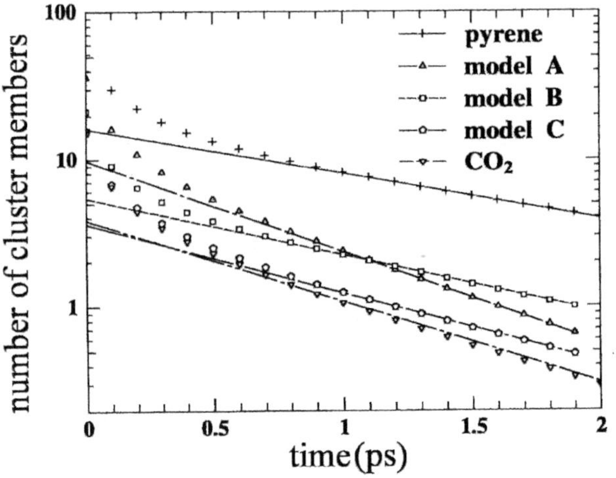

Fig. 1.5.5 Time evolution of the number of solvating molecules which were tagged at $t = 0$ and remain tagged after a period t [12]

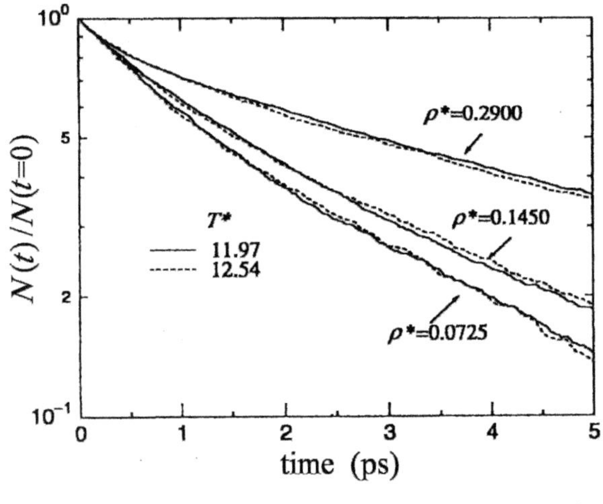

Fig. 1.5.6 Time evolution of the number of solvating molecules which were tagged around a benzene at $t = 0$ and remain tagged after a period t [13] Reprinted from: H. Inomata, S. Saito, P.G. Debenedetti: Molecular dynamics simulation of infinitely dilute solutions of benzene in supercritical CO_2, *Fluid Phase Equilibria* (Elsevier Science) **116**, 282 (1996)

1.5.2 Diffusion and Viscosity

(a) Self-Diffusion Coefficient

Calculation Method

The self-diffusion coefficient of the species 1 is calculated by the mean square displacement (MSD) vs. time plot from the molecular dynamics simulations:

$$D_1 = \frac{1}{6}\lim_{t\to\infty}\frac{\partial}{\partial t}\left\langle \frac{1}{n_1}\sum_{i=1}^{n_1}\delta R_{i1}^2(t)\right\rangle, \tag{1.5.5}$$

Here, the vector $\delta R_{i1}(t)$ means the displacement vector of the i-th molecule in the species 1 at time t. When there are n_1 molecules that belong to the same species, the MSD is averaged with respect to n_1 molecules like. (1.5.5). For this reason, the average value is stable [14].

Equation of State of the Self-Diffusion Coefficient in a Lennard–Jones Fluid

There are two sets of the equation of state for the self-diffusion coefficient in the Lennard–Jones fluid [15, 16]. These are derived by molecular dynamics simulations. These are consistent with each other. The calculated equations of state are in agreement with the experimental results on CO_2 [15] and Ar [16].

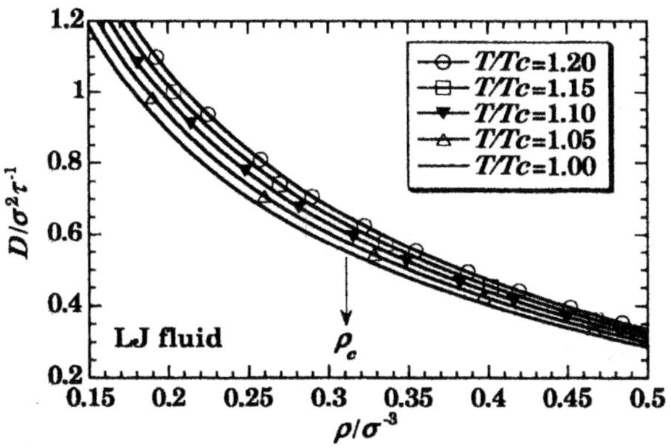

Fig. 1.5.7 Density dependence of the self-diffusion coefficient of the Lennard-Jones fluid around the critical point

The calculated self-diffusion coefficient shows no significant enhancement nor depression near the critical point as plotted in Fig. 1.5.7. The critical parameters of the Lennard–Jones fluid are $T_C = 1.3396\,\varepsilon / k_B$, $P_C = 0.1405\,\varepsilon / \sigma^3$, $\rho_C = 0.3108\,\sigma^{-3}$ [17]. The number of particles in the system was 108 [15] or 256 [16]. These numbers were too small to obtain the anomaly around the critical point.

The Self-Diffusion Coefficient in Supercritical Solution

In a supercritical fluid, solvent molecules form a characteristic cluster around solute molecules. The cluster structures are investigated by molecular dynamics simulations using a carbon dioxide fluid containing a naphthalene molecule [18]. The diffusion coefficient of the naphthalene molecule shows a large fluctuation when the trajectory of the solute is treated with a time scale of 100 ps.

(b) Viscosity

Equilibrium molecular dynamics simulations were performed for the Lennard–Jones fluid for 171 conditions spanning the range $0 \le \rho\sigma^3 \le 1$ and $0.8 \le k_B T / \varepsilon \le 4$ [16]. The modified mean square displacement (MSD) formula

$$\left\langle [A(t) - A(0)]^2 \right\rangle = \int_0^\infty \left\langle \left.\frac{\mathrm{d}A(t)}{\mathrm{d}t}\right|_{t=t} \left.\frac{\mathrm{d}A(t)}{\mathrm{d}t}\right|_{t=0} \right\rangle \mathrm{d}t, \tag{1.5.6}$$

$$\frac{\mathrm{d}A(t)}{\mathrm{d}t} = \sum_i^N \frac{p_{\alpha i} p_{\beta i}}{m_i} + \sum_i^N F_{\alpha i} r_{\beta i} \qquad (\alpha \ne \beta) \tag{1.5.7}$$

is used to compute the shear viscosity at each condition.

An analytical equation for the viscosity coefficient is fitted to the simulated data as a polynomial function of density and temperature. The resultant smoothing equation correlates the simulated data quite well and agrees with argon experimental data within the uncertainty of the data [16]. The enhancement near the critical point, like in supercritical CO_2, [19] was not observed.

(c) Mutual Diffusion Coefficient

Theory

The recently reported mutual diffusion coefficient in the solution of methane and *n*-decane is very small compared to the self-diffusion coefficient in the supercritical region [20]. The diffusion coefficient of acetone in supercritical carbon dioxide approaches zero near the critical pressure of carbon dioxide which is close to that of the mixture [21].

The mutual diffusion coefficient D is written as a product of the thermodynamic factor Q and the kinematic factor L [22]:

$$D = QL. \tag{1.5.8}$$

The thermodynamic factor Q is zero at the critical point because of the following relation [22].

$$Q = \frac{x_1}{k_B T}\left(\frac{\partial \mu_1}{\partial x_1}\right)_{T,P}. \tag{1.5.9}$$

The quantity μ_1 is the chemical potential of component 1 and x_1 is the mole fraction of component 1. Temperature and pressure are shown as T and P, respectively. The quantity k_B is the Boltzmann constant.

The two-component mixture of the N-molecule system is considered in three-dimensional space as an example. The kinematic factor L in (1.5.8) is given as follows [22]:

$$L = \frac{1}{3Nc_1c_2}\left[\left(\frac{m_1}{m_2}\right)\cdot c_1 + c_2\right]^2 \int_0^\infty \left\langle \left(\sum_{i=1}^{n_1} v_{i1}(t)\right)\cdot\left(\sum_{i=1}^{n_1} v_{i1}(0)\right)\right\rangle \mathrm{d}t. \tag{1.5.10}$$

The quantity m_1 is the mass of component 1. The velocity of the i-th molecule in the species 1 is written as $v_{i1}(t)$. The number density of species 1 is described as c_1. There are n_1 molecules of species 1 in the system. The equation shows that the velocity correlation function of the center of mass of species 1 determines the essential part of the factor L. The thermodynamic factor Q is calculated by the radial distribution functions $g_{ab}(R)$ according to the Kirkwood–Buff theory [22–24].

The kinematic factor L is also written as the mean square displacement of the center of mass of component 1, $\langle \delta R_{1\mathrm{cm}}^2(t)\rangle$.

$$L = \frac{1}{6Nc_1c_2}\left[\left(\frac{m_1}{m_2}\right)\cdot c_1 + c_2\right]^2 \lim_{t\to\infty}\frac{\partial}{\partial t}\langle \delta R_{1\mathrm{cm}}^2(t)\rangle. \tag{1.5.11}$$

The kinematic factor L can be expressed in terms of the joint diffusion coefficient D_{ab} ($a, b = 1$ or 2) [24, 25].

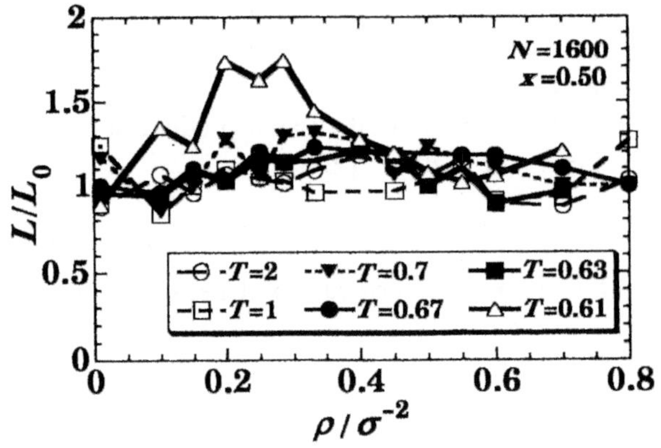

Fig. 1.5.8 Plot of the kinematic factor relative to the self-diffusion coefficient, L/L_0, vs. density, ρ

$$D_{ab} = \frac{N}{3}\int_0^\infty \left\langle v_{ia}(t)\cdot v_{jb}(t)\right\rangle \mathrm{d}t\,, \tag{1.5.12}$$

$$L = L_0 + x_1 x_2 (D_{11} + D_{22} - 2D_{12})\,, \tag{1.5.13}$$

$$L_0 = x_2 D_1 + x_1 D_2\,. \tag{1.5.14}$$

In (1.5.12), $\left\langle v_{ia}(t)\cdot v_{jb}(t)\right\rangle$ is the velocity cross-correlation function of one molecule i of species a and a distinct molecule j of species b (the same or different from the species a). The self-diffusion coefficient of species 1 is written as D_1 in (1.5.14).

The second term in (1.5.13) expresses the correlation of the motion in the solution. In the normal solution L is not very different from L_0 because the factor Q is about one and the mutual diffusion coefficient D is close to L_0. For this reason, we are going to use L_0 as the base line of the mutual diffusion coefficient D in the discussion of the density dependence of D in the critical region hereafter.

Density Dependence of the Mutual Diffusion Coefficient

The concentration fluctuates to a large extent in the solution near the critical points. For this reason, the thermodynamic factor Q must be estimated carefully. The length of the unit cell should be large enough to obtain the Kirkwood–Buff parameters. A two-dimensional unit cell is adopted for this reason. The formulas shown above are transformed to the two-dimensional version.

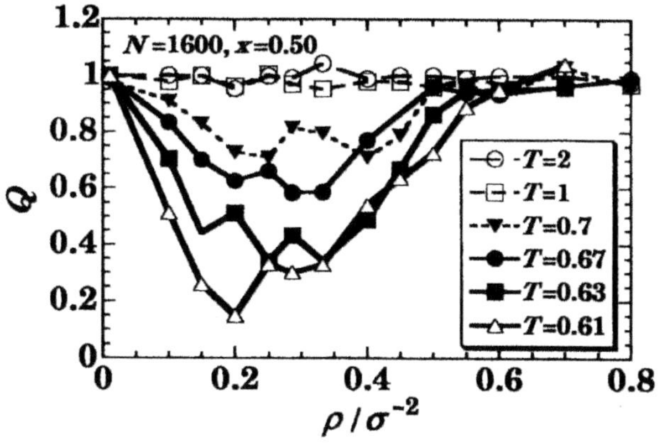

Fig. 1.5.9 Plot of thermodynamic factor, Q, vs. density, ρ

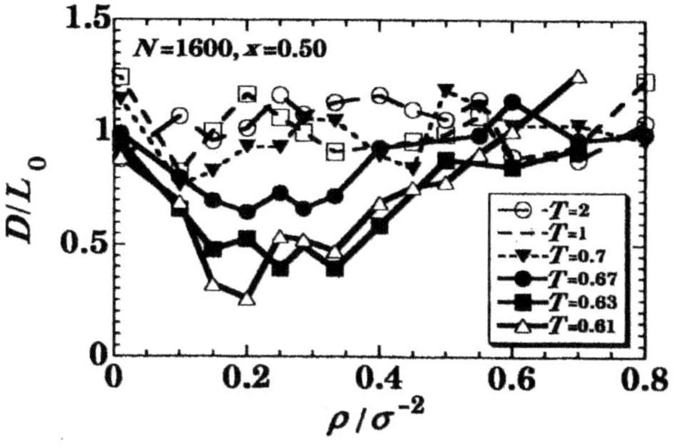

Fig. 1.5.10 Plot of mutual diffusion coefficient relative to the self-diffusion coefficient, D/L_0, vs. density, ρ

Their mixture consists of Ar and Kr, and the interaction is assumed to be Lennard–Jones (LJ) potential for simplicity. Component 1 is Ar. The mole fraction is 0.50, where the concentration has large fluctuation. The unit of energy is the LJ potential for Ar, ε. The unit of length is the size parameter of Ar, σ in the LJ model. The number of molecules in the unit cell is 1600.

The density dependencies of L/L_0, Q and D/L_0 are shown in Figs. 1.5.8, 1.5.9 and 1.5.10, respectively [26]. The thermodynamic factor Q is about 1 far from the critical point of the mixture ($T_C \approx 0.61\varepsilon/k_B$, $r_C \approx 0.25\sigma^{-2}$). The factor Q has small values near the critical density ($0.25\,\sigma^{-2}$) at low temperatures ($0.61\,\varepsilon/k_B$), which gives the large concentration fluctuation. The ratio L/L_0 is also about 1 in the state far from the critical region. Near the critical point, L / L_0 has a value larger than 1. However, it is not large enough, so the mutual diffusion coefficient relative to the self diffusion coefficient D / L_0 is small compared to 1 in the critical region. This result is qualitatively consistent with the recently reported mutual diffusion coefficients [20, 21].

References

1.2 Solvation Structure from Spectroscopy

[1] O. Kajimoto, *Chem. Rev.*, **99**, 355 (1999).
[2] P. Villard, *Comptes Rendus*, **120**, 182 (1895).
[3] J. A. Hyatt, *J. Org. Chem.*, **49**, 5097 (1984).
[4] S. Akimoto and O. Kajimoto, *Chem. Phys. Lett.*, **209**, 263 (1993).
[5] L. Onsager, *J. Am. Chem. Soc.* 1936, **58,** 1486.
[6] A. D. Buckingham, *Proc. Roy. Soc.,* **A248,** 169 (1958).
[7] E. G. McRae, *J. Phys. Chem.*, **61**, 562 (1957).
[8] "*Supercritical Fluid Science and Technology*", ed. K. P. Johnston and J. M. L. Penninger, American Chemical Society monograph **406**, Washington, DC (1989).
[9] R. Noyori, ed., *Chem. Rev.*, **99**, 2 (1999).
[10] M. E. Sigman, S. M. Lindley and J. E. Leffler, *J. Am. Chem. Soc.*, **107**, 1471 (1985).
[11] C. R. Yonker, S. L. Frye, D. R. Kalkwarf and R. D. Smith, *J. Phys. Chem.*, **90**, 3022 (1986).
[12] C. R. Yonker and R. D. Smith, *J. Phys. Chem.*, **92**, 235 (1988).
[13] C. R. Yonker and R. D. Smith, *J. Phys. Chem.*, **92**, 2374 (1988).
[14] C. R. Yonker and R. D. Smith, *J. Phys. Chem.*, **93**, 1261 (1989).
[15] S. Kim and K. P. Johnston, *Ind. Eng. Chem. Res.*, **26**, 1206 (1987).
[16] O. Kajimoto, M. Futakami, T. Kobayashi and K. Yamasaki, *J. Phys. Chem.*, **92**, 1347 (1988).
[17] O. Kajimoto, K. Yamasaki and K. Honma, *Faraday Disc. Chem. Soc.*, **85**, 65 (1989).
[18] Y.-P. Sun, M. A. Fox, and K. P. Johnston, *J. Amer. Chem. Soc.*, **114**, 1187 (1992).
[19] J. N. Wilson, *Chem. Rev.*, **25**, 377 (1939).
[20] Y. Ooshika, *J. Phys. Soc. Japan*, **9**, 594 (1954).
[21] N. Mataga and T. Kubota, "*Molecular Interactions and Electronic Spectra*", Marcel Dekker, New York (1970).
[22] H. Luo and S. C. Tucker, *J. Phys. Chem.*, **100**, 11165 (1996).
[23] M. T. Klein, Y. G. Mentha and L. A. Torry, *Ind. Eng. Chem. Res*, **31**, 182 (1992).
[24] H. Luo, H. and S. C. Tucker, *J. Phys. Chem.* **B 101**, 1063 (1997).

[25] L. W. Flanagin, P. B. Balbuena, K. P. Johnston and P. J. Rossky, *J. Phys. Chem.*, **99**, 5196 (1995).
[26] P. B. Balbuena, K. P. Johnston and P. J. Rossky, *J. Am. Chem. Soc.*, **116**, 2689 (1994).
[27] R. Ishii, S. Okazaki, I. Okada, M. Furusaka, N. Watanabe, M. Misawa and T Fukunaga, *J. Chem. Phys.*, **105**, 7011 (1996).
[28] S. Okazaki, M. Matsumoto, I. Okada, K. Maeda, and Y. Kataoka, *J. Chem. Phys.*, **103**, 8594 (1995).
[29] I. B. Petsche and P. G. Debenedetti, *J. Chem. Phys.*, **91**, 7075 (1989).
[30] P. G. Debenedetti, *Chem. Eng. Sci.*, **42**, 2203 (1987).
[31] P. G. Debenedetti, I. B. Petsche and R. S. Mohamed, *Fluid Phase Equilibria*, **52,** 347 (1989).
[32] C. Reichardt, "Solvents and Solvent Effects in Organic Chemistry", VCH, New York (1990).
[33] J. G. Kirkwood, W. West and R. T. Edwards, *J. Chem. Phys.*, **5**, 14 (1937).
[34] H. Sato, Thesis, Kyoto University (1996).
[35] E. U. Franck and K. Roth, *Discuss. Faraday Soc.*, **43**, 108 (1967).
[36] A. P. Luck, *Discuss. Faraday Soc.*, 43, 115 (1967).
[37] T. W. Zerda, B. Wiegrand and J. Jonas, *J. Chem. Eng. Data*, **31**, 274 (1986).
[38] J. P. Blitz, C. R. Yonker and R. D. Smith, *J. Phys. Chem.*, **93**, 6661 (1989).
[39] N. Wada, M. Saito, D. Kitada, R. L. Smith, Jr., H. Inomata, K. Arai and S. Saito, *J. Phys. Chem. B*, **101**, 10918 (1997).

1.3 Solvation Structure from Molecular Simulations

[1] C. S. Murthy, K. Singer and I. R. McDonald, *Mol. Phys.* **44**, 135 (1981).
[2] W. A. Steels and H. A. Posch, *J. Chem. Soc. Faraday Trans.*, **238**, 1843 (1987).
[3] L. C. Geiger, B. M. Ladanyi and M. E. Chapin, *J. Chem. Phys.*, **93**, 4533 (1990).
[4] D. Moeller, "Thermodynamik Fluider Mischungen, Kalorimetrie und Computersimulationen", VDI-Verlag, Düsseldorf (1991).
[5] D. Moeller and J. Fischer, *Fluid Phase Equilibria*, **100**, 35 (1994).
[6] S. Tsuzuki, T. Uchimaru, K. Tanabe, S. Kuwajima, N. Tajima and T. Hirano, *J. Phys. Chem.*, **100**, 4400 (1996).
[7] S. Tsuzuki, T. Uchimaru, M. Mikami, K. Tanabe, T. Sako and S. Kuwajima, *Chem. Phys. Lett.*, **255**, 347 (1996).
[8] S. Tsuzuki, T. Uchimaru, M. Mikami and K. Tanabe, *J. Chem. Phys.*, **109**, 1 (1998).
[9] M. Welker, G. Steinebrunner, J. Solca and H. Huber, *Chem. Phys.*, **213**, 253 (1996).
[10] G. Steinebrunner, A. J. Dyson, B. Kirchner and H. Huber, *J. Chem. Phys.*, **109**, 3153 (1998).

[11] A. Szabo and N. S. Ostlund, "Modern Quantum Chemistry, Introduction to Advanced Electronic Structure Theory", Macmillan, London (1982).
[12] A. J. Stone, "*The Theory of Intermolecular Forces*", Oxford University Press, Oxford (1996).
[13] S. F. Boys and F. Bernardi, *Mol. Phys.*, **19**, 553 (1970).
[14] M. A. Walsh, T. H. England, T. R. Dyke and B. J. Howard, *Chem. Phys. Lett.*, **142**, 265 (1987).
[15] K. W. Jucks, Z. S. Huang, R. E. Miller, G. T. Fraser, A. S. Pine, and W. J. Lafferty, *J. Chem. Phys.*, **88**, 2185 (1988).
[16] B. H. Besler, K. M. Merz Jr. and P.A. Kollman, *J. Comp. Chem.*, **11**, 431 (1990).
[17] M. P. Allen and D. J. Tildesley, "*Computer Simulation of Liquids*", Clarendon Press, Oxford (1987). The programs referred to in this book are available at http://www.dl.ac.uk/CCP/CCP5/main.html.
[18] C. G. Gray and K. E. Gubbins, "*Theory of Molecular Fluids*", Clarendon Press, Oxford (1984).
[19] J. Kolafa and I. Nezbeda, *Fluid Phase Equilibria,* **100**, 1 (1994).
[20] "*Monte Carlo Methods in Statistical Physics*", ed. K. Binder, Springer-Verlag, Berlin (1979).
[21] "Applications of the Monte Carlo Method in Statistical Physics", ed. K. Binder, Springer-Verlag, Berlin (1984).
[22] "*The Monte Carlo Method in Condensed Matter Physics*", ed. K. Binder, Springer-Verlag, Berlin (1992).
[23] J. J. Binney, N. J. Dowrick, A. J. Fisher and M. E. J. Newnan, "*The Theory of Critical Phenomena*", Oxford University Press, Oxford (1992).
[24] J. M. Yeomans, "*Statistical Mechanics of Phase Transitions*", Oxford University Press, Oxford (1992).
[25] I. B. Petsche and P. G. Debenedetti, *J. Chem. Phys.*, **91**, 7075 (1989).
[26] P. Postorino, R. H. Tromp, M. A. Ricci, A. K. Soper and G. W. Neilson, *Nature*, **366**, 668 (1993).
[27] R. H. Tromp, P. Postorino, G. W. Neilson, M. A. Ricci and A. K. Soper, *J. Chem. Phys.*, **101**, 6210 (1994).
[28] C. A. Eckert, D. H. Ziger, K. P. Johnston and S. J. Kim, *J. Phys. Chem.*, **90**, 2738 (1986).
[29] P. T. Cummings, H. D. Cochran, J. M. Simonson, R. E. Mesmer and S. Karaborni, *J. Chem. Phys.*, **94**, 5606 (1991).
[30] H. D. Cochran, P. T. Cummings and S. Karaborni, *Fluid Phase Equilibria*, **71**, 1 (1992).
[31] P. T. Cummings, A. A. Chialvo and H. D. Cochran, *Chem. Eng. Sci.*, **49**, 2735 (1994).
[32] S. T. Cui and J. G. Harris, *Chem. Eng. Sci.*, **49**, 2749 (1994).
[33] S. T. Cui and J. G. Harris, *J. Phys. Chem.*, **99**, 2900 (1995).
[34] J. Gao, *J. Am. Chem. Soc.*, **115**, 6893 (1993).
[35] S. Okazaki, M. Matsumoto, I. Okada, K. Maeda and Y. Kataoka, *J. Chem. Phys.*, **103**, 8594 (1995).
[36] R. Ishii, S. Okazaki, I. Okada, M. Furusaka, N. Watanabe, M. Misawa and T. Fukunaga, *J. Chem. Phys.*, **105**, 7011 (1996).

[37] L. C. Geiger, B. M. Ladanyi and M. E. Chapin, *J. Chem. Phys.*, **93**, 4533 (1990).
[38] Y. Kataoka, *J. Chem. Phys.*, **87**, 589 (1987).
[39] R. D. Mountain, *J. Chem. Phys.*, **90**, 1866 (1989).
[40] A. G. Kalinichev and J. D. Bass, *Chem. Phys. Lett.*, **231**, 301 (1994).
[41] J. J. de Pablo and J. M. Prausnitz, *Fluid Phase Equilibria*, **53**, 177 (1989).
[42] A. A. Chialvo and P. T. Cummings, *J. Chem. Phys.*, **101**, 4466 (1994).
[43] Y. Guissani and B. Guillot, *J. Chem. Phys.*, **98**, 8221 (1993).
[44] R. D. Mountain, *NISTIR*, 6028 (1997).
[45] T. I. Mizan, P. E. Savage and R. M. Ziff, *J. Phys. Chem.*, **100**, 403 (1996).
[46] C. C. Liew, H. Inomata and K. Arai, *Fluid Phase Equilibria*, **144**, 287 (1998).
[47] E. S. Fois, M. Sprik and M. Parrinello, *Chem. Phys. Lett.*, **223**, 411 (1994).
[48] J. Ortega, J. P. Lewis and O. F. Sankey, *J. Chem. Phys.*, **106**, 3696 (1997).
[49] R. D. Mountain, *J. Chem. Phys.*, **103**, 3084 (1995).
[50] I. M. Svishchev, P. G. Kusalik, J. Wang and R. J. Boyd, *J. Chem. Phys.*, **105**, 4742 (1996).
[51] N. Yoshii, H. Yoshie, S. Miura and S. Okazaki, *J. Chem. Phys.*, **109**, 4873 (1998).
[52] N. Yoshii, S. Miura and S. Okazaki, *Bull. Chem. Soc. Jpn.*, **72**, 151 (1999).
[53] K. Yamanaka, T. Yamaguchi and H. Wakita, *J. Chem. Phys.*, **101**, 9830 (1994).
[54] N. Matubayasi, C. Wakai and M. Nakahara, *Phys. Rev. Lett.*, **78**, 2573 (1997).
[55] N. Matubayasi, C. Wakai and M. Nakahara, *J. Chem. Phys.*, **107**, 9133 (1997).
[56] M. M. Hoffmann and M. S. Conradi, *J. Am. Chem. Soc.*, **119**, 3811 (1997).
[57] Y. Ikushima, K. Hatakeda, N. Saito and M. Arai, *J. Chem. Phys.*, **108**, 5855 (1998).
[58] A. K. Soper, F. Bruni and M. A. Ricci, *J. Chem. Phys.*, **106**, 247 (1997).
[59] P. Jedlovszky, J. P. Brodholt, F. Bruni, M. A. Ricci, A. K. Soper and R. Vallauri, *J. Chem. Phys.*, **108**, 8528 (1998).
[60] L. X. Dang, *J. Chem. Phys.*, **97**, 2659 (1992).
[61] Y. E. Gorbaty and Y. N. Demianets, *Chem. Phys. Lett.*, **100**, 450 (1983).

1.4 Static Properties of Supercritical Fluids

[1] A. Kido, O. Kitao and K. Nakanishi, *Chem. Phys. Lett.*, **199**, 403 (1992).
[2] R. Yamamoto, O. Kitao and K. Nakanishi, *Mol. Simulation*, **12**, 383 (1994). or R. Yamamoto, H. Tanaka, K. Nakanishi and X. C. Zeng, *Chem. Phys. Lett.*, **231**, 401 (1994).

[3] J. J. Nicolas, K. E. Gubbins, W. B. Street and D. J. Tildesley, *Mol. Phys.*, **37**, 1429 (1979).
[4] J. K. Johnson, J. A. Zollweg and K. E. Gubbins, *Mol. Phys.*, **78**, 591 (1993).
[5] J. Kolafa and I. Nezbeda, *Fluid Phase Equilibria*, **100**, 1 (1994).
[6] E. A. Guggenheim, "*Mixtures*", Clarendon Press, Oxford (1952).
[7] J. H. Hildebrand, J. M. Prausnitz and R. L. Scott, "*Regular and Related Solutions*", Van Nostrand Reinhold, New York (1970).
[8] C. J. Wormald and J. M. Eyears, *J. Chem. Thermodynamics*, **20**, 323 (1988).
[9] M. S. Gruszkiewicz and J. B. Ott, *Fluid Phase Equilibria*, **124**, 81 (1996).
[10] I. R. McDonald, *Mol. Phys.*, **23**, 41(1972).
[11] I. Fujihara and K. Nakanishi, *J. Chem. Phys.*, **107**, 3121 (1997).
[12] B. Widom, *J. Chem. Phys.*, **39**, 2808 (1963).
[13] J. G. Kirkwood, *J. Chem. Phys.*, **3**, 300 (1935).
[14] K. S. Shing and S. T. Chung, *J. Phys. Chem.*, **91**, 1674 (1987).
[15] H. Eya, Y. Iwai, T. Fukuda and Y. Arai, *Fluid Phase Equilibria*, **77**, 39 (1992).
[16] Yu.V. Tsekhanskaya, M. B. Iomtev and E. V. Mushkina, *Russ. J. Phys. Chem.*, **38**, 1173 (1964).
[17] Y. Iwai, H. Uchida, Y. Koga, Y. Arai and Y. Mori, *Ind. Eng. Chem. Res.*, **35**, 3782 (1996).
[18] Y. Iwai, H. Uchida, Y. Arai and Y. Mori, *Fluid Phase Equilibria*, **144**, 233 (1998).
[19] R. A. Van Leer and M. E. Paulaitis, *J. Chem. Eng. Data*, **25**, 257 (1980).
[20] Y. Iwai, H. Yamamoto, Y. Tanaka and Y. Arai, *J. Chem. Eng. Data*, **35**, 174 (1990).
[21] Y. Mori, T. Shimizu, Y. Iwai and Y. Arai, *J. Chem. Eng. Data*, **37**, 317 (1992).
[22] Y. Koga, Y. Iwai and Y. Arai, *J. Chem. Phys.*, **101**, 2283 (1994).
[23] Y. Iwai, Y. Koga and Y. Arai, *Fluid Phase Equilibria*, **116**, 267 (1996).
[24] Y. Koga, Y. Iwai, M. Yamamoto and Y. Arai, *Fluid Phase Equilibria*, **131**, 83 (1997).
[25] M. P. Allen and D. J. Tildesley, "*Computer Simulation of Liquids*", Clarendon Press, Oxford (1987).
[26] D. Frenkel and B. Smit, "*Understanding Molecular Simulation, From Algorithm to Applications*", Academic Press, Sandiego (1996).
[27] T. Nitta, and J. Yoneya, *J. Chem. Eng. Jpn*, **28**, 31 (1995).
[28] W. A. Steele, "*The Interaction of Gases with Solid Surfaces*", Pergamon Press, Oxford (1974).
[29] H. Shojibara, Y. Sato, S. Takishima and H. Masuoka, *J. Chem. Eng. Jpn*, **28**, 245 (1995).
[30] T. Shigeta, J. Yoneya and T. Nitta, *Mol. Sim.*, **16**, 291 (1996).
[31] T. Nitta and T. Shigeta, *Fluid Phase Equilibria*, **144**, 245 (1998).
[32] T. Shigeta and T. Nitta, *J. Chem. Eng. Jpn*, **29**, 516 (1996).
[33] T. Shigeta and T. Nitta, *J. Chem. Eng. Jpn*, **32**, 150 (1999).

1.5 Dynamic Properties of Supercritical Fluids

[1] S. Saito, *J. Supercrit. Fluids*, **8**, 177 (1995).
[2] F. V. Bright and M. E. McNally, "Supercritical Fluid Technology - Theoretical and Applied Approaches to Analytical Chemistry", ACS Symposium Series, **488** (1992).
[3] T. J. Bruno and J. F. Ely, "*SCF Technology-Review in Modern Theory and Application*", CRC Press, Boca Raton (1991).
[4] C. A. Eckert, D. H. Ziger, K. P. Johnston and T. K. Ellison, *Fluid Phase Equilibria*, **14**, 167 (1983).
[5] S. Kim and K. P. Johnston, *AIChE J.*, **33**, 1603 (1987).
[6] I. B. Petsche and P. G. Debenedetti, *J. Chem. Phys.*, **91**, 7075 (1989).
[7] H. D. Cochran and L. L. Lee, ACS Symposium Series, **406**, 27 (1989).
[8] P. T. Cummings, A. A. Chialvo, H. D. Cochran, *Chem. Eng. Sci.*, **49**, 2735 (1994).
[9] T. L. Hill, J. Chem. Phys., **23**, 617 (1955); "Statistical Mechanics - Principle and Selected Applications", McGraw-Hill, New York (1956).
[10] E. M. Sevick, P. A. Monson and J. M. Ottino, *J. Chem. Phys.*, **88**, 1198 (1988).
[11] I. B. Petsche and P. G. Debenedetti, *J. Chem. Phys.*, **95**, 386 (1991).
[12] C. C. Liew, H. Inomata and S. Saito, *Fluid Phase Equilibria,* **104**, 317 (1995).
[13] H. Inomata, S. Saito and P. G. Debenedetti, *Fluid Phase Equilibria*, **116**, 282 (1996).
[14] M. P. Allen and D. J. Tildesley, "*Computer Simulation of Liquids*", Clarendon Press, Oxford (1987).
[15] Y. Kataoka and M. Fujita, *Bull. Chem. Soc. Jpn.*, **68**, 152 (1995).
[16] R. L. Rowley and M. M. Painter, *Int. J. Thermophysics,* **18**, 1109 (1997). According to the author, constant $b_{3,2}$ for viscosity in Table V should be read as 1067.97. Constant ω_2 for viscosity in Table V should be read as –2.0265.
[17] J. Kolafa and I. Nezbeda, *Fluid Phase Equilibria,* **100**, 1 (1994).
[18] O. Kitao, K. Tanabe, S. Ono, S. Kumakura and K. Nakanishi, *Fluid Phase Equilibria,* **144**, 279 (1998).
[19] H. Iwasaki and M. Takahashi, *J. Chem. Phys.*, **74**, 1930 (1981).
[20] D. K. Dysthe and B. Hafskjold, *Int. J. Thermophysics*, **16**, 1213 (1995).
[21] H. Nishiumi, M. Fujita and K. Ago, *Fluid Phase Equilibria*, **117**, 356 (1996).
[22] D. L. Jolly and R. J. Bearman, *Mol. Phys.*, **41**, 137 (1980).
[23] A. Ben-Naim, "Water and Aqueous Solution – Introduction to a Molecular Theory", Plenum Press, New York (1974).
[24] T. Kato, *J. Phys. Chem.*, **89**, 5750 (1985).
[25] H. L. Freedman and R. Mills, *J. Solution Chem.*, **10**, 395 (1981).
[26] Y. Kataoka, *Fluid Phase Equilibria*, **144**, 257 (1998).

2 Phase Equilibria and Static Properties

2.1 Thermophysical Properties of Supercritical Fluids

2.1.1 Behavior of Supercritical Fluids

The state condition of a fluid is described as a function of two intensive properties and the properties of the fluid can be represented also as a function of two independent properties. Figure 2.1.1 [1] shows the *PVT* surface of water in which the relation of pressure, volume and temperature of water is described. The critical point is shown as an open circle labeled C.P. The critical temperature of water is T_C = 647.096 K, the critical pressure is P_C = 22.064 MPa and the critical volume is V_C = 0.003106 m^3/kg. A list of critical and basic constants for selected fluids is given in Appendix A. The saturated liquid line and the saturated vapor line meet at the critical point. The dome-shaped region under the saturated liquid and vapor lines is the two-phase coexistence region of liquid and vapor. The region above the dome-shaped region is a single-phase homogeneous region which may be divided into three regions by the critical isotherm and critical isobar. The region at high pressure and low temperature is the compressed liquid region and the region at high temperature and low pressure is the superheated vapor or gas region. The supercritical fluid (SCF) region covers the states at which temperature is higher than the critical

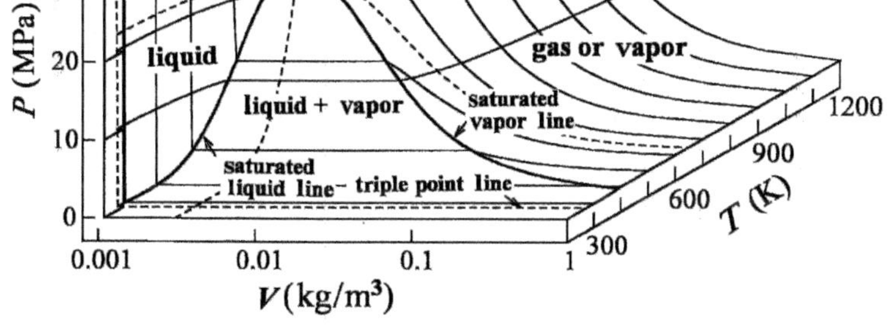

Fig. 2.1.1 *PVT* surface of water: C.P.= critical point

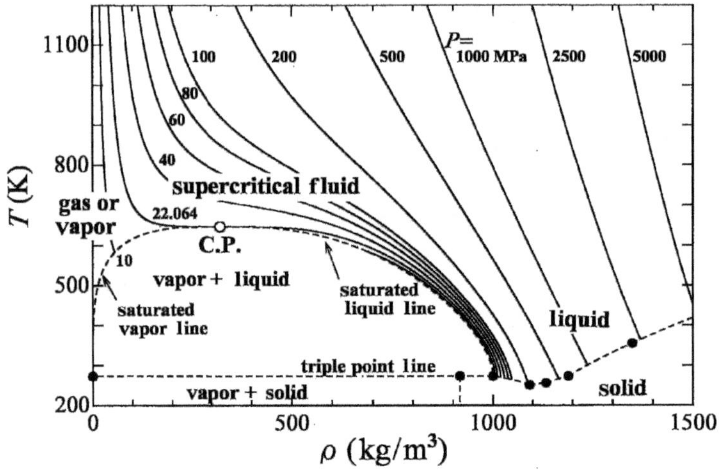

Fig. 2.1.2 Constant pressure lines of water in T–ρ plane

temperature, and pressure is higher than the critical pressure. In the SCF region, the phase separation can no longer be observed by compression of the vapor or expansion of the liquid.

The critical isotherm shown as a broken line labeled T_C in Fig. 2.1.1 has the characteristic nature of the horizontal point of inflection at the critical point, where the critical isotherm meets the peak of the dome, in the P–V plane. The critical point of a pure fluid may be defined by the equations

$$\left(\frac{\partial P}{\partial V}\right)_{T_C} = 0 \quad \text{and} \quad \left(\frac{\partial^2 P}{\partial V^2}\right)_{T_C} = 0 \tag{2.1.1}$$

at P_C and V_C along the critical isotherm.

A projection of the PVT surface in the $T-\rho$ plane is shown in Fig. 2.1.2 [1] with several isobars including the critical isobar. ρ in kg/m^3 denotes the fluid density that is the reciprocal of the specific volume. Here, the critical point is also shown as an open circle labeled C.P, whereas the triple points of water with various phases of ice are shown as solid circles. As shown in Fig. 2.1.2, the critical isobar also has the characteristic nature of the horizontal point of inflection at the critical point in the $T-\rho$ plane. In the vicinity of the critical point, significant change of density of fluids may be observed for a small difference of temperature for isobaric processes or for a small difference of pressure for isothermal processes. When the temperature or pressure is higher than critical, the density changes continuously without

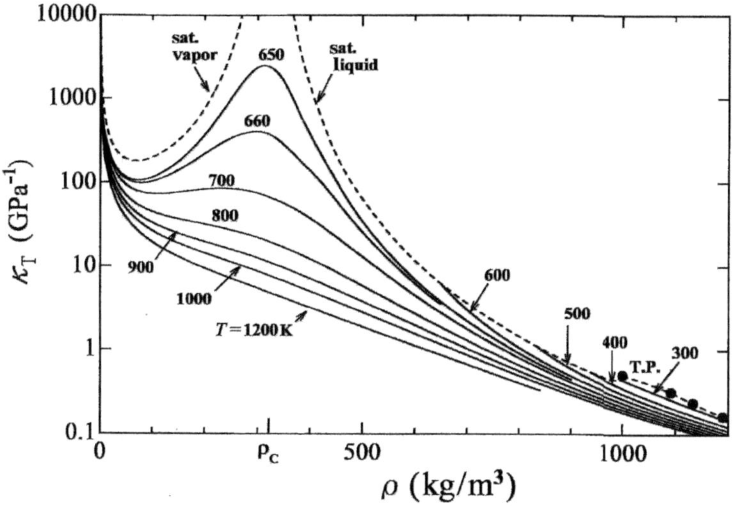

Fig. 2.1.3 Behavior of the isothermal compressibility of water as a function of density: T.P.= triple point

phase separation. If the temperature or pressure is lower than critical, phase separation is observed. It may be recognized that the density difference between the saturated liquid and saturated vapor at a temperature or pressure slightly lower than critical is significant, because the top of the dome is very flat.

In the isothermal process, the relative change in fluid density due to the pressure change is represented by the isothermal compressibility κ_T defined by

$$\kappa_T = \frac{1}{\rho}\left(\frac{\partial \rho}{\partial P}\right)_T = -\frac{1}{V}\left(\frac{\partial V}{\partial P}\right)_T . \tag{2.1.2}$$

Equation (2.1.1) shows that the derivative of the pressure with respect to volume along the critical temperature is zero at the critical point. This means that the isothermal compressibility diverges as it approaches the critical point and becomes infinite at the critical point. Figure 2.1.3 [1] shows the behavior of κ_T of water at constant temperatures as a function of density. κ_T decreases with increasing density from an ideal gas state to a liquid state except in the SCF region, where κ_T enhances as it approaches the critical point.

In the isobaric process, the relative change in fluid density due to the temperature change is represented by the thermal expansion coefficient α_P defined by

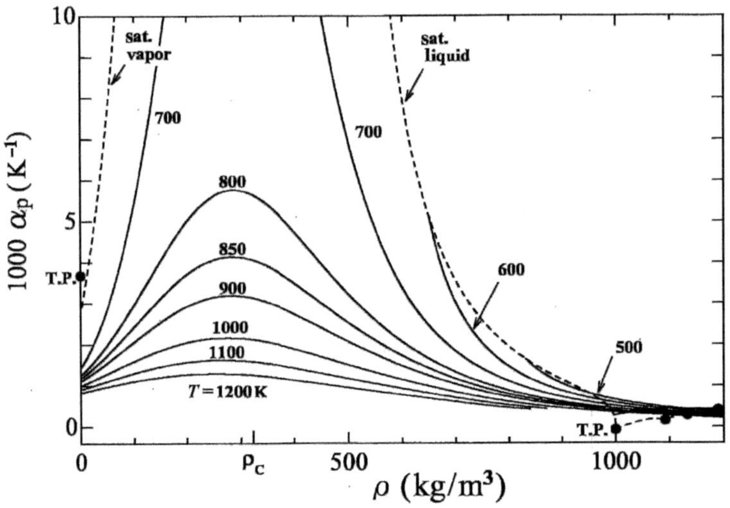

Fig. 2.1.4 Behavior of the thermal expansion coefficient of water as a function of density

$$\alpha_P = -\frac{1}{\rho}\left(\frac{\partial \rho}{\partial T}\right)_P = \frac{1}{V}\left(\frac{\partial V}{\partial T}\right)_P . \tag{2.1.3}$$

Figure 2.1.4 [1] shows the behavior of α_P of water at constant temperatures as a function of density. α_P also diverges as it approaches the critical point and becomes infinite at the critical point.

Direct measurements of the specific heat at constant volume, C_V, have shown that it diverges weakly close to the critical point and becomes infinite at the critical point [2]. Figure 2.1.5 [1] shows the behavior of C_V of water at constant temperatures in the SCF region as a function of density. C_V at $T <$ 700 K shows enhancement as it approaches the critical point.

The specific heat at constant pressure C_P is related to C_V by

$$C_P = C_V + \frac{T}{\rho^2}\cdot\frac{(\partial P/\partial T)_\rho^2}{(\partial P/\partial \rho)_T} = C_V + \frac{T}{\rho}\cdot\frac{\alpha_P^2}{\kappa_T} . \tag{2.1.4}$$

C_P increases more rapidly than C_V as it approaches the critical point and becomes infinite at the critical point. Figure 2.1.6 [1] shows the constant C_P lines as solid lines in the $T-\rho$ plane, where the broken lines denote isobars and saturated liquid and vapor lines. In this figure, the extent of the critical enhancement of C_P in magnitude and area in the SCF region can be seen.

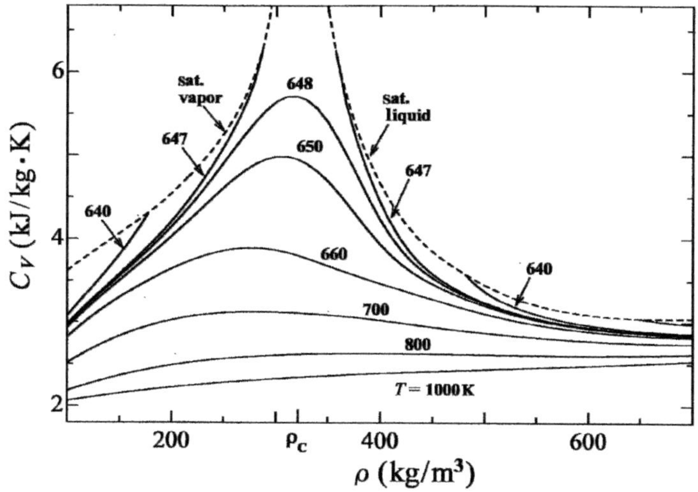

Fig. 2.1.5 Behavior of the specific heat at constant volume of water in the SCF region as a function of density

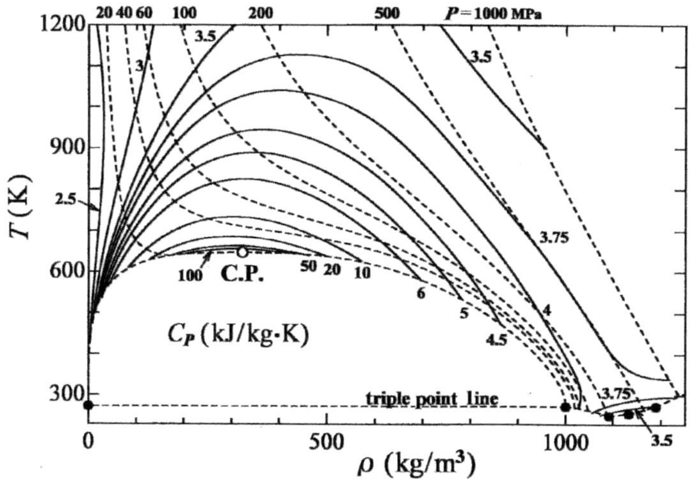

Fig. 2.1.6 Constant lines of specific heat at constant pressure of water in the T–ρ plane

The speed of sound w of fluids can be calculated from V, κ_T, C_V and C_P by

$$w = \sqrt{\left(\frac{\partial P}{\partial \rho}\right)_S} = \sqrt{\frac{C_P}{C_V} \cdot \left(\frac{\partial P}{\partial \rho}\right)_T} = \sqrt{V \cdot \frac{C_P}{C_V \cdot \kappa_T}} . \tag{2.1.5}$$

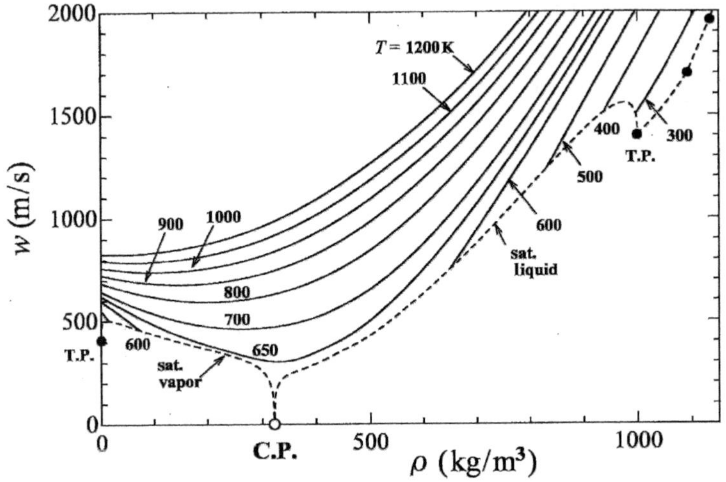

Fig. 2.1.7 Behavior of the speed of sound of water as a function of density

As it approaches the critical point, κ_T, C_V and C_P increase, but w decreases to zero at the critical point. Figure 2.1.7 [1] shows the behavior of w of water at constant temperatures as a function of density.

Figures 2.1.8 and 2.1.9 [1] show the constant lines of dielectric constant ε and ion product K_W of water as solid lines in the $T-\rho$ plane, respectively. ε and K_W increase with increasing density in the SCF region.

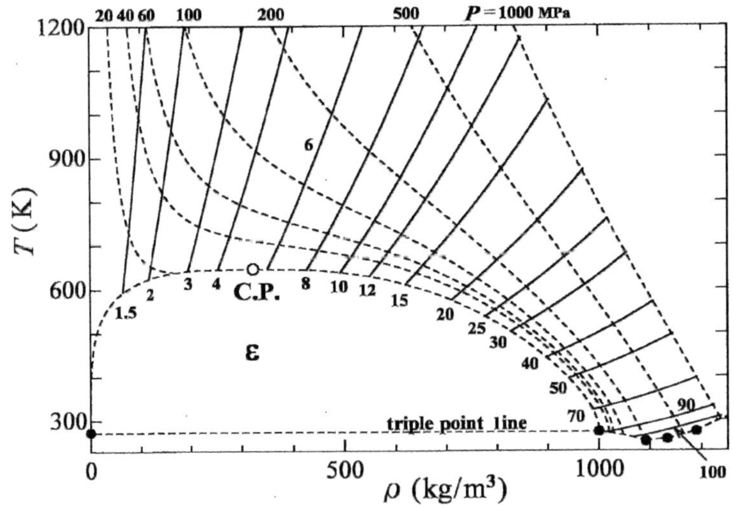

Fig. 2.1.8 Constant lines of the dielectric constant of water in T–ρ plane

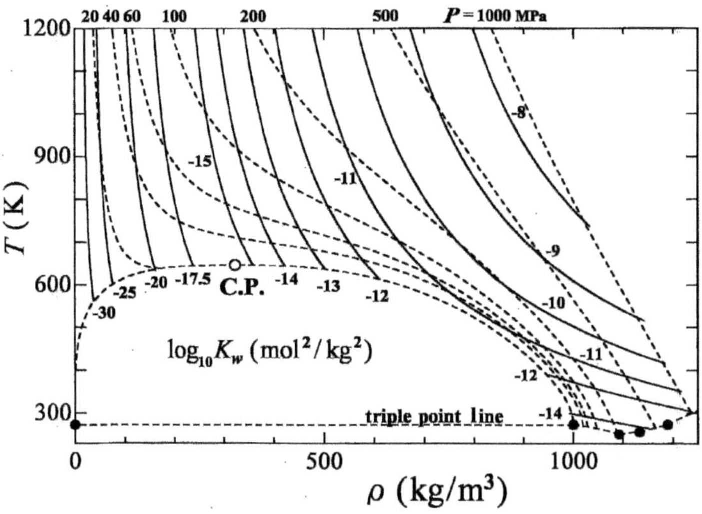

Fig. 2.1.9 Constant lines of the ion product of water in T–ρ plane

2.1.2 Measurement of the Critical Point of Liquid and Vapor

The critical point of liquid and vapor can be determined by visual observation of the meniscus at the vapor–liquid interface. Figure 2.1.10 shows the apparatus for meniscus observation and simultaneous measurements of *PVT* properties [3].

Close to the critical point, strong scattering of light is observed throughout the meniscus observation and this leads to complete opacity at the critical point [4]. This phenomenon is called critical opalescence. Near the critical point, an indistinct meniscus is observed with critical opalesence if $T < T_C$. If $T > T_C$, there is no meniscus, and only critical opalescence is observed. The critical temperature T_C is the temperature at which critical opalescence is observed most intensely. Keeping the temperature constant, observation continues with variation of density until the critical density ρ_C is found. At the state of T_C and ρ_C, the critical pressure P_C is measured.

On the other hand, the critical point can be found by analysis of *PVT* measurements [5]. T_C is found by the requirements of a horizontal point of inflection in (2.1.1) on an isotherm in the $P-V$ plane. P_C is found as the vapor pressure at T_C by analysis of the vapor pressure measurements in the $(\log P)-(1/T)$ plane. ρ_C is found by extrapolating the mean of the saturated liquid and vapor densities up to T_C. This extrapolation is called the law of rectilinear diameters.

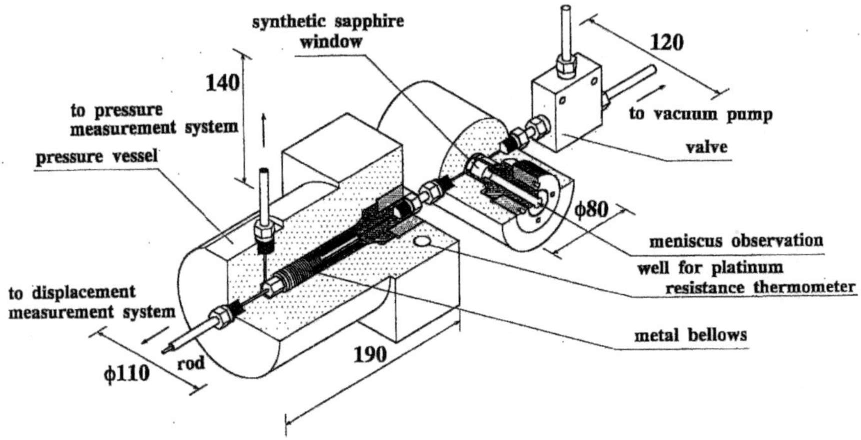

Fig. 2.1.10 Apparatus for meniscus observation and *PVT* measurements at the critical point [3] Reprinted from: Y. Suehiro, M. Nakajima, K. Yamada, M. Uematsu: Critical parameters of $\{xCO_2 + (1\text{-}x)CHF_3\}$ for x=(1.0000, 0.7496, 0.5013, and 0.2522), *Int. J. Chem. Thermodynamics* (Academic Press) **28**, 1153 (1996)

2.2 Phase Equilibria for a Mixture with a Supercritical Fluid

To improve the application of supercritical fluid technology, information on the phase equilibrium relation with a supercritical fluid is important [1]. In some cases of scientific and technological interest, the pressures employed for a mixture of fluid are above the critical point of the gas component, so this equilibrium is between the condensed phase and the supercritical fluid. In this section, some specific phase separations with a supercritical fluid are shown with pressure–composition diagrams.

2.2.1 Phase Rule

The Gibbs phase rule relates the number of phases presented at equilibrium to the number of components at specified composition (x), temperature (T), and pressure (P). The number of degrees of freedom of the system becomes

$$f\,(\text{degree of freedom}) = c\,(\text{number of components}) + 2 - p\,(\text{number of phases}). \tag{2.2.1}$$

Table 2.2.1 Degrees of freedom and number of phases for a binary system

f	p	
3	1	(solid–liquid /vapor)
2	2	(liquid–vapor/solid–liquid / liquid–liquid/ solid–vapor)
1	3	(solid-liquid-vapor/solid-solid–liquid /liquid–liquid–vapor)
0	4	(solid-solid–liquid–vapor/solid–liquid–liquid–vapor)

* number of components $c = 2$, degrees of freedom f, number of phases p

For example, for a binary vapor–liquid equilibrium, the number of components is $c = 2$ and the number of phases is $p = 2$. The degrees of freedom are $f = 2$ by (2.2.1).

Vapor–liquid equilibrium (VLE) is most frequently encountered in systems made up of two or more components that are completely miscible in the liquid phase. Vapor–liquid equilibrium relations of ternary and multicomponent systems cannot be represented by two-dimensional graphs, so considerations are limited to binary systems in this section.

For a system containing two components ($c = 2$), the phase rule becomes $f = 2 + 2 - p$. Since one phase ($p = 1$) must exist at least, the maximum number of independent phase-rule variables required to specify the intensive state of a stable system is three. If the independent variables are taken to be T, P, and one composition, all equilibrium states of the system can be represented in three-dimensional $P - T - x$ space. Within this space, the states of pairs of phases, coexisting at equilibrium ($f = 4 - 2 = 2$) define surfaces. The identifications are summarized in Table 2.2.1 [2].

2.2.2 Phase Equilibria and Phase Diagrams

A typical three-dimensional diagram for vapor–liquid equilibrium is shown in Fig. 2.2.1, which shows schematically the $P - T - x$ surface that represents equilibrium states of the saturated vapor and saturated liquid for a binary system. The lower surface contains the saturated vapor states. The upper surface contains the saturated liquid states. These surfaces intersect along the lines UBHC1 and KAC2, which represent the vapor pressure–temperature curves for pure components 1 and 2. The lower and upper surfaces form a continuous rounded surface across the top of the diagram between C1 and C2. The points C1 and C2 are the critical points of pure components 1 and 2, and the critical points of mixtures of 1 and 2 lie along a line on the rounded edge of the surface between C1 and C2. This is called the critical locus.

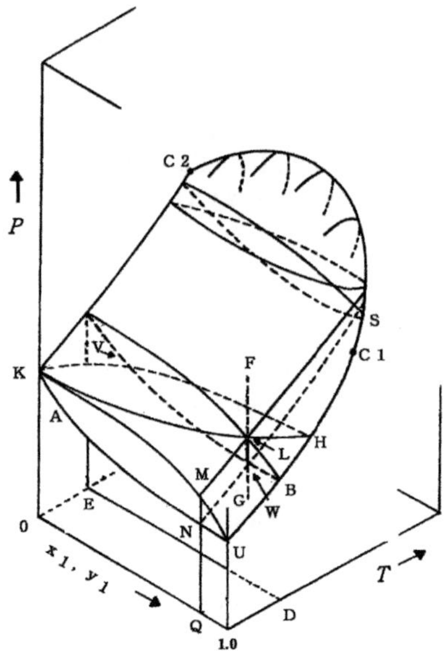

Fig. 2.2.1 Pressure–temperature–composition diagram for vapor–liquid equilibrium

The region lying above the upper surface of Fig. 2.2.1 is the subcooled-liquid region, and the region below the lower surface is the superheated-vapor. The interior space between the two surfaces is the coexistence of liquid and vapor phases. Reducing the pressure of the liquid at F at constant temperature and composition, the first bubble of vapor appears at point L, which lies on the upper surface. For this reason, L is called the bubble point. The state of the vapor with a bubble in equilibrium with the liquid at L is represented by a point on the lower surface at the temperature and pressure of L. This point is indicated by V. The line VL is an example of a tie line, which connects points representing phases in equilibrium.

Reducing the pressure along the line FG, liquid vaporizes until at W the process is complete. W lies on the lower surface and represents the state of saturated vapor with the mixture composition. So W is the point at which the last drop of liquid disappears; it is called the dew point, and the lower surface is the dew-point surface. Continued reduction of pressure merely leads into the superheated-vapor region. Because of the complexity of Fig. 2.2.1, the detailed characteristics of binary vapor–liquid equilibrium are usually depicted by two-dimensional graphs that display what is seen on various planes that cut the three-dimensional diagram. The three principal planes, each perpendicular to one of the coordinate axes, are illustrated in Fig. 2.2.1. Thus a vertical plane perpendicular to the temperature axis is outlined as ALBDEA. The lines on this plane represent a pressure–composition phase diagram at constant temperature. Each is a tangent point at which a horizontal line touches the curve. This is so because all tie lines connecting phases in equilibrium are horizontal, and the tie line connecting identical phases (the definition of a critical point) must therefore be the last such line to cut the diagram. A horizontal plane passed through Fig. 2.2.1 perpendicular to the P-axis is identified by HIJKLH.

The typical types of vapor–liquid behaviors are summarized in Figs 2.2.2 and 2.2.3. The popular kinds of diagrams are $T-x$ and $P-x$. In Fig. 2.2.2, types I and II involve liquid phases that are immiscible and partially miscible, respectively. In the less common systems of type III, immiscible liquid phases exist at temperatures intermediate to the condensation temperatures of the pure substances.

Figure 2.2.3 shows six typical types of phase separation behaviors on the pressure–temperature diagram [1]. C_α and C_β denote critical points of substances. Type 1 shows a system without liquid–liquid separation. Types 2–6 show systems with liquid–liquid separation. On an end point of the liquid-liquid critical locus in type 2, the upper critical end point (U, UCEP) is illustrated. In type 3, the ends of the critical locus of a and b will lead to an end point of solid–liquid–liquid equilibrium or UCEP of vapor–liquid–liquid equilibrium. The loci of c and d have unique vapor–liquid separations. Types 4–6 involve lower critical end points (L, LCEP).

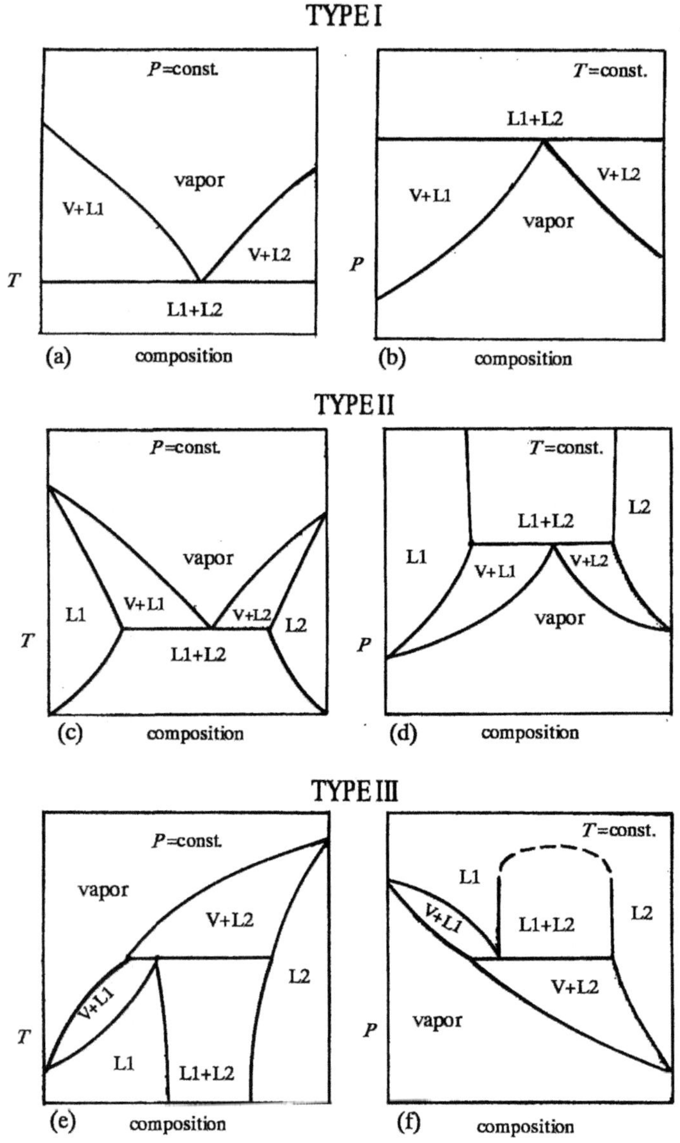

Fig. 2.2.2 Typical separation behaviors of temperature–composition and pressure–composition diagrams: TYPE I: systems with immiscible liquid phases; TYPE II: systems with partially miscible liquid phases; TYPE III: systems with immiscible liquid phases at temperatures between the condensation temperatures of the pure substances

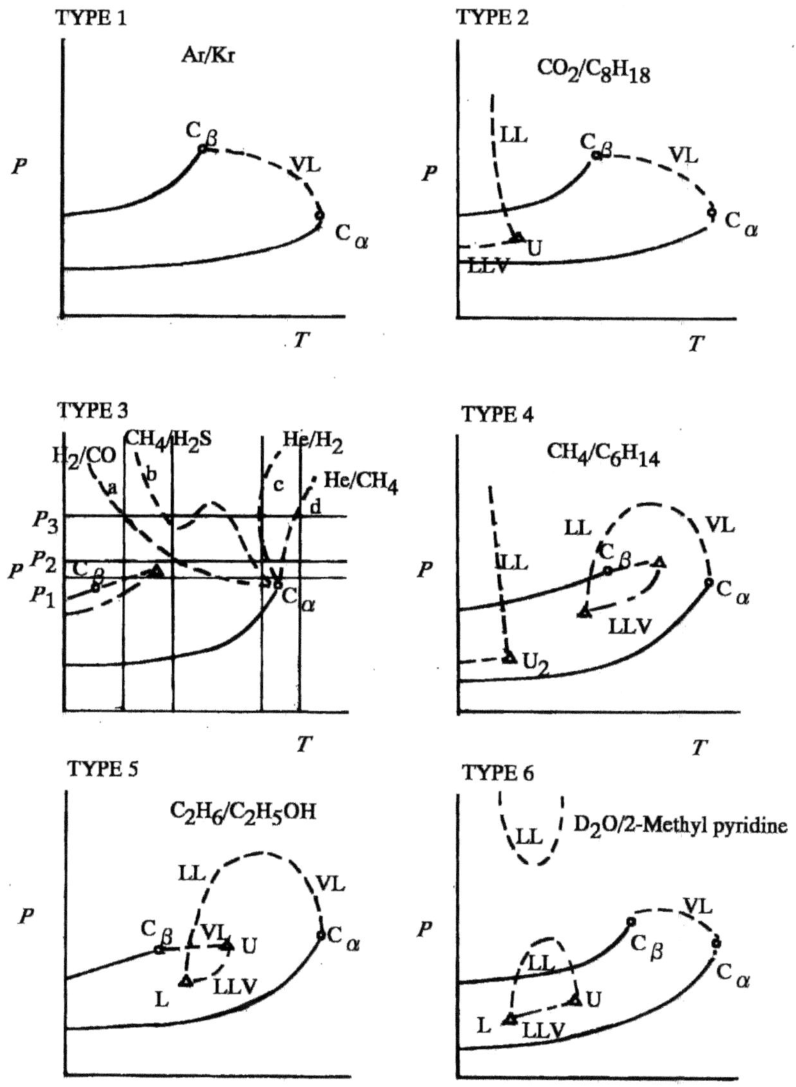

Fig. 2.2.3 Typical separation behaviors of pressure–temperature diagram; TYPE 1: system without liquid–liquid separation, TYPE 2 to TYPE 6: systems with liquid–liquid separation

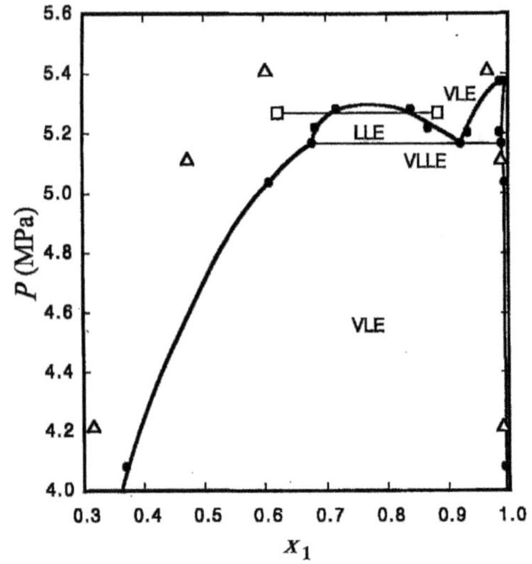

Fig. 2.2.4 Phase equilibrium for ethane + ethanol at 311.15 K: ●, Kato et al. [3]; △, Suzuki et al. [4] at 313.4 K; □, McHugh et al. [5], LLE along VLL loci at 311.25 K

2.2.3 Experimental Phase Diagrams for a Binary Mixture

In some cases of fluid mixtures, three-phase separation is observed for binary system. Figure 2.2.4 shows the $P-x$ diagram for ethane + ethanol at 311.15 K [3] including the vapor–liquid equilibrium at 313.4 K reported by Suzuki et al. [4] and liquid–liquid equilibrium (LLE) data of McHugh et al. [5] at 311.25 K along the vapor–liquid–liquid (VLL) line. The VLL phase separation is observed at 5.169 MPa. Figure 2.2.5 gives the saturated density behaviors [3] including the saturated liquid density data of McHugh et al. [5] along the VLL loci.

For a binary system consisting of two liquid phases and a vapor phase in equilibrium, the degree of freedom is one. At a fixed temperature, the pressure and compositions of all three phases are determined. If the three-phase equilibrium is maintained, the compositions of the three phases must remain the same. However, the relative amounts of the phases will change to reflect the alteration in total composition.

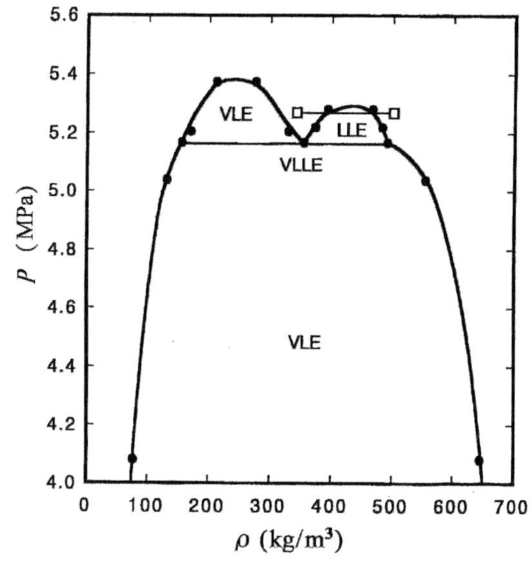

Fig. 2.2.5 Saturated density behaviors for ethane + ethanol at 311.15 K: ●, Kato et al. [3]; □, McHugh et al. [5], LLE along VLL loci at 311.25 K

2.3 Measurement of Phase Equilibria

Measurement methods about the phase equilibria of supercritical fluids have been studied and developed only recently. The first investigation of the behavior of the supercritical state was reported by Toriumi et al. [1, 2]. Experimental instruments and apparatus for vapor–liquid equilibrium relation measurements of a system containing supercritical fluids were then devised by many researchers, with remarkable achievements.

The measurement method is principally divided into two methods: a static method and a dynamic one, both of which are presented briefly in this section. The systematic classification of the measuring method on the phase equilibrium of supercritical fluids has already been described elsewhere [3, 4].

2.3.1 Static Method

The static method can be subdivided into the static analytical method and the static nonanalytical method. Each characteristic illustration is shown in Fig. 2.3.1 (a) and (b), respectively.

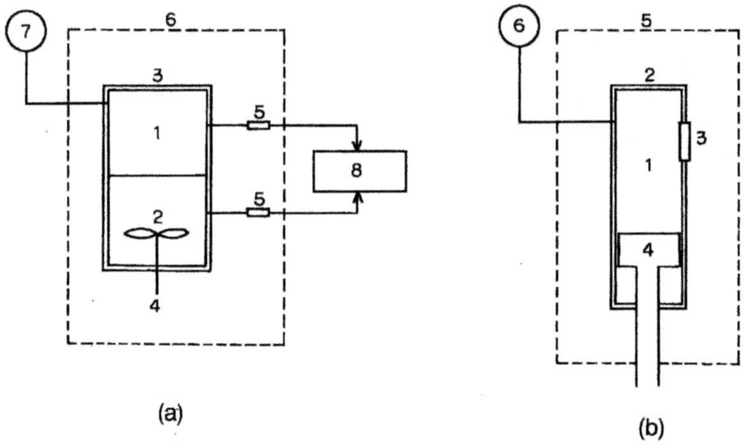

Fig. 2.3.1 Static experimental method. (**a**): 1, vapor; 2, liquid; 3, equilibrium cell; 4, agitator; 5, sampler; 6, thermostat; 7, pressure gauge; 8, analysis; (**b**): 1, mixture; 2, equilibrium cell; 3, window; 4, free piston; 5, thermostat; 6, pressure gauge

(a) Static Analytical Method

Figure 2.3.1 (a) represents a typical example of the static analytical measurement. Both the vapor and liquid used in the investigation are currently introduced into the evacuated equilibrium cell by an intensifier or a pump for the liquid and directly or by the use of a heater for the liquefied vapor. In order to reach the equilibrium state, the vapor and liquid in the equilibrium cell are stirred by various methods of agitation.

After equilibrium is reached, the temperature and pressure of the substances in the cell are precisely measured and a constant volume of the samples of either vapor or liquid or both samples are taken out from the phases in the cell to analyze their composition. To keep a uniform equilibrium state, the temperature, pressure and composition distribution are always checked after sampling from both phases. Some examples used practically in the measurements will be given here for illustration.

A schematic diagram of the experimental apparatus designed by Sagara et al. [5] is shown in Fig. 2.3.2. They measured the vapor–liquid equilibria of binary and ternary systems containing hydrogen and light hydrocarbons at temperatures from 103.15 K (−170 °C) to 248.15 K (−25 °C) and pressures below 10.8 MPa. The equilibrium cell 2 and sample lines are made of stainless steel. Equilibrium was indicated when the system pressure reached a constant value. Samples were withdrawn through the stainless steel tubes and then

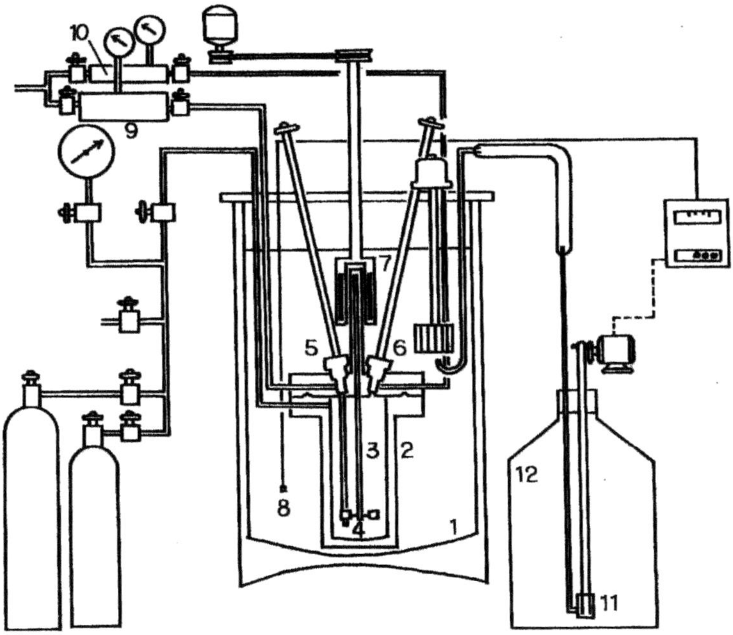

Fig. 2.3.2 Schematic diagram of the experimental apparatus of Sagara et al. [5]: 1, vacuum cryostat; 2, equilibrium cell; 3, magnetic stirrer; 4, capillary tube; 5, liquid sample valve; 6, vapor sample valve; 7, external magnet; 8, thermocouple; 9, liquid sampler; 10, vapor sampler; 11, plunger pump; 12, liquid N_2 container

flushed into the samplers, 9 and 10, respectively. The samplers were attached to prevent the separation of the sample into vapor and liquid in the sample lines. After flushing, the expanded vapor samples were then introduced individually to the gas chromatograph. The first sample taken was disposed of without analysis because it probably contained nonequilibrium portions in the dead space. Pressure change during a run could be neglected for the large content in the cell.

Bae et al. [6] designed and constructed a visual type of static apparatus which consists of an equilibrium cell with two 17 mm thick reinforced glass windows, a sampling device for the liquid phase and associated equipment for measurements of temperature and pressure, in order to measure the binary phase equilibrium containing C_2H_4 in the temperature range from 223.15 to 323.15 K at pressures up to 10.1 MPa. Figure 2.3.3 [6] shows a glass-windowed cell containing a specially designed liquid sampling device and a magnetically driven impeller, 3, designed to disperse the vapor phase continuously into an underlying liquid phase. The visibility afforded by the large glass

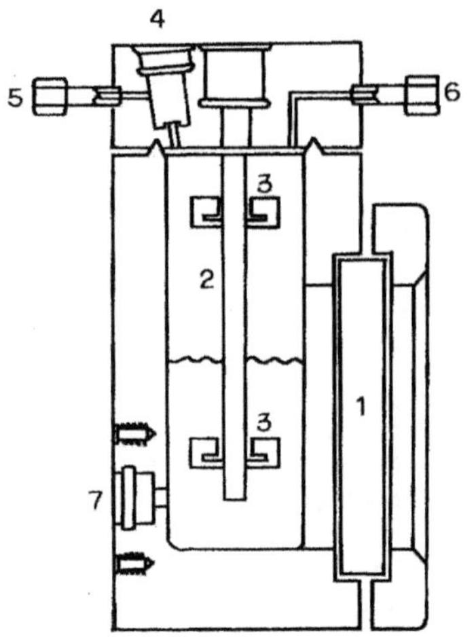

Fig. 2.3.3 Equilibrium cell and agitator used by Bae et al. [6]: 1, glass window; 2, agitator; 3, metallic impeller; 4, valve; 5, charge line; 6, pressure measurement system; 7, liquid withdrawal position

windows, 1, of the cell is useful in the detection of liquid or solid phases, dew and bubble points, and critical points. Vapor is introduced in the small upper holes of the cell. After the hollow shaft, 2, is descended, vapor flows out through the lower holes to disperse into the liquid phase. The vapor bubbles are continuously drawn into the liquid phase at any condition, to reduce the time required to attain equilibrium.

Binary phase equilibria and ternary liquid–liquid equilibria (LLE) were measured by Nakayama et al. [7], to extract ethanol from ethanol aqueous solutions by the use of their apparatus designed for measurements up to 450 K and 20 MPa. A schematic diagram of the static method to obtain high-pressure liquid–liquid equilibrium data is shown in Fig. 2.3.4 [7]. The equilibrium cell, 1, of volume about 270 cm^3 is connected to a buffer tank, 4, through a stainless steel diaphragm, 3. Samples can be withdrawn from the equilibrium cell by sample devices, 2. The volume and pressure decrease that occurs during sampling is automatically compensated for by the action of the diaphragm connected to the buffer tank. Two types of sampling device, a sliding rod and a capillary tube sampler, were used. The sliding rod sampler, with some improvements, based on a sampling valve reported by Fredenslund et al. [8], Ng et al. [9, 10] and Bae et al. [6], could be moved back and forth horizontally between the cell and the sampling line. The capillary tube sampler was simpler in design and operation and required no special maintenance. It is necessary to

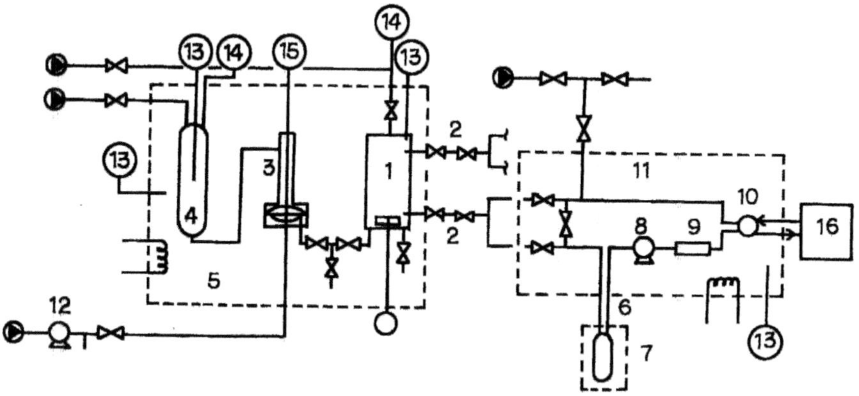

Fig. 2.3.4 Schematic diagram of experimental apparatus of Nakayama et al.[7]: 1, equilibrium cell; 2, sampler; 3, diaphragm; 4, buffer tank; 5, liquid constant temperature bath; 6, trap; 7, low temperature bath; 8, magnetic pump; 9, static mixer; 10, eight-way cock; 11, air constant temperature bath; 12, pump; 13, thermocouple; 14, pressure transducer; 15, displacement meter; 16, gas chromatograph

compare the data obtained from both devices; the capillary tube sampler had a large volume decrease than the sliding rod sampling device.

(b) Static Nonanalytical Method

Figure 2.3.1 (b) shows a static nonanalytical method or a static synthetic method for vapor–liquid equilibrium measurement. A known composition mixture is introduced into the equilibrium cell equipped with window, and is pressurized at a constant temperature until the sample forms the homogeneous phase of the mixture. The mole fractions of the sample mixture can be precisely determined from the feed amount of each substance made up of it. The sample volume is expanded by diminished pressure after the stability of the experimental temperature and pressure is obtained. As soon as the homogeneous mixture starts to separate the vapor–liquid coexistence phase, the temperature and pressure are continuously recorded. Then, an inflection point on the isothermal relation between the pressure and inside volume of the cell is recognized, since the compressibility between the liquid and vapor–liquid coexistence phases represents a large difference. The bubble point pressure is determined from the inflection point on the isothermal line. The advantages of this method are a simple apparatus, no analysis of the sample, and critical state measurement, but the disadvantages are, for multi-component system, measurements away from the critical state, and especially difficult visual observations.

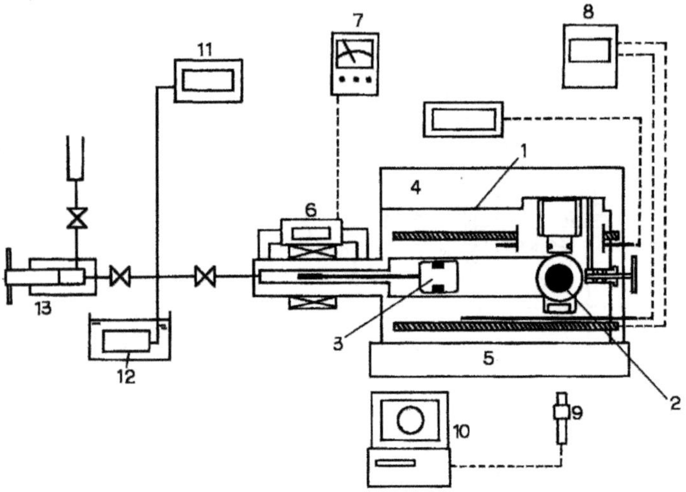

Fig. 2.3.5 Schematic diagram of experimental apparatus of Sato et al. [11]: 1, high-pressure cell; 2, sapphire window; 3, free piston; 4, heat insulator; 5, hotplate stirrer; 6, linear variable displacement meter; 8, temperature controller; 9, CCD camera; 10, display and VTR; 11, pressure gauge; 12, pressure damper; 13, high-pressure pump. Reprinted from: Y. Sato, S. Takishima, H. Masuoka: Solubility of carbon dioxide in eicosane, docosane, tetracosane, and octacosane at temperatures from 323 to 473 K and pressures up to 40 MPa, *Fluid Phase Equilibria* (Elsevier Science) **147**, 181 (1998)

A practical example of this method employed by Sato et al. [11] is shown in Fig. 2.3.5. They measured bubble point pressures for N_2 + decane and N_2 + hexadecane systems at temperatures from 323.2 to 473.2 K and pressures up to 30 MPa to develop a gas injection process. This apparatus consists of a high-pressure cell of variable volume, increasing pressure part, displacement measurement part of the free piston, and temperature controlling part. The high-pressure cell, 1, is equipped with a sapphire window, 2. The phase behavior in the cell is observed through the window by use of a CCD camera, 9, and display, 10. The pressure damper, 12, is cooled by liquid N_2 to recover N_2 gas.

2.3.2 Dynamic Method

The dynamic types of experimental methods can be subdivided into two methods, as well as the static method. They are the phase recirculation method and the flow type method. The characteristic of each typical dynamic method is illustrated in Fig. 2.3.6 (a) and (b), respectively.

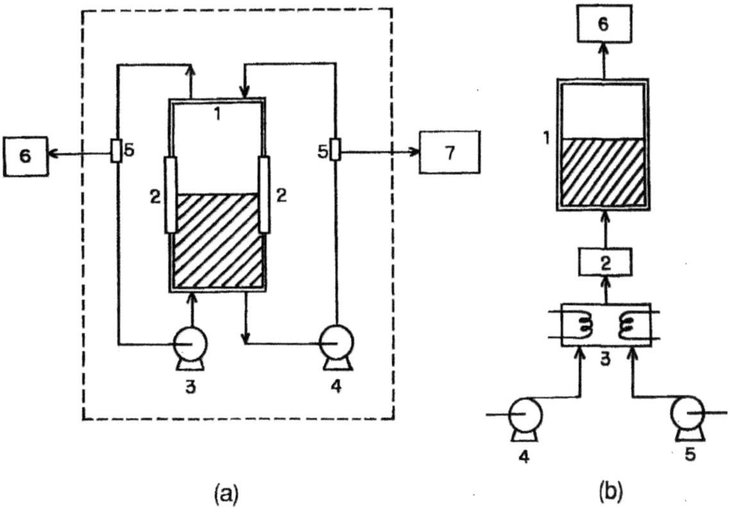

Fig. 2.3.6 Dynamic experimental method. (**a**): 1, equilibrium cell; 2, window; 3, recirculation pump for vapor; 4, recirculation pump for liquid; 5, sampler; 6, analysis for vapor; 7, analysis for liquid; (**b**): 1, equilibrium cell; 2, mixer; 3, preheater; 4, feed pump for vapor; 5, feed pump for liquid; 6, analysis system

(a) Phase Recirculation Method

In one phase recirculation type, either vapor or liquid, mainly the vapor phase is circulated through the equilibrium cell to obtain the vapor–liquid equilibrium at low or cryogenic temperatures.

Generally a two-phase recirculation method is used and is represented in Fig. 2.3.6 (a). Both vapor and liquid phases circulate countercurrently through the equilibrium cell. The liquid is introduced at the top of the cell and flows down through the vapor phase, while the vapor is introduced at the bottom of the cell and bubbles up through the liquid phase by using two recirculation pumps. Phase equilibrium in the cell can be obtained fairly rapidly by this method. It is not so difficult to operate the recirculation pump and the instruments in the system for the vapor–liquid equilibrium measurement over a wide range of temperature and pressure. Sampling of the vapor and liquid phase in the recirculation system has been devised by many investigators.

Kubota et al. [12] reported the vapor–liquid equilibrium data for the $C_2H_4 + C_3H_6$ system at temperatures from 283.15 to 348.15 K and pressures up to the critical pressure of the mixture, by using a vapor–liquid circulating type apparatus shown in Fig. 2.3.7. In the middle of the equilibrium cell, 1, a pair of Pyrex glass windows is placed to observe mixing and circulating

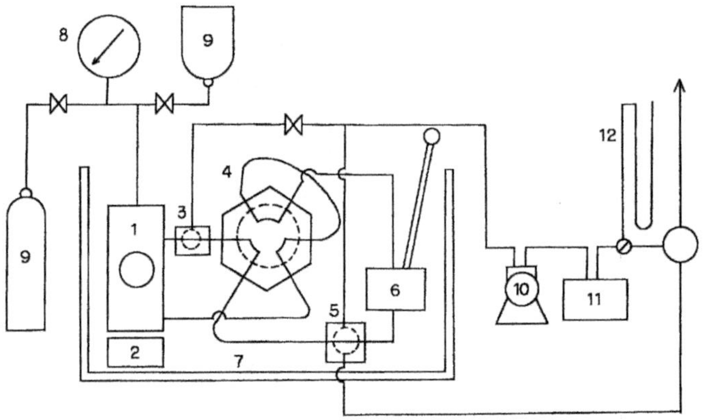

Fig. 2.3.7 Schematic layout of experimental apparatus of Kubota et al. [12]: 1, equilibrium cell; 2, magnet mixer; 3, two-way ball valve; 4, six-way ball valve; 5, two-way ball valve; 6, high-pressure pump; 7, temperature bath; 8, bourdon gauge; 9, sample cylinder; 10, low-pressure pump; 11, gas chromatograph; 12, mercury manometer

conditions of the sample in the cell and the behavior of the interface in the critical region. Three ball valves, 3, 4, and 5, are used in this apparatus. The change from the vapor circulating path to the liquid one is carried out by valve, 4. The sample trapped in valve 5 is introduced into the low-pressure analysis system. The sampler of the gas chromatograph is included in the low-pressure circulating line.

As shown in Fig. 2.3.8, the apparatus used by Takishima et al. [13] is capable of measuring three-phase equilibria and consists of the four main parts: feed system, dual-window equilibrium cell, 1, recycling system, and sampling system. Except for the feed system, each part of the apparatus is located in a temperature-controlled air bath 13–15. To avoid the condensation of heavier components in the recirculation line for the vapor phase, the temperature of air bath 14 is maintained at a slightly higher value than the equilibrium temperature, whereas the temperature in the recirculation line for the liquid phase is slightly lower than the equilibrium temperature to avoid the vaporization of CO_2. When no further change is detected in pressure and interfacial levels, recirculation is stopped and a small amount of each phase is enclosed in sampler 6. To vaporize the liquid sample and to prevent condensation of the vapor sample, the temperatures of air bath 14 are elevated to the same value as that of the sampling systems 15. After the evacuation of the flash tank, 11, the liquid sample is expanded into the tank and recirculated through the magnetic pump, 7, to homogenize the sample throughout the sampling system.

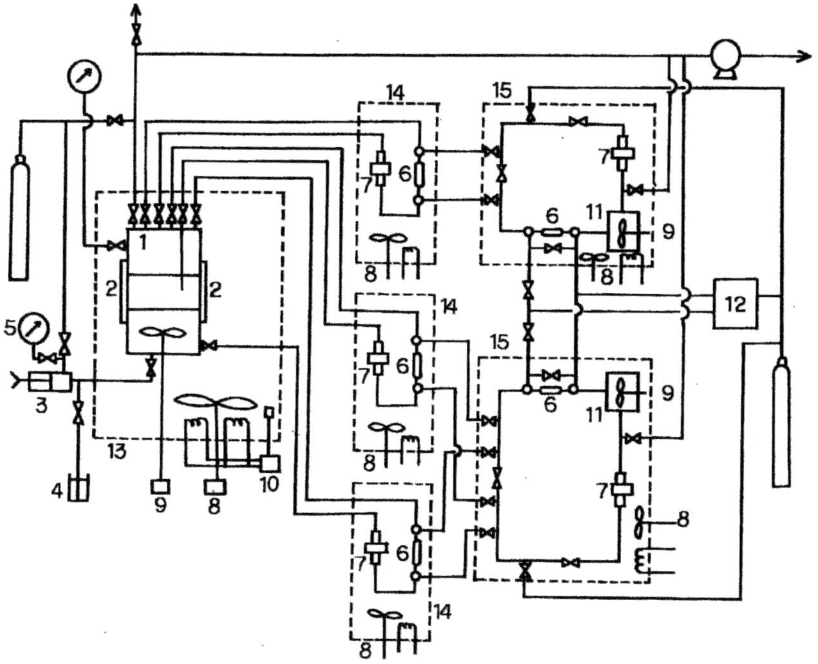

Fig. 2.3.8 Schematic diagram of apparatus of Takishima et al. [13]: 1, equilibrium cell; 2, window; 3, sample charging pump; 4, liquid sampler; 5, pressure gauge; 6, sampler; 7, magnetic pump; 8, stirrer; 9, magnetic stirrer; 10, temperature controller; 11, flash tank; 12, gas chromatograph; 13–15, air bath

Suzuki et al. [14] reported isothermal high-pressure vapor–liquid equilibrium data for seven binary systems by a recirculating type apparatus shown in Fig. 2.3.9. The apparatus is made up of two sections: an equilibration section that consists of an equilibrium cell, 1, enclosed in an air bath, 11, and an analysis section that consists of a homogenizing system enclosed by an air bath, 11, kept at a temperature much greater than the equilibrium temperature. The equilibrium cell, consisting of a two-piece autoclave, is designed to carry a maximum pressure of 25 MPa at 453 K. Three pairs of diametrically opposed windows allow viewing of the sample and liquid level. Sampling valves consist of two pairs of two-way valves placed at 90° angles from each other and two holes in the body that allow for heating with cartridge heaters. The internal volume of the liquid sampler is 0.75 cm^3 and that of the vapor sampler can be varied from 1 to 10 cm^3 by manually changing a stainless steel insert.

The experiment starts by evacuation of the apparatus overnight to pressure of below 0.1 Pa. Materials are supplied into the cell through a high-

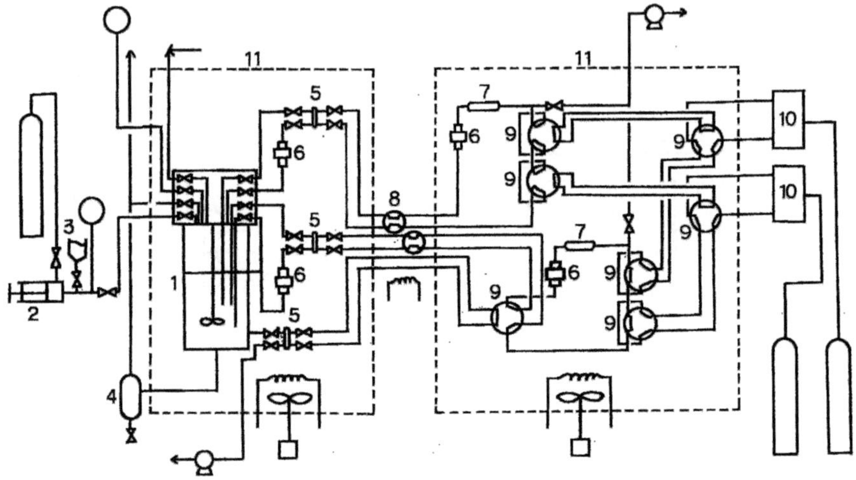

Fig. 2.3.9 Schematic diagram of apparatus of Suzuki et al. [14]: 1, equilibrium cell; 2, charge pump; 3, liquid supply; 4, drain tank; 5, sampler; 6, magnetic pump; 7, flash tank; 8, four-way valve; 9, six-way valve; 10, gas chromatograph; 11, air bath

pressure piston pump. The vapor and liquid phases are recirculated through the sampling valves by pneumatically controlled magnetic pumps and are vigorously stirred by the magnetic coupled stirrer. Samples are isolated in the sampling valves and flashed to the low-pressure section, and all line heaters and sampling valve heaters are activated. Low-pressure magnetic pumps are used for stirring action, and surge tanks provide additional system volume.

The schematic diagram of a circulation type apparatus constructed by Iwai et al. [15] is shown in Fig. 2.3.10. They measured a CO_2 + linalool system at temperatures from 313.2 to 333.2 K and pressures up to the critical region of the mixture by this method. The equilibrium cell equipped with a dual glass window, 1, is evacuated. CO_2 is supplied from a gas cylinder, liquefied through a cooling unit, compressed, and directed to the equilibrium cell. The heavy component is introduced into the equilibrium cell by a charging pump. An impeller, 2, is used to promote the equilibrium condition. After equilibrium is obtained, stirring by the impeller is stopped and circulation of the vapor and liquid phases is started by the magnetic pump, 4. A small amount of each phase is introduced into sampling loops, 6. The six-way valves, 5, are switched in order to isolate the sampling loops from the equilibrium cell. The samples in the loops are decompressed through expansion valves, and then introduced into glass tubes, 9, in which CO_2 and the heavy component are separated. Ethanol and hexane are used to trap decane and

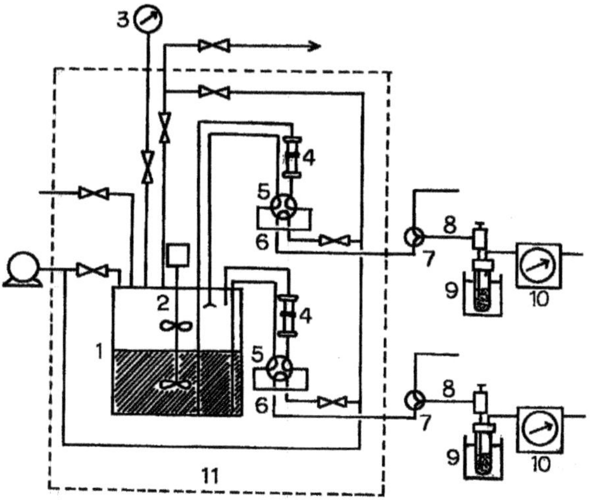

Fig. 2.3.10 Circulation type apparatus of Iwai et al. [15]: 1, equilibrium cell; 2, impeller; 3, pressure gauge; 4, magnetic pump; 5, six-way valve; 6, sampling loop; 7, three-way valve; 8, line heater; 9, cold trap; 10, dry test meter; 11, water bath

linalool, respectively, and have already been saturated with CO_2 by using a bypass line before the sampling operation to avoid dissolution of CO_2 in the sampling loops into ethanol or hexane. The amount of CO_2 is measured by the dry test meter, 10. A small amount of heavy component remaining in the tubing and the expansion valve is removed and trapped by using supercritical CO_2 through the bypass line and ethanol or hexane. The amount of heavy component is determined by means of a gas chromatograph.

Hiaki et al. [16] measured isothermal vapor–liquid equilibria for CO_2 with 1-butanol and 2-butanol systems by using a high-pressure view-cell with circulation of both vapor and liquid phases. The measurement was carried out at a temperature of 313.2 K and pressures up to 8 MPa, respectively. A schematic diagram of the apparatus is illustrated in Fig. 2.3.11. As shown in the figure, the experimental apparatus consists of a sample loading part, an equilibrium cell connected to magnetic circulating pumps, and sampling and analysis parts of both vapor and liquid phase components. An equilibrium cell, 1, equipped with three windows, 2, that is, top, middle, and bottom of the cell, and sampling system are each placed in the temperature-controlled air bath, 11. The sampling of both phases is undertaken by using magnetic circulating pumps, 4, and a surge tank, 5, for holding some of the phase in the small compartment. After equilibrium is reached, the samples of both phases in the surge tank are analyzed with a gas chromatograph. As the components of the sample

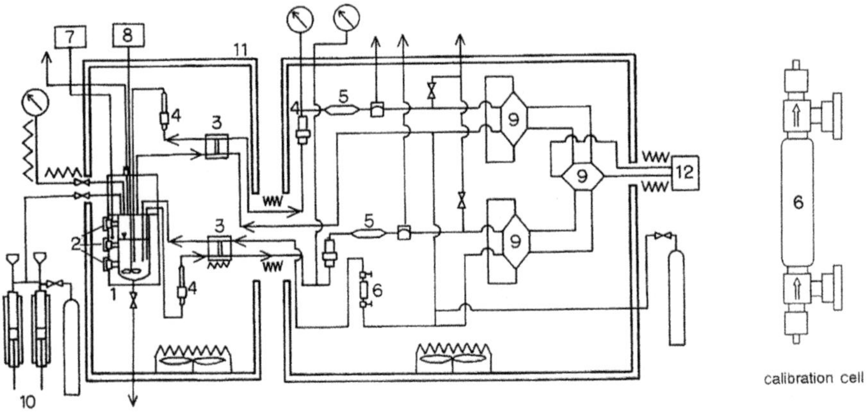

Fig. 2.3.11 Schematic diagram of the experimental apparatus of Hiaki et al. [16]: 1, equilibrium cell; 2, windows; 3, sampler; 4, circulating pump; 5, surge tank; 6, calibration cell; 7, Pt resistance thermometer; 8, agitator; 9, six-way valve; 10, sample loader; 11, air bath; 12, gas chromatograph. Reprinted from: T. Hiaki, H. Miyagi, T. Tsuji, M. Hongo: Vapor-liquid equilibria for supercritical carbon dioxide + butanol system at 313.2 K, *J. Supercritical Fluids* (Elsevier Science) **13**, 23 (1998)

are supercritical CO_2 and liquid butanol, we have to prepare sample mixtures in the high-pressure calibration cell shown in Fig. 2.3.11. The calibration line between the peak area and composition is determined from analysis of samples of known composition.

(b) Flow Type Method

A typical illustration of the dynamic flow type method or single vapor and liquid pass method is shown in Fig. 2.3.6 (b). Vapor and liquid components are separately introduced into the preheater and injected into the mixer at a controlled temperature and pressure. It may be possible to replace the preheater and mixer according to the condition of the sample. The sample mixture passes into an equilibrium cell where the mixture separates into the vapor and liquid phases. It is possible to measure the phase equilibrium containing the heavy liquid and solid. This method is often applied to supercritical fluid extraction, and is well suited for a large quantity treatment because of its simplicity and rapidity. A practical example will be described below.

Inomata et al. [17] designed a flow type apparatus using an overflow-type liquid level self-control system and a back pressure regulator. The static method has a disadvantage that it requires measurement at high temperatures,

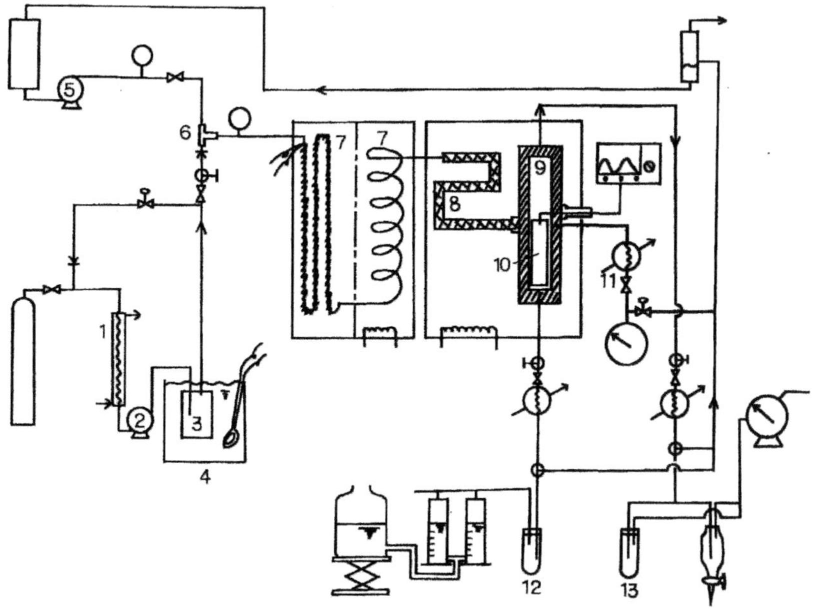

Fig. 2.3.12 Schematic diagram of flow-type apparatus by Inomata et al. [17]: 1, water heat exchanger; 2, plunger pump; 3, buffer tank; 4, water bath; 5, nonpulsating pump; 6, tee; 7, preheater tubing; 8, static mixer; 9, equilibrium cell; 10, liquid level sensor; 11, back pressure regulator; 12, liquid sampler; 13, vapor sampler

where thermal decomposition of the components may occur. This disadvantage can be overcome by minimizing the residence time of the sample in the high-temperature range, and this apparatus is used to collect seven sets of binary vapor–liquid equilibrium data in the temperature range from 340 to 710 K and pressure up to 23 MPa. The flow type apparatus represented in Fig. 2.3.12 [17] is composed of a feed system, preheater, equilibration system and sampling system. A vapor component, such as CO_2, is cooled and liquefied in the water heat exchanger, 1. The liquefied component is pressurized through a plunger pump, 2, and supplied to a buffer tank, 3, immersed in a water bath, 4, which allows the component to be heated to its supercritical temperature. The pressure fluctuation caused by the plunger pump is reduced with the buffer tank. A liquid component is provided through a nonpulsating pump, 5, and joined in a tee zone, 6, with the vapor component supplied from the buffer tank.. The mixture is immediately heated when passing through a preheater tubing, 7, fitted with an electric line heater. In the following tubing, 7, immersed in an air bath, the temperature of the mixture is adjusted to the desired equilibrium temperature. Then it flows into a static mixer, 8, and is well mixed in order to ascertain the attainment of the equilibrium state in a short time before the separation of vapor and liquid phases in an equilibrium cell, 9. In the cell, the liquid level is automatically controlled at the position of the outlet nozzle connected to the overflow line, while the pressure is maintained at a fixed value by the back pressure regulator, 11. The liquid level is also detected by a capacitor sensor, 10, and the signal is very useful for steady-state operation. The compositions of the equilibrium phases are determined from the mole number of each component evaluated from the weighed mass of liquid and vapor volume of each component.

2.4 Prediction of Phase Equilibria

2.4.1 Fundamental Equations

For every component i in a mixture, the thermodynamic condition of phase equilibria at constant temperature and pressure is that the fugacity or chemical potential of component i of each phase is equal, as follows:

$$f_i^{\mathrm{I}} = f_i^{\mathrm{II}} = f_i^{\mathrm{III}} = \cdots \qquad \text{or} \qquad \mu_i^{\mathrm{I}} = \mu_i^{\mathrm{II}} = \mu_i^{\mathrm{III}} = \cdots, \tag{2.4.1}$$

where f is the fugacity, μ is the chemical potential and superscripts I, II and III denote the phases I, II and III. Therefore, phase equilibria can be calculated if the fugacity or the chemical potential of component i in the mixture is obtained with an appropriate equation.

In order to calculate the fugacity, the following thermodynamic relationship combined with the equation of state and appropriate mixing rules, is generally used:

$$\ln \phi_i = \ln\left(\frac{f_i}{x_i P}\right) = \frac{1}{RT}\int_V^{\infty}\left[\left(\frac{\partial P}{\partial n_i}\right)_{T,V,n_{j\neq i}} - \frac{RT}{V}\right]\mathrm{d}V - \ln z\,, \tag{2.4.2}$$

where the subscripts i and j denote the components i and j, respectively, ϕ is the fugacity coefficient, x is the mole fraction, P is the pressure, R is the gas constant, T is the absolute temperature, n is the amount of substance, V is the volume and z $(=PV/nRT)$ is the compressibility factor.

Table 2.4.1 Constants for van der Waals, Redlich–Kwong, Soave–Redlich–Kwong and Peng–Robinson equations of state [1]

equation	u	w	b	a
van der Waals	0	0	$\frac{RT_c}{8P_c}$	$\frac{27}{64}\frac{R^2T_c^2}{P_c}$
Redlich–Kwong [2]	1	0	$\frac{0.08664RT_c}{P_c}$	$\frac{0.42748R^2T_c^{2.5}}{P_cT^{1/2}}$
Soave–Redlich–Kwong [3]	1	0	$\frac{0.08664RT_c}{P_c}$	$\frac{0.42748R^2T_c^2}{P_c}\left[1+f_\omega\left(1-T_r^{1/2}\right)\right]^2$ $f_\omega = 0.48+1.574\omega-0.176\omega^2$
Peng–Robinson [4]	2	−1	$\frac{0.07780RT_c}{P_c}$	$\frac{0.45724R^2T_c^2}{P_c}\left[1+f_\omega\left(1-T_r^{1/2}\right)\right]^2$ $f_\omega = 0.37464+1.54226\omega-0.26992\omega^2$

2.4.2 Empirical Equation of State

Equations of state are classified as theoretical and empirical equations of state. Empirical equations of state are commonly used and they are classified as van der Waals (vdW) type and virial expansion type equations of state. It is easy to use the vdW type equation of state because only a few parameters are necessary. However, it is not sufficient to represent the *PVT* relationship in the liquid phase by using the vdW type equation of state. On the other hand, the virial expansion type equation of state can be applied in the high-density region containing the liquid phase region, where many parameters are necessary. Many vdW type equations of state have been proposed as shown in Table 2.4.1 [1–4], and Soave–Redlich–Kwong (SRK) [3] and Peng–Robinson (PR) equations of state [4] are widely used in industry. The SRK equation can predict the vapor–liquid equilibria of several hydrocarbons and the *PVT* relationship in the gas phase, while the equation cannot well predict the *PVT* relationship in the liquid phase. The PR equation is improved to give good representation of liquid densities.

The vdW type equation of state can generally be expressed by the following equation;

$$P = \frac{RT}{v-b} - \frac{a}{v^2+ubv+wb^2}, \tag{2.4.3}$$

where v is the molar volume, u and w are integer values whose values are shown in Table 2.4.1, and a and b are the parameters shown in Table 2.4.1.

The Benedict–Webb–Rubin (BWR) equation of state [5] is usually used as the virial type equation of state. The BWR equation of state is more complicated than the vdW type equations of state, and it is applicable over broader temperature and pressure ranges.

The generalized BWR equation of state is given by the following equation:

$$\begin{aligned}P &= RT\rho + \left(B_0RT - A_0 - C_0/T^2 + D_0/T^3 - E_0/T^4\right)\rho^2 \\ &+ \left(bRT - a - d/T - e/T^4 - f/T^{23}\right)\rho^3 \\ &+ \alpha\left(a + d/T + e/T^4 + f/T^{23}\right)\rho^6 \\ &+ \left(c/T^2 + g/T^8 + h/T^{17}\right)\rho^3\left(1 + \gamma\rho^2\right)\exp\left(-\gamma\rho^2\right),\end{aligned} \tag{2.4.4}$$

$$\begin{aligned}B_0 &= (0.443690 + 0.115449\omega)/\rho_C, \\ A_0 &= (1.28438 - 0.920731\omega)RT_C/\rho_C, \\ C_0 &= (0.356306 + 1.70871\omega)RT_C^3/\rho_C, \\ D_0 &= (0.0307452 + 0.179433\omega)RT_C^4/\rho_C, \\ E_0 &= [0.006450 - 0.022143\omega\exp(-3.8\omega)]RT_C^5/\rho_C, \\ b &= (0.528629 + 0.349261\omega)/\rho_C^2, \\ a &= (0.484011 + 0.754130\omega)RT_C/\rho_C^2, \\ d &= (0.0732828 + 0.463492\omega)RT_C^2/\rho_C^2, \\ \alpha &= (0.0705233 - 0.044448\omega)/\rho_C^3, \\ c &= (0.504087 + 1.32245\omega)RT_C^3/\rho_C^2, \\ \gamma &= (0.544979 - 0.270896\omega)/\rho_C^2, \\ e &= \left[4.65593\times10^{-3} - 3.07393\times10^{-2}\omega + 5.58125\times10^{-2}\omega^2\right. \\ &\left.- 3.40721\times10^{-3}\exp\left(-7.72753\omega - 45.3152\omega^2\right)\right]RT_C^5/\rho_C^2, \\ f &= \left[0.697\times10^{-13} + 8.08\times10^{-13}\omega - 16.0\times10^{-13}\omega^2\right. \\ &\left.- 0.363078\times10^{-13}\exp\left(30.9009\omega - 283.680\omega^2\right)\right]RT_C^{24}/\rho_C^2, \\ g &= \left[2.20\times10^{-5} - 1.065\times10^{-4}\omega + 1.09\times10^{-5}\exp(-26.024\omega)\right]RT_C^9/\rho_C^2, \\ h &= \left[-2.40\times10^{-11} + 11.8\times10^{-11}\omega - 2.05\times10^{-11}\exp(-21.52\omega)\right]RT_C^{18}/\rho_C^2,\end{aligned} \tag{2.4.5}$$

where ρ is the density of the fluid, ω is Pitzer's acentric factor listed in Appendix A and subscript C means the critical property. The constants D_0, E_0 and d were introduced by Starling and Han [6], and the constants e, f, g and h were added by Nishiumi and Saito [7]. Equation (2.4.4) becomes the original BWR equation of state when the constants D_0, E_0, d, e, f, g and h are zero, and (2.4.4) is the modification of the BWR equation of state proposed by Starling and Han [6] (BWRS equation of state) when the constants e, f, g and h are zero.

2.4.3 Mixing Rules

In order to apply the equation of state to mixtures, the mixing rules for the constants a and b of (2.4.3) are required. Several mixing rules have been proposed and they are classified into two categories: (1) coventional mixing rules, and (2) local mole fraction mixing rules. The local mole fraction mixing rules, such as Huron–Vidal mixing rule [8], are useful for correlating phase equilibria of complex mixtures (asymmetric mixtures) containing polar components or small and large molecules. However, the local mole fraction mixing rules are not necessarily useful for calculation of phase equilibria containing supercritical fluids because the solubility of the solute in a supercritical fluid are very low [9]. The conventional mixing rules are rather useful to correlate the phase equilibria containing a supercritical fluid by using the vdW type equation of state as follow:

$$a = \sum_i \sum_j y_i y_j a_{ij} \, , \; a_{ij} = \left(1 - k_{ij}\right)\sqrt{a_i a_j} \, , \qquad (2.4.6)$$

$$b = \sum_i \sum_j y_i y_j b_{ij} \, , \; b_{ij} = \left(1 - l_{ij}\right)\frac{b_i + b_j}{2} \, , \qquad (2.4.7)$$

where k_{ij} and l_{ij} are the interaction parameters between unlike molecules i and j, and the parameters are empirically determined to give good representation of the experimental data. Recently, attempts are also being made to generalize the interaction parameters so that they can be calculated solely from the physical properties of pure substances [10, 11].

2.4.4 Correlation of Vapor–Liquid Equilibria

The vapor–liquid equilibria can be calculated by the following equation because the fugacity of component i in the vapor phase is equal to that in the liquid phase:

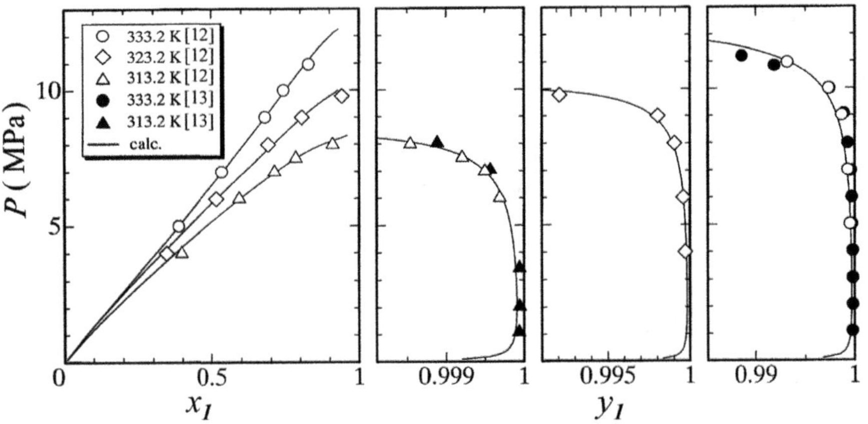

Fig. 2.4.1 Correlation results of vapor-liquid equilibria for CO_2 (1) + linalool (2) [12]. Reprinted from: High-pressure vapor–liquid equilibria for carbon dioxide + linalool: Y. Iwai, N. Hosotani, T. Morotomi, Y. Koga, Y. Arai, *J. Chem. Eng. Data* (American Chemical Society) **39**, 900 (1994).

$$\phi_i^{\mathrm{V}} P y_i = \phi_i^{\mathrm{L}} P x_i , \tag{2.4.8}$$

where x_i and y_i are the mole fractions of component i in the liquid phase and vapor phase, respectively, and superscripts V and L denote the vapor phase and liquid phase, respectively. The vapor-liquid equilibria can be calculated if ϕ_i can be evaluated by an appropriate equation of state.

Figure 2.4.1 shows the correlation results of vapor-liquid equilibria for CO_2 (1) + linalool (2) by the PR equation of state using the conventional mixing rules with two interaction parameters, k_{12} and l_{12} [12, 13]. The properties of linalool are estimated by the method of Lydersen and Edmister [14], a kind of group contribution method. The interaction parameters, k_{12} and l_{12}, are treated as the adjustable parameters and they are determined to give good representation of experimental data. The experimental data are correlated with good agreement by using the interaction parameters ($k_{12} = 0.053$, $l_{12} = -0.035$; 313.2 K), ($k_{12} = 0.046$, $l_{12} = -0.038$; 323.2 K), and ($k_{12} = 0.051$, $l_{12} = -0.027$; 333.2 K).

2.4.5 Correlation of Solubility of the Solid Component in a Supercritical Fluid

Several empirical methods using the fluid density or solubility parameter have been proposed as correlation or prediction methods of solubilities of the solid component in the supercritical fluid. The relationship between solubility, y_2

(mole fraction), and fluid density, ρ_1, is often represented by the following empirical equations [15, 16].

$$\log y_2 = A + B\rho_1 , \tag{2.4.9}$$

$$\ln y_2 = C \ln \rho_1 + D , \tag{2.4.10}$$

where A, B, C and D are constants. The relationship between solubility, y_2 (mole fraction), and the solubility parameter of the fluid, δ_1, is also represented by the following empirical equation [17].

$$\log y_2 = E\delta_1^2 + F\delta_1 + G , \tag{2.4.11}$$

where E, F and G are constants. These empirical equations can easily be applied to the prediction of solubility if the fluid density or solubility parameter is given. However, the parameters in these empirical equations have no clear physical meaning, and it would be difficult to apply the empirical equations to unknown systems.

The equations of state are often used to correlate the solubility of the solid component in the supercritical fluid. The fugacity of solid component 2 in the supercritical phase, f_2^{G}, is given by the following equation,

$$f_2^{\mathrm{G}} = P y_2 \phi_2^{\mathrm{G}} , \tag{2.4.12}$$

where ϕ_2^{G} is the fugacity coefficient of solid component 2 in the supercritical fluid and can be calculated by using the appropriate equation of state and (2.4.2). On the other hand, the fugacity, f_2^{S}, of solid component 2 in the solid phase can be approximated by the following equation,

$$f_2^{\mathrm{S}} = f_2^{\mathrm{oS}} \exp\left(\int_{P_2^{\mathrm{sat}}}^{P} \frac{v_2^{\mathrm{oS}}}{RT} \mathrm{d}P \right) = P_2^{\mathrm{sat}} \phi_2^{\mathrm{oS}} \exp\left(\int_{P_2^{\mathrm{sat}}}^{P} \frac{v_2^{\mathrm{oS}}}{RT} \mathrm{d}P \right), \tag{2.4.13}$$

where P_2^{sat} is the saturation vapor pressure of solid component 2, v_2^{oS} is the solid-state molar volume of solid component 2, superscript S means the solid state and ϕ_2^{oS} is the fugacity coefficient of pure solid component 2 under the saturation vapor pressure at equilibrium temperature and can be approximated by unity because the saturation vapor pressure, P_2^{sat} is very low. From (2.4.12) and (2.4.13), the solubility, y_2, of solid component 2 in a supercriti-

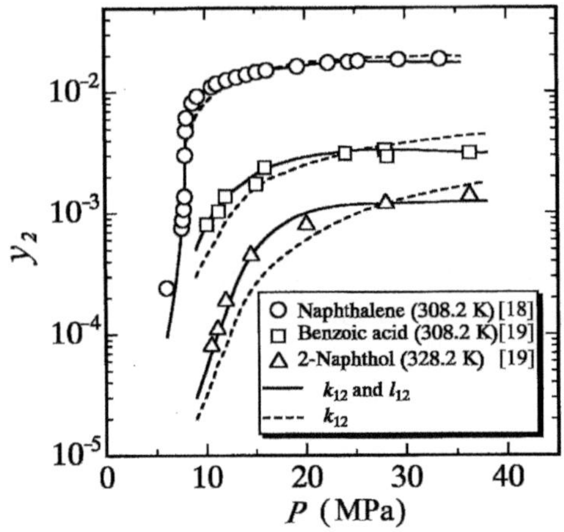

Fig. 2.4.2 Correlation results for solubilities of naphthalene (2), benzoic acid (2) and 2-naphthol (2) in supercritical CO_2 (1) [9]

cal fluid is given by the following equation when the pressure effect on the solid-state molar volume is negligible,

$$y_2 = \frac{P_2^{\text{sat}}}{P} \frac{1}{\phi_2^{\text{G}}} \exp\left[\frac{v_2^{\text{oS}}\left(P - P_2^{\text{sat}}\right)}{RT}\right]. \tag{2.4.14}$$

The values of P_2^{sat} and P_2^{oS} can be given from the physical properties of a pure component, and they can be obtained through measurement or by using an appropriate correlation equation. Therefore, the solubility, y_2, can be calculated if ϕ_2^{G} is obtained with an appropriate calculation method. Usually, ϕ_2^{G} can be calculated by using an appropriate equation of state. Figure 2.4.2 shows the correlation results for solubilities of naphthalene [18], benzoic acid and 2-naphthol [19] in supercritical CO_2 by the equation of state proposed by Yu et al. [20, 21] using the conventional mixing rules with two interaction parameters, k_{12} and l_{12} [9]. The equation of state proposed by Yu et al. is developed to represent a good correlation with the *PVT* relationships in a wide range from low-boiling components such as CO_2 to high-boiling components. The equation of state proposed by Yu et al. is shown by the following equation [20, 21],

$$P = \frac{RT}{v-b} - \frac{a}{v(v+c)+b(3v+c)}, \tag{2.4.15}$$

where a, b and c are pure component parameters which can be calculated with the critical properties and Pitzer's acentric factor. The conventional mixing rules are used for the parameters, a, b and c as follows:

$$a=\sum_i\sum_j y_i y_j a_{ij}\ ,\ a_{ij}=\left(1-k_{ij}\right)\sqrt{a_i a_j}\ , \tag{2.4.16}$$

$$b=\sum_i\sum_j y_i y_j b_{ij}\ ,\ b_{ij}=\left(1-l_{ij}\right)\frac{b_i+b_j}{2}\ , \tag{2.4.17}$$

$$c=\sum_i\sum_j y_i y_j c_{ij}\ ,\ c_{ij}=\left(1-l_{ij}\right)\frac{c_i+c_j}{2}\ . \tag{2.4.18}$$

Figure 2.4.2 shows that the calculated results give reasonable agreement with the experimental data by using the interaction parameters as adjustable parameters (naphthalene: — k_{12}= 0.059 and l_{12}= −0.090; --- k_{12}= 0.105), (benzoic acid: — k_{12}= −0.039 and l_{12}= −0.232, --- k_{12}= 0.055) and (2-naphthol: — k_{12}= −0.051 and l_{12}= −0.400, --- k_{12}= 0.112). In particular, the experimental data can be correlated with good agreement by applying the interaction parameter l_{12} to size parameters b and c.

2.4.6 Correlation of Entrainer Effect

The solubility, y_2, of the solid component in the supercritical fluid with an entrainer can be calculated based on the calculation of phase equilibria of a three-component system when the system is thought to be a supercritical fluid (1) + solid component (2) + entrainer (3) system. The equations of state or the solution models are usually used as the correlation method for the entrainer effect. Figure 2.4.3 shows the correlation results for solubilities of acridine (2) in supercritical CO_2 (1) with methanol (3) [22] by using the equation of state proposed by Yu et al. as shown in (2.4.15) and the conventional mixing rule as shown in (2.4.16)–(2.4.18) [9]. In the calculation, the interaction parameters, k_{12} and l_{12}, are determined to give good representation of solubilities of acridine in supercritical CO_2, and the interaction parameters, k_{13} and l_{13}, are determined to give good representation of vapor−liquid equilibria of CO_2 + methanol at 313.2 K [23]. The interaction parameters, k_{23} and l_{23}, are determined to give good representation of solubilities of acridine in supercritical CO_2 with methanol. The entrainer effects can be correlated with good agreement by using the equation of state proposed by Yu et al. with the conventional mixing rules (k_{12}= 0.089, l_{12}= −0.091; k_{13}= 0.088, l_{12}= 0.101; k_{23}= −0.333, l_{12}= −0.241) as shown in Fig. 2.4.3.

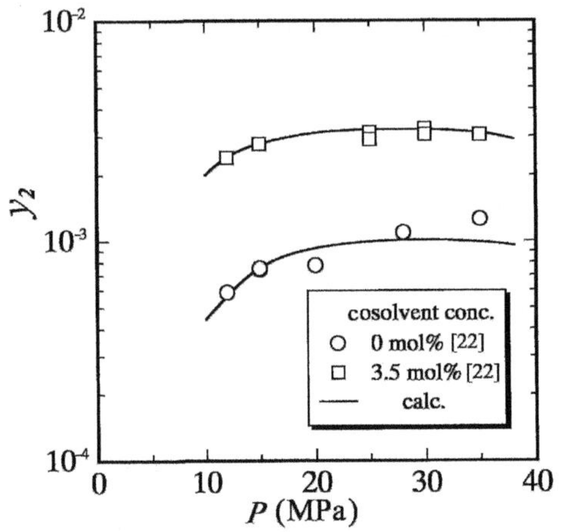

Fig. 2.4.3 Correlation results for solubilities of acridine (2) in supercritical CO_2 (1) with methanol (3) at 308.2 K [9]

On the other hand, the solution model is also useful to correlate the solubilities of solid components in the supercritical fluid mixed with the entrainer. Based on the solution model, the solubility of the solid component in a supercritical fluid mixture at infinite dilution is given by the following equation [24],

$$\ln y_2 = \frac{\Delta h_2^{\mathrm{m}}}{RT}\left(\frac{T}{T_2^{\mathrm{m}}} - 1\right) - \frac{v_2}{RT}[(\delta_1 - \delta_2)^2\phi_1 + (\delta_2 - \delta_3)^2\phi_3 - (\delta_1 - \delta_3)^2\phi_1\phi_3] - \ln\frac{\phi_2}{y_2} - 1 + \frac{\phi_2}{y_2}, \quad (2.4.19)$$

where T is the absolute temperature, R is the gas constant, Δh_2^{m} and T_2^{m} are the molar heat of fusion and the melting point for pure solid component 2, respectively, and v_i and δ_i are the molar volume and the solubility parameter of component i in the supercritical fluid phase, respectively. While the solubility parameter for supercritical CO_2, δ_1, is calculated by the method proposed by Giddings et al. [25], those for components 2 and 3 are estimated by the following equation [24],

$$\delta_i = \sqrt{\frac{\Delta U_i}{v_i}} \qquad (i = 2 \text{ and } 3), \tag{2.4.20}$$

where ΔU_i is the molar heat of vaporization, and the value can be calculated by the group contribution method proposed by Fedors [26]. The relationship between the molar volume, v_i, and the density of supercritical CO_2, ρ_1, can be approximated by the following equation [27],

$$\ln v_i = \alpha_i \ln \rho_1 + \beta_i . \tag{2.4.21}$$

When the values of α_i are fixed to -1.34, the values of β_i for components 2 and 3 are calculated by the following linear equation [24],

$$\beta_i = 0.001129 v_{\mathrm{C},i} - 0.7633 \qquad (i = 2 \text{ and } 3), \tag{2.4.22}$$

where $v_{\mathrm{C},i}$ is the critical molar volume of component i calculated by the group contribution method proposed by Lydersen [1]. A solute molecule in the supercritical fluid phase may be apt to cluster with entrainer molecules. Therefore, the volume of solute is approximated by the following linear equation [24],

$$v_2 = v_2^{(1)} + \theta \Delta v_2 , \tag{2.4.23}$$

where θ is the local composition of entrainer around the solute, $v_2^{(1)}$ and $v_2^{(3)}$ are the volumes in the pure supercritical fluid phase (component 1) and pure solvent (entrainer, component 3), respectively, and Δv_2 is written by [24].

$$\Delta v_2 = v_2^{(3)} - v_2^{(1)} . \tag{2.4.24}$$

Yonker and Smith [28] studied the solvent structure by using UV spectroscopy, and consequently found that the local composition of the entrainer approaches asymptotically to the bulk composition with increasing pressure under supercritical conditions. Therefore, θ is empirically assumed to be given by the following equation [24],

$$\theta = \frac{y_3}{(P/P^{\mathrm{bulk}}) + [1 - (P/P^{\mathrm{bulk}})] y_3} , \tag{2.4.25}$$

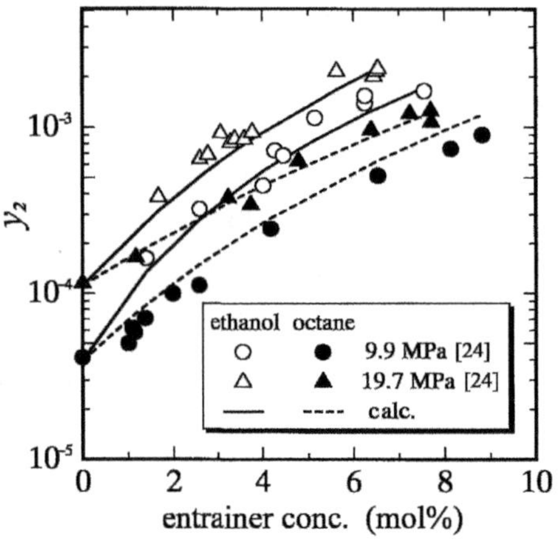

Fig. 2.4.4 Correlation results for solubilities of stearic acid (2) in supercritical CO_2 (1) with entrainers (3) at 308.2 K [24]

where P is the pressure, and P^{bulk} means the pressure at the point where the local composition of the entrainer is equal to the bulk composition, y_3. Figures 2.4.4 and 2.4.5 show the correlation results for solubilities of stearic acid and stearyl alcohol in supercritical CO_2 with entrainers, respectively. In the calculation, Δv_2 and P^{bulk} are used as adjustable parameters (stearic acid: $\Delta v_2 \times 10^6 = 84.44$ and $P^{bulk} \times 10^{-6} = 178.5$ (ethanol), 103.2 and 70.73 (octane); stearyl alcohol: 80.12 and 52.38 (ethanol), 68.80 and 80.38 (octane)) to give good agreement between the experimental solubilities and the calculated results in supercritical CO_2 mixed with the entrainer. The calculated results give reasonable agreement with the experimental data, as shown in Figs. 2.4.4 and 2.4.5.

2.5 Group Contribution Method

The phase equilibria of supercritical fluids have usually been correlated with good accuracy using the two binary interaction parameters in the conventional mixing rule. In many cases, however, we cannot find the experimental data in order to determine the interaction parameters. In these cases, we will make use of a predictive method in order to get the phase equilibria.

In the predictive method of phase equilibria, the group contribution (GC) methods are those for predicting the equilibria from the group pair parameters relating to the groups composed of the molecules in the mixture and have been

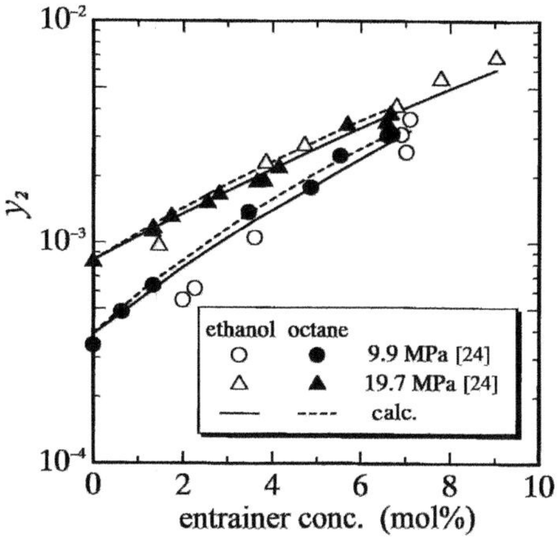

Fig. 2.4.5 Correlation results for solubilities of stearyl alcohol (2) in supercritical CO_2 (1) with entrainers (3) at 308.2 K [24]

widely used from the practical point of view. In general, the number of molecules treated in the chemical industry is marvelous, but the number of groups used as a constituent unit of a component is comparatively small and the number of fundamental important groups is about 50. If the group pair parameters are available, the phase equilibria for many mixtures can be predicted [1, 2].

One of the group contribution models used for predicting phase equilibria at high pressures are EOS-group contribution methods that combine the group contribution method of G^E with the cubic equations of state (EOS). The mixture parameters in the van der Waals (vdW) equation of state, like the Soave–Redlich–Kwong (SRK) and Peng–Robinson (PR) equations of state, have been expressed using the mixing rules named by G^E mixing rule and have been predicted using ASOG and UNIFAC group contribution methods. The other is the GC-EOS which obtains the pure and mixture energy parameters in the perturbed equation of state using group-based group interaction parameters.

Firstly ASOG and UNIFAC are introduced simply. Second, the vapor–liquid equilibria and solubilities for systems containing supercritical fluids are predicted using EOS-group contribution method and GC-EOS.

2.5.1 Prediction of Excess Gibbs Free Energy Using the Group Contribution Method (ASOG and UNIFAC)

The molar excess Gibbs free energy G^{E} is related to the activity coefficient γ_i of each component obtained from vapor−liquid equilibria data:

$$\frac{G^{\mathrm{E}}}{RT} = \sum_i x_i \ln \gamma_i \,, \tag{2.5.1}$$

where R is the gas constant, T is the absolute temperature and x_i is the mole fraction of component i. Figure 2.5.1 shows the vapor−liquid equilibrium, activity coefficients and G^{E}/RT for acetone + water and ethanol + water systems. ASOG and UNIFAC are group contribution methods for predicting activity coefficients or G^{E} using the following rules [1,2]:

(i) The logarithmic activity coefficient $\ln \gamma_i$ is expressed by the summation of the combinatorial part $\ln \gamma_i^{\mathrm{C}}$ and the residual part $\ln \gamma_i^{\mathrm{R}}$.

$$\ln \gamma_i = \ln \gamma_i^{\mathrm{C}} + \ln \gamma_i^{\mathrm{R}} \,. \tag{2.5.2}$$

(ii) $\ln \gamma_i^{\mathrm{R}}$ is given by the group activity coefficient Γ_k as

$$\ln \gamma_i^{R} = \sum \nu_k^i (\ln \Gamma_k - \ln \Gamma_k^{(i)}) \,, \tag{2.5.3}$$

where ν_k^i denotes the number of group k in molecule i. ASOG and UNIFAC are classified by the difference of the representative equations of $\ln \gamma_i^{\mathrm{C}}$ and $\ln \Gamma_k$.

The group pair parameters of ASOG and UNIFAC have been determined using vapor−liquid equilibrium data at low pressures. The presence of the parameters is shown in Figs. 2.5.2 and 2.5.3. It is worth mentioning that the predictive accuracy of both methods is similar for low-pressure vapor−liquid equilibria. ASOG can predict the phase equilibria for systems containing the following substances:

Hydrocarbons, water, alcohols, phenols, poly-alcohols, ketones, ethers, aldehydes, esters, fatty acids, formic acids, amides, nitriles, amines, aromatic amines, nitro compounds, aromatic nitro compounds, chlorides, fluorides, iodides, carbon disulfide, pyridines, furfural, acryronitrile, dimethylsulfoxide, *N*-methylpyrolidone, thiols, dimethylformamide, ethanediol, diethylenegrycol, sulforane.

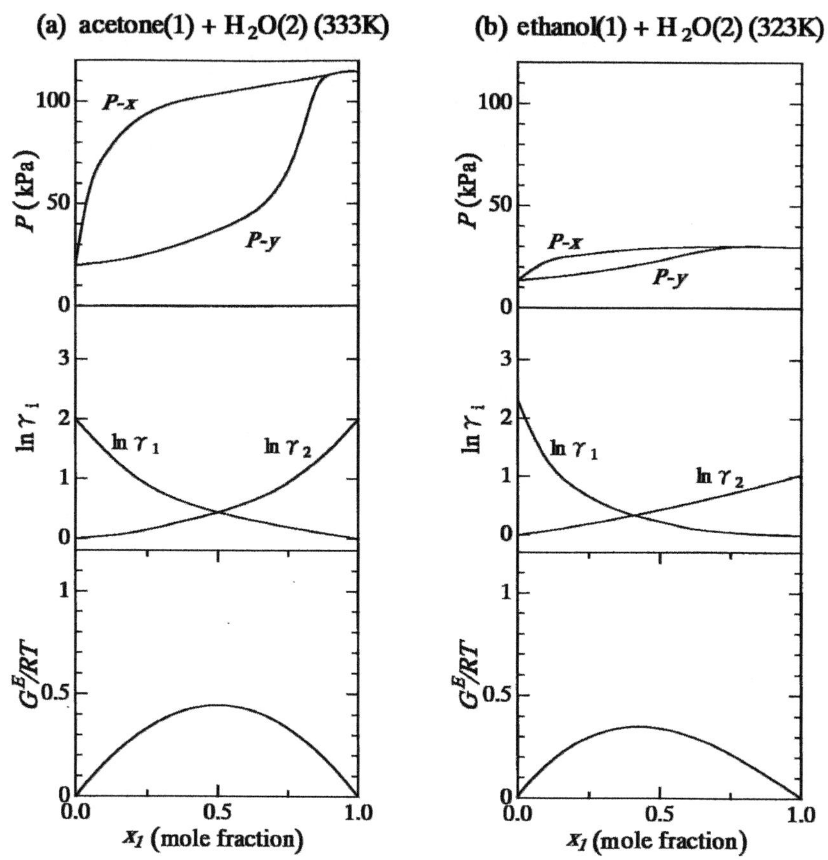

Fig. 2.5.1 Vapor–liquid equilibria, activity coefficients and excess free energies for binary systems

2.5.2 EOS-Group Contribution Method

(a) Huron–Vidal Mixing Rule

The excess Gibbs free energy is given by the fugacity coefficients ϕ.

$$\frac{G^{\mathrm{E}}}{RT} = \ln \phi^{\mathrm{m}} - \sum_i x_i \ln \phi_i^0 . \tag{2.5.4}$$

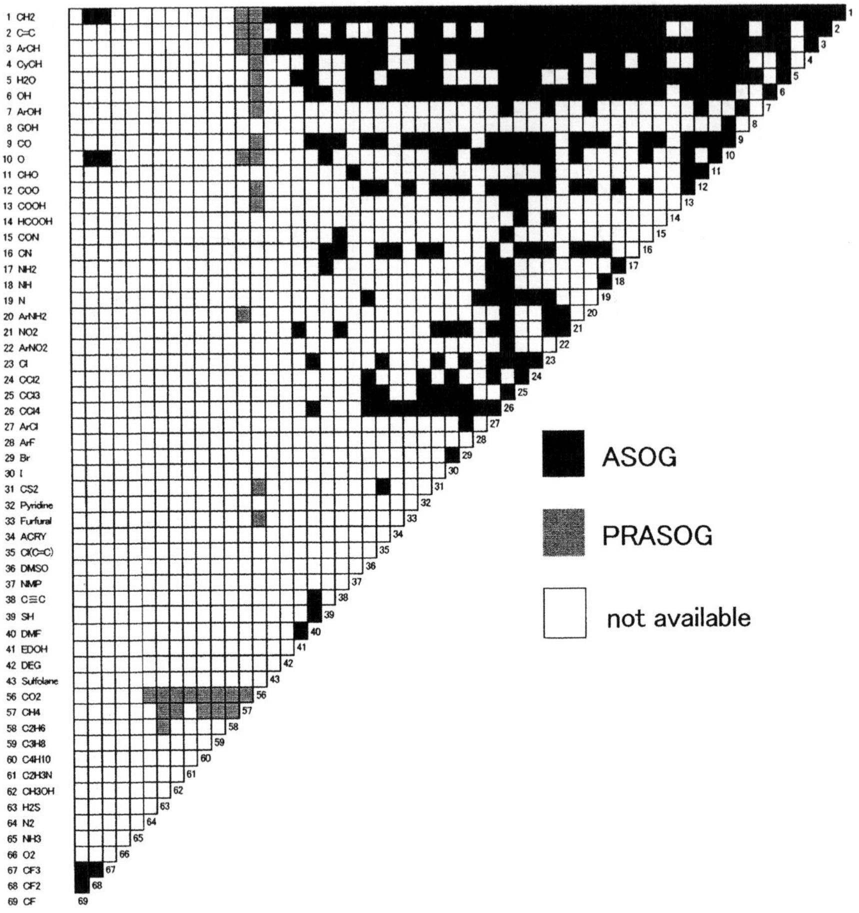

Fig. 2.5.2 ASOG and PRASOG group pair parameters

Here superscripts 0 and m show the properties in the pure component and mixture, respectively. When the fugacity coefficient is given by the SRK equation, G^{E} is given by the following equation:

$$\frac{G^{\mathrm{E}}}{RT}=\frac{a}{RTb}\ln\left(\frac{v}{v+b}\right)-\sum_i x_i\frac{a_{ii}}{RTb_i}\ln\left(\frac{v_i}{v_i+b_i}\right)$$
$$+\sum_i x_i\ln\left(\frac{v_i-b_i}{v-b}\right)+\frac{P}{RT}\left(v-\sum_i x_i v_i\right), \tag{2.5.5}$$

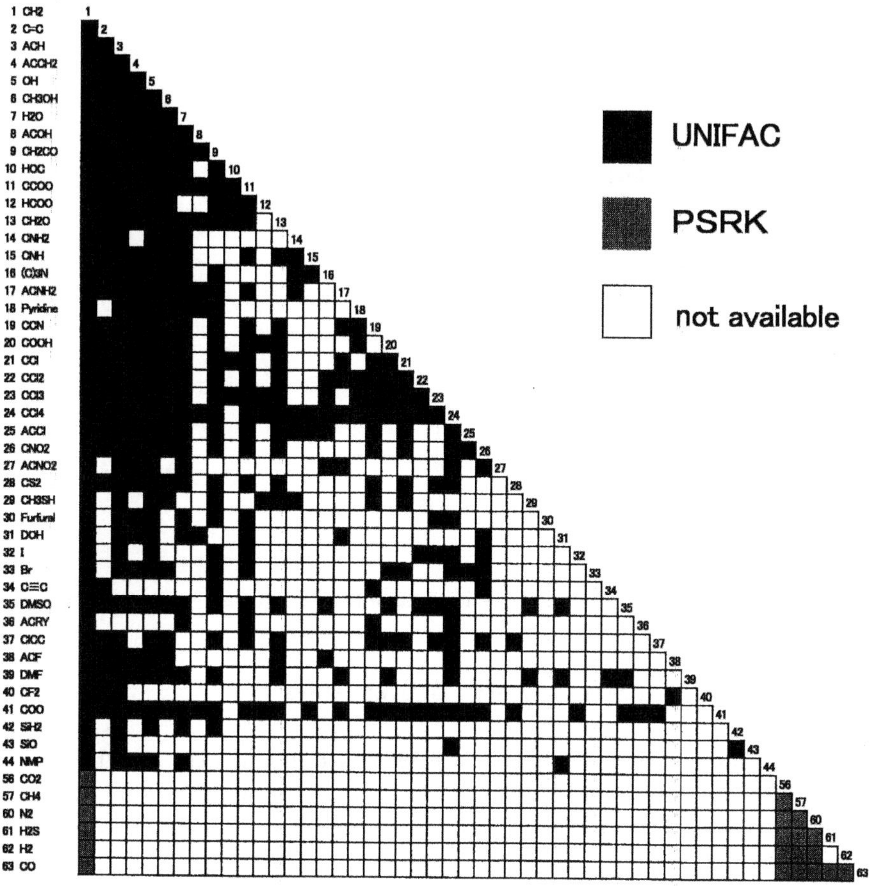

Fig. 2.5.3 UNIFAC and PSRK group pair parameters

where v is the molar volume. Huron and Vidal applied the infinite pressure condition (∞) and proposed the first G^{E} mixing rule, called the Huron-Vidal mixing rule [3] for parameters a and b in the equation of state.

$$\frac{a}{b} = \sum_i x_i \frac{a_{ii}}{b_i} - \frac{1}{\ln 2} G_{\infty}^{\mathrm{E}}, \tag{2.5.6}$$

$$b = \sum_i x_i b_i . \tag{2.5.7}$$

This mixing rule is useful for correlating phase equilibria, although it needs the re-determined ASOG or UNIFAC group pair parameters [4] in order to predict phase equilibria at high pressure.

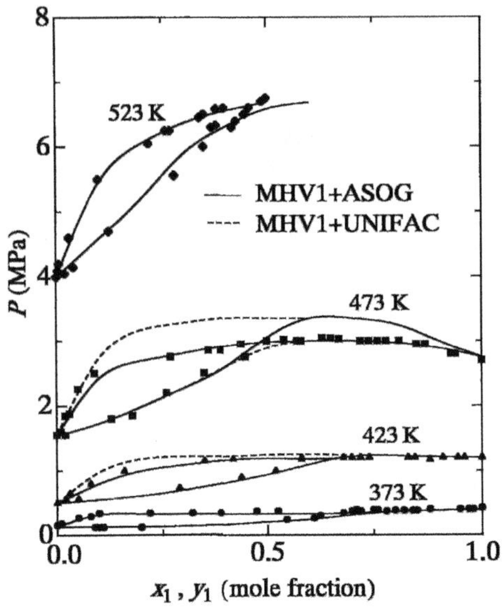

Fig. 2.5.4 Predicted high-pressure vapor–liquid equilibria for acetone + water system

(b) MHV1, MHV2, PSRK and MR1 Mixing Rules

Since the Huron–Vidal mixing rule was proposed, it was expected that a predictive method of high-pressure vapor–liquid equilibria with good accuracy using low-pressure ASOG and UNIFAC group pair parameters would be developed. Firstly Mollerup proposed a mixing rule of zero pressure condition. The useful ones are MHV1 and MHV2 [5, 6]:

$$\frac{a}{b} = \sum_i x_i \frac{a_{ii}}{b_i} + \frac{1}{(-0.593)} \left\{ \frac{G_0^{\mathrm{E}}}{RT} + \sum_i x_i \ln \frac{b}{b_i} \right\}, \quad \text{(MHV1)} \qquad (2.5.8)$$

$$-0.4783 \cdot \left(\frac{a}{b} - \sum_i x_i \frac{a_{ii}}{b_i} \right) - 0.0047 \left\{ \left(\frac{a}{b} \right)^2 + \sum_i x_i \left(\frac{a_{ii}}{b_i} \right)^2 \right\}$$
$$= \frac{G_0^{\mathrm{E}}}{RT} + \sum_i x_i \ln \frac{b}{b_i}. \qquad \text{(MHV2)} \qquad (2.5.9)$$

Figure 2.5.4 shows the predicted results [7] of high-pressure vapor–liquid equilibrium for the acetone + water system that obtained G_0^{E} in MHV1 with low-pressure ASOG and UNIFAC parameters. For MHV2, the modified UNIFAC has been combined and the group pair parameters have been deter-

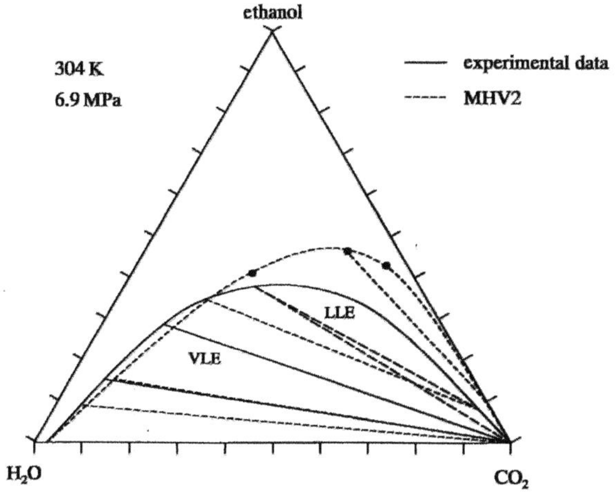

Fig. 2.5.5 Predicted high-pressure phase equilibria for CO_2 + ethanol + water system

mined for nine solvents and 13 gases (H_2, O_2, H_2. CO, CO_2, H_2S, CH_4, C_2H_2, C_2H_4, C_2H_6, C_3H_6, iC_4H_{10}). Figure 2.5.5 shows the predicted results for the CO_2 + ethanol + water system. However, it should be mentioned that group pair parameters between gas and gas are zero.

Gmehling et al. has proposed the following PSRK model [8] similar to MHV1.

$$\frac{a}{b} = \sum_i x_i \frac{a_{ii}}{b_i} + \frac{1}{(-0.64663)} \left\{ \frac{G_0^E}{RT} + \sum_i x_i \ln \frac{b}{b_i} \right\}. \tag{2.5.10}$$

The merit of PSRK is the keeping of the wide applicability of UNIFAC, because the UNIFAC parameters are used directly and the group pair parameters are determined between eight gases (CH_4, C_2H_6, C_3H_6, C_4H_{10}, CO_2, N_2, H_2, CO) and solvent groups. The presence of PSRK parameters is given in Fig. 2.5.3. Figures 2.5.6 and 2.5.7 show the predicted results of high-pressure vapor–liquid equilibria for the ethane + toluene system and of the critical locus for CO_2 + paraffins [9].

Zhong et al. have proposed the following mixing rule, named MR1 [10], in which a correlation factor, f, to G^E from UNIFAC is introduced in MHV1 and correlated with the carbon number:

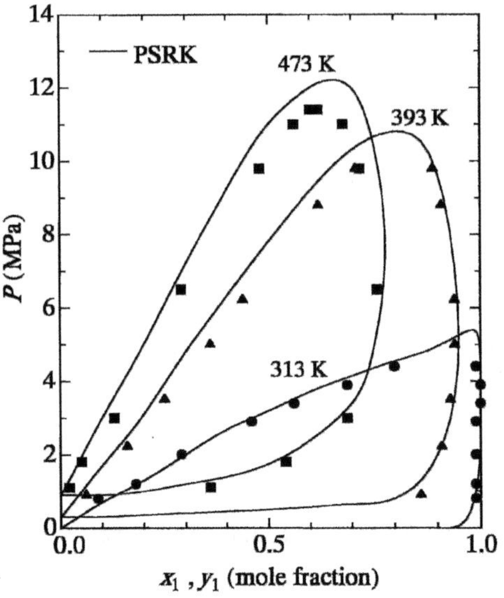

Fig. 2.5.6 Predicted vapor–liquid equilibria for ethane + toluene system

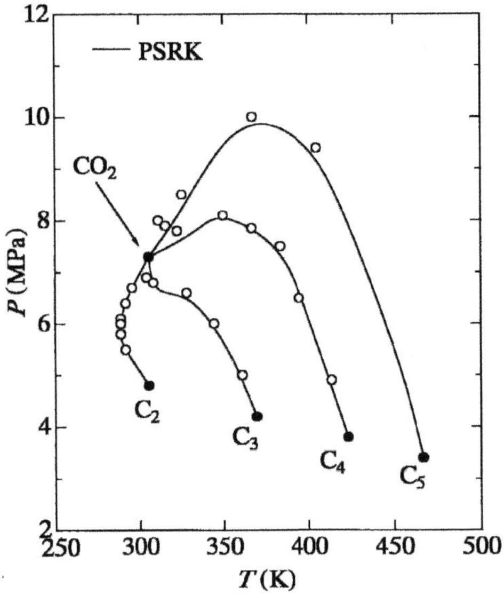

Fig. 2.5.7 Predicted critical locus for CO_2 + n-paraffin system

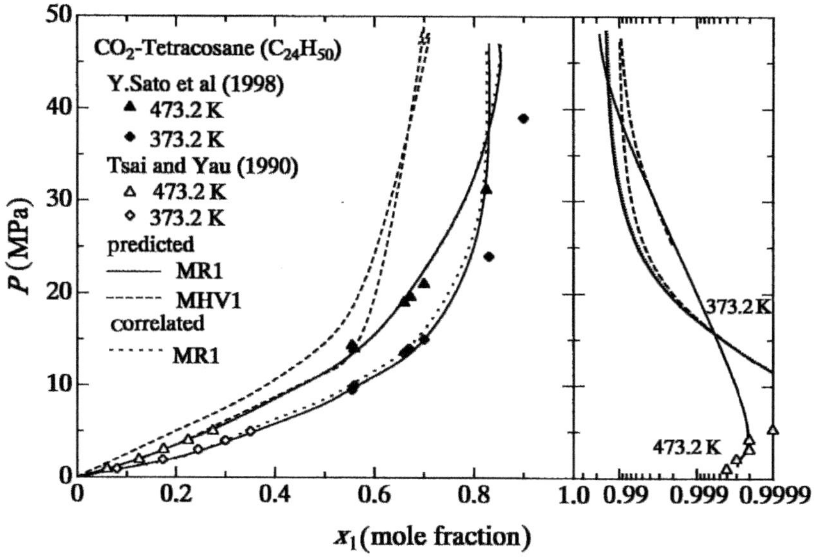

Fig. 2.5.8 Predicted solubilities of CO_2 in tetracosane system

$$\frac{a}{b} = \sum_i x_i \frac{a_{ii}}{b_i} + \frac{1}{(-0.64663)} \left\{ \frac{G_0^E}{RT}(1+f) + \sum_i x_i \ln \frac{b}{b_i} \right\}. \tag{2.5.11}$$

It is stated that MR1 gives very good predictive accuracy for gas/large alkane systems. Figure 2.5.8 shows the predicted phase equilibria for the CO_2 + tetracosane system. PSRK, MHV1, MHV2 and MR1 make use of (2.5.7) in order to calculate the size parameter b.

(c) Wong–Sandler Mixing Rule

One of the problems of G^E mixing rules from the proposal of Huron and Vidal is the nonsatisfaction of the condition for the second virial coefficients B,

$$B = \sum_i \sum_j x_i x_j B_{ij} . \tag{2.5.12}$$

Wong and Sandler proposed a new type of mixing rule.

$$B = \sum_i \sum_j \left(\frac{B_{ii} + B_{jj}}{2} \right) (1 - k_{ij}) . \tag{2.5.13}$$

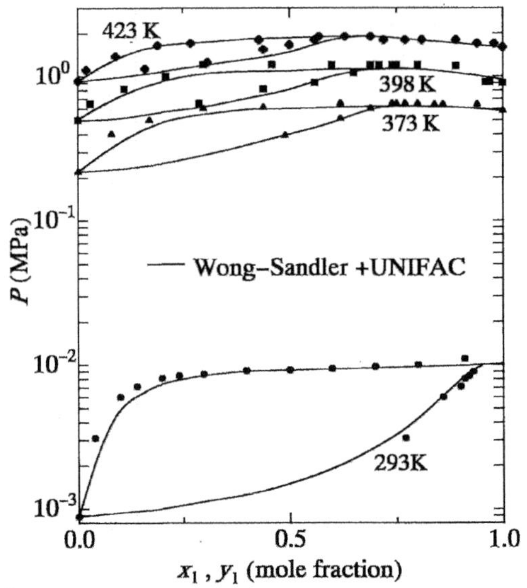

Fig. 2.5.9 Predicted vapor–liquid equilibria for *n*-pentane + ethanol system

As the contribution of PV^{E} $(= P(v - 2x_i v_i))$ is small at low pressure, the equality of the excess Gibbs free energy G^E at low pressure and the excess Helmholtz free energy $A^{\mathrm{E}} (= G^E - PV^{\mathrm{E}})$ at infinite pressure was assumed:

$$A^{\mathrm{E}}(P = \infty) \approx G^{\mathrm{E}}(P = \text{low pressure}). \tag{2.5.14}$$

The Wong–Sandler mixing rule [11, 12] was given by the following equation for the Peng–Robinson equation of state:

$$\frac{a}{b} = \sum_i x_i \frac{a_{ii}}{b_i} - \frac{1}{(-1.762)} A_\infty^{\mathrm{E}}. \tag{2.5.15}$$

Equations (2.5.13) and (2.5.15) are the Wong–Sandler mixing rule and the prediction needs the determination of binary interaction parameters k_{ij} from low-pressure vapor–liquid equilibria predicted using UNIFAC. Figure 2.5.9 [11] shows the high-pressure vapor–liquid equilibria for the pentane + ethanol system. The accuracy is good.

(d) PRASOG Mixing Rule

As we have mentioned in b. and c., it will be interesting theoretically and practically if high-pressure vapor−pressure equilibria can be predicted by the mixing rule that satisfies the condition of the second virial coefficient and from low-pressure group pair parameters. Tochigi et al. have proposed the PRASOG [13, 14] given by the following equations that satisfy both conditions:

$$\frac{a}{b}=\sum_i x_i \frac{a_{ii}}{b_i}\left[\frac{\ln\{(1+\sqrt{2}+v_{i0}/b_i)/(1-\sqrt{2}+v_{i0}/b_i)\}}{\ln\{(1+\sqrt{2}+v_0/b)/(1-\sqrt{2}+v_0/b)\}}\right] \tag{2.5.16}$$

$$-\left\{\frac{G_0^{\mathrm{E}}}{RT}+\sum_i x_i \ln\frac{b}{b_i}+\sum_i x_i \ln\left(\frac{v_0/b-1}{v_{0i}/b_i-1}\right)\right\}$$

$$\times\left[\frac{2\sqrt{2}RT}{\ln\{(1+\sqrt{2}+v_0/b)/(1-\sqrt{2}+v_0/b)\}}\right],$$

$$b-\frac{a}{RT}=\sum_i\sum_j x_i x_j\left\{\frac{b_i+b_j}{2}-\frac{\sqrt{a_{ii}a_{jj}}}{RT}\right\}. \tag{2.5.17}$$

The predicted results using low-pressure ASOG group pair parameters agree with experimental data for the acetone + water and ethanol + water systems [13] . The group pair parameters were then determined for eight gases (CO_2, CH_4, C_2H_6, C_3H_8, C_4H_{10}, H_2S, N_2, O_2) and solvent groups. The presence of group pair parameters is given in Fig. 2.5.2. Figures 2.5.10 and 2.5.11 show the predicted results for CO_2 + methyl palmitate and CO_2 + fatty acids systems [14] .

2.5.3 Group Contribution Equation of State (GC-EOS)

The GC-EOS [15] is given by the free volume term and group interaction term.

$$A^{\mathrm{R}}=A^{\mathrm{R}}(\text{repulsive})+A^{\mathrm{R}}(\text{interaction})\,. \tag{2.5.18}$$

Here the repulsive term is given by the Carnahan−Starling equation derived by Mansoori and Leland.

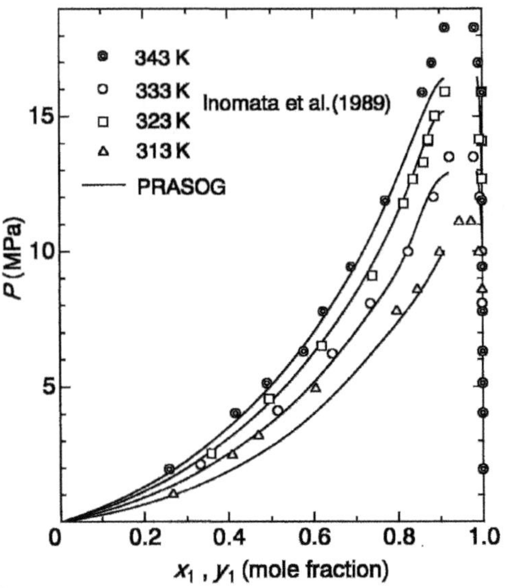

Fig. 2.5.10 Predicted vapor−liquid equilibria for the CO_2 + methyl palmitate system

$$\frac{A^{\mathrm{R}}(\text{repulsive})}{RT} = 3\frac{\lambda_1\lambda_2}{\lambda_3}(Y-1) + \frac{\lambda_2^3}{\lambda_3^2}(-Y+Y^2-\ln Y) + n\ln Y,$$

where

$$\lambda_k = \sum_j n_j d_j^k \quad \text{and} \qquad Y = (1-\pi\lambda_3/6V)^{-1}. \tag{2.5.19}$$

Here d_j is the diameter of the rigid sphere of component j.

From the representation of the attractive term by the van der Waals equation,

$$A^{\mathrm{R}}(\text{interaction}) = -a\rho\,, \tag{2.5.20}$$

the following equation is used:

$$A^{\mathrm{R}}(\text{interaction}) = -\frac{Z}{2}\sum_i N_i \sum_j v_j^i q_j \sum_k \theta_k (g_{kj}\tilde{q}\,\tau_{kj}\rho) / \sum_i \theta_i \tau_{ij}\,,$$

where

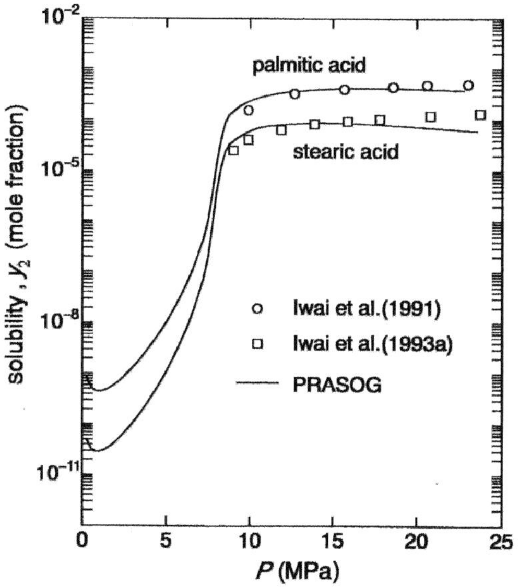

Fig. 2.5.11 Predicted solubilities of fatty acids in CO_2 at 308 K

$$\tau_{kj} = \exp[-\alpha_{kj}(g_{kj} - g_{jj})\tilde{q}\rho / RT], \tag{2.5.21}$$

$$\tilde{q} = \sum_i N_i \sum_j v_i^j q_j \qquad \text{and} \qquad \theta_j = (q_j / \tilde{q}) \sum_i N_i v_i^j .$$

Here Z is the number of partitions, v_j^i is the number of group j in molecule i, q_j is the surface parameter of group j, g_{jj} is the energy parameter between the same groups, g_{kj} is the energy parameter between different groups, and α_{kj} is a nonrandom parameter.

The parameters in GC-EOS have been determined for the groups relating to 13 solvent groups (CH_3, CH_2, CH, CyCH, ACH, AC, $ACCH_3$, $ACCH_2$, CH_3OH, CH_2OH, CH_3CO, CH_2CO, H_2O) and nine gases (H_2, N_2, CO, O_2, CH_4, C_2H_4 CO_2, C_2H_6, H_2S) [16]. Figure 2.5.12 shows the predicted results for the CO_2 + n-propanol + water system [17]. This may be a model which can solve isomer problems by consideration of the effect of neighboring groups.

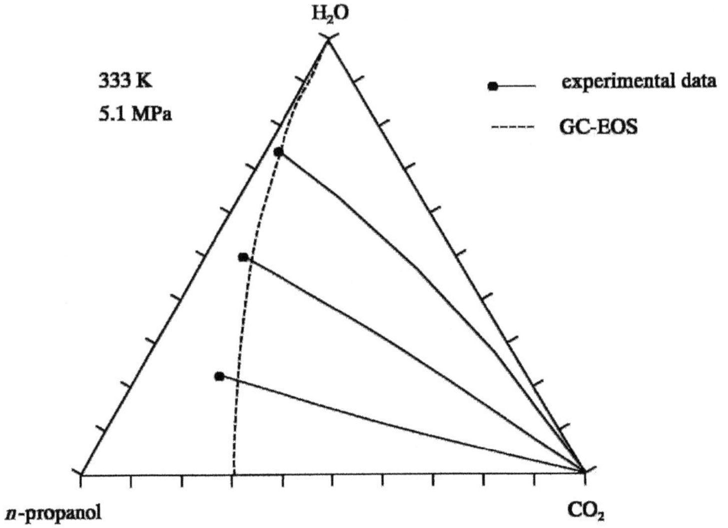

Fig. 2.5.12 Predicted phase equilibria for the CO_2 + *n*-propanol + water system

References

2.1 Thermophysical Properties of Supercritical Fluids

[1] M. Sato and M. Uematsu, "*Physico-Chemical Properties for Chemical Engineering*", Vol. 19, Kagaku Kogyosya, Tokyo, p. 3 (1999).

[2] M. L. McGlashan, "*Chemical Thermodynamics*", Academic Press, London, p. 131 (1979).

[3] Y. Suehiro, M. Nakajima, K. Yamada and M. Uematsu, *J. Chem. Thermodynamics*, **28**, 1153 (1996).

[4] J. S. Rowlinson and F. L. Swinton, "*Liquids and Liquid Mixtures*", 3rd ed., Butterworth Scientific, London, p. 70 (1982).

[5] D. Ambrose, in "*Experimental Thermodynamics, Volume II: Experimental Thermodynamics of Non-reacting Fluids*", ed. B. Le Neindre and B. Vodar, Butterworth, London, p. 621 (1975).

2.2 Phase Equilibria for a Mixture with a Supercritical Fluid

[1] M. A. McHugh and V. J. Krukonis, "*Supercritical Fluid Extraction (Principles and Practice)*", Butterworths, Stoneham, MA, Chap. 3, p. 23 (1986).

[2] S. Saito, et al., "*Science and Technology of Supercritical Fluid*", Sankyo Business, Sendai, Miyagi, Japan, Chap. 3.2 (1996).
[3] M. Kato, H. Tanaka and H. Yoshikawa, *J. Chem. Eng. Data*, **44**, 116 (1999).
[4] K. Suzuki, H. Sue, M. Itou, R. L. Smith, H. Inomata, K. Arai and S. Saito, *J. Chem. Eng. Data*, **35**, 63 (1990).
[5] M. A. McHugh, M. W. Mallett and J. P. Kohn, in "*Chemical Engineering at Supercritical Fluid Condition*", ed. M. E. Paulaitis, J. M. Penninger, R. D. Gray Jr., P. Davidson, Ann Arbor Science, Stoneham, MA, Chap. 5, pp. 113–138 (1983).

2.3 Measurement of Phase Equilibria

[1] T. Toriumi and R. Hara, "*Reports of the Asahi Glass Foundation for Industrial Technology*", Asahi Glass Co. Ltd., Tokyo, **8**, 66 (1941, 1942).
[2] T. Toriumi, T. Kawakami, J. Sakai, D. Ogawa and N. Azuma, *J. Chem. Soc. Jpn. Chemistry and Industrial Chemistry*, **49**, 1 (1946).
[3] J. D. Raal and A. L. Muhlbauer, "*Phase Equilibria Measurement and Computation*", Taylor & Francis, Washington D.C. (1997).
[4] K. Nagahama, *Fluid Phase Equilibria*, **116**, 361 (1996).
[5] H. Sagara, Y. Arai and S. Saito, *J. Chem. Eng. Jpn.*, **5**, 339 (1972).
[6] H. K. Bae, K. Nagahama and M. Hirata, *J. Chem. Eng. Jpn.*, **14**, 1 (1981).
[7] T. Nakayama, H. Sagara, K. Arai and S. Saito, *Fluid Phase Equilibria*, **38**, 109 (1987).
[8] A. Fredenslund, J. Mollerup and L. J. Christiansen, *Cryogenics*, *July*, 414 (1973).
[9] H. J. Ng and D. B. Robinson, *J. Chem. Eng. Data*, **23**, 325 (1978).
[10] H. J. Ng and D. B. Robinson, *Fluid Phase Equilibria*, **2**, 283 (1979).
[11] Y. Sato, S. Takishima and H. Masuoka, *Fluid Phase Equilibria*, **147**, 181 (1998).
[12] H. Kubota, H. Inatome, Y. Tanaka and T. Makita, *J. Chem. Eng. Jpn.*, **16**, 99 (1983).
[13] S. Takishima, K. Saiki, K. Arai and S. Saito, *J. Chem. Eng. Jpn.*, **19**, 48 (1986).
[14] K. Suzuki, H. Sue, M. Itou, R. L. Smith, H. Inomata, K. Arai and S. Saito, *J. Chem. Eng. Data,* **35**, 63 (1990).
[15] Y. Iwai, N. Hosotani, T. Morotomi, Y. Koga and Y. Arai, *J. Chem. Eng. Data*, **39**, 900 (1994).
[16] T. Hiaki, H. Miyagi, T. Tsuji and M. Hongo, *J. Supercritical Fluids*, **13**, 23 (1998).
[17] H. Inomata, K. Tuchiya, K. Arai and Saito, *J. Chem. Eng. Jpn.*, 19, 386 (1986).

2.4 Prediction of Phase Equilibria

[1] R. C. Reid, J. M. Prausnitz and B. E. Poling, "*The Properties of Gases & Liquids*", 4th ed., McGraw-Hill, New York (1988).
[2] O. Redlich and J. N. S. Kwong, *Chem. Rev.*, **44**, 233 (1949).
[3] G. Soave, *Chem. Eng. Sci.*, **27**, 1197 (1972).
[4] D.-Y. Peng and D. B. Robinson, *Ind. Eng. Chem. Fundam.*, **15**, 59 (1976).
[5] M. Benedict, G. B. Webb and L. C. Rubin, *J. Chem. Phys.*, **8**, 334 (1940).
[6] K. E. Starling and M. S. Han, *Hydrocarbon Process.*, **51**, 129 (1972).
[7] H. Nishiumi and S. Saito, *J. Chem. Eng. Jpn.*, **8**, 356 (1975).
[8] M.-J. Huron and J. Vidal, *Fluid Phase Equilibria*, **3**, 255 (1979).
[9] Y. Iwai, B. C.-Y. Lu, H. Yamamoto and Y. Arai, *Kagaku Kogaku Ronbunshu*, **15**, 676 (1989).
[10] Y. Iwai, T. Shimizu, H. Yamamoto and Y. Arai, *J. Chem. Eng. Jpn.*, **24**, 801 (1991).
[11] Y. Iwai, H. Yamamoto, T. Shimizu and Y. Arai, in "*Solvent Extraction 1990, Part B*", ed. T. Sekine, Elsevier, Amsterdam, p. 1695 (1992).
[12] Y. Iwai, N. Hosotani, T. Morotomi, Y. Koga and Y. Arai, *J. Chem. Eng. Data*, **39**, 900 (1994).
[13] J. Suzuki and K. Nagahama, *Kagaku Kogaku Ronbunshu*, **22**, 195 (1996).
[14] R. C. Reid, J. M. Prausnitz and T. K. Sherwood, "*The Properties of Gases & Liquids*", 3rd ed., McGraw-Hill, New York (1977).
[15] S. Robin and B. Vodar, *Discuss. Faraday Soc.*, **15**, 233 (1953).
[16] E. Stahl, W. Schilz, E. Schütz and E. Willing, *Angew. Chem. Int. Ed. Engl.*, **17**, 731 (1978).
[17] J. J. Czubryt, M. N. Myers and J. C. Giddings, *J. Phys. Chem.*, **74**, 4260 (1970).
[18] Yu. V. Tsekhanskaya, M. B. Iomtev and E. V. Mushkina, *Russ. J. Phys. Chem.*, **38**, 1173 (1964).
[19] W. J. Schmitt and R. C. Reid, *J. Chem. Eng. Data*, **31**, 204 (1986).
[20] J.-M. Yu and B. C.-Y. Lu, *Fluid Phase Equilibria*, **34**, 1 (1987).
[21] J.-M. Yu, B. C.-Y. Lu and Y. Iwai, *Fluid Phase Equilibria*, **37**, 207 (1987).
[22] J. M. Dobbs, J. M. Wong, R. J. Lahiere and K. P. Johnston, *Ind. Eng. Chem. Res.*, **26**, 56 (1987).
[23] K. Ohgaki and T. Katayama, *J. Chem. Eng. Data*, **21**, 53 (1976).
[24] Y. Koga, Y. Iwai, Y. Hata, M. Yamamoto and Y. Arai, *Fluid Phase Equilibria*, **125**, 115 (1996).
[25] J. C. Giddings, M. N. Myers, L. McLaren and R. A. Keller, *Science*, **162**, 67 (1968).
[26] R. F. Fedors, *Polym. Eng. Sci.*, **14**, 147 (1974).
[27] Y. Iwai, Y. Koga, T. Fukuda and Y. Arai, *J. Chem. Eng. Jpn.*, **25**, 757 (1992).
[28] C. R. Yonker and R. D. Smith, *J. Phys. Chem.*, **92**, 2374 (1988).

2.5 Group Contribution Method

[1] H. K. Hansen, P. Rasmussen, Aa. Fredenslund, M. Schiller and J. Gmehling, *Ind. Eng. Chem. Res.*, **30**, 2355 (1991).
[2] K. Tochigi, D. Tiegs, J. Gmehling and K. Kojima, *J. Chem. Eng. Jpn.*, **23**, 453 (1990).
[3] M. Huron and J. Vidal, *Fluid Phase Equilibria*, **3**, 255 (1979).
[4] K. Tochigi, K. Kurihara and K. Kojima, *Ind. Eng. Chem. Res.*, **29**, 2142 (1990).
[5] S. Dahl and M. L. Michelsen, *AIChE J.*, **36**, 1829 (1900).
[6] M. L. Michelsen, *Fluid Phase Equilibria*, **60**, 213 (1990).
[7] K. Tochigi, T. Iizumi, P. Kolar and K. Kojima, *Sekiyu Gakkaishi*, **36**, 398 (1993).
[8] T. Holderbaum and J. Gmehling, *Fluid Phase Equilibria*, **70**, 251 (1991).
[9] P. Kolar and K. Kojima, *Fluid Phase Equilibria*, **118**, 175 (1996).
[10] C. Zhong and H. Masuoka, *J. Chem. Eng. Jpn.*, **29**, 315 (1997).
[11] H. Orbey, S. I. Sandler and D. S. H. Wong, *Fluid Phase Equilibria*, **85**, 41 (1993).
[12] D. S. Wong and S. I. Sandler, *AIChE J.*, **38**, 671 (1992).
[13] K. Tochigi, *Fluid Phase Equilibria*, **144**, 59 (1998).
[14] K. Tochigi, T. Iizumi, H. Sekigawa, K. Kurihara and K. Kojima, *Ind. Eng. Chem. Res.*, **37**, 3731 (1999).
[15] S. Skjold-Jorgensen, *Fluid Phase Equilibria.*, **16**, 317 (1984).
[16] E. A. Brignole, S. Skjold-Jorgensen and Aa. Fredenslund, "*Supercritical Fluid Technology*", ed. J. M. L. Penninger et al., Elsevier, Amsterdam, p. 87 (1985).
[17] H. Inomata, A. Kondo, K. Arai and S. Saito, *J. Chem. Eng. Jpn.*, **23**, 199 (1990).

3 Transport Properties of Supercritical Fluids

Along with the increase in the use of supercritical fluids in versatile industrial fields, transport properties for various supercritical fluids have become more important to either design reactors or analyze the transport phenomena taking place in the reactors. In fact, measurements of these properties have been increasingly made. In this chapter, the measurements and predictive methods for the viscosity η, thermal conductivity λ and binary diffusion coefficient D_{12} for various fluids in the near-critical and critical regions are described. In addition, the anomalies in these properties in the near-critical region are also discussed.

3.1 Viscosity

In the chemical industry, the viscosity of pure components and their mixtures is an important property in hydraulics calculations for pipeline systems and flow through porous media. With the increased popularity of process simulators, there is a need for a consistent, reliable and accurate predictive method for viscosity calculations. Accurate knowledge of transport properties of supercritical fluids, and of their mixtures, is essential both for the optimum design of the different items of chemical process plants for supercritical fluid technology and for development of accurate theories of transport phenomenon in a dense fluid medium.

The gaseous viscosity at atmospheric pressure has been measured by many researchers so far. Therefore it is rather easy to find viscosity data of a particular substance and a desired temperature condition. However, even at present, only few research groups are active in measuring the viscosity of high-pressure gases and supercritical fluids due to the experimental difficulty. In this section, we describe the experimental apparatus and procedures for the viscosity measurement of pure supercritical fluids and mixtures containing a supercritical fluid. Some of the experimental results and comparison with prediction are also shown. Readers can refer to many references [1 –8] to study in more details the viscosity of supercritical fluids and/or dense fluids.

3.1.1 Definition and General Behavior of Viscosity

The viscosity of a fluid measures its tendency to dissipate energy when disturbed from equilibrium by the imposition of a flow field. The viscosity is defined by the linear, phenomenological relationship known as Newton's law. When a shear stress is applied to a fluid, motion of the fluid occurs and a velocity gradient is established in the fluid. If the shear stress per unit area at any point is divided by the velocity gradient, the ratio obtained can be defined as the viscosity of the fluid. If the shear stress is produced by a force F applied tangent to a plane of area A, then Newton's law postulates,

$$F = -\eta A \frac{\mathrm{d}v}{\mathrm{d}x}, \tag{3.1.1}$$

where η is the viscosity and v is the velocity.

The viscosity of dilute gases is independent of pressure (or density) and can be expressed as a function of temperature. In this case, the viscosity is proportional to the square root of the absolute temperature T, and dilute gases become more viscous at higher temperatures. This behavior can be certified by simple kinetic theory. That is, the gaseous viscosity of spherically symmetric monatomic molecules at low pressure is expressed by the following equation,

$$\eta = \frac{\rho v l}{3}, \tag{3.1.2}$$

where ρ is the density of the gas, v is the average velocity, and l is the mean free path. Since the pressure dependency of the density is offset by that of the mean free path, the viscosity is almost constant over the low-pressure region. Since the average velocity is proportional to $T^{1/2}$, the viscosity is proportional to $T^{1/2}$.

However, as the pressure increases further, the viscosity also becomes greater. This difference in the pressure dependence of viscosity between a dilute gas and a dense gas arises from the difference of the mechanism of momentum transfer. In a dilute gas, it is the molecules that transport the momentum, while in a dense gas transport of momentum occurs over nonzero distance on collision. The molecular interaction plays an important role in momentum transport in dense fluids.

The liquid viscosity decreases exponentially with increasing temperature. This can be explained by the formation of a hole or the increase of the free volume. In order to represent this viscosity behavior, many models have

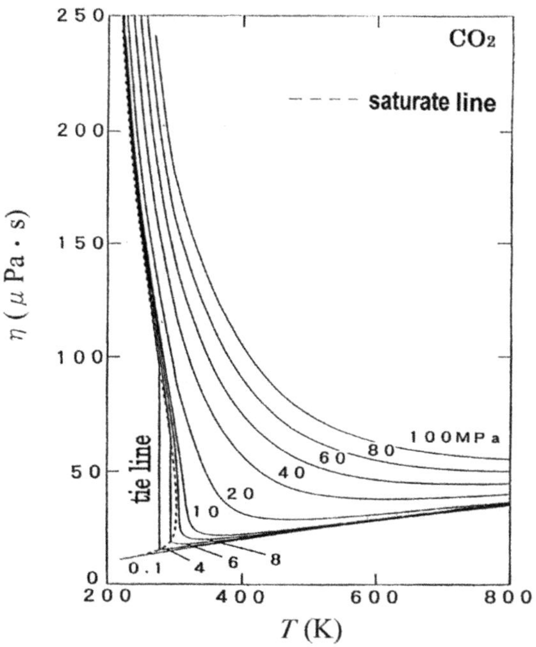

Fig. 3.1.1 Viscosity of CO_2

been proposed, since a rigorous theoretical treatment is hardly possible for liquid viscosity analysis.

Figure 3.1.1 shows the viscosity of CO_2 [9]. The dotted lines indicate the viscosity of saturated liquid and that of saturated vapor. Both lines coincide at the critical point. The viscosities of the saturated liquid and the saturated vapor at the same temperature can be connected by a tie line. The solid lines in the figure represent isobars. As can be seen, the viscosity near the critical region shows complicated temperature and/or density dependence. Figure 3.1.2 shows the pressure dependence of the five isotherms of CO_2 viscosity. It can be seen that the pressure dependence of each isotherm is not similar. It is characteristic that the isotherms at 320 K, which is close to the critical temperature of CO_2, increase rapidly near the critical pressure (7.8 MPa). The density dependence of the same isotherms in Fig. 3.1.2 is given in Fig. 3.1.3. In this case, as can be seen, every isotherm shows very similar density dependency. From the comparison between Figs 3.1.2 and 3.1.3, it can be concluded that the complicated pressure dependency of the viscosity arises from the pressure dependency of the density. Therefore the viscosity data over a wide range of temperature and pressure should be expressed as a function of temperature and

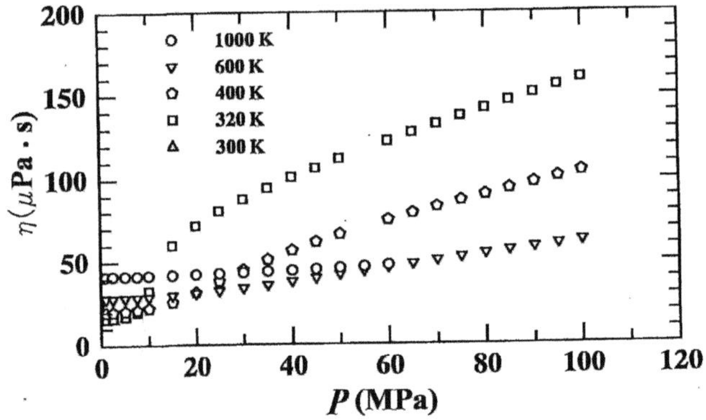

Fig. 3.1.2 Pressure dependence of the viscosity of CO_2

density. In order to do this, an accurate equation of state is usually required to calculate the density from the experimental temperature and pressure data at the viscosity measurements.

In general, the gas viscosity under the constant density condition increases at almost the same rate as the temperature increases. However, the rate of increase of the gas viscosity at constant pressure with temperature changes remarkably in the low-pressure region and high-pressure region (as shown in Fig. 3.1.2). Figure 3.1.4 is the reduced *PVT* diagram of supercritical CO_2. At the higher-density region above the dotted line, the viscosity of CO_2 at constant pressure decreases as the temperature increases. Therefore the viscosity in this low-temperature and high-density region behaves like a usual liquid. On the other hand, in the lower-density region below the dotted line, the vis-

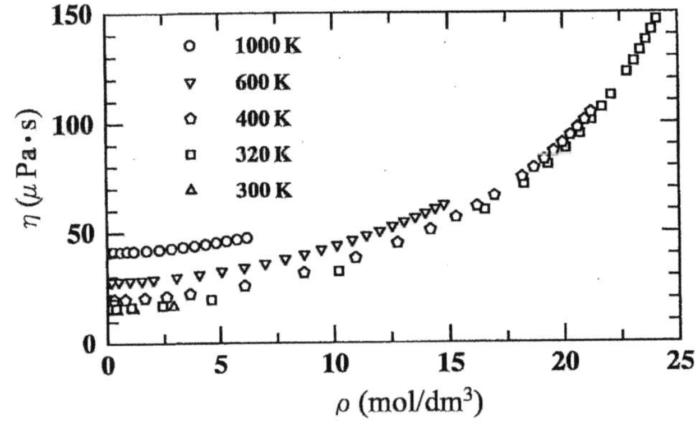

Fig. 3.1.3 Density dependence of the viscosity of CO_2

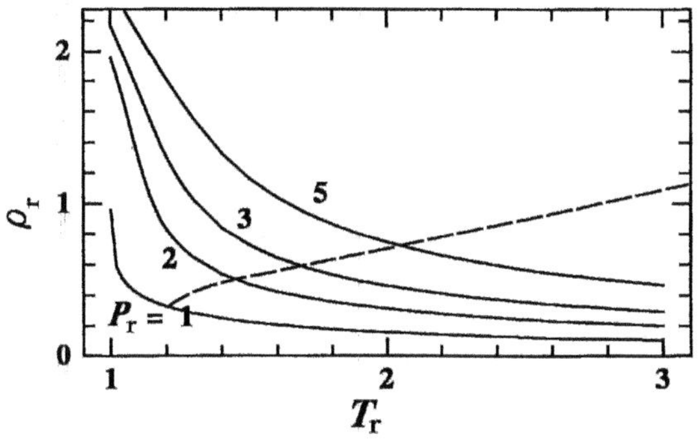

Fig. 3.1.4 *PVT* diagram of supercritical CO_2

cosity of CO_2 at constant pressure increases as the temperature increases. In this high-temperature and low-density region, the viscosity of CO_2 shows gas-like behavior. These behaviors are general to the other supercritical fluids.

3.1.2 Measurement of Viscosity

(a) Viscometer

The viscometers which have often been employed for accurate work for dense fluids fall into two categories: oscillating-body and capillary-flow viscometers. For the measurements of dense fluids near the critical region, the oscillating-body viscometer may be the most reliable instrument. On the other hand, the viscosity of liquids under ambient conditions can be measured accurately and routinely by capillary-tube viscometers. In this section, we describe the oscillating-disk viscometer developed for the purpose of measuring the critical viscosity anomaly by Iwasaki and Takahashi [10–13].

The oscillating-body viscometer consists of an axially symmetric body which performs torsional oscillations in a fluid. The oscillating system is suspended from an elastic wire and gently rotated to start the motion. The changes in frequency and damping decrement caused by the fluid, compared with those in a vacuum, depend only upon the density and viscosity of the fluid and on the characteristic properties of the suspension systems. In principle, one could determine the viscosity and density simultaneously from measurements of the frequency and decrement. In general cases, the density can be measured directly with greater accuracy by other methods so that only the viscosity is of interest.

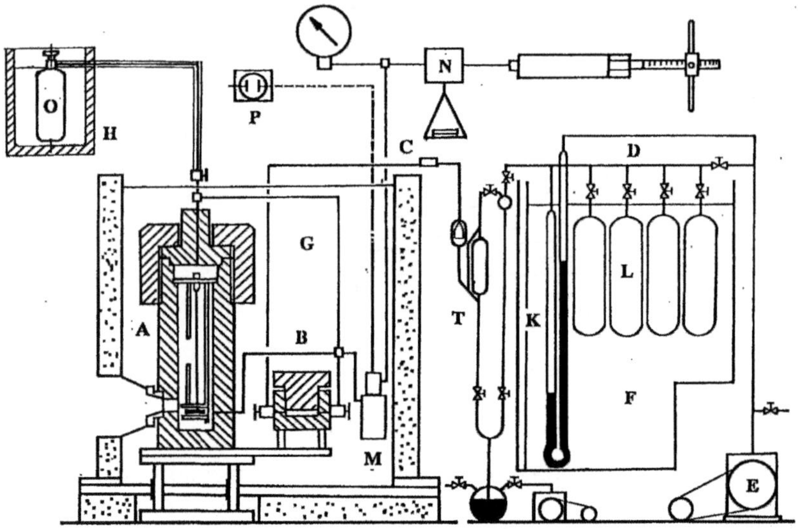

Fig. 3.1.5 Schematic diagram of apparatus for viscosity and density measurements. A: viscometer; B: high-pressure pipette; C: connecting tube; D: connecting capillary; E: vacuum pump; F, G, H: thermostat; K: mercury manometer; L: glass cylinder; M: pressure difference detector; N: dead weight gauge; O: sample cylinder; P: pilot lamp; T: Toepler pump; U: hand pump; V: pressure gauge

Experimental apparatus for viscosity and density measurements is shown in Fig. 3.1.5. The apparatus consists of an oil bath, water bath, Toepler pump and pressure measurement system. A viscometer, high-pressure pipette, and pressure difference detector are contained in the oil bath of about 70 l volume. The oil bath is fixed to the outer, solid stone wall, which was built separately from the basement of the building, to ensure rigidity and freedom from vibrations. A mercury manometer and glass cylinders are set in the water bath. The temperature of the water bath is regulated to 298.15 K. The volumes of the glass cylinders and glass tube connecting the cylinders are calibrated by the Burnett method with the use of nitrogen gas whose density data are known accurately.

The Maxwell type oscillating viscometer is shown schematically in Fig. 3.1.6. Figure 3.1.7 shows a picture of the high-pressure cell, in which the suspension system is included. The optical window, through which observations of the oscillation periods and damping ratio are made, can be seen at the bottom part of the high-pressure cell body. The main part is the suspension system shown in Fig. 3.1.8, which is composed of an oscillating stainless steel

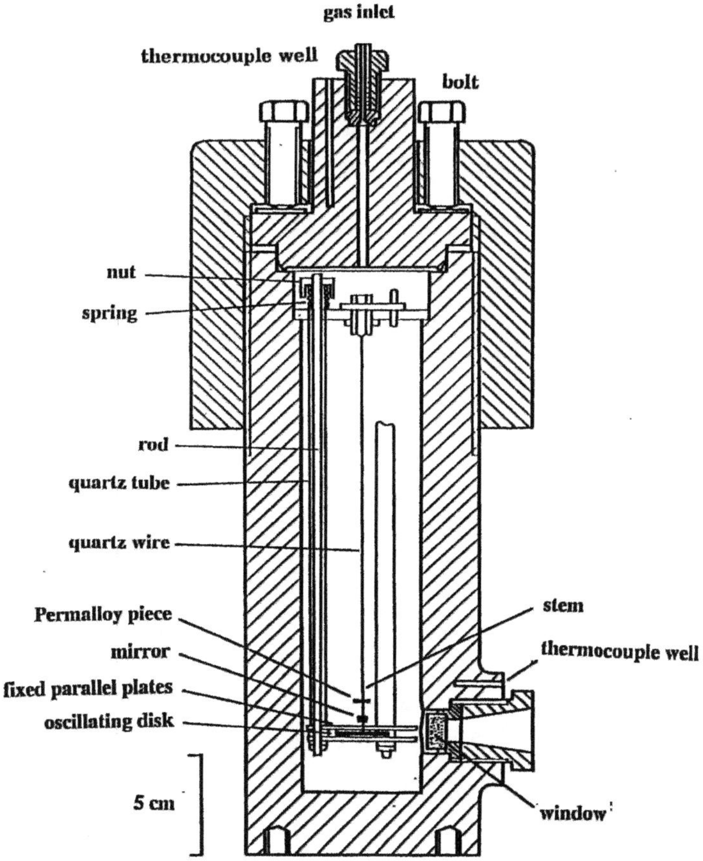

Fig. 3.1.6 Maxwell type viscometer

disk, a fine suspension quartz wire, two fixed stainless steel disks, and a stainless steel stem with a nickel-plated glass mirror and a Permalloy piece.

The size of the oscillating disk is 28.104 mm in diameter and 0.975 mm thickness at 298.15 K. The stainless steel stem of 3 mm long and 1 mm in diameter is soldered at the center of the oscillating disk. The size of the fixed disk is 52.000 mm in diameter and 2.000 mm in thickness. At the center of the fixed disk, a hole of 3 mm in diameter is drilled. The two fixed disks are placed between three spacer cylinders (10.0 mm in outer diameter, 3.5 mm in inner diameter, and 2.000 mm in length) fixed by the three stainless steel rods and nuts. The three rods pass through the hole of each of the three spacer cylinders and fix the cylinders. The whole suspension system is fixed to the high-pressure vessel with the two screws in a symmetric position.

Fig. 3.1.7 Picture of the viscometer cell body

Great care is taken in aligning the suspension system itself by ensuring good parallelism of the fixed disks and the oscillating disk and, moreover, the change of the relative position of the oscillating disk between the fixed disks with the change of temperature is made as small as possible. For accurate data analysis based on the mathematical model of the flow process, the distance between the oscillating disk and the upper fixed disk should be equal to the distance between the oscillating disk and the lower fixed disk at any ambient temperatures. In order to satisfy this requirement, the three stainless steel rods are covered with the three quartz tubes. The quartz tubes are inserted between the upper fixed disk and the fixing disk, and the length of the rods is adjusted.

The high-pressure vessel made from stainless steel has about 1000 cm^3 volume. The vessel has an optical window made of quartz (18 mm in diameter

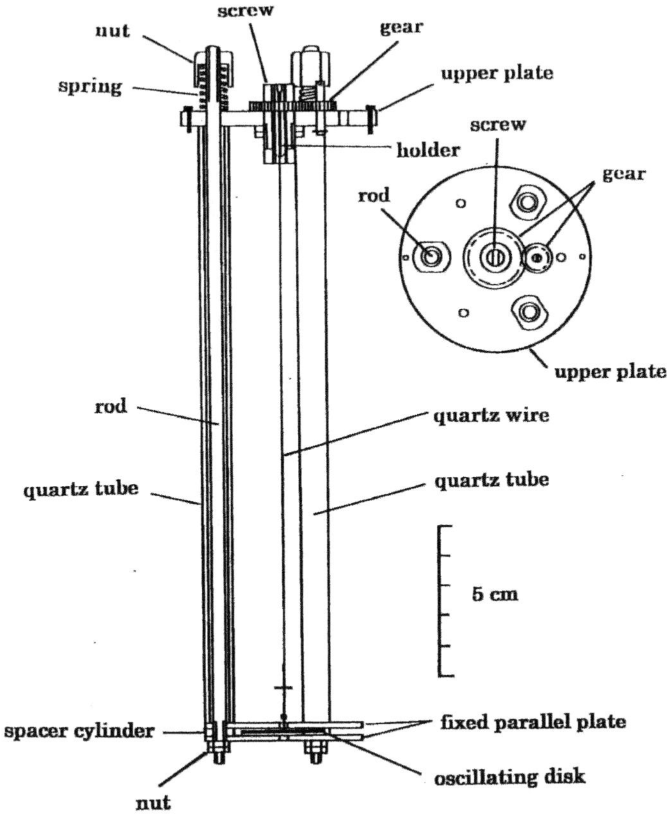

Fig. 3.1.8 Suspension system

and 10 mm thickness). The vessel is sealed with the plug and nut. The plug is suppressed evenly with equal power with six bolts. The copper packing is used as a seal. The two holes are used to insert the thermocouple. The sample temperature is measured by the thermocouple in the hole near the optical window.

The quartz wire is prepared as follows. First, a quartz rod, 20 cm long and about 0.3 mm in diameter, is drawn from a piece of clear, fused quartz, care being taken that its diameter is uniform throughout the length of the rod. Then the quartz rod is treated with hydrogen fluoride to obtain a uniformly thin wire, 18.5 cm long and about 35 μm in diameter, keeping both ends of the rod away from contact with the hydrogen fluoride solution. Both ends of the quartz strand prepared thus are plated with silver and with copper by turns. One end of the quartz strand, 5 mm long, is soldered to the thin stainless steel rod, 12 mm long, which is screwed perpendicularly on the oscillating disk.

The other end, 10 mm long, is also soldered to the stainless steel holder fixed to the screw, which is fixed to the upper plate with the gear and nut.

In front of the oil bath, at a distance of approximately 3 meters from the axis of oscillation, a precision glass scale is placed. The amplitudes and period of oscillation are measured with the aid of the scale, three phototransistors, and a clock circuit connected to a personal computer. The three slits (1.00 ± 0.01 mm in width, about 5 mm long) are set up to a regular interval (29.00 ± 0.01 mm). Three phototransistors are attached in the back of each slit. The light from a high-pressure Hg lamp goes to the suspension system passing through the optical window, then is reflected by the mirror. The picture of the reflected light performs a round trip vibration due to the oscillation of the oscillating disk. The position of the scale is adjusted so that the reflected light from the mirror hits the central phototransistor in a condition at which the oscillating disk is halted. When the reflected light passes the phototransistor, a current occurs in the phototransistor. Since the current is sent to a personal computer as a signal, the passing times of the reflected light on the phototransistors can be measured by a clock circuit in the computer. From a continual series of passing time data, the oscillation period and damping rate can be determined. The details of this analysis of oscillation are described by Takahashi et al. [14, 15].

(b) Density Measurement

The density value should be measured to determine the viscosity of supercritical fluids and especially of fluids near the critical region very accurately, since the supercritical fluid is quite compressible and the local density distribution due to gravity or temperature fluctuation can easily occur. The density measurements are made by the use of the constant volume type apparatus, shown in Figure 3.1.5, developed by Date [16]. The apparatus can be divided into two main parts. One is the high-pressure system in which the *PVT* measurements for certain amounts of fluid sample are made at the same conditions as the viscosity measurements. Another is the low-pressure system in which the molar quantities of the fluid sample employed above are determined from the measured values of P, V and T in the expanded state near to 0.1 MPa. Almost the same type of low-pressure system was used by Kaminishi et al. [17, 18].

The high-pressure system consists of a high-pressure gas pipette and a pressure difference detector. As shown in Fig. 3.1.5, the pipette is set horizontally and connected with the bottom part of the viscometer at the level of the oscillating disk so that its inner space is kept at the same level as the oscillating disk. The tubing connecting the pipette and viscometer takes the same level also to ensure the rapid establishment of the equalization of density in

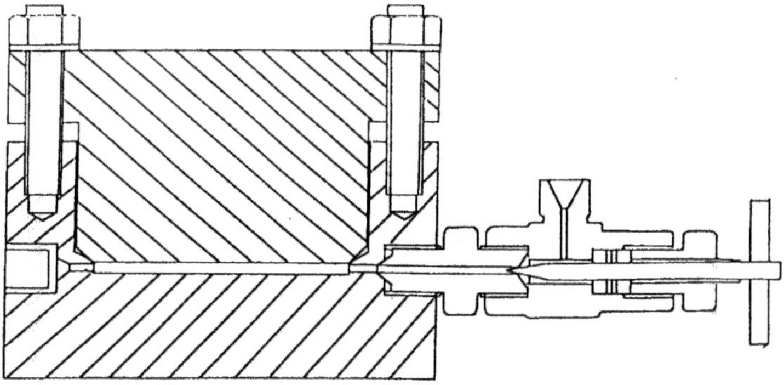

Fig. 3.1.9 High-pressure gas pipette

the pipette with that of the position of the oscillating disk in the viscometer. In this way, the gravity effect on the density value can be neglected and the local density of the fluid near the oscillating disk can be determined accurately. The details of the pipette are shown in Fig. 3.1.9. The internal volume of the pipette is 6.837 cm^3 at 298.15 K and the height of the room in the pipette is 2 mm. The pipette equips two valves; one is connecting to the viscometer and the other to the Toepler pump. The pipette is set in the oil bath, while the glass cylinder is set in the water bath. The water bath is thermostated at 298.15 ± 0.01 K. The sample fluid is introduced into the pipette at the same temperature and pressure at which the viscosity is measured, and then introduced into the low-pressure system. The low-pressure system is evacuated before introducing the fluid. The residual gas in the pipette is transferred by using the Toepler pump. The quantity (moles) of the fluid is determined from the volume, pressure, and temperature of the system. The density of the fluid is determined from the above-determined fluid quantity and the volume of the pipette. The volume of the pipette is determined as follows: the nitrogen (99.99% pure) is sampled into the pipette at 298.15 K and 12 MPa, and then introduced into the low-pressure system. The quantity of nitrogen is determined by the same procedure described above. Then, the volume of the pipette is calculated from the above-determined fluid quantity and the compressibility factor data of nitrogen.

For the measurement of pressure, the mercury U-tube pressure difference detector (shown in Fig. 3.1.5) is used. One leg of the U tube is linked with the viscometer. The other leg is filled with oil and connected to the dead weight gauge. The viscometer, the mercury U-tube detector, and the dead weight gauge are carefully set with each other so that the oscillating disk, the head of mercury in the U tube, and the lower end of the piston of the dead weight gauge in its working state are aligned at the same level. The effect of the temperature on the thermal expansion of mercury and oil are considered precisely throughout the experiment. The atmospheric pressure is measured by a standard mercury manometer with an accuracy of ± 0.1 mmHg. The accuracy of the pressure measurements is estimated to be ± 0.1 kPa.

The temperatures of the oil bath and the high-pressure cell body are measured with a platinum resistance thermometer calibrated against a Leeds and Northrup platinum resistance thermometer. The temperature of the oil bath is kept constant to about ± 3 mK over the periods of measurements. The temperature of the high-pressure cell body is kept to within ± 1 mK. The accuracy of the temperature measurements is estimated to be ± 3 mK.

(c) Data Analysis

The results of the measurements are evaluated with the aid of Newell's theory [19] in a manner similar to that used by Iwasaki and Kestin [20]. The correction factor C_n, which characterizes the effect of the edges and corners of the disk upon the oscillation, is determined experimentally with nitrogen as the standard fluid at the same experimental temperature and pressure conditions. The viscosity values of nitrogen recommended by Stephan et al. [21] are used and the density of nitrogen is calculated from a BWR equation of state developed by Jacobsen and Stewart [22]. The error in the viscosity determination is estimated to be less than ± 0.3% considering the error of the logarithmic decrement (± 0.25%) and the error of the period of oscillation (± 0.01%).

3.1.3 Correlation

The viscosity values of trifluoromethane (CHF_3; HFC-23) [23] near the critical temperature are shown in Figs. 3.1.10 and 3.1.11. The viscosity vs. pressure isotherms increase rapidly near the critical pressure as shown in Fig. 3.1.10. The viscosity vs. density isotherms are shown to lie on a single curve, since the experimental temperatures are close to the critical temperature. The density dependence of the viscosity shown in Figure 3.1.11 indicates that the viscosity equation expressed as a function of temperature and density has a much simpler functional form than that expressed as a function of temperature and pressure.

The correlations for the viscosity of high-pressure gases and supercritical fluids have been developed based on the corresponding state principle and the modified Enskog theory. Here, the correlations developed by Lee and Thodos [24] and Chung et al. [25, 26] are tested for CHF_3. Figures 3.1.12 and 3.1.13 shows the deviation plots of correlations by Lee−Thodos and Chung et

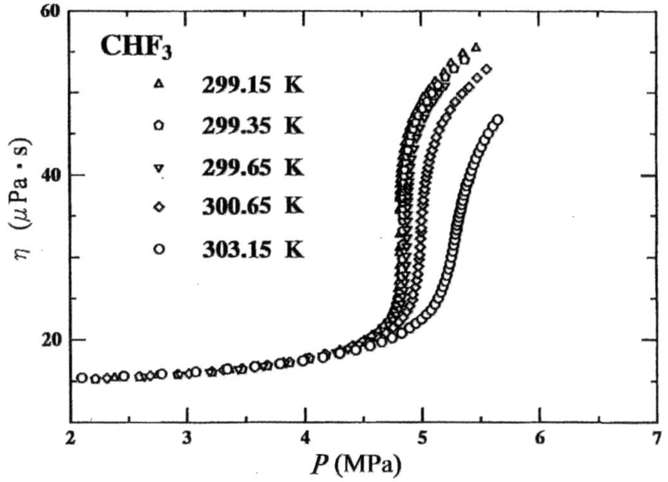

Fig. 3.1.10 Viscosity vs. pressure plots for CHF_3

al., respectively. It can be seen the Lee–Thodos correlation can show better agreement with experiment. The correlation of Chung et al. gives larger deviation in the higher-density region.

However, the Lee–Thodos correlation was developed only for the purpose of representing the viscosity behavior of pure components over a wide range of temperature and pressure cannot apply to their mixtures. Other corre-

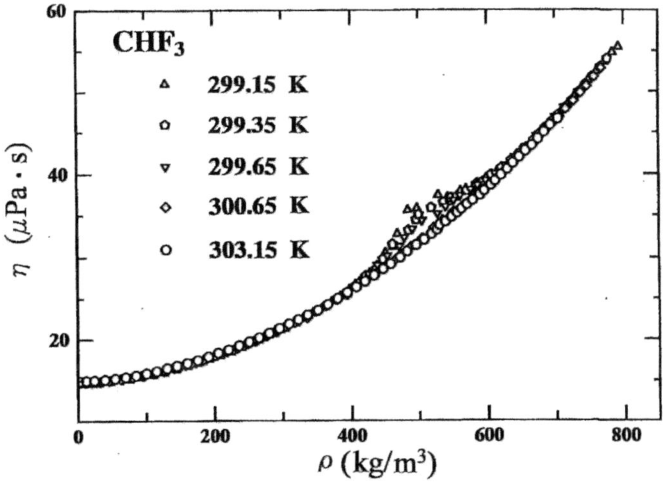

Fig. 3.1.11 Viscosity vs. density plots for CHF_3

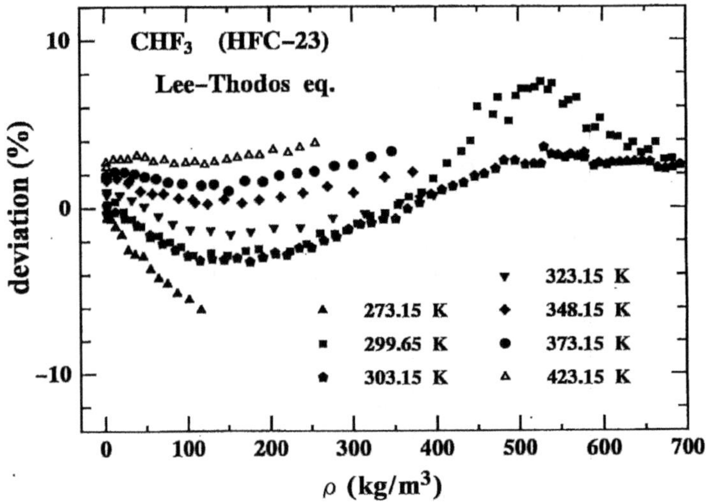

Fig. 3.1.12 Deviation plots between experimental values from those calculated from Lee–Thodos correlation for CHF_3

lations based on the corresponding-state principle need the critical parameters of a mixture, which is rarely available in the literature or any databases. Therefore Yokoyama et al. have tried to combine the Lee–Thodos correlation for pure fluids and a modified kinetic theory for dense fluid mixtures. Originally, DiPippo et al. [27] proposed a modified kinetic theory to calculate the viscosity of dense fluid mixtures. Later, Vesovic and Wakeham [28, 29] modified DiPippo's theory to improve the applicability of the original version. The modified kinetic theory is the most powerful tool for predicting the mixture viscosity under the condition that the pure component viscosity as a function of temperature and density over a whole range of interest are well known. The reader is referred to the original publications [28, 29] for the details of the Vesovic–Wakeham (VW) version.

Figure 3.1.14 shows the calculated results of the viscosity correlation for the CO_2 + CHF_3 mixture obtained from the VW version. It should be noted that the mixture viscosity at normal pressure is fitted with the extended corresponding state principle of Kestin et al. [30, 31]. It is found that the deviations for the mixtures are almost the same order of magnitude as those of the pure components. This result suggests that the mixture viscosity can be predicted from a knowledge of the pure components' viscosity over a whole range of interest and mixture viscosity at normal pressure.

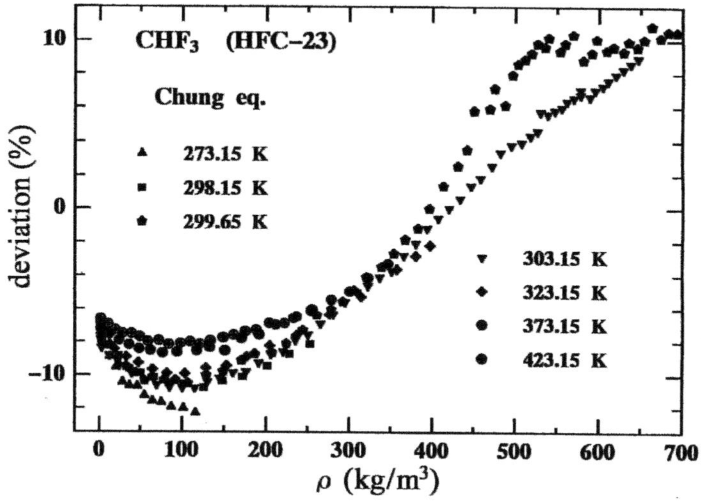

Fig. 3.1.13 Deviation plots between experimental values from those calculated from Chung et al. correlation for CHF_3

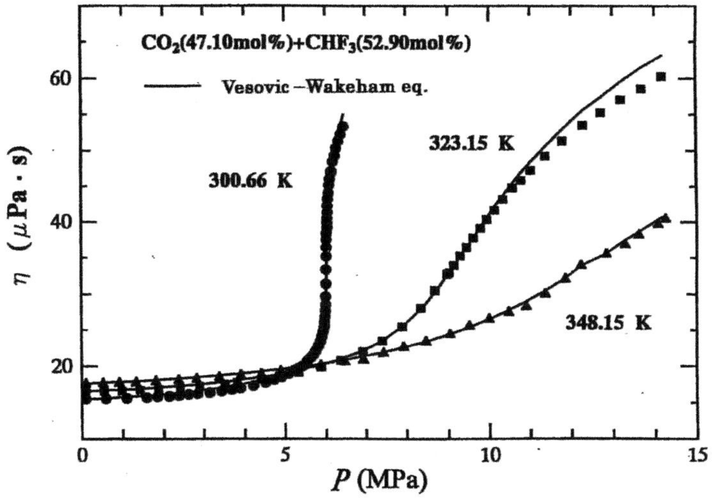

Fig. 3.1.14 Comparison between experimental viscosity values and those calculated from Vesovic–Wakeham theory for a CO_2 + CHF_3 mixture

3.1.4 Critical Anomaly of Viscosity

Near a critical point, the viscosity of a fluid exhibits an anomalous increase. Such critical anomaly behavior has been investigated theoretically and experimentally. In accordance with the practical importance of supercritical fluid technology in the last two decades, studies on transport properties of nearcritical and supercritical fluids have significantly increased. To develop and examine the theoretical models for the critical anomaly of transport properties, we need accurate data for the viscosity of fluids in the critical region. Investigations of the viscosity of the pure component near the critical point has been investigated for various substances such as CO_2 [12, 32–37], ethane [12, 38], ethylene [13], xenon [37, 38], nitrogen [39], sulfur hexafluoride [40], water [41], nitrous oxide (N_2O) [42], and trifluoromethane (CHF_3) [23]. For binary mixture system, Wang and Meyer [43] measured the critical anomaly of the ^3He + ^4He systems and Yokoyama et al. [44] studied the critical anomaly of the CO_2 + CHF_3 systems. Theoretical models for the transport properties of binary mixtures has been developed by Luettmer-Strathmann and Sengers [45, 46] and Kiselev and Kulikov [47]. However, only limited experimental data are available for the transport properties of supercritical fluid mixtures. In future, experimental efforts to accumulate the thermophysical properties of multicomponent mixtures near the critical point should be more important to test the theory. In this subsection, we discuss some of the experimental studies for the critical viscosity anomaly.

(a) Measurements of Viscosity Near the Critical Region

Measurements of the transport properties of supercritical fluids near the critical region are hampered by the fact that they are nonequilibrium quantities. Accurate measurements should be carried out by introducing macroscopic gradients into the system under the equilibrium condition. For the purpose of obtaining the transport properties the hydrodynamic equations must be solved under certain imposed boundary conditions. Exact solutions of these equations are known only for fluids showing stable thermophysical properties in idealized geometrical situations for theoretical analysis. However, in the critical region a number of thermophysical properties such as local density and specific heat vary appreciably with position as a results of gradients.

If the temperature is not uniform in a fluid near its critical point, the gravity force can easily generate convection currents and convective heat flow. If the shear stress applied to an oscillating body is large enough, the secondary flow becomes sufficiently pronounced and sometimes the flow becomes turbulent. The convective heat flow and mass flow must be distinguished from the irreversible fluxes associated with the thermodynamic forces by linear laws. Therefore great care should be taken to maintain not only a near-equilibrium state, but also hydrodynamic, thermal and mechanical stability over the time needed to complete a measurement.

(b) Data Analysis and Correlation of the Critical Anomaly of Viscosity

The details of the experimental apparatus and procedure are given in Sect. 3.1.2. The unique point of the measurement by Yokoyama et al. is the direct measurement of the density of sample fluid at the same level as the oscillating disk. This procedure eliminates the need to apply corrections for significant variations of density with elevation in the viscometer cell in the neighborhood of the critical point. Figures 3.1.15 and 3.1.16 show the viscosity of N_2O and CHF_3, respectively, in the critical region as a function of density. It can be seen that the viscosity indicates an anomalous increase near the critical point. These results show almost the same tendency observed previously for CO_2 and other gases.

In general, the experimental viscosity $\eta(\rho,T)$ can be separated into an anomalous part $\Delta\eta(\rho,T)$ and a background viscosity $\eta_b(\rho,T)$, i.e.,

$$\eta(\rho,T)=\eta_b(\rho,T)+\Delta\eta(\rho,T). \tag{3.1.3}$$

The background viscosity is obtained empirically by extrapolating the behavior of the viscosity outside the critical region, and it may be written as

$$\eta_b(\rho,T)=\eta_0(T)+\eta_e(\rho,T), \tag{3.1.4}$$

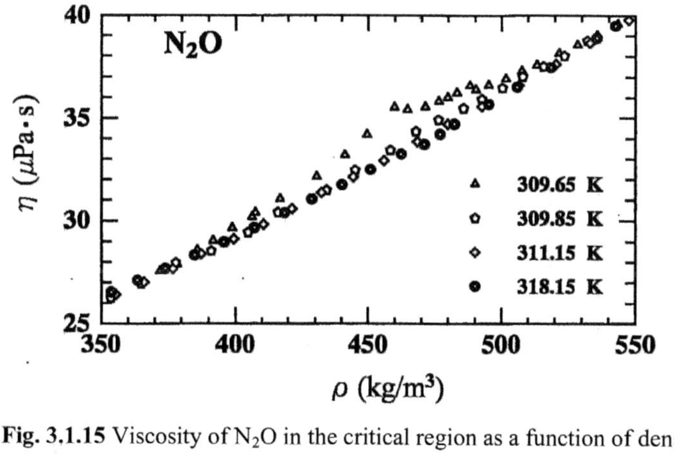

Fig. 3.1.15 Viscosity of N_2O in the critical region as a function of density

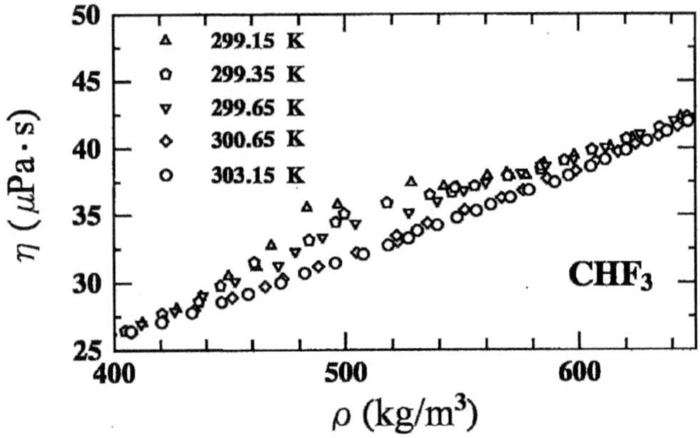

Fig. 3.1.16 Viscosity of CHF_3 in the critical region as a function of density

where $\eta_0(T)$ is the viscosity in the limit of low densities and $\eta_e(\rho,T)$ is the excess viscosity. Here, $\eta_0(T)$ is assumed to be equal to the viscosity at atmospheric pressure. So, the excess viscosity can be determined from the viscosity data outside the critical region and the viscosity data at atmospheric pressure. In the critical region, the anomalous viscosity term $\Delta\eta(\rho,T)$ can be obtained from (3.1.3). The values of $\Delta\eta$ for N_2O and CHF_3 are shown in Figs. 3.1.17 and 3.1.18, respectively. As can be seen, $\Delta\eta$ has a maximum near the critical density.

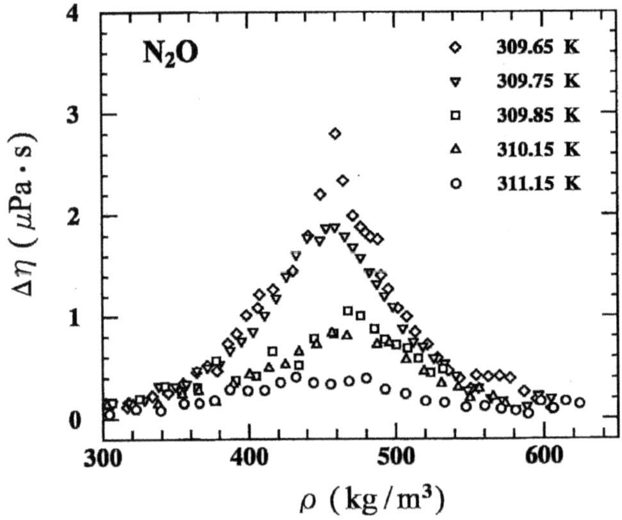

Fig. 3.1.17 Anomalous viscosity $\Delta\eta$ of N_2O

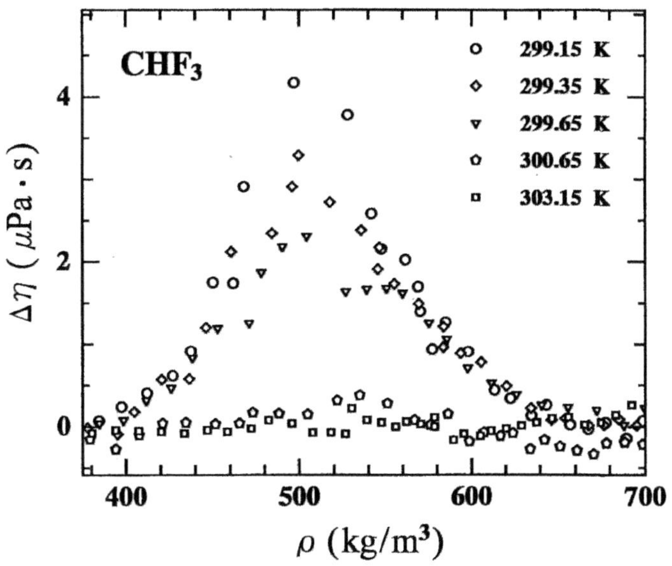

Fig. 3.1.18 Anomalous viscosity $\Delta\eta$ of CHF_3

Close to the critical region, a simplified theoretical treatment based on mode-coupling theory yields the following power law [48, 49],

$$\frac{\eta}{\eta_b} = (q\xi)^{\phi}, \tag{3.1.5}$$

where q is a constant, ξ is the correlation length, and ϕ is a universal exponent. The original estimate for the value of ϕ from mode-coupling theory is 0.054. Recently, this has been shown to be inconsistent with most experimental data for the critical enhancement of the viscosity and should be replaced by a value closer to 0.065 [50–52]. Sengers et al. [53] suggested that (3.1.5) can be rewritten in terms of the symmetrized isothermal compressibility as follows:

$$\frac{\eta}{\eta_b} = \left(\frac{\chi_T^*}{\chi_0^*}\right)^{\nu\phi/\gamma} \quad \text{for} \quad \chi_T^* > \chi_0^*, \tag{3.1.6}$$

$$\frac{\eta}{\eta_b} = 1 \quad \text{for} \quad \chi_T^* < \chi_0^*,$$

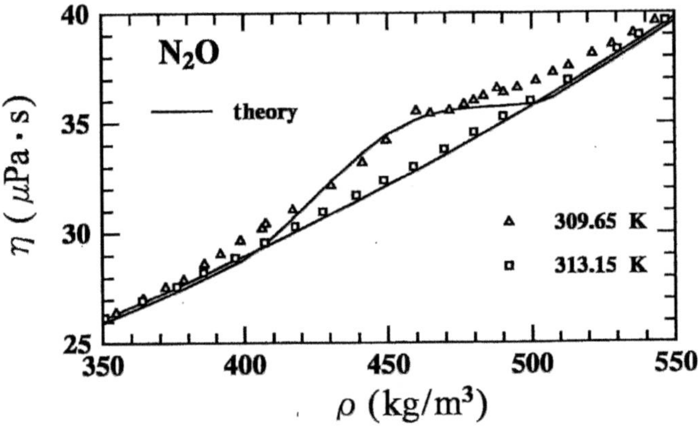

Fig. 3.1.19 Comparison between the calculated values and the experimental viscosity of N_2O at 309.650 K and 318.150 K

where χ_T^* is the reduced symmetrized compressibility, $\chi_0^* = \Gamma/(q\xi_0)\gamma\nu$, γ and ν are critical exponents, Γ is the critical amplitude of χ_T^*, and ξ_0 is the critical amplitude of ξ. Yokoyama et al. used (3.1.5) and (3.1.6) for the correlation of their viscosity data in the critical region. χ_T^*, ξ_0 and ξ are calculated with the use of the linear model of the scaled equation of state. The value of q^{-1} can be determined from the double logarithmic plot of the viscosity ratio η/η_b and ξ and a prerequisite condition of $\phi = 0.065$. The q^{-1} value obtained is 81×10^{-10} m for N_2O and 28×10^{-10} m for CHF_3. The comparison between the calculated results and the experimental data of N_2O at 309.650 and 318.150 K is shown in Fig. 3.1.19. It can be seen that a fairly good agreement between theory and experiment was obtained. They studied the critical anomaly also for SF_6 and binary mixtures of $CO_2 + CHF_3$. In addition, the theoretical analysis based on the theory of the global behavior of the transport properties in the critical region developed by Olchowy and Sengers [54] is under investigation.

3.1.5 Remarks

In this section, we have described mainly experimental studies of the viscosity of dense fluids. It should be noted that a very accurate equation of state (*PVT* relationships) and accurate specific heat data (or equation) are required for the theoretical analysis of viscosity especially in the critical region. For dilute and moderately dense fluids, the oscillating body viscometer is the most suitable instrument at this point. There is still room for further study to improve the reproducibility and accuracy of the measurements with oscillating body viscometers, such as the method of detecting the periods and damping ratio with the use of optical and/or electrical instruments. Newly designed viscometers should be investigated in future to improve the accuracy and applicability of the measurements.

As for the theory, the modified kinetic theory of Vesovic and Wakeham is recommended for the mixture viscosity calculation. Further study should go to develop the prediction method for mixture viscosity at normal pressure and also for the viscosity of pure components over a wide range of temperature and pressure (density).

3.2 Thermal Conductivity

Transport properties, especially viscosity and thermal conductivity, are essential in the analysis of heat and mass transfer processes within heat exchangers, compressors, flow meters, distillation columns, extractors, and so on. They have a major impact on equipment design and thus influence whether it is economically feasible to build a machine which can actually approach the theoretical efficiency of the thermodynamic cycle.

The thermal conductivity of fluids varies considerably from dilute gas to compressed liquid, from vapor to supercritical fluid. It is, therefore, impossible to predict the behavior of the thermal conductivity, especially of supercritical fluids, without a detailed experimental and theoretical study. Knowledge of the thermal conductivity is also very important in the development of molecular theories of fluids and fluid mixtures.

3.2.1 Definition and Units of Thermal Conductivity

When a temperature gradient $\mathrm{d}T/\mathrm{d}x$ exists in a physical system, transfer of a quantity of heat J_q in the direction of decreasing gradient occurs. The quantity of heat J_q that is transported over an area A perpendicular to the x-axis in a time interval t is proportional to the temperature gradient $\mathrm{d}T/\mathrm{d}x$:

$$J_q = -\lambda A t \frac{\mathrm{d}T}{\mathrm{d}x}. \tag{3.2.1}$$

The above equation is a form of Fourier's law of heat conduction and the coefficient λ is the thermal conductivity. Thermal conductivity is a measure of heat conduction and is the amount of heat energy that is transported over a unit area at a unit time by a unit temperature gradient.

At nonstationary conditions the temperature T changes with time and the following equation holds true:

$$\frac{\partial T}{\partial t} = a \frac{\partial^2 T}{\partial x^2}, \tag{3.2.2}$$

where a is the thermal diffusivity and is defined by the equation:

$$a = \frac{\lambda}{C_P \rho}. \tag{3.2.3}$$

Here ρ is the density and C_P is the specific heat at constant pressure.

Thermal conductivity is the measure of the transfer of heat energy which is due to a molecule's motion. Effects due to group motion like convection, and radiation due to electromagnetic waves are not included in this property.

The units used for thermal conductivity are W m^{-1} K^{-1}. To convert English or cgs units to these units:

Btu hr^{-1} ft^{-1} R^{-1} × 1.73074 = W m^{-1} K^{-1},
kcal m^{-1} hr^{-1} K^{-1} × 1.16222 = W m^{-1} K^{-1},
cal cm^{-1} s^{-1} K^{-1} × 418.400 = W m^{-1} K^{-1}.

3.2.2 Measurement of Thermal Conductivity

When a temperature gradient is imposed on a fluid, three kinds of heat transfer occur, namely, conduction, convection and radiation. To measure the thermal conductivity, it is necessary to avoid the onset of free convection and the presence of the radiative contribution. The thermal conductivity of a fluid measures its propensity to dissipate energy, when disturbed from equilibrium, by the imposition of a temperature gradient. The methods of thermal conductivity measurement are divided into two main groups: direct (or thermodynamic) methods and indirect (or nonthermodynamic) methods [1]. By direct methods, we mean those that involve the direct measurement of temperature gradients by means of thermometers or thermocouples. As examples of the first group we have the transient hot-wire method, the concentric cylinders method, and the parallel plates method. As indirect methods, we have the laser-flash method, the forced-Rayleigh-scattering method, and so on. In this section, only the direct methods are discussed briefly due to space limitations.

(a) Transient Hot-Wire Method

As shown in Fig. 3.2.1(a), the mathematical basis of this method is that of an infinite-line heat source (H) immersed in an infinite isotropic fluid (X) with temperature-independent properties. This method involves the use of a very thin wire at the center of a cylindrical sample fluid. The wire is used both as a heating element for the sample and as a resistance thermometer. The wire is subjected to an electric pulse current (AC or DC) and its temperature history is monitored by a personal computer after this pulse.

The working equation is given as follows.

$$\Delta T = T - T_0 = \frac{q}{4\pi\lambda} \ln\left(\frac{4at}{r_0^2 \exp\gamma}\right), \tag{3.2.4}$$

where T is the temperature of the wire, T_0 is the initial temperature of the wire, q is the heat input per unit length of the wire, r_0 is the radius of the wire, γ is the Euler constant, λ is the thermal conductivity of the fluid, and a is the thermal diffusivity of the fluid defined by (3.2.3). The thermal conductivity is obtained from the slope of the plot of the temperature rise, ΔT, of the wire as a function of the elapsed time $\ln t$.

The following assumptions are usually made: (a) the line source is infinitely long and vertical, (b) λ of the wire is infinite, (c) C_P of the wire is zero, (d) the line source is in an isotropic infinite fluid, (e) the thermophysical properties of the fluid such as λ, a, ρ and C_P are constant during the

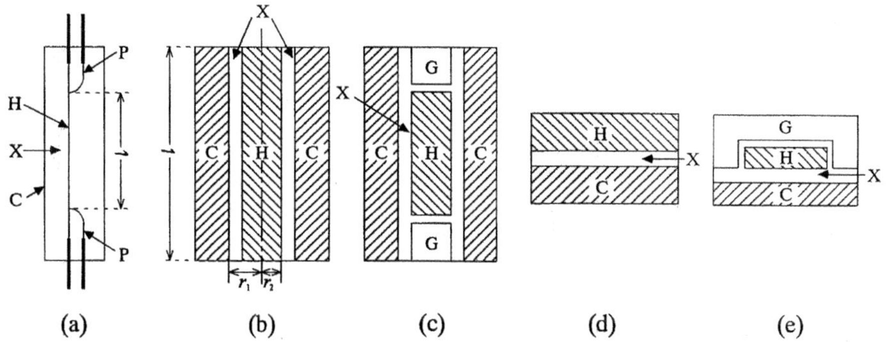

Fig. 3.2.1 Schematic diagrams of thermal conductivity measuring systems. (**a**) transient hot-wire method, (**b**) and (**c**) concentric cylinders method, (**d**) and (**e**) parallel plates method. H: emitter; C: receiver; X: sample fluid; G: guard heater; l: effective length; P: potential lead; r_1: internal radius of receiver, r_2: external radius of emitter

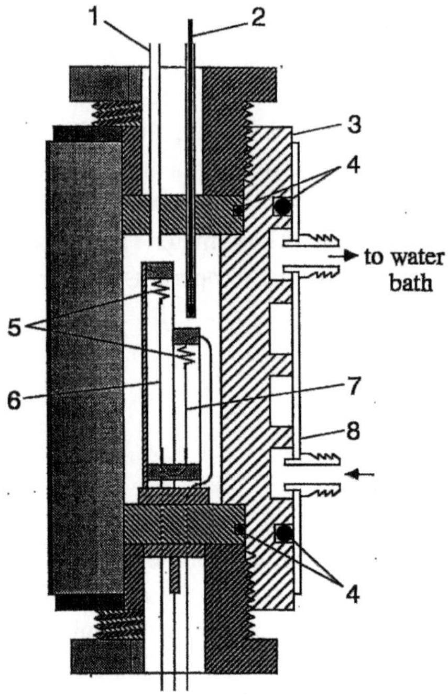

Fig. 3.2.2 Cross-sectional view of the transient hot-wire apparatus used by Yamamoto et al. [4] 1: sample inlet; 2: thermistor; 3: high-pressure vessel; 4: O-ring; 5: copper spring; 6: main Pt-wire ($D = 5$ µm, $L = 50$ mm); 7: compensating Pt-wire ($D = 5$ µm, $L = 25$ mm); 8: water jacket

measurement, (f) the line source is in thermodynamic equilibrium with the fluid at $t \leq 0$, (g) energy transfer from the wire is purely conductive.

A complete description of this technique, including the mathematical modeling of the instruments and their operation, can be found in refs. [2, 3]. Experimental methods based on this technique can be classified into the compensated-lead method and the potential-lead method. The cross-sectional view of the hot-wire cell developed by Yamamoto et al. [4] is illustrated schematically in Fig. 3.2.2. To compensate for the heat flux in the axial direction through the sample fluid at the ends of the wire, two platinum wires (5 µm in diameter) of different length (main wire: 50 mm, compensating wire: 25 mm) are used. The main and compensating wires are arranged in opposing arms of a bridge circuit and the deviation from the balance point owing to the initiation of the heating cycle is measured. This apparatus is not specifically designed to measure thermal conductivity in the critical region. In the measurement of electrically conducting liquids, bare wires with DC currents cannot be used because secondary electric currents are produced. To solve those problems, an

secondary electric currents are produced. To solve those problems, an AC current and wire electrically insulated by a non-conduction coating are used.

(b) Concentric Cylinders Method

As shown in Fig. 3.2.1(b), the ideal model of this method consists of two cylinders of infinite length and of different radii, at two different temperatures. Heat is conducted from the inner hot cylinder (H) to the outer cold cylinder (C) through the sample fluid (X). The fluid is constant in an annulus between two infinitely long coaxial cylinders. The thermal conductivity is computed from the parameters: the power, Q, developed in the inner cylinder by a specified heater; the temperature difference, ΔT, between the two cylinders; and a geometrical factor relating Q and ΔT through the geometry of the heat flow pattern. The working equation for this method is

$$\lambda = \frac{Q \ln(r_1/r_2)}{2\pi l \Delta T}, \tag{3.2.5}$$

where r_1 is the internal radius of the outer cylinder, r_2 is the external radius of the inner cylinder, and l is the length of the cylinders.

This technique has the advantage of a simple mathematical model. However, difficulties arise at the experimental level because of heat losses at the top and the bottom of the instrument. Therefore two guard heaters (G) or thermal insulators are often used as shown in Fig. 3.2.1 (c). As the cylinders must be of finite length, heat losses by convection and by radiation are the other problems to be taken into account. Besides the problem of the determination of the exact amount of heat that flows through the specimen, there are other sources of error: heat leakage in the leads outside the potential taps, the position of the potential taps, ΔT and T measurements, and the determination of the geometrical factor of the apparatus. A cross-sectional view of the concentric cylinders apparatus used by Tanaka et al. [5] is shown in Fig. 3.2.3. The apparatus consists of three concentric copper cylinders fixed vertically in a brass high-pressure vessel. The sample fluid is introduced into the annular gap of about 0.5 mm between the emitter (H) and the receiver (C). To minimize axial conduction, thermal insulators are provided on both ends of the emitter. This apparatus is not equipped with guard heaters, which prevent axial heat losses. Therefore the cell was calibrated using the most probable values of the thermal conductivity of reference fluids such as Ar, Ne and N_2. A similar concentric cylinders apparatus was also used by Tanaka et al. [6] for measurements of the thermal conductivity of environmentally acceptable refrigerants.

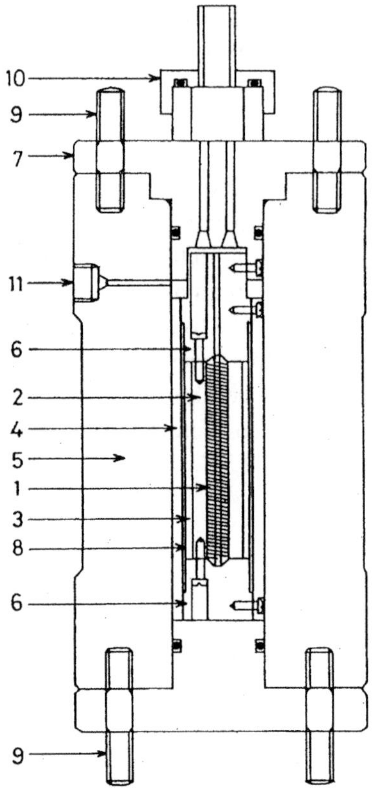

Fig. 3.2.3 Cross-sectional view of the concentric cylinders apparatus used by Tanaka et al. [5] 1: heater ; 2: inner cylinder, copper; 3: emitter, copper (O.D.= 25 mm, L= 72 mm); 4: receiver, copper (O.D.= 34 mm, I.D.= 26 mm, L= 120 mm); 5: high-pressure vessel, brass; 6: insulator, phenol resin; 7: flange, brass; 8: annular gap; 9: bolt, brass; 10: opening for leads; 11: sample inlet

(c) Parallel-Plates Method

As shown in Fig. 3.2.1 (d), the mathematical model of this apparatus consists of two infinite parallel plates at different temperatures; heat flows from the top hot plate (H) at a higher temperature to the bottom cold plate (C). The sample fluid is located in the small gap (X) between these parallel plates. The working equation is

$$\lambda = \frac{Qd}{A\Delta T}, \tag{3.2.6}$$

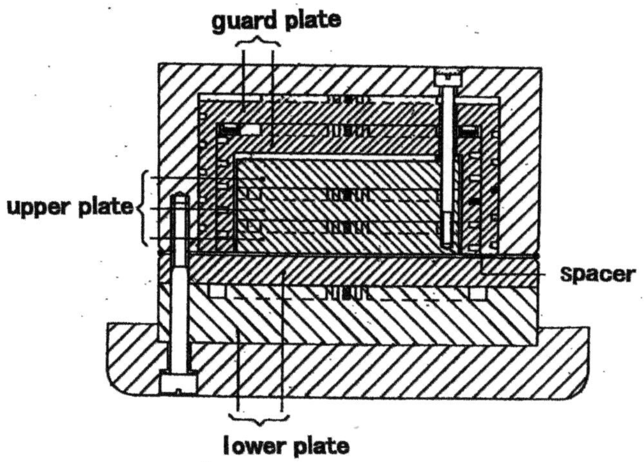

Fig. 3.2.4 Cross-sectional view of the parallel plates apparatus used by Mostert et al. [7]

where Q is the power developed in the upper plate, A is the effective area of the upper plate, d is the width of the gap, and ΔT is the temperature difference across the fluid layer. The cross-sectional view of the parallel-plates apparatus used by Mostert et al. [7, 8] is illustrated schematically in Fig. 3.2.4. The sample fluid is located in a small gap between the lower and upper plate, the latter being surrounded by a guard plate and guard ring. A known amount of heat is produced in the upper plate and transported through the sample fluid to the lower plate. To prevent net heat losses to the environment, the guard plates and guard rings are kept continuously at the same temperature as the upper plate by supplying a regulated amount of heat to the guard plates and rings as well. The height of the cell is 47 mm and its diameter is 67 mm. The lower and upper plates and the guard plates and rings are made of electrolytic copper in view of the high thermal conductivity of this material.

The experimental arrangement of this type of apparatus presents several problems, namely, temperature uniformity of the plate, temperature stability during the time interval needed to establish steady-state conditions, and the inconvenience of pretending to work with steady-state apparatus. However, the principal advantage of this kind of apparatus is the possibility of realizing measurements near the critical point of fluids.

It is widely accepted that the guarded parallel-plate method is the most accurate method for measuring the thermal conductivity of fluids close to the critical point. Two other methods mentioned above are less suitable in the critical region, since convection will develop more easily in those configurations. Moreover, the transient temperature rise in hot-wire instruments or the temperature difference across the gap of a concentric cylinders apparatus is of the order of 1 K, which is much too large to be useful very close to the critical point. Recently, a new relative method of thermal conductivity measurement for fluids has been proposed by Zaihua et al. [9] using a thermoelectric module consisting of a BiTe semiconductor and the Peltier effect. It was found that this method is very convenient and effective to measure the thermal conductivity of fluids near the critical point and in the supercritical state.

Direct (or thermodynamic) measurements of the thermal conductivity near the critical point of fluids are rare, and generally do not approach the critical point on near-critical isochores closer than $\Delta T_r = (T - T_C)/T_C = 0.01$. With light-scattering techniques [10], on the other hand, measurements of the thermal diffusivity can be obtained down to $\Delta T_r = 10^{-5}$. Thus the thermal diffusivity can be an excellent complement to thermal conductivity data near the critical point, when the values of the density and the specific heat capacity at constant pressure are available. However, this method cannot be applied far from the critical point because of the low intensity of the scattered light there. Consequently, the two techniques almost always have little or no overlap in temperature.

3.2.3 General Behavior of Thermal Conductivity

The thermal conductivity–temperature–pressure surface of nitrogen generated by Stephan et al. [11] is shown in Fig. 3.2.5. The behavior of the viscosity (see Sect. 3.1) and the thermal conductivity of a fluid with respect to temperature and pressure are quite similar qualitatively. Figure 3.2.5 can be considered to represent the general behavior of the thermal conductivity for many fluids because of the law of corresponding states. The corresponding states principle

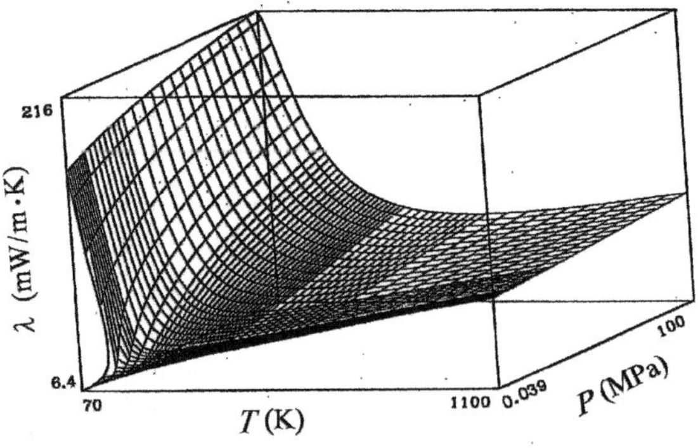

Fig. 3.2.5 Thermal conductivity–temperature–pressure surface of N_2 determined by Stephan et al. [11]

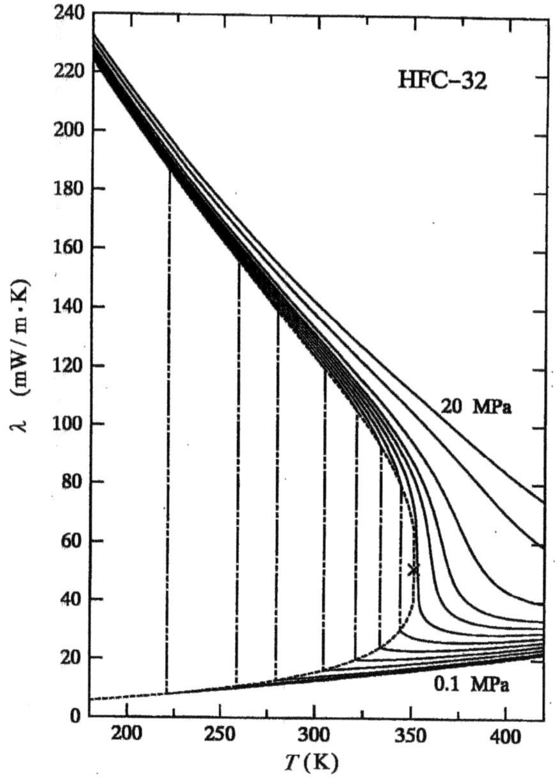

Fig. 3.2.6 Thermal conductivity of HFC-32 (CH_2F_2) as functions of temperature and pressure determined by Yata [12]

holds true for the thermal conductivity of fluids by reducing λ with λ_C, the thermal conductivity at the critical point, or with λ_{0C}, the dilute gas thermal conductivity at the critical temperature T_C.

The thermal conductivity of most gases at standard temperature and pressure ranges from 5 to 30 mW/m·K except for H_2, He and Ne. For these light gases, the thermal conductivities range from 50 (Ne) to 180 (H_2) mW/m·K. According to molecular kinetic theory, the thermal conductivity of dilute gases increases with increasing temperature approximately in proportion to $T^{1/2}$. The pressure effect on λ of dilute gases is comparatively small in the pressure range between 0.01 and 0.5 MPa. At a constant temperature below T_C, the thermal conductivity generally increases with increasing pressure in the pressure range above 0.5 MPa. At the saturation pressure, the thermal conductivity increases remarkably and discontinuously due to liquefaction as shown in Fig. 3.2.6 for HFC-32 (CH_2F_2) [12]. Although a negative pressure effect is sometimes reported

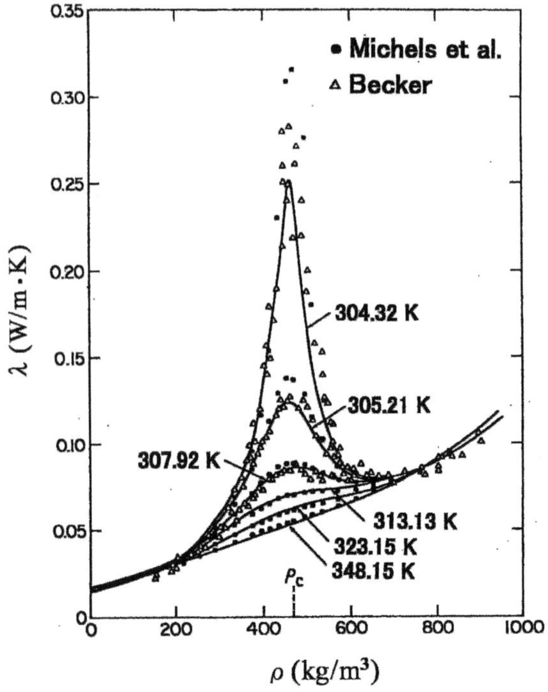

Fig. 3.2.7 Thermal conductivity of CO_2 in the critical region as a function of density along isotherms

for the viscosity of dilute gases in a low-temperature region, that is, $(\partial\eta/\partial P)_T <$ 0, the thermal conductivity has always been observed and reported to increase with pressure, $(\partial\lambda/\partial P)_T > 0$. The temperature coefficients of gaseous thermal conductivity, $(\partial\lambda/\partial T)_P$, are positive at low pressures. However, these values decrease gradually with increasing pressure and become negative at higher pressures. The temperature coefficients of liquid thermal conductivity are negative. Thus, the thermal conductivity isotherms intersect mutually in a moderate pressure region when plotted in λ versus P coordinates. On the other hand, the thermal conductivity isotherms increase regularly with increasing density without intersecting mutually in λ versus ρ coordinates except in the critical region. As shown in Fig. 3.2.7 for CO_2, the λ isotherms slightly above T_C show an abnormal enhancement in the vicinity of ρ_C. The critical enhancement of thermal conductivity reaches 200–300% while that of viscosity reaches about 15%.

As seen in Fig. 3.2.6, the thermal conductivity of the liquid is approximately 10–100 times larger than that of the gas state. The liquid thermal conductivity decreases linearly with increasing temperature, while the gaseous thermal conductivity increases with increasing temperature. The behavior of both η and λ in the gas phase and the liquid phase are quite different, which may be due to the difference of the transport mechanisms of momentum and heat energy in the gas and liquid phases. In dilute gases, transport of these physical quantities is made mainly through molecular motion and molecular collisions. On the other hand, in the liquid phase transport is made primarily through molecular interaction. Thus, the transport properties of liquids are affected to a great extent by molecular structure and shape, polarity, aggregation state of molecules, and the like. The liquid phase viscosity at standard conditions ranges from 1 to 10^7 times that of a simple molecular liquid, depending on the nature of the substance and molecular structure. The magnitude of the liquid phase thermal conductivity is much narrower and ranges from 1 to 10 times that of simple molecular liquids for most substances.

The effect of pressure on η and λ is also quite different. The liquid viscosity increases with pressure with a positive curvature ($[\partial^2\eta/\partial P^2]_T > 0$). On the other hand, the effect of pressure on the liquid thermal conductivity is not so large. The isotherm increases with pressure with a negative curvature ($[\partial^2\lambda/\partial P^2]_T < 0$). The pressure coefficient of λ, $(\partial\lambda/\partial P)_T$, becomes smaller at higher pressures.

3.2.4 Correlation

In this section, we will discuss how the most probable thermal conductivity surface is determined in the case when a great amount of experimental data exists for a given substance, and we will also expound on the method of data analysis, evaluation, and correlation.

(a) Evaluation of Thermal Conductivity Data

Selection of the primary and secondary data is an important step in data evaluation and correlation. Primary data have a high degree of confidence and are used in the development of a correlation. Secondary data, on the other hand, are not used in correlation but are rather used in comparisons with the most probable values determined. The details of the data evaluation criteria are described in refs. [13, 14]. It is convenient if the analysis, comparison and evaluation of data are performed for whole physical states such as: (a) dilute gases, (b) saturated vapor, (c) compressed gases, (d) saturated liquids, (e) liquids under pressure, and the like [14]. In order to compare data among investigators, a suitable empirical equation which acts as a basis is employed and in addition to the magnitude and scatter of the thermal conductivity data, the dependences of λ on factors such as temperature, pressure and density are also critically compared.

(b) Standard Correlation Procedure for Thermal Conductivity

It is generally accepted that the thermal conductivity should be correlated in terms of temperature and density rather than temperature and pressure, because over a wide range of experimental conditions the behavior of thermal conductivity is determined by its density dependence. In view of the theory of molecular motion, and for practical purposes, the dependence of thermal conductivity on temperature and density, $\lambda(\rho,T)$, is normally expressed as a sum of three independent terms:

$$\lambda(\rho,T)=\lambda_0(T)+\Delta\lambda(\rho,T)+\Delta_C\lambda(\rho,T). \tag{3.2.7}$$

The first term $\lambda_0(T)=\lambda(0,T)$ is the dilute gas term at temperature T as the density ρ approaches the limit zero. To obtain $\lambda_0(T)$ from the experiment we must extrapolate the measurements at low densities to zero density, usually with a low-order polynomial or graphically. This value is for a dilute gas where only two-body molecular interactions occur. $\Delta_C\lambda(\rho,T)$ is the critical enhancement and arises from the clustering and density fluctuations that occur in a fluid near the critical point. As a fluid approaches its critical point, the thermal conductivity increases when compared to a fluid at a temperature T and density ρ far from the critical point as shown in Figs. 3.2.8 and 3.2.9. $\Delta\lambda(\rho,T)$ is the excess (or residual) thermal conductivity and represents all the other contributions besides $\lambda_0(T)$ and $\Delta_C\lambda(\rho,T)$ to the thermal conduc-

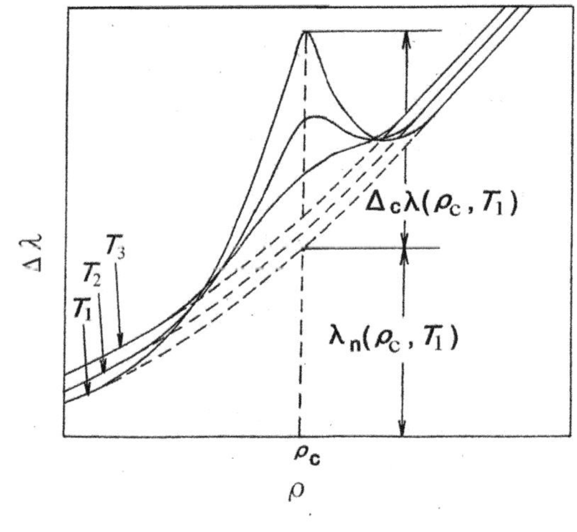

Fig. 3.2.8 Schematic diagram of the critical enhancement of thermal conductivity in the critical region as a function of density at several temperatures. $T_C < T_1 < T_2 < T_3$

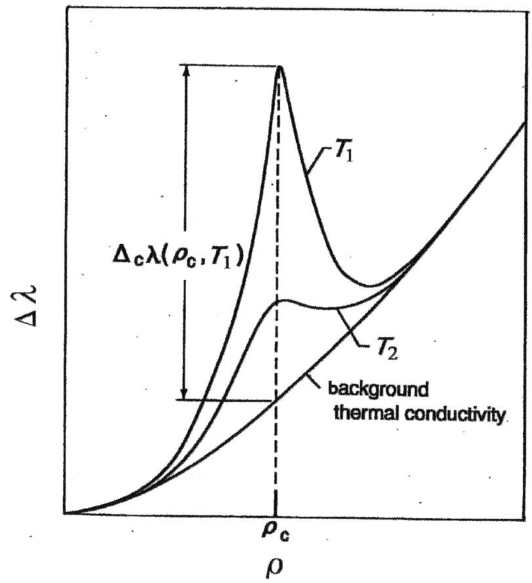

Fig. 3.2.9 Schematic diagram of the residual (or excess) thermal conductivity in the critical region as a function of density at several temperatures. $T_C < T_1 < T_2$

tivity of the fluid at high densities, and these include many-body collisions, molecular-velocity correlations, and collisional transfer, among others. The first two terms in (3.2.7) taken together are sometimes called the background thermal conductivity, $\bar{\lambda}(\rho, T)$.

$$\bar{\lambda}(\rho, T) = \lambda_0(T) + \Delta\lambda(\rho, T). \tag{3.2.8}$$

It should be noted that this correlation method requires an equation of state to translate measured pressures into equivalent densities.

The expression of the thermal conductivity consisting of the three independent contributions is useful, enabling us to treat both $\lambda_0(T)$ and $\Delta_C\lambda(\rho, T)$ independently and theoretically. Experimental data for $\lambda_0(T)$ are usually abundant and it is possible to accurately show $\lambda_0(T)$ based on both theory and experimental values. On the other hand, there is no reliable theoretical guidance concerning the excess property $\Delta\lambda(\rho, T)$, so there is no other choice but to perform a correlation based entirely on experimental data. Therefore, in order to perform a correlation based on (3.2.7), we should first critically determine $\lambda_0(T)$ from both the theoretical and experimental sides, and then use this to get the background thermal conductivity $\lambda(\rho, T)$ with experimental data far from the critical region wherein the contribution of the critical enhancement $\Delta_C\lambda(\rho,T)$ is negligible. Then, by subtracting $\lambda(\rho,T)$ from all the experimental data, we get the first-iterate values of $\Delta_C\lambda(\rho,T)$, and using this we can get the first-iterate values for the parameter within $\Delta_C\lambda(\rho,T)$. Using these values, a more accurate estimate of $\lambda(\rho,T)$ is performed and the second-iterate value for $\Delta_C\lambda(\rho,T)$ is derived. By repeating this process until convergence, the complete final form of $\lambda(\rho,T)$ can be achieved.

(c) Thermal Conductivity of Dilute Gases

According to the theory of molecular motion, thermal conductivity is ascribed not only to a particle's translational motion, but also to contributions from the internal energy due to rotational and vibrational motions. Therefore λ_0 is related to the various collision cross-sections, which are in turn related to the relaxation and diffusion of the translational and internal energies.

According to the Chapman–Enskog theory, λ_0 can be expressed as follows:

$$\lambda_0(T) = \frac{0.177568(T/M)^{1/2} C_P^0/R}{\sigma^2 \Omega_T(T^*)}, \tag{3.2.9}$$

where $\Omega_T(T^*)$ is the reduced effective collision cross-section for thermal conductivity, C_P^0 is the isobaric heat capacity for ideal gases, R is the gas constant and λ_0 is the thermal conductivity in mW/m·K. T^* is the reference temperature defined by the following equation:

$$T^* = \frac{k_B T}{\varepsilon}, \tag{3.2.10}$$

where ε and σ are the energy and length scaling parameters, respectively. Preselected primary data for λ_0 having a high degree of confidence are used to determine $\Omega_T(T^*)$ in the following equation:

$$\Omega_T(T^*) = \sum_{i=0}^{n} b_i/(T^*)^i. \tag{3.2.11}$$

The ε and σ values needed for correlation are determined from analysis of viscosity data η_0 while b_i values are determined using experimental λ_0 data.

For more practical correlations simple polynomial equations are sufficient to represent the experimental values of λ_0 as a function of temperature:

$$\lambda_0(T) = \sum_{i=0}^{n} A_i T^i . \tag{3.2.12}$$

(d) Critical Enhancement of Thermal Conductivity

The thermal conductivity of fluids exhibits a strong enhancement in the critical region as originally demonstrated by Michels, Sengers and van der Gulik [15, 16] for CO_2 as shown in Fig. 3.2.7. There are two separate critical regions where the critical enhancement is important. The first region is the "near-critical region" or "critical region proper" where a nonanalytical equation of state, namely, a scaled fundamental equation, must be used. According to Hanley et al. [17] and Sengers [18], this region is defined in terms of the reduced coordinates

$$\Delta T_\mathrm{r} = (T - T_\mathrm{C})/T_\mathrm{C} \text{ , } \Delta\rho_\mathrm{r} = (\rho - \rho_\mathrm{C})/\rho_\mathrm{C} \text{ ,} \tag{3.2.13}$$

as

$$|\Delta T_\mathrm{r}| \le 0.03 \text{ and } |\Delta\rho_\mathrm{r}| \le 0.25 . \tag{3.2.14}$$

This is a very limited zone in temperature and pressure, where the density is changing very rapidly. It is in the near-critical region that the thermal conductivity increases ten times or more than the dilute gas values. The second region designated the "extended critical region" covers the zone in densities and temperatures for which the experimental measurements reveal an anomalous increase in the thermal conductivity. This region extends to considerably high temperatures up to $T/T_\mathrm{C} = 2$. In this region an analytical equation of state can be used. Most of the applications in supercritical fluid technology are implemented in the extended critical region.

The thermal conductivity enhancement in the critical region $\Delta_\mathrm{C}\lambda(\rho,T)$ can be expressed using the critical enhancement $\Delta_\mathrm{C} D_\mathrm{T}$ of the thermal diffusivity $D_\mathrm{T} = \lambda / \rho C_P$ as follows [13, 14, 17, 18]:

$$\Delta_\mathrm{C} D_\mathrm{T} = \frac{\Delta_\mathrm{C}\lambda}{\rho C_P} , \tag{3.2.15}$$

where C_P is the specific heat at constant pressure. The theory of dynamic critical phenomena predicts asymptotic behavior for transport properties in the critical region, and this can be expressed by the following equations [13, 17, 18]:

$$\Delta_C D_T = \frac{\Delta_C \lambda}{\rho C_P} = \frac{R k_B T}{6\pi\eta\xi}, \tag{3.2.16}$$

$$\eta = \Delta_C \eta + \bar{\eta} = \bar{\eta}(Q\xi)^z, \tag{3.2.17}$$

where η is the viscosity, k_B is Boltzmann's constant, ξ is the correlation length of the density fluctuations in the critical region and Q is an amplitude dependent on the kind of system. R and z are universal quantities and as such have almost the same values for all fluids as shown below.

$$R = 1.01, \ z = 0.06. \tag{3.2.18}$$

To accurately express the critical enhancements of the transport properties, crossover equations are needed, which are equations incorporating the asymptotic behavior in the critical region and the background behavior of $\Delta\lambda(\rho,T)$ far away from the critical point. Vesovic et al. [13] and Olchowy and Sengers [19] have proposed such crossover equations and they are as follows:

$$\Delta_C D_T = \frac{\Delta_C \lambda}{\rho C_P} = \frac{R k_B T}{6\pi\eta\xi}(\Omega - \Omega_0), \tag{3.2.19}$$

$$\eta = \bar{\eta}\exp(zH). \tag{3.2.20}$$

Here Ω, Ω_0 and H are functions of density and temperature. They are quite complex however and the theoretical background is too long to be given here. We encourage interested readers to consult the related literature [13, 17–19] for an in-depth discussion. As the critical point is approached, clustering occurs and since $\xi \to \infty$, we can get the asymptotic behavior predicted by (3.2.15) and (3.2.16) from (3.2.18) and (3.2.19). Far from the critical point, $\xi \to 0$, and as a consequence $\Omega = \Omega_0$ and $H \to 0$. Because of this, $\Delta_C D_T$ vanishes and the value of η approaches $\bar{\eta}$, that is $\eta \to \bar{\eta}$. The underlying cause of the infinite singularity in the thermal conductivity at the critical point lies in the divergence of the length scale over which molecular motions are correlated.

In the case when sound experimental data for $\Delta_C \lambda(\rho,T)$ exist, it is more useful to provide empirical representations for industrial use. Roder and coworkers [20–22] used the following equations for the representation of the thermal conductivity enhancement of methane, ethane, and their binary mixtures. To represent the critical enhancement, a Gaussian function was used:

$$\Delta_C\lambda(\rho,T)=\Delta_C\lambda^0\exp\left(-x^2\right). \tag{3.2.21}$$

The maximum value of $\Delta_C\lambda$ at $\rho=\rho_C$ is the amplitude $\Delta_C\lambda^0$ of the function for the particular temperature under consideration. The amplitude is chosen to be a simple function of temperature, as follows

$$\Delta_C\lambda^0=\frac{C_1}{T+C_2}+C_3+C_4T. \tag{3.2.22}$$

The variable x is a simple function of density chosen to be zero at the maximum value of the critical enhancement, that is, at the density ρ_C:

$$x=C\left(\rho-\rho_C\right). \tag{3.2.23}$$

If necessary, the small asymmetry of the critical enhancement is achieved by providing different coefficients in (3.2.23) above and below the critical density:

$$x=C_5\left(\rho-\rho_C\right)\qquad \text{or}\qquad \rho<\rho_C, \tag{3.2.24}$$

$$x=C_6\left(\rho-\rho_C\right)\qquad \text{for}\qquad \rho>\rho_C. \tag{3.2.25}$$

Laesecke et al. [23] used the following empirical equation for the critical enhancement of the thermal conductivity of HCFC-123 ($CHCl_2CF_3$):

$$\Delta_C\lambda(\delta,\tau)=a_1\exp[a_2(\tau-1)^m+a_3(\delta-1)^n], \tag{3.2.26}$$

where a_1, m and n are constants determined from experimental data and where m and n are integers. δ and τ are defined as follows:

$$\delta=\frac{\rho}{\rho_C},\ \tau=\frac{T_C}{T}. \tag{3.2.27}$$

(e) Excess (or Residual) Term

The excess thermal conductivity translates the increase of thermal conductivity over its zero density limit, due to the increase in density, free from critical enhancement contributions.

As described previously in Sect. 3.1, the viscosity of a gas at low pressure (low density) is greatly affected by changes in temperature and pressure (density). There exists a certain temperature boundary in which at temperatures above this boundary viscosity increases with increasing pressure (density). At temperatures below this boundary, the viscosity slightly decreases as the pressure (density) increases.

In contrast to the case of viscosity, there is no known temperature region where the initial density effect on thermal conductivity is negative. However, the critical enhancement of thermal conductivity is quite large and so we should be very careful when separating $\Delta\lambda$ from $\Delta_C\lambda$.

The term $\Delta\lambda(\rho,T)$ shows the thermal conductivity as a function of ρ and T in the region adequately distant from the critical point and where there are no effects due to critical enhancement. $\Delta\lambda(\rho,T)$ differs significantly from the previously stated $\lambda_0(T)$ and $\Delta_C\lambda(\rho,T)$ because up to now there is no clear-cut theory explaining its behavior. Therefore, empirical fitting based on experimental data is the fundamental basis for $\Delta\lambda$. There has been much discussion about the dependence of the excess thermal conductivity on temperature. However, most of the previous attempts have been temperature-dependent excess functions $\Delta\lambda$.

$$\Delta_C\lambda(\rho,T)=\lambda(\rho,T)-\lambda_0(T)-\Delta\lambda(\rho,T), \tag{3.2.28}$$

$$\bar{\lambda}(\rho,T)=\lambda_0(T)+\Delta\lambda(\rho,T). \tag{3.2.29}$$

To do this, we must first completely separate $\Delta\lambda$ from λ_0 and $\Delta_C\lambda$. Separating λ_0 from the other contributions is relatively simple as it can be determined from experimental data. In contrast, separating the critical enhancement $\Delta_C\lambda$ is not an easy task. The performance of a process of repetition based on the trial and error method is necessary. By performing these equations and determining the excess property data, an evaluation of the experimental measurement data and selection of the primary data is done, and based on these a correlation with ρ and T can be performed. Generally, a polynomial function of ρ is used for $\Delta\lambda$ correlations as follows [13]:

$$\Delta\lambda(\rho,T)=\sum_{i=1}^{n}C_i(T)\rho^i, \tag{3.2.30}$$

where $C_i(T)$ is a function of temperature.

Roder and coworkers [20–22] used the following excess functions:

$$\Delta\lambda(\rho,T)=\alpha(T)\rho+\beta(T)\rho^{n(T)}, \tag{3.2.31}$$

where the parameters α , β , and n are functions of temperature:

$$\alpha(T)=\alpha_1+\alpha_2 T , \tag{3.2.32}$$

$$\ln\beta(T)=\beta_1+\beta_2 T+\beta_3 T^2+\beta_4 T^3 , \tag{3.2.33}$$

$$n(T)=n_1+n_2 T+n_3 T^2 . \tag{3.2.34}$$

In recent years, the transient hot-wire method has become the mainstream method of thermal conductivity measurement and because of this it would seem that measurement of thermal conductivity would be much easier than viscosity measurements. However, the data for gases at moderate densities measured by various investigators does not coincide well. It has been shown that a deviation of up to ± 20% exists, showing the difficulty involved in the selection of primary data. In the dense fluid region sufficiently removed from the critical point or the compressed liquid region, the measurement becomes relatively easier and all the reported data have been shown to coincide within ± 5%. In these regions, plotting $\Delta\lambda(\rho,T)$ versus ρ will greatly help in the evaluation of experimental data. Generally speaking, however, the data measured and reported by various investigators do not coincide well and in most cases there is no other choice but to approximate $\Delta\lambda$ as a function of density only, independent of temperature:

$$\Delta\lambda(\rho,T)\approx\Delta\lambda(\rho), \tag{3.2.35}$$

$$\Delta\lambda(\rho)=\sum_{i=1}^{n}a_i\rho^i . \tag{3.2.36}$$

3.2.5 Critical Anomaly of Thermal Conductivity

In 1962, Michels, Sengers and van der Gulik [15, 16] measured the thermal conductivity of CO_2 in the critical region with a parallel-plate apparatus. Their measurements originally established the existence of a pronounced critical enhancement in the thermal conductivity. As mentioned in Sect. 3.2.2, it is generally accepted that the guarded parallel-plate method is the most accurate method for measuring the thermal conductivity of fluids very close to the critical point. The transient hot-wire method and the concentric cylinders method are less suitable for measurements in the critical region, since convection will develop more easily in their vertically long configurations. In addition, the transient temperature rise in the hot-wire method or the temperature difference across the annular gap of the concentric cylinders apparatus is around or over 1 K, which is much too large to be useful very close to the critical point. In view of the difficulties encountered in the critical region it is not surprising to find that reliable direct measurements of the critical enhancement of λ at or near critical densities in reduced temperatures close to T_C are quite scarce in the literature. Thus, this section presents a brief review of the situation of thermal conductivity measurements both on the critical enhancement behavior of supercritical fluids and on the environmentally acceptable refrigerants in dilute and moderately dense vapors.

Mostert et al. [7, 8] measured the thermal conductivity of ethane and (ethane + CO_2) mixtures in the critical region with a parallel-plate apparatus originally developed by Michels and coworkers [15, 16] and further improved by Mostert et al. [7]. Any horizontal temperature inhomogeneity or any deviation of the fluid layer from a strictly horizontal position will cause convection. Convective heat transfer increases with increasing values of the Rayleigh number, which becomes extremely large in the critical region. To reduce the Rayleigh number a small gap distance $d = (155 \pm 0.5)$ μm was used, while the temperature difference ΔT varied from 100 mK far away from the critical point to values as small as 2–3 mK very close to the critical point. The absence of any measurable contribution from convective heat transfer was checked experimentally by repeating measurements at the same average density and temperature with different values of ΔT. The thermal conductivity of ethane was measured as a function of temperature along eleven isochores to obtain detailed information on the critical enhancement of the thermal conductivity. The experimental results were found to be consistent with the theoretical equation proposed by Olchowy and Sengers [19].

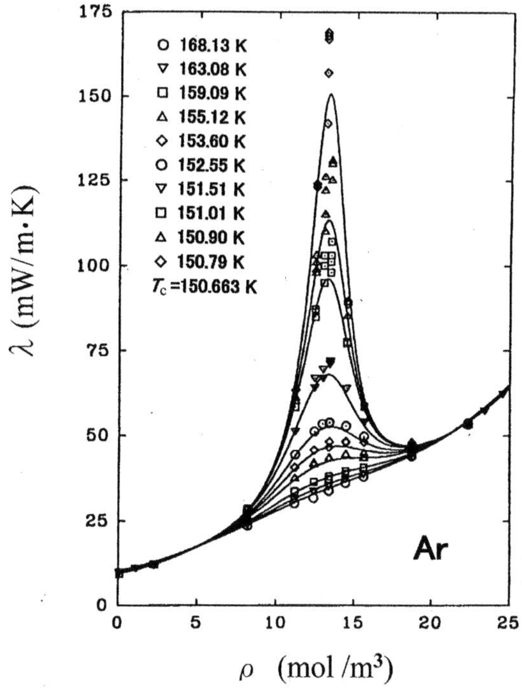

Fig. 3.2.10 Thermal conductivity of Ar in the critical region as a function of density along isotherms measured by Tiesinga et al. [24]

Tiesinga et al. [24] measured the thermal conductivity of Ar at temperatures from 302 down to 150.8 K and at densities up to 25 mol/*l* by the same apparatus. The experimental data are plotted in Fig. 3.2.10 as a function of density along ten isotherms.

The transient hot-wire method for measuring the thermal conductivity of gases was first used in its modern form by Haarman [2] and developed by Kestin and his coworkers.

Clifford et al. [25] measured the thermal conductivity of Ar, N_2 and H_2 using this method between 300 and 400 K and at pressures up to 25 MPa. Scott et al. [26] tried to measure the thermal conductivity of CO_2 in the supercritical region with a transient hot-wire instrument. The use of apparatus of this type near the critical point was discussed. The wires used were of 7 μm diameter and 93 and 164 mm in length. Temperature rises of 6–8 K occurred during the experiments. The experimental results were compared with the expression for the critical enhancement in λ given by Hanley, Sengers and Ely [17]. According to the data analysis by Vesovic et al. [13], excellent agreement was obtained between the experimental data of Scott et al. [26] and the most probable values determined by Vesovic et al. [13] at 45 K above T_C as well as at lower densities at all temperatures. However, closer to the critical point, the transient hot-wire apparatus yielded an apparent critical enhancement of λ which became too large, presumably because convection became significant.

Johns et al. [27, 28] also measured the thermal conductivity of Ar, N_2, CO_2, and mixtures of N_2 and CO_2 at elevated temperatures from 302 to 470 K and pressures up to 30 MPa. The experimental results for (N_2 + CO_2) mixtures were used to test several prediction methods for the thermal conductivity of gas mixtures under pressure. Good agreement was found with the prediction using the corresponding-states method of Ely and Hanley [29].

Roder and coworkers have measured the thermal conductivity of methane [20], ethane [21], and (methane + ethane) mixtures [22] at elevated temperatures and pressures with a transient hot-wire apparatus after measuring N_2, He, Ar, O_2, H_2 and propane.

Yamamoto et al. [4] measured the gaseous thermal conductivity of CFC alternative refrigerants HFC-134a (CF_3CH_2F), HCFC-123 (CF_3CHCl_2), and HCFC-141b (CH_3CCl_2F) in the temperature range 273–373 K at pressures up to saturation using a new improved transient hot-wire apparatus. The uncertainty of the experimental data was estimated to be within 1%. Tanaka et al. [30] also measured the gaseous thermal conductivity of HFC-32 (CH_2F_2), HFC-125 (CF_3CHF_2), and their binary mixtures with the same apparatus in the temperature range 283–333 K at pressures up to saturation. The thermal conductivity as a function of composition of the mixtures at constant pressure and temperature was found to have a small maximum near 0.3–0.4 mole fraction of HFC-32 as shown in Fig. 3.2.11. The composition dependence of the thermal conductivity at a constant temperature was correlated with the aid of the Wassiljewa equation.

Although a number of hydrofluorocarbons (HFCs) and hydrochlorofluorocarbons (HCFCs) have been developed as possible substitutes for the restricted chlorofluorocarbons (CFCs), they have not always satisfied all the environmental or industrial requirements. In the case of blowing agents for polyurethane foams, some HFCs are expected to be promising alternatives for CFCs and HCFCs but their global warming potential (GWP) are often beyond permissible limits. Therefore, the development of a new type of alternative compounds in which chlorine atoms are replaced by those with low toxicity and low flammability such as oxygen, sulfur or silicon should be undertaken. Matsuo et al. [31] have measured the gaseous thermal conductivity of the following new HFCs and fluorinated ethers (HFEs) which are expected to be

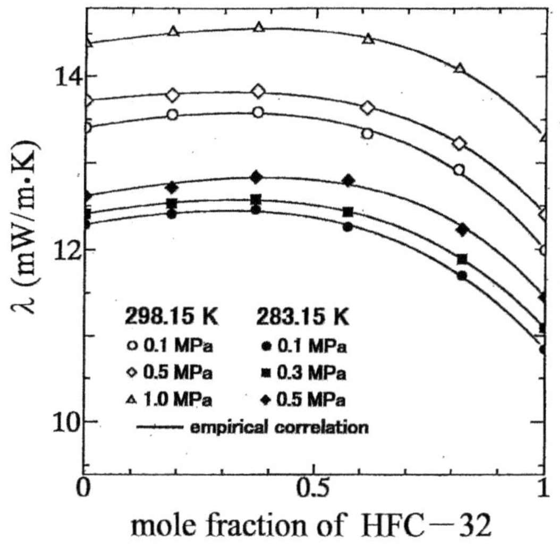

Fig. 3.2.11 Composition dependences of the thermal conductivity of gaseous (HFC-32 + HFC-125) mixture at 283.15 and 298.15 K at various pressures [30]

promising replacements for traditional CFC-11 (CCl_3F) and temporary HCFC-141b blowing agents by the transient hot-wire method:

HFC-236ea	($CHF_2-CHF-CF_3$),
HFC-245ca	($CH_2F-CF_2-CHF_2$),
HFC-245fa	($CHF_2-CH_2-CF_3$),
HFC-365mfc	($CF_3-CH_2-CF_2-CH_3$),
HFE-356mmz	($(CF_3)_2CH-O-CH_3$),
HFE-347pc-f	($CHF_2CF_2-O-CH_2CF_3$),
HFE-347mcc	($CF_3CF_2CF_2-O-CH_3$),
HFE-329mc-c	($CF_3CF_2-O-CF_2CHF_2$),
HFE-254pc	($CHF_2CF_2-O-CH_3$),
HFE-236me	($CHF_2-O-CHFCF_3$),
HFE-227me	($CF_3CHF-O-CF_3$),
HFE-245mc	($CF_3CF_2-O-CH_3$),
HFE-134p	($CHF_2-O-CHF_2$).

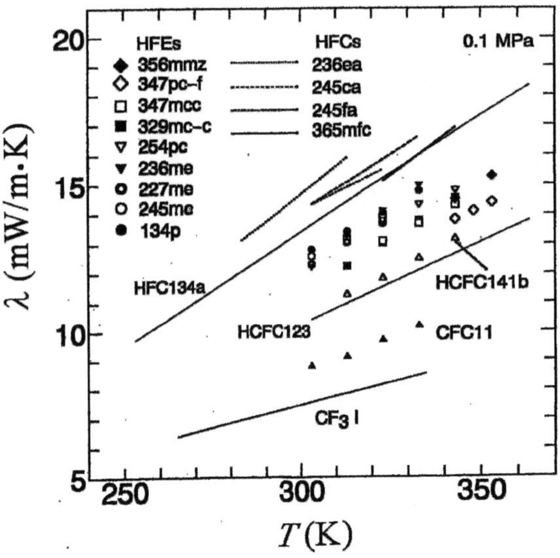

Fig. 3.2.12 Comparison of gaseous thermal conductivity of fluorinated ethers with several fluorocarbon refrigerants at atmospheric pressure measured by Matsuo et al. [31]

Figure 3.2.12 shows the comparison of thermal conductivities of these fluorinated ethers with some CFCs, HCFCs, and HFCs [31].

Tufeu and coworkers [32–34] have measured the critical thermal conductivity enhancement of ethane, ethylene, propane, *n*-butane, isobutane, *n*-pentane, CO_2, H_2O, D_2O, NH_3, and SF_6 along the critical isochore by the steady state concentric-cylinders apparatus made of silver and set up in a vertical position. The annular gap between the cylinders could be varied from 0.2 to 0.4 mm. The convective heat flow was estimated from the Rayleigh number. Near the critical point, the smallest gap, 0.2 mm, was used. The temperature of the sample was measured with a 0.02 K uncertainty and the pressure with a 0.05% uncertainty. The experimental data were analyzed based on the corresponding-state principle. A predictive method to calculate $\Delta_C\lambda$ along the critical isochore was developed employing the critical constants, the slope of the critical isochore, and the viscosity along the critical isochore. The proposed equation can be used to estimate $\Delta_C\lambda$ along the critical isochore within 12%.

Tanaka and coworkers measured the gaseous thermal conductivity of methane, SF_6 [5], CFC-12 (CCl_2F_2), CFC-13 ($CClF_3$), HCFC-22 ($CHClF_2$), HFC-23 (CHF_3) [35], four binary mixtures of HCFC-22 with CFC-12, CFC-13, HFC-23, and CFC-114 ($CClF_2CClF_2$) [36], HFC-134a (CH_2FCF_3),

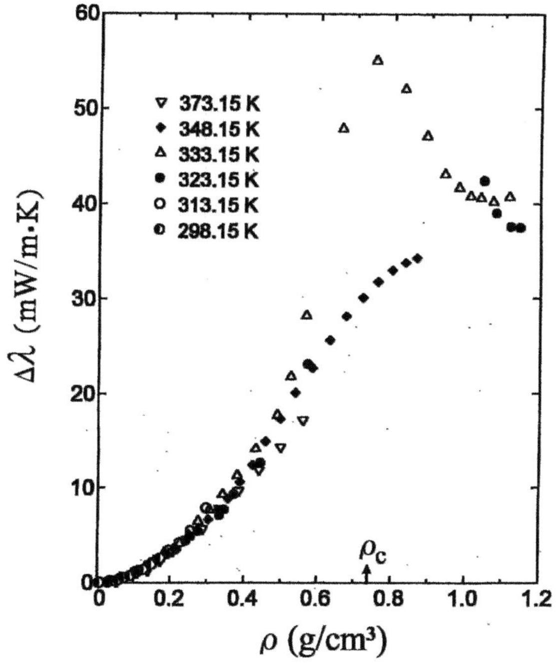

Fig. 3.2.13 Residual (or excess) thermal conductivity of SF_6 in the critical region as a function of density measured by Tanaka et al. [5]

HFC-143a (CH_3CF_3), HCFC-141b, and HCFC-142b (CH_3CClF_2) [6] with a steady-state coaxial cylinders apparatus. The measurements were carried out on a relative basis with the reference thermal conductivity values of He, Ne, Ar, Kr, Ne, and methane and SF_6. The uncertainty of the experimental data obtained was estimated to be within 2% except for the uncertainty associated with the reference thermal conductivity values. The density dependence of the thermal conductivity of SF_6 is illustrated in Fig. 3.2.13 [5].

Takada et al. [37] have also undertaken thermal conductivity measurements for fluorinated ethers HFEs mentioned above by a new steady-state coaxial cylinders apparatus, and their results are in excellent agreement with those of Matsuo et al. [31]. It is found that the fluorinated ethers investigated have comparatively lower λ values among the alternative substances in the following order:

$$\text{CFCs} < \text{HCFCs} < \text{HFEs} < \text{HFCs}.$$

Lee [68] developed a generalized estimation method for $\Delta_C\lambda$ for pure fluids. The power-law hypothesis and scaled equation of state representing the thermodynamic behavior near the critical point were adopted to analyze the experimental thermal conductivity data for 15 fluids. This predictive method required only critical constants. A comparison of calculated values with corresponding $\Delta_C\lambda$ data for 16 fluids yielded a mean deviation of 12% within the critical region bounded by $1<T_r<1.03$ and $0.7<\rho_r<1.3$.

The data sources of the most probable values of thermal conductivity of fluids correlated up to now are listed in Table 3.2.1 [11,13, 38–67].

Table 3.2.1 Data sources of the most probable values of thermal conductivity of fluids

$\lambda_0(T)$
^3He [38], ^4He [38, 39], Ne [38, 39], Ar [38–40], Kr [38–40], Xe [38– 40], nH_2 [41], N_2 [42–45], O_2 [42, 43, 45], CO [42, 44, 45], NO [42, 45], CO_2 [42, 45], N_2O [42, 45], CH_4 [42, 45–47], CF_4 [42, 45, 46], C_2H_4 [42], C_2H_6 [42], SF_6 [42, 45], Air [48]
$\lambda(\rho, T)$
Ar [49, 50], Kr [49], Xe [49], F_2 [51], N_2 [11, 49], O_2 [49, 52], D_2O [53], H_2O [54–56], CO_2 [13], Air [57], CH_4 [47], C_2H_4 [58], C_2H_6 [59, 60], C_3H_8 [61], C_6H_6 [62], $C_6H_5CH_3$ [63], CFC12 [64], CFC113 [64], CFC114 [64], RC318 [64], HCFC123 [65], HFC134a [66], HFC152a [67]

3.3 Binary Diffusion Coefficient

Measurements of binary diffusion coefficients for various organic compounds in supercritical fluids, mainly CO_2, have been made by many workers in the last three decades, as Liong et al. reviewed extensively [1]. However, the data are not sufficient enough to be covered in industrial use for reactor design because the data for the various systems between solutes and solvents are needed in wide ranges of temperature, pressure and compositions. Furthermore, the prediction methods are not well established theoretically and empirically for valuable substances. The objective compounds applied for supercritical fluid extraction or reactions are commonly those having relatively high molecular weights, and the physical properties needed for predicting the binary diffusion coefficients are not available in most cases. Moreover, most data on diffusion coefficients are limited to infinite dilution binary or pseudo-binary conditions although not only the concentration dependence but also the values in multi-component systems are actually required for industrial purposes. Such data are scarcely available because of experimental difficulties in measurements. In this section, first, the theories and experimental apparatus of the Taylor dispersion method [2–5], which is mostly employed, are described. Secondly, the data sources are listed, and the useful predictive correlations proposed are demonstrated. The Schmidt number correlation [6], probably the most practically predictive correlation for infinite dilution binary diffusion coefficients so far, is mentioned, and the accuracy of the literature data is examined. Thirdly, some data for pseudo-binary diffusion coefficients in mixture solvents, assuming a solvent consisting of multicomponents as a single component, and its correlations are discussed. Finally, studies on the anomaly in the binary diffusion coefficients in the near-critical region are described.

3.3.1 Definition of Binary Diffusion Coefficients

The phenomena of momentum, heat and mass transfer in fluids are analogous. The transferred quantities are proportional to the gradients of momentum, heat and concentration, and the phenomenological relationships are referred to as Newton's law in (3.1.1), Fourier's law in (3.2.1) and Fick's law, respectively. Similarly to momentum and heat transfer, the flux of the diffusion mass J can be given as (3.3.1) when a solute 1 diffuses into a solvent 2 in the z-direction:

$$J_1 = -D_{12}\frac{\mathrm{d}C_1}{\mathrm{d}z}, \tag{3.3.1}$$

where C is the concentration of the solute and D_{12} is the binary diffusion coefficient. Generally speaking, binary diffusion coefficients vary with composition, but often the values are assumed to be independent of concentration. This assumption can be verified when the solute concentration is at infinite dilution or is low.

Diffusion coefficients in gaseous systems at ambient or low pressures are well represented by the Chapman–Enskog kinetic theory [7]. At higher pressures, the modified theories, or the rough hard sphere theory [8, 9], can give a good representation under limited conditions, so that the expression does not fully cover supercritical fluids, or liquid-like dense fluids. The theory has been developed from dilute gaseous collision theory. On the other hand, the Stokes–Einstein equation [10], given by (3.3.2), can represent the diffusion in the liquid phase when the fluid does not slip at the surface of the diffusion molecule, correspondingly for diffusion of large spherical molecules.

$$D_{12} = \frac{RT}{6\pi\eta_2 R_1}, \tag{3.3.2}$$

where R is the gas constant, η_2 is the solvent viscosity, and R_1 is the radius of the diffusion molecule. Based on this equation, various predictive correlations for liquid phase diffusion have been proposed such as the Wilke–Chang equation [10]. However, the applicability of this model to supercritical fluids or dense gases is also limited.

3.3.2 Measurement of Binary Diffusion Coefficient

(a) Taylor Dispersion Method

The Taylor dispersion technique, developed by Taylor [2–4] and Aris [5], has been widely employed to measure diffusion coefficients for low-pressure-gaseous, supercritical, and ambient to high-pressure liquid systems. Most data for binary diffusion coefficients in supercritical fluids have been measured by this technique, whereas the diaphragm cell method, which is common for liquid systems, has not been used. Since the exact basic mathematical equation cannot be analytically solved, it is usually simplified under some assumptions. This approximation restricts both the design of the apparatus and the measurement conditions to meet the assumptions. The various experimental errors ascribed to this method and the design of the experimental apparatus are carefully described [11, 12]. This method is substantially based on the infinite dilution of solute concentration while a relatively large amount of the solute was injected [13, 14].

When a tracer species is imposed as a delta shot to a fully developed laminar flow moving in circular cross-sectional tubing, the tracer concentration is described by Taylor and Aris as:

$$D_{12} \frac{1}{r}\frac{\partial}{\partial r}\left(r \frac{\partial C}{\partial r} \right) + D_{12} \frac{\partial^2 C}{\partial z^2} = \frac{\partial C}{\partial t} + 2u_{\mathrm{a}}\left(1 - \frac{r^2}{R^2} \right) \frac{\partial C}{\partial z}, \tag{3.3.3}$$

where D_{12} is the binary diffusion coefficient of the tracer species, R is the tubing radius, u_a is the average velocity, t is the time, and r and z are the radial and axial distances, respectively. The initial and boundary conditions are:

$$C = \frac{m}{\pi R^2}\delta(z) \quad \text{at } t = 0, \tag{3.3.4}$$

$$\frac{\partial C}{\partial r} = 0 \quad \text{at } r = 0 \text{ and } r = R, \tag{3.3.5}$$

$$C = 0 \quad \text{at } z = \pm\infty, \tag{3.3.6}$$

where m is the injected amount of the tracer species. By defining the average concentration per cross-sectional area of tubing given by

$$C_a = \frac{2}{R^2}\int_0^R C r \, \mathrm{d}r . \tag{3.3.7}$$

(3.3.3) can approximately reduce to (3.3.8):

$$K\frac{\partial^2 C_a}{\partial z^2} - u_a\frac{\partial C_a}{\partial z} = \frac{\partial C_a}{\partial t}, \tag{3.3.8}$$

where

$$K = D_{12} + \frac{u_a^2 R^2}{48 D_{12}} . \tag{3.3.9}$$

The initial and boundary conditions lead to

$$C_a = \frac{m}{\pi R^2}\delta(z) \quad \text{at } t = 0, \tag{3.3.10}$$

$$C_a = 0 \quad \text{at } z = \pm\infty, \tag{3.3.11}$$

The solution of (3.3.8) is given by

$$C_a = \frac{m}{\pi R^2 (4\pi K t)^{1/2}} \exp\left[-\frac{(z - u_a t)^2}{4Kt}\right]. \tag{3.3.12}$$

In this study the D_{12} value is obtained by the method of curve fitting in the time domain (FTD) so as to minimize the root-mean-square error ε defined by

$$\varepsilon = \left(\frac{\int_{t_1}^{t_2} [C_{\mathrm{a,exp}}(t) - C_{\mathrm{a,cal}}(t)]^2 dt}{\int_{t_1}^{t_2} [C_{\mathrm{a,exp}}(t)]^2 dt} \right)^{1/2}, \tag{3.3.13}$$

where a fitting period between t_1 and t_2 is chosen so that the response curve calculated is fitted to that measured experimentally at higher than 10% peak heights.

Figure 3.3.1 shows the experimental set-up for the Taylor dispersion method. The apparatus consists of a pump, an injector, a diffusion column, a detector and a back-pressure controller. The pump is commonly a syringe pump. The accurate measurement requires a constant flow rate, a uniform solvent velocity profile in the diffusion column, and quite stable and constant pressure in the course of the measurement. The diffusion column is a capillary tube which should be installed horizontally in the oven because a gravity effect on the diffusion coefficients was reported [13–15]. The detector is usually a UV detector. Funazukuri et al. [15–19] employed a multi-UV detector capable of scanning from 195 to 350 nm, and Silva and Macedo [20] used a flame ionization detector, which has a wide linearity range, although the solute concentration was measured after decreasing pressure at atmospheric pressure. The back-pressure controller is also essential to obtain accurate data.

Figure 3.3.2 shows the effect of wavelength on diffusion coefficient and fitting error. As depicted, Funazukuri et al. [15] found that the measurement wavelength should be carefully selected when a UV detector is used. In most cases the response curves are analyzed by the moment method, or from the half peak widths (variances). Funazukuri et al. [19] demonstrated that the curve fit-

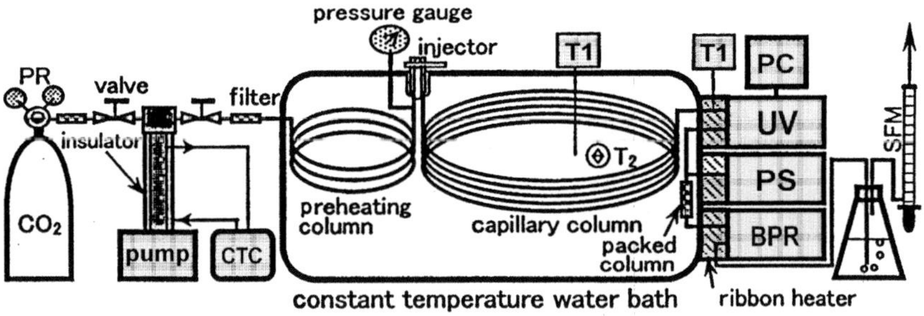

Fig. 3.3.1 A schematic diagram of the apparatus for the Taylor dispersion measurement: PR, pressure regulator; CTC, constant temperature circulator; T_1, thermometer; T_2, thermocouple; PC, personal computer; UV, UV detector; PS, pressure sensor; BPR, back pressure regulator; SFM, soap-bubble flow meter [19]

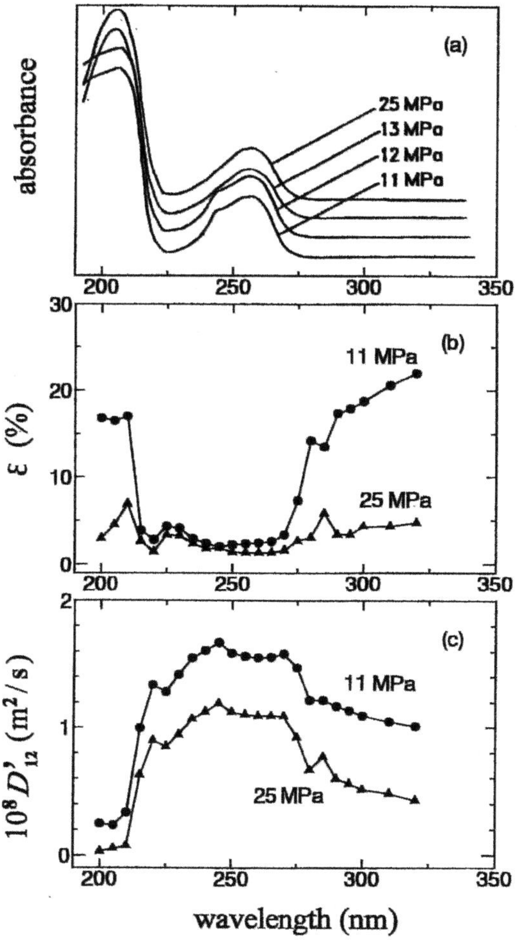

Fig. 3.3.2 The effects of wavelength on (**a**) peak absorbance intensity at the top of the response curves, (**b**) fitting error ε , and (**c**) apparent diffusion coefficients D'_{12} of benzene in CO_2 at 313.2 K and various pressures [15]

ting method was more accurate than the methods with the variances of the response curves such as the moment method. Figure 3.3.3 (a) shows the response curves of acetone obtained experimentally, and calculated by the FTD method and the moment method, and Figure 3.3.3 (b) shows the deviations for the both methods. Figure 3.3.4 shows the error contour map for the FTD method at 308.2 K and 10.30 MPa when the binary diffusion coefficient, D_{12}, and solvent velocity, u_a, are determined as fitting parameters. It is found that both values are determinable accurately, and the parameter sets for the best fit (denoted +) is dif-

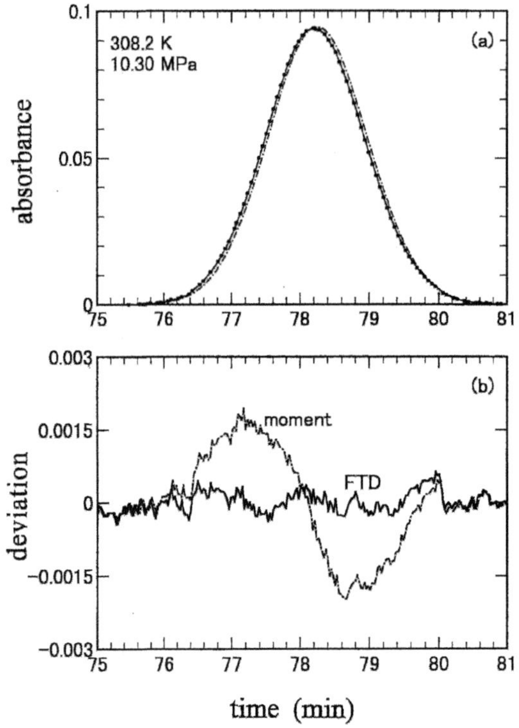

Fig. 3.3.3 (**a**) The response curves measured experimentally (●), and calculated by curve fitting in the time domain (——) and the moment method (— · —), and (**b**) deviations for both methods from the response curve measured

ferent from that (×) for the moment method. Figure 3.3.5 shows the fitting error vs. pressure for the D_{12} data of acetone in CO_2 at 308.2 and 313.2 K in the pressure range from 7.9 to 40 MPa. The error for the moment method is much higher than the curve fitting method, and both errors increase with decreasing pressure. In the near-critical region of the solvent, the error abruptly increases with decreasing pressure. Figure 3.3.6 shows binary diffusion coefficients of acetone measured at 308.2 K and 313.2 K and pressures from 7.9 to 40 MPa. The values at both temperatures decrease with increasing pressures.

The solute measurable is liquid which can be injected directly through a common high-pressure liquid chromatography (HPLC) injector. In most initial studies various solutes having relatively large molecular weights dissolved with common organic solvents such as n-hexane were introduced into the diffusion column. Strictly speaking, this case is not a binary system, and the effect of the organic solvent on the diffusion coefficients is ambiguous. How-

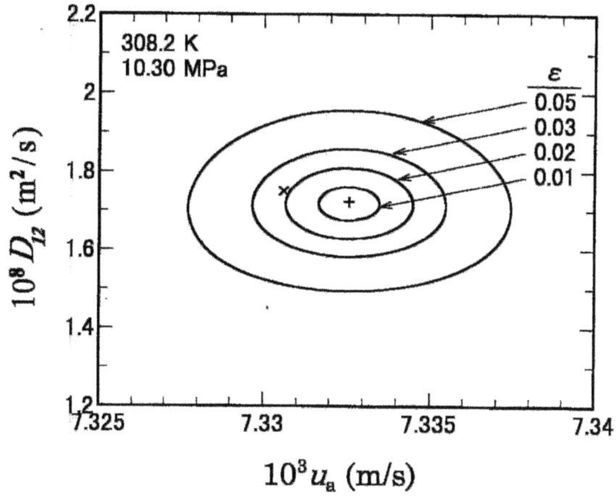

Fig. 3.3.4 Error contour map for the curve fitting method at 308.2 K and 10.30 MPa; +: best fit; ×: the moment method

ever, the effect seems to be small when the molecular weight of a solute is much higher than that of a solvent dissolving because of the great difference in the diffusion coefficient values. In order to overcome this drawback, the solute dissolved into supercritical CO_2 was imposed directly to the diffusion column [21–24], or a diffusion column coated with thin polymer film was employed [25, 26]. In the latter case the parameters involved in the system are binary diffusion coefficient and the partition coefficient of solute between the polymer layer and the supercritical solvent.

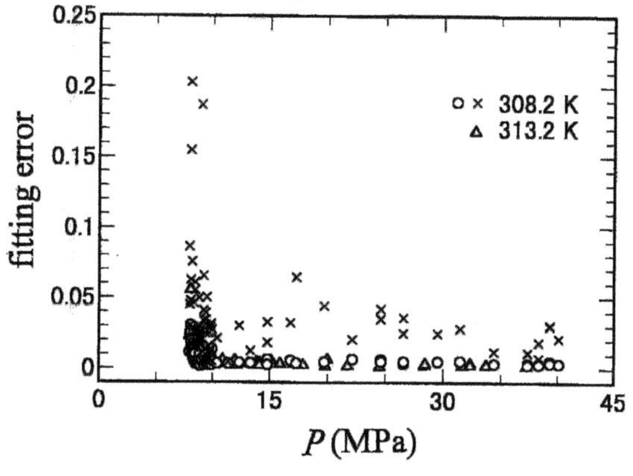

Fig. 3.3.5 Comparison of fitting error between the curve fitting method (O, △) and the moment method (×)

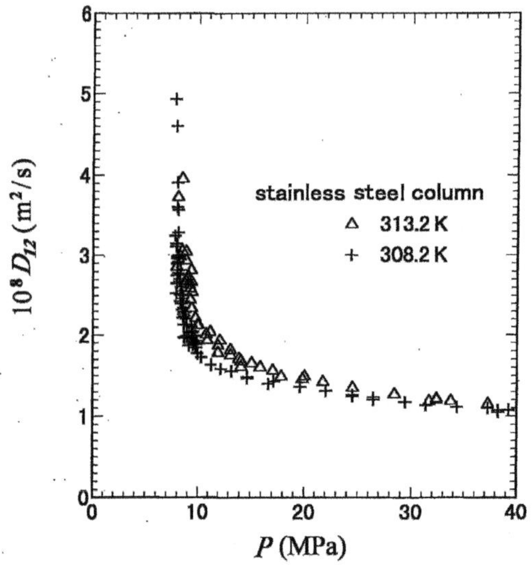

Fig. 3.3.6 Measured binary diffusion coefficients of acetone in CO_2 at 308.2 K and 313.2 K and pressures from 7.9 to 40 MPa

(b) Capillary Tube Method

The capillary tube method is adequate for liquid and solid solutes. For supercritical fluids a couple of studies with this method [27–30] have been reported. The driving force for the diffusion is the concentration difference between zero and saturation solute concentration at the contacting surface with the supercritical fluid when no solute is assumed to exist in the bulk flow. Thus, the diffusion coefficient obtained is not the infinite dilution diffusion coefficient, and an averaged value from zero to the saturation solute concentration.

(c) Other Methods

An NMR spin-echo technique has commonly been applied to measurements of self-diffusion coefficients of various monatomic compounds and multiatomic compounds in supercritical states. Lamb et al. [31] employed this technique to measure binary diffusion coefficients of naphthalene in supercritical CO_2. Russian workers [32–34] determined binary diffusion coefficients from transient concentration profiles.

(d) Literature Data of Binary Diffusion Coefficients in Pure Supercritical Fluids

The binary diffusion coefficients for various solutes in supercritical CO_2 in the literature [13–52] are listed in Table 3.3.1. Since the measurement conditions are not fully identical, the accuracy in the data cannot be directly examined. The molecular weights of measured solutes range from 24 for ethylene to 885 for glycerol trioleate. The binary diffusivities in other supercritical fluids: sulfurhexafluroine [53], ethane [54], propane, butane and pentane [55] are also available.

3.3.3 Correlations

(a) Empirical Correlation

A predictive method for binary diffusion coefficients in supercritical fluids has not been well established theoretically and empirically. While some efforts have been made to develop the method theoretically, the methods are limited to a hard-sphere fluid [56], and a Lennard-Jones fluid [57, 58]. For the practical purposes the most interesting solutes are nonideal fluids, and the parameters involved in these models cannot be estimated. Some empirical correlations based on theories such as the Stokes–Einstein relationship [25, 38, 39], the rough hard sphere model [56, 59, 60], and the Dymond free volume type equation [38, 59, 61, 62] have been proposed. The Schmidt number correlation [6, 19, 26] is also available.

The Wilke–Chang equation has been claimed to be less accurate for predicting the binary diffusion coefficients [17, 39]. Some modification, including a solvent density term, has been made [25]. The rough hard sphere model can successfully describe the transport properties for dense fluids including the liquid state [56, 59]. The model involves two parameters: the roughness factor (translational–rotational coupling parameter) and the diameter of the closest packed molecules. Unfortunately, two parameters are dependent on each other when both values are determined adjustably [60], and these are commonly unavailable for real molecules. The Dymond free volume type equation in (3.3.14) is useful in the limited range of solvent density.

Table 3.3.1 Data sources of binary diffusion coefficients of various organic compounds in supercritical carbon dioxide

ref.	investigators	method	no. of data	temperature (K)
32	Iomtev and Tsekhanskaya (1964)	SD	19	308.2-328.2
33	Tsekhanskaya (1971)	SD	5	308.2
34	Vinkler and Morozov (1975)	Dynamic	13	293.2-313.2
35	Swaid and Schneider (1979)	TD	63	308.2-328.2
37	Lauer et al. (1983)	TD	14	288.3-328.1
38	Debenedetti and Reid (1986)	DRC	9	308.2-328.2
39	Sassiat et al. (1987)	TD	50	303-333
27	Knaff and Schlünder (1987)	CD	16	308-333
31	Lamb et al. (1989)	NMR	25	308.2, 328.2
16	Funazukuri et al. (1989)	TD	5	313.2
41	Dahmen et al. (1990)	TD	48	314
42	Dahmen et al. (1990)	TD	177	313.5, 314.5
17	Funazukuri et al. (1991)	TD	26	308.2-323.2
22	Liong et al. (1991)	TD	85	308-318
23	Liong et al. (1992)	TD	81	308-318
24	Wells et al. (1992)	TD	31	308.2-318.2
18	Funazukuri et al. (1992)	TD	24	313.2
46	Umezawa and Nagashima (1992)	TD	55	299-308

pressure (bar)	solute
82-300	Naphthalene
72-84	Naphthalene
24	Naphthalene
80-160	Benzene, 1-Phenylpropane, 1,3,5-Trimethylbenzene
67.2-351	Naphthalene
160-250	Benzoic acid, 2-Naphthol
110-265	Acetone, Benzene, Naphthalene, 1,3,5-Trimethylbenzene, Phenanthrene, Pyrene, Chrysene
118-226	Naphthalene, Caffeine
100-1000	Naphthalene
160	Benzene, Naphthalene, C18:1-ME[1)], C18:2-ME[2)], C18:3-ME[3)]
100-180	Cyclopentanone, Cycloheptanone, Cyclononanone, 3-Pentanone, 3-Heptanone, 5-Nonanone
95-180	2-Propanone, 2-Butanone, 2-Pentanone, 3-Pentanone, 3-Heptanone, 3-Heptanone, 2-Nonanone, 5-Nonanone, 6-Undecanone, Adamantanone, DMP[4)], TMP[5)], MBBA[6)], Stearic acid, Oleic acid, 5-OCB[7)], Linoleic acid, Squalene
140-336	C16:1-ME[8)], C18:1-ME[1)], C18:2-ME[2)], C18:3-ME[3)] trans-C18:1-ME[9)], C20:5-ME[10)], C22:6-ME[11)], C24:1-ME[12)]
98.7-211	C18:0-EE[13)], C22:0-EE[14)], C22:6-EE[15)], C20:5-ME[10)], C22:6-ME[11)]
96.7-211	C4:0-EE[16)], C8:0-EE[17)], C10:0-EE[18)], C14:0-EE[19)],
96.7-211	C16:0-EE[20)], Phenylacetic acid, Vanillin
160	C14:1-ME[21)], C20:1-ME[22)], C22:1-ME[23)], Vitamin K_3 Vitamin A acetate, Vitamin K_1, Vitamin E dl-Limonene, cis-Jasmone, Indole, Benzene, Naphthalene
299-308	Acetone, Benzene, Pentane, Hexane, Heptane, Octane Nonane, Decane, Undecane, Tetradecane

Table 3.3.1 continued

ref.	investigators	method	no. of data	temperature (K)
47	Bueno et al. (1993)	TD	24	313-333
48	Sengers et al. (1993)	TD	17	306.1-314.1
49	Shenai et al. (1993)	TD	18	308-328
50	Suárez et al. (1993)	TD	75	313.2-333.2
29	Catchpole and King (1994)	TD & CD	16	298.1-318.7
25	Lai and Tang (1995)	TRPC	75	308-328
15	Funazukuri and Nishimoto (1996)	TD	31	313.2
51	Funazukuri (1996)	TD	20	313.2
28	Higashi et al. (1998)	CD	21	308.2
13	Nishiumi et al. (1996)	TD	130	314.3
21	Akgerman et al. (1996)	TD	45	308-328
20	Silva and Macedo (1998)	TD	45	313.15-333.15
52	Tuan et al. (1999)	SSPP	9	313.2-333.2
30	Higashi et al. (1999)	CD	35	308.2, 318.2
19	Funazukuri et al. (2000)	TD	121	308.2,313.2
26	Funazukuri et al. (2000)	TRPC	106	308.2-328.2

SD: Sublimation dissolution; TD: Taylor dispersion; DRC: Diffusion in rectangular channel; CD: Capillary diffusion; NMR: NMR spin echo; TRPC: Tracer response with polymer coated column; SSPP: Steady-state parallel plates

pressure (bar)	solute
150-350	Benzene, Toluene, o-,m-,p-Xylene, Ethylbenzene 1-Phenylpropane, (1-Methylethyl)benzene 1,3,5-Trimethylbenzene
9.68-13.5	Benzene, Toluene
172, 276	Benzoic acid, Phenanthrene, Acridine
150-350	Benzene, Toluene, Ethylbenzene 1-Phenylpropane, (1-Methylethyl)benzene
71-251	Naphthalene, Benzoic acid, Oleic acid, Vitamin E Glycerol trioleate
75-144.2	Toluene, Ethyl acetate, Phenol
90-300	Benzene
160	Benzene,Toluene, o-,m-,p-Xylene, Ethylbenzene 1-Phenylpropane, (1-Methyethyl)benzene, Benzyl ether 1,3,5-Trimethylbenzene, Benzaldehyde, Acetone, 2-Butanone, 1,4-Dioxane, Ethyl acetate, Diethyl ether 1-Octene, Dipentene, Dichloromethane, Chloroform
72.1-200	Naphthalene, 2,6-Dimethylnaphthalene, 2,7-Dimethylnaphthalene
92.2-225.6	Acetone
96-247	Naphthalene, Phenanthrene, Hexachlorobenzne
81-162	Diethyl ether, Diisopropyl ether, Tetrahydrofuran
10.6-14.0	Methyl oleate
77-200	Naphthalene, 2,6-Dimethylnaphthalene, 2,7-Dimethylnaphthalene
79-400	Acetone
87.2-303.2	Phenol, α-Tocopherol, β-Carotene

1) cis-9-octadecenoic (oleic) acid methyl ester; 2) cis-9,12-octadecadienoic (linoleic) acid methyl ester; 3) cis-6,9,12-octadecatrienoic (γ-linolenic) acid methyl ester; 4) 2,4-dimethyl-3-pentanone; 5) 2,2,4,4-tetramethyl-3-pentanone; 6) N-(4-methoxybenzylidene)-4-n-butylaniline; 7) 4-cyano-4'-n-pentoxybiphenyl; 8) cis-9-hexadecenoic (palmitoleic) acid methyl ester; 9) trans-9-octadecenoic (elaidic) acid methyl ester; 10) cis- 5,8,11,14,17-eicosapentaenoic acid (EPA) methyl ester; 11) cis-4,7,10,13, 16,19-docosahexaenoic acid (DHA) methyl ester; 12) cis-15-tetracosenoic (nervonic) acid methyl ester; 13) octadecanoic (stearic) acid methyl ester; 14) docosanoic (behenic) acid methyl ester; 15) DHA-ethyl ester; 16) butanoic (butyric) acid ethyl ester; 17)octanoic (caprylic) acid ethyl ester; 18) decanoic (capric) acid ethyl ester; 19) tetradecanoic (myristic) acid ethyl ester; 20) hexadecanoic (palmitic) acid ethyl ester; 21) cis-9-tetradecenoic (myristoleic) acid methyl ester; 22) cis-11-eicosenoic acid methyl ester; 23) cis-13-docosenoic (erucic) acid methyl ester.

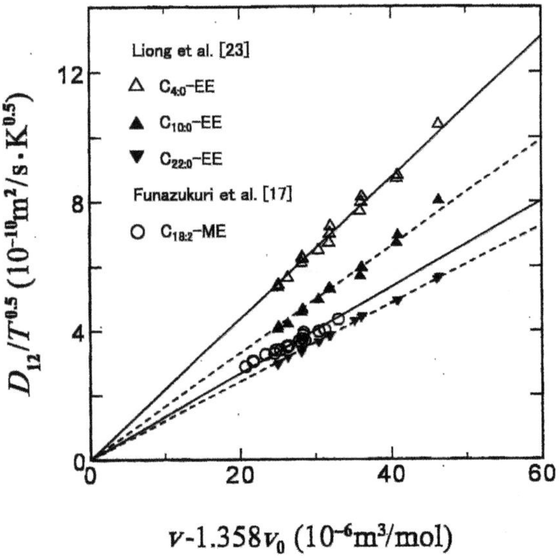

Fig. 3.3.7 The Dymond free volume type equation for fatty acid methyl ester (ME) and ethyl ester (EE) in CO_2

$$D_{12} / \sqrt{T} = \alpha(v - \beta v_0), \tag{3.3.14}$$

where α and β are constants specific to the system, and β is 1.358 [59] or an adjustable parameter, and v_0 is the hard-sphere closest packed volume, equal to $N_A\sigma^3 / \sqrt{2}$: N_A is the Avogadro's number and σ is the hard-sphere diameter. Figure 3.3.7 shows plots for various solutes in CO_2 at 308–323 K in the pressure range from 9.7 to 34 MPa. As depicted, in this limited pressure range (3.3.14) represents the measured data well. However, the deviation increases with decreasing pressure as shown in Fig. 3.3.8 for the literature D_{12} data of benzene [15, 35, 36, 39, 43, 46, 48, 50] in CO_2.

(b) Schmidt Number Correlation

Funazukuri and Wakao [6] proposed the Schmidt number correlation based on the rough hard-sphere model. The rough hard-sphere model represents well the transport properties such as thermal conductivity, viscosity, and self-diffusion coefficient for dense fluids. When the model is applied to prediction, the hard-sphere diameters for the solutes, in particular those for large molecular weight compounds, are not available in many cases. In order to overcome this drawback Funazukuri and Wakao assumed that the hard-sphere diameter

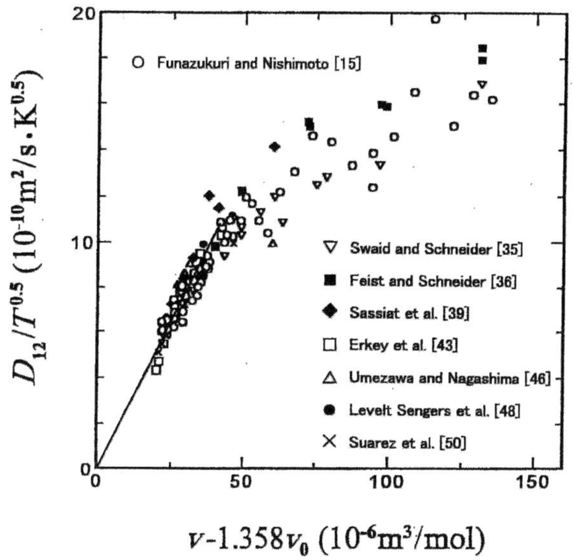

Fig. 3.3.8 The Dymond free volume type equation for benzene in CO_2

ratio of solute to solvent was equal to the van der Waals diameter ratio of solute to solvent. Moreover, Funazukuri et al. [18] found that the ratio of Schmidt number ($=\eta/\rho D_{12}$) at high pressure to that at atmospheric pressure was almost independent of pressure at each temperature. Based on the above assumption and the evidence, the following correlation was developed [6, 19].

The Schmidt number correlation for predicting binary diffusion coefficients as well as self-diffusion coefficients from low to high pressure regions is expressed as follows:

$$Sc^+ - 1 = \frac{Sc}{Sc^*} - 1 = \exp\left[\sum_{i=0}^{5} a_i \left(\frac{v_0}{v}\right)^i\right], \tag{3.3.15}$$

where

$$Sc^* = \frac{5}{6}\left[\frac{\sigma_1 + \sigma_2}{2\sigma_2}\right]^2 \left[\frac{2M_1}{M_1 + M_2}\right]^{1/2} \quad \text{for binary diffusion,}$$

$$Sc^* = \frac{5}{6} \quad \text{for self-diffusion,}$$

and the coefficients a_i, obtained from the accurate self-diffusion coefficient data [63–67], are listed in Table 3.3.2, and Sc and Sc^* are Schmidt numbers at high pressure and at atmospheric pressure at the same temperature, respectively, v is the molar volume of solvent, and v_0 is the hard-sphere closest packed volume of solvent molecules, and σ_1 and σ_2 are hard-sphere diameters for solute and solvent, respectively. The σ_2 value for CO_2 was obtained from the v_0 correlation with temperature [18]. The coefficients are listed in Table 3.3.3. When the σ_1 value is not available, the following assumption is made:

$$\frac{\sigma_1}{\sigma_2} = \frac{\sigma_{\mathrm{vw},1}}{\sigma_{\mathrm{vw},2}}, \tag{3.3.16}$$

where $\sigma_{\mathrm{vw},1}$ and $\sigma_{\mathrm{vw},2}$ are van der Waals diameters of solute and solvent molecules, respectively, obtained from the method of Bondi [68].

Figure 3.3.9 shows the plot for the Schmidt number correlation [19]. It is found that the D_{12} data having low fitting errors ($\varepsilon < 0.01$ with blank key) are well expressed with the correlation at both temperatures in a wide pressure range (from 10 to 40 MPa), while the data having higher fitting errors ($\varepsilon >$ 0.01 with solid key) deviate from the correlation. Figure 3.3.10 compares the literature data with the correlation. The deviations of most literature data [13, 15, 39, 42, 46, 51] are more than that of Funazukuri et al. [19]. The data of

Table 3.3.2 Correlation of Schmidt number

i	a_i
0	-4.92519817
1	5.45529385×10^1
2	-2.45231443×10^2
3	6.07893924×10^2
4	-7.08884016×10^2
5	3.29611433×10^2

$$Ln\left(Sc^+ - 1\right) = \sum_0^5 a_i \left(\frac{v_0}{v}\right)^i,$$

where $v_0 = N\sigma^3/\sqrt{2}$,

N is Avogadro's number, and

σ is the effective hard-sphere diameter.

Table 3.3.3 Correlation of effective hard sphere packed volume for CO_2

j	b_j
0	$+4.452 \times 10^{-5}$
1	-1.152×10^{-7}
2	$+2.749 \times 10^{-10}$
3	-3.073×10^{-13}
4	$+1.290 \times 10^{-16}$

$$v_0\,(\mathrm{m^3/mol}) = \frac{1}{1.384}\sum_0^4 b_j \left(T[K]\right)^j,$$

where $v_0 = N\sigma^3/\sqrt{2}$,

N is Avogadro's number, and

σ is the effective hard-sphere diameter.

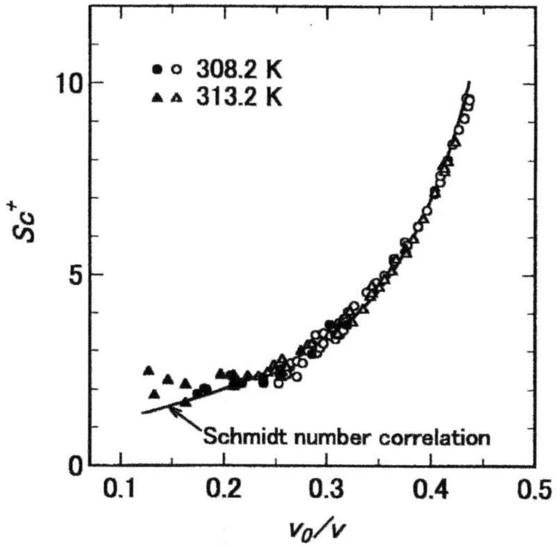

Fig. 3.3.9 Schmidt number correlation for acetone at 308.2 and 313.2 K in the pressure ranges from 7.9 to 40 MPa

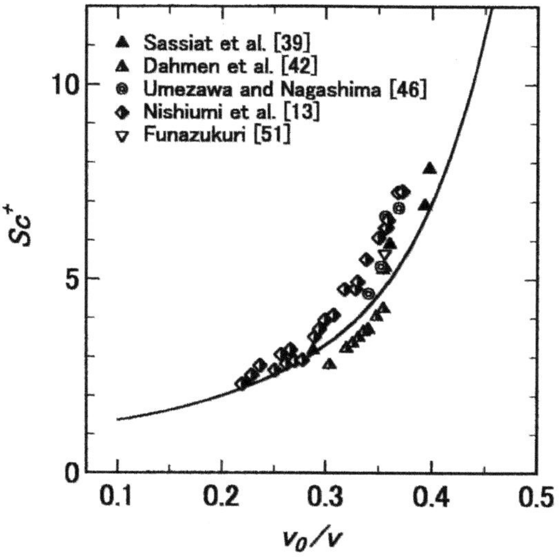

Fig. 3.3.10 Schmidt number correlation for acetone in the literature

Sassiat et al. [39] and Umezawa and Nagashima [46] at lower pressures are roughly consistent with those from the correlation. This correlation is also valid for the data of self-diffusion coefficients [63–67], as shown in Fig. 3.3.11. Figure 3.3.12 shows the same plot for the self-diffusion coefficients obtained from molecular dynamics calculations [69].

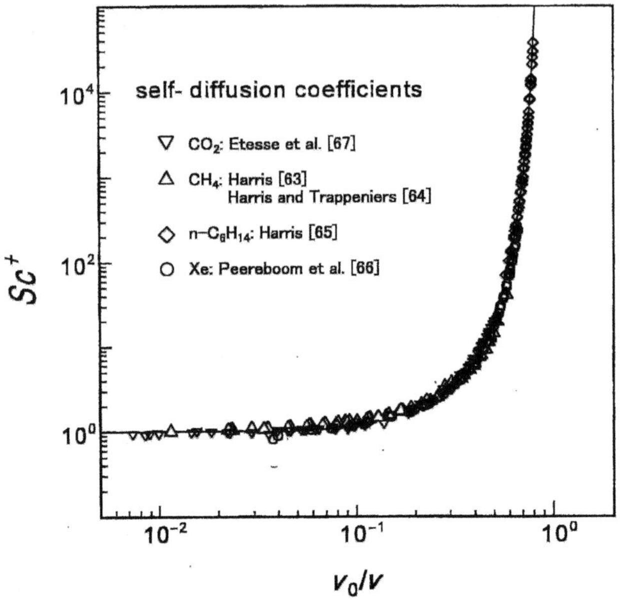

Fig. 3.3.11 Schmidt number correlation for self-diffusion coefficients of CO_2 [67], CH_4 [63, 64], *n*-hexane [65] and Xe [66]

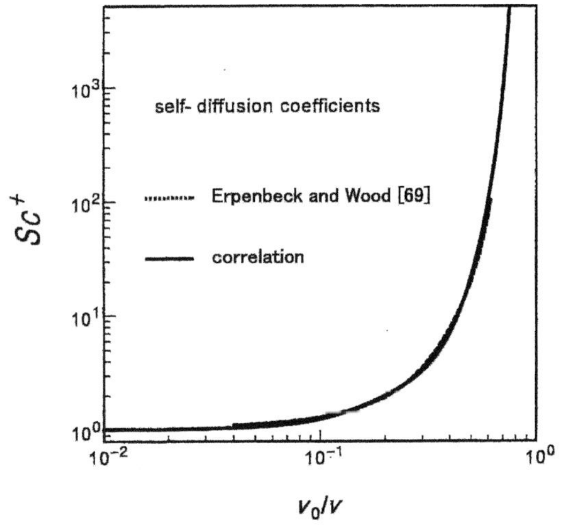

Fig. 3.3.12 Comparison of Schmidt number ratio Sc^+ $(= Sc / Sc^*)$ for self-diffusion coefficients obtained from molecular dynamics calculations of Erpenbeck and Wood [69] with those from the correlation

3.3.4 Pseudo-Binary Diffusion Coefficients in Mixture Solvents

Supercritical fluid extraction/chromatography has widely been used to separate/analyze valuable, in particular, heat labile compounds. A small amount of entrainer has often been added to a supercritical solvent to enhance solubilities of objective compounds or resolution in the separation. In order to design or optimize the reactors of extraction or chromatography the transport properties such as molecular diffusion coefficients in supercritical fluids are required. Binary diffusion coefficients for common organic compounds have been measured extensively. However, few studies on molecular diffusion coefficients in multicomponent systems are available under supercritical conditions while the actual operating systems often consist of many components. In order to improve accuracy in reactor design, the accurate prediction of diffusion coefficients in multicomponents is needed.

Molecular diffusion coefficients D_{ij} of a solute i in a mixture solvent having multi(n)-components are given [70] by

$$\mathbf{J}_i = -\sum_{j=1}^{n-1} D_{ij} \nabla C_j , \tag{3.3.17}$$

where $\mathbf{J}_i$ and C_i are the diffusion flux and concentration for component i, respectively. While (3.3.17) can describe multicomponent diffusion, it is difficult to relate the diffusion coefficients D_{ij} to the mass transfer parameters such as mass transfer coefficient, axial fluid dispersion, partition coefficient, adsorption rate constant, intra-particle diffusion coefficient, etc., in the mathematical models for extraction or chromatographic processes. These parameters are often defined on the assumption of a pseudo-binary system: the objective component is a solute and the mixture solvent is regarded as a single component. Furthermore, the diffusion coefficients D_{ij} are difficult to measure in a wide range of temperature, pressure and compositions.

Although the pseudo-binary diffusion coefficients $D_{1\mathrm{m}}$ have been measured [39, 44, 49, 71–73] by the Taylor dispersion technique [2–5], the number of data points and the measurement conditions are limited. Due to the lack of data, the prediction method has not been well established. Olesik and Woodruff [44] measured diffusion coefficients of benzene and *m*-cresol in mixtures of solvent and a low-concentration entrainer. They found that the conventional correlations for diffusion in the liquid phase did not represent their measured diffusion data. Some workers [49, 71, 72] have pointed out that $D_{1\mathrm{m}}$ was correlated with the mixture solvent viscosity η_m. However, the correlations employed are slightly different: Shenai et al. [49] found the D_{1m} was ex-

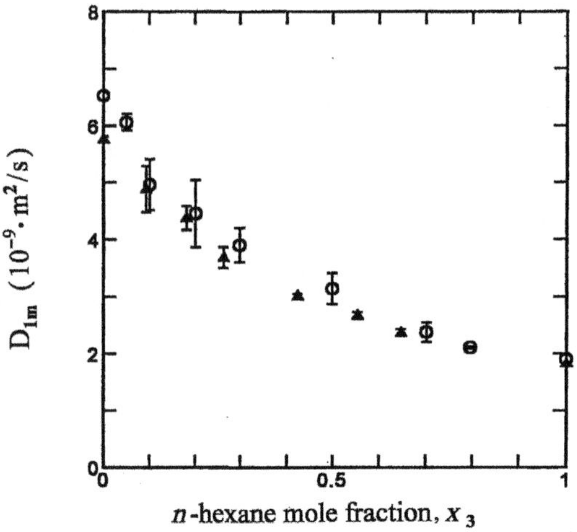

Fig. 3.3.13 Pseudo-binary diffusion coefficients of linoleic acid methyl ester in a mixture of CO_2 and *n*-hexane at 313.2 K, and pressures of 16 (O) and 25 (▲) MPa [73]

pressed linearly with T/η_m for benzoic acid, phenanthrene and acridine in a mixture of CO_2 and methanol (5 mol%). Olesik and coworkers [71, 72] also found that D_{1m} was proportional to $1/\eta_m$, not including the temperature explicitly, for benzene, anthracene, *m*-cresol and *p*-nitrophenol in methanol + water + CO_2 and ethanol + water + CO_2. They pointed out that the Erying rate theory, $\ln D_{1m} = A + B/T$, was also valid.

Figure 3.3.13 shows pseudo-binary diffusion coefficients of linoleic acid methyl ester at 313.2 K and pressures of 16 and 25 MPa in a mixture of CO_2 and *n*-hexane [73]. The values change with compositions by three times at most. Figure 3.3.14 shows a logarithmic plot of D_{12}/T or D_{1m}/T vs. viscosity for linoleic acid methyl ester in supercritical CO_2 at 308.2 to 328.2 K and pressures of 14 to 33.6 MPa [17], in a mixture of CO_2 and *n*-hexane at 313.2 K and pressures of 16 and 25 MPa [73], and in liquid organic solvents at temperatures of 303.2–333.2 K and pressures of 16–25 MPa [74]. Although the data for mixtures deviate somewhat, Eq. (3.3.18) is valid for mixture solvents as well as pure solvents when a solute is specific:

$$D/T = \delta\eta^{\gamma}, \tag{3.3.18}$$

where δ and γ are constants specific to the system. The plots for vitamin K_3 and indole as solutes are also found to be valid in mixture solvents as well as these organic solvents: the average absolute deviation and the maximum deviation are 5.99% and 17.7% at *n*-hexane mole fraction $x_3 = 0.702$ and

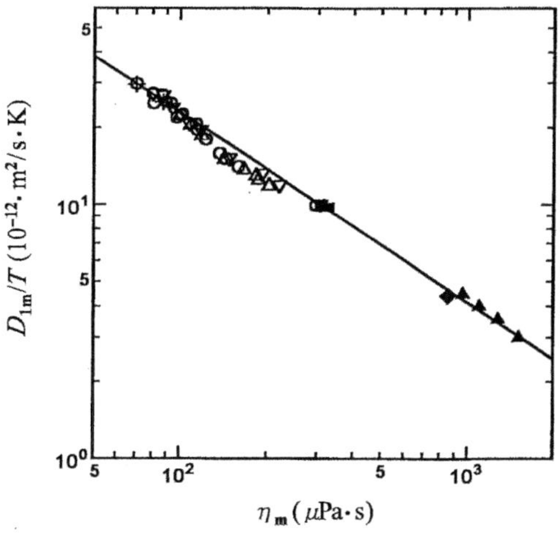

Fig. 3.3.14 D_{1m}/T vs. viscosity η_m for linoleic acid methyl ester in a mixture of CO_2 and *n*-hexane at 313.2 K and 16 (△) and 25 (▽) MPa [73], together with those in supercritical CO_2 at 303.2–333.2 K and 14–33.6 MPa (○) [17] and in liquid organic solvents of *n*-hexane (■), *n*-dodecane (▲) and cyclohexane (◆) at 303.2–333.2 K and 16–25 MPa [74]

16.0 MPa for linoleic acid methyl ester, 5.14% and 12.1% at $x_3 = 0.499$ and 16.0 MPa for vitamin K_3, and 5.60% and 12.3% at $x_3 = 0.554$ and 25.0 MPa for indole, respectively. Equation (3.3.18) seems to be a possible correlation for prediction.

3.3.5 Critical Anomaly of Diffusion Coefficient

As compared with viscosity and thermal conductivity, the anomaly in binary diffusion coefficients in the near-critical region has not been well clarified. Some recent experimental studies are available [13, 14, 28, 78].

(a) Measurement of Diffusion Coefficient Near the Critical Region

Effect of Small Changes in Solute Composition

Generally, diffusion coefficients estimated from the Taylor dispersion method are considered to be in infinitely dilute solutions. As shown in Fig. 3.3.15, however, Nishiumi et al. [13] found that even small changes in acetone composition had a strong effect on the diffusion coefficients obtained near the critical point of CO_2. They found that the diffusion coefficients had a maxi-

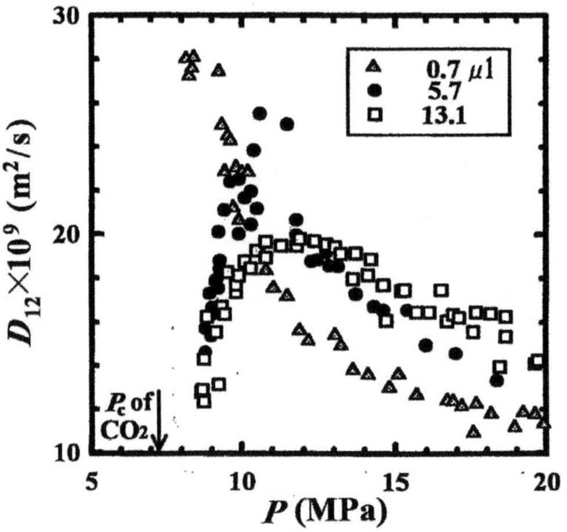

Fig. 3.3.15 Effect of injection volumes on the diffusion coefficient of acetone in supercritical CO_2 at 313.15 K [13]. Reprinted from: H. Nishiumi, M. Fujita, K. Agou: Diffusion of acetone in supercritical carbon dioxide, *Fluid Phase Equilibria* (Elsevier Science) **117**, 356 (1996)

mum depending on the injection volumes and approached zero for 5.7 and 13.1 μl injection volumes near the critical pressure of CO_2, although their values decreased monotonically with pressure for 0.7–4.3 μl injection volumes. They observed the same qualitative behavior in the case of benzene [78]. Below 8 MPa, the decay in concentration distribution due to fluctuations in the critical region was measured with a flame ionization detector since light scattering prevented the use of the optical method.

Effect of Column Configurations and Column Types

In the Taylor dispersion method, most researchers used a coiled type diffusion tube. Some used a straight diffusion tube to avoid secondary flow effects caused by gravity, centrifugal force and buoyancy. Nishiumi ct al. [13] compared the values of diffusion coefficients measured with a straight tube, a horizontal and a vertical position coiled tubes. The effect of tube configuration for the diffusion coefficient of acetone in supercritical CO_2 is shown in Fig. 3.3.16. The results show that a vertical-loop configuration should be avoided because of complicated buoyancy corrections especially near the critical point.

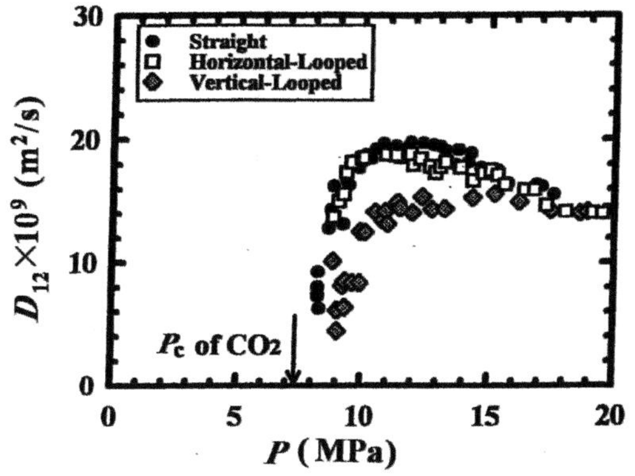

Fig. 3.3.16 Effect of column configurations on diffusion coefficient of benzene in supercritical CO_2 at 314.25 K (13 μl) [13]

(b) Experimental Data on Diffusion Coefficient Near the Critical Point

The critical and consolute anomalous behavior of diffusion coefficients has been reported by researchers since 1957. Umezawa and Nagashima [46] showed a slight anomaly of the diffusion coefficient of nonane in supercritical CO_2 by the Taylor dispersion method. Using the Mach-Zehnder interferomtery, Dysthe et al. [79] observed composition dependence of the diffraction coefficient in the critical region for the system of methane and decane at 303 K. Nishiumi et al. [13, 78] observed strong effects on the diffusion coefficients obtained near the critical point of CO_2 as stated above. Higashi et al. [28, 30] showed the critical slowing down of the diffusion coefficients of naphthalene and dimethylnaphthalene isomers in supercritical CO_2 at 308.2 K by the extended solid dissolution technique.

In spite of the scientific interest in anomalous behavior for over a century, this topic is generally not covered in engineering textbooks on properties. One reason may be that it is only recently that there has been a keen interest in applications of supercritical fluids to industry.

Recently, Funazukuri et al. [75] recalculated the response signals using the data of Nishiumi et al. [13] and the critical properties measured by Reaves et al. [76]. Figure 3.3.17 shows a three-dimensional response curve reproduced by using the diffusivity data and the parameters reported in the paper of Nishiumi et al. [13]. The L axis is the axial distance of the column. The $t - L/u$ axis is the time difference of the mean residence time. The T_C axis is the tracer concentration in the mixture of acetone and CO_2 in terms of the

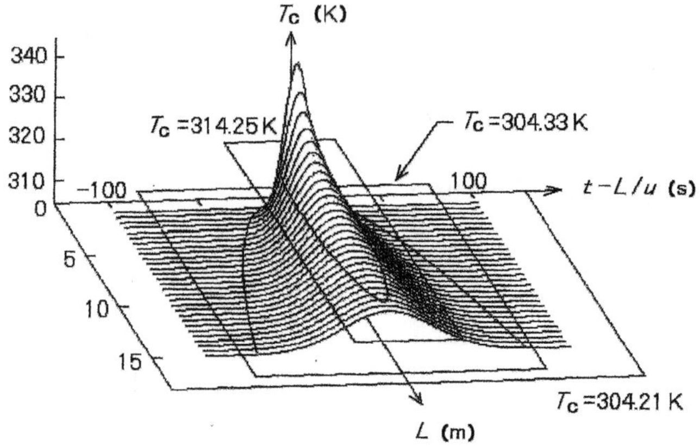

Fig. 3.3.17 Three-dimensional response curve reproduced under conditions of Nishiumi et al. [13] when an amount 13.1 μl is injected and the CO_2 solvent flows at the lowest possible velocity (0.01 ms^{-1}) [75]

critical temperature of the mixture. The upper portion cut with the plane at T_C = 314.25 K is the region where the critical temperature of the mixture is higher than the measurement temperature of 314.25 K, that is, the subcritical state. Thus, it is found that the mixture of the solute and CO_2 solvent does not attain to the supercritical state at most axial positions of the diffusion column under their measurement conditions showing the anomaly, as shown in Fig. 3.3.17.

(c) Theoretical Treatment of Anomalous Behavior of Diffusion Coefficients

From Equations of State

According to irreversible thermodynamics based on the phenomenological equation, the chemical potential gradient provides the driving force of diffusion instead of the concentration gradient near the critical point. The mutual diffusion coefficient D_{12} is expressed as,

$$D_{12} = D_0 \frac{x_1}{RT}\left(\frac{\partial \mu_1}{\partial x_1}\right)_{P,T} \tag{3.3.19}$$

where D_0 is a reference value, and x_1 and μ_1 are the composition and chemical potential of the solute, respectively.

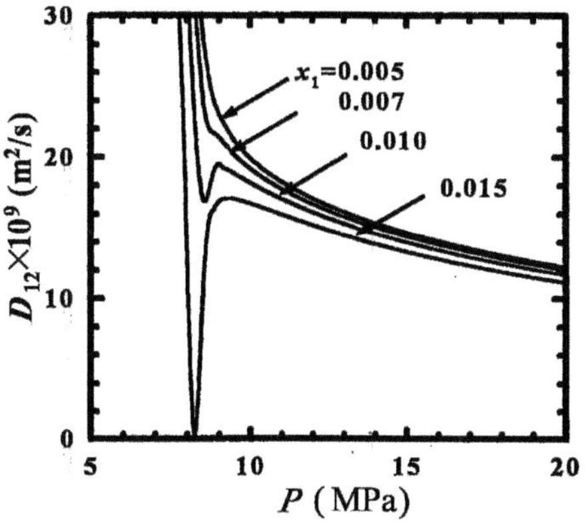

Fig. 3.3.18 Effect of solute composition on diffusion coefficients calculated from the Peng–Robinson equation ($\delta_{ij} = 0.08$) [14]

As is well known for the critical point of vapor–liquid equilibria, the chemical derivative term in (3.3.19) predicts that the diffusion coefficient at the critical point becomes zero. If we know the functional form of μ in terms of composition, pressure and temperature, we can calculate the diffusion coefficient at an arbitrary state. At present, having excellent equations of state, we can apply them to calculate diffusion coefficients. As shown in Fig. 3.3.18, Ago et al. [14, 78] revealed that a small increase in composition causes a decrease in the diffusion coefficient near the critical point. The behavior of the diffusion coefficient with composition depends on a maximum point that was consistent qualitatively with experimental values, although minimum points could not be observed because of the critical fluctuation.

From Molecular Dynamics

For a binary mixture, the mutual diffusion coefficient, D_{12}, is expressed by the following equation.

$$D_{12} = L \cdot Q \tag{3.3.20}$$

where L is the correlated kinetic factor and Q is the thermodynamic factor.

Ago et al. [77] applied molecular dynamics (MD) simulation to a two-dimensional system composed of an equimolar Ar–Kr mixture. The scheme of this MD simulation is described in Sec. 1.5.2. The results are shown in Fig. 3.3.19, where L_0 is the self-diffusion coefficient. Although the normalized kinetic factor L/L_0 reaches a maximum, the thermodynamic factor Q

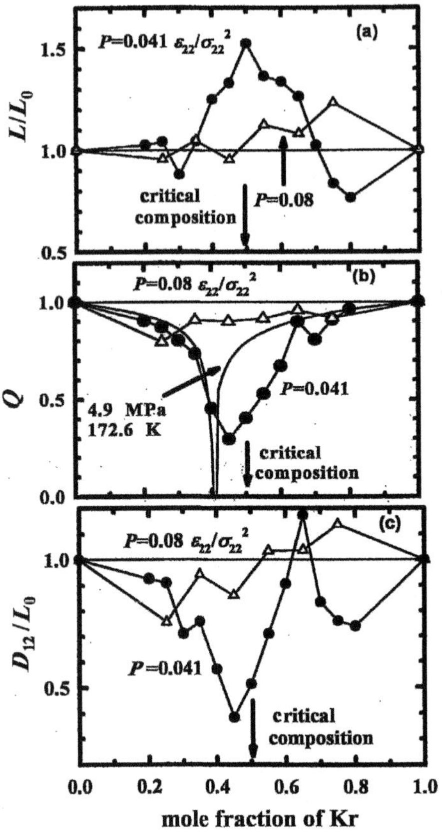

Fig. 3.3.19 Composition dependence of (**a**) normalized kinetic factor, (**b**) thermodynamic factor and (**c**) normalized diffusion coefficient from molecular dynamics [77]

decreases significantly near the critical point. Therefore, the normalized mutual diffusion coefficients D_{12} / L_0 exhibits a minimum near the critical point. Both molecular dynamics and equation of states show the effect of large-concentration fluctuations and exhibit the same qualitative behavior as observed in more complicated experimental systems.

3.3.6 Remarks

While binary diffusion coefficients of various organic compounds in supercritical solvents are available in the literature, the data are not sufficient enough for the design of reactors. Pseudo-binary diffusion coefficients in a supercritical fluid mixture, assuming the mixture solvent to be a single component, are limited. While the effort to develop theoretical prediction methods has been made, the solutes applicable are limited. The Schmidt number correlation is a possible prediction method for infinite dilution binary diffusion coefficients. The correlation of D/T with solvent viscosity in (3.3.18) is also valid for both pure and mixture solvents when a solute is specific. The coefficients δ and γ involved should be determined for each solute. The anomaly in binary diffusion coefficients in the near-critical region has not been well clarified. Nishiumi et al. [13, 14] claimed the anomaly, but Funazukuri et al. [75] criticized their conclusions because of experimental shortcomings. Some further work on this topic is necessary.

References

3.1 Viscosity

[1] J. O. Hirshfelder, G. F. Curtiss and R. B. Bird, "*Molecular Theory of Gases and Liquids*", John Wiley & Sons, New York (1954).

[2] "*Transport Phenomena in Fluids*", ed. H. J. M. Hanley, Marcel Dekker, New York (1969).

[3] S. Chapman and T. G. Cowling, "*The Mathematical Theory of Non-Uniform Gases*", Cambridge University Press, Cambridge (1970).

[4] J. H. Ferziger and H. G. Kaper, "*Mathematical Theory of Transport Processes in Gases*", North-Holland, New York (1972).

[5] T. Makita, "*Nendo To Netsudendouritu*", Baifu-Kan, Tokyo (1975).

[6] R. A. Reid, J. M. Prausnitz and B. J. Poling, "*The Properties of Gases and Liquids*", McGraw-Hill, New York (1987).

[7] "*Experimental Thermodynamics Vol. III: Measurements of the Transport Properties of Fluids*", ed. W. A. Wakeham, A. Nagashima and J. V. Sengers, Blackwell Scientific Publications, London (1991).

[8] J. Millat, J. H .Dymond and C. A. Nieto de Castro, "*Transport Properties of Fluids*", Cambridge University Press, Cambridge (1996).

[9] V. Vesovic, W. A. Wakeham, G. A. Olchowy, J. V. Sengers and J. T. R. Watson, *J. Phys. Chem. Ref. Data*, **19**, 763 (1990).

[10] H. Iwasaki and H. Takahashi, *Bull. Chem. Res. Inst. Non-Aqueous Solutions, Tohoku University*, **6**, 61 (1956).

[11] H. Iwasaki and M.Takahashi, *Rev. Phys. Chem. Japan*, **38**, 18 (1968).

[12] H. Iwasaki and M.Takahashi, *J. Chem. Phys.*, **74**, 1930 (1981).

[13] H. Iwasaki and M. Takahashi, *Proceedings of the 4th International Conference on High Pressure*, Kyoto, p. 523, (1974).

[14] M. Takahashi, C. Yokoyama and S. Takahashi, *Trans. JAR*, **4**, 25 (1987).

[15] M. Takahashi, C. Yokoyama and S. Takahashi, *Trans. JAR*, **6**, 57 (1989).

[16] K. Date, *Rev. Phys. Chem. Japan*, **43**, 1 (1973).
[17] G. Kaminishi and T. Toriumi, *Bull. Chem. Res. Inst. Non-Aqueous Solutions, Tohoku University*, **10**, 51 (1961).
[18] G. Kaminishi, Y. Arai, S. Saito and S. Maeda, *J. Chem. Eng. Japan*, **1**, 109 (1968).
[19] G. F. Newell, *ZAMP*, **10**, 160 (1959).
[20] H. Iwasaki and J. Kestin, *Physica*, **29**, 1345 (1963).
[21] K. Stephan, R. Krauss and A. Laesecke, *J. Phys. Chem. Ref. Data*, **16**, 993 (1987).
[22] R. T. Jacobsen and R. T. Stewart, *J. Phys. Chem. Ref. Data*, **2**, 757 (1973).
[23] C. Yokoyama and M. Takahashi, *Int. J. Thermophys.*, **18**, 1369 (1997).
[24] H. Lee and G. Thodos, *Ind. Eng. Chem. Res.*, **29**, 1404 (1990).
[25] T. H. Chung, L. L. Lee and K. E. Starling, *Ind. Eng. Chem. Fundam.*, **23**, 8 (1984).
[26] T. H. Chung, L. L. Lee and K. E. Starling, *Ind. Eng. Chem. Res.*, **27**, 671 (1988).
[27] R. DiPippo, J. R. Dorfman, J. Kestin, H. E. Khalifa and E. A. Mason, *Physica*, **86A**, 205 (1977).
[28] V. Vesovic and W. A. Wakeham, *Int. J. Thermophys.*, **10**, 125 (1989).
[29] V. Vesovic and W.A. Wakeham, *Chem. Eng. Sci.*, **44**, 2181 (1989).
[30] J. Kestin, S. T. Ro and W. A. Wakeham, *Physica*, **58**, 165 (1972).
[31] J. Kestin, K. Knierim, E. A. Mason, B. Najafi, S. T. Ro and M. Waldman, *J. Phys. Chem. Ref. Data*, **13**, 229 (1984).
[32] S. N. Naldrett and O. Maass, *Can. J. Res.*, **18B**, 322 (1940).
[33] A. Michels, A. Botzen and W. Schuurman, *Physica*, **23**, 95 (1957).
[34] J. Kestin, J. H. Whitelaw and T. F. Zien, *Physica*, **30**, 161 (1964).
[35] W. Herreman, A. Laattenist, W. Grevendonk and A. De Bock, *Physica*, **52**, 489 (1971).
[36] L. Bruschi and G. Torzo, *Phys. Lett.*, **98A**, 265 (1983).
[37] R. F. Berg and M. R. Moldover, *J. Chem. Phys.*, **93**, 1926 (1990).
[38] H. J. Strumpf, A. F. Collings and C. J. Pings, *J. Chem. Phys.*, **60**, 1309 (1974).
[39] V. N. Zozulya and Yu. P. Blagoi, *Sov. Phys., JETP*, **39**, 99 (1974).
[40] J. H. B. Hoogland and N. J. Trappeniers, *Proceedings of the 8th Symposium on Thermophysical Properties*, ed. J. V. Sengers, AIME, New York, **1**, p. 440 (1982).
[41] S. L. Rivkin, A. J. Levin, L. B. Izrailevsky and K. G. Kharitonov, in *Proceedings of the 8th International Conference on the Properties of Water and Steam*, Paris, p. 153 (1975).
[42] C. Yokoyama, M. Takahashi and S. Takahashi, *Int. J. Thermophysics*, **15**, 603 (1994).
[43] S. Wang and H. Meyer, *J. Low-Temp. Phys.*, **69**, 377 (1987).
[44] M. Takahashi, N. Titakawa and C. Yokoyama, *"Proceedings of the 5th Japan Symposium on Thermophysical Properties"*, Toyama, pp. 255-258 (1994).
[45] J. Luettmer-Strathmann and J. V. Sengers, *Int. J. Thermophys.*, **15**, 1241 (1994).

[46] J. Luettmer-Strathmann and J. V. Sengers, *J. Chem. Phys.*, **104**, 3026 (1996).
[47] S. B. Kiselev and V. D. Kulikov, *Int. J. Thermophys.*, **15**, 283 (1994).
[48] T. Ohta, *Progr. Theor. Phys.*, **54**, 1566 (1975).
[49] T. Ohta and K. Kawasaki, *Progr. Theor. Phys.*, **55**, 1348 (1976).
[50] J. C. Nieuwoudt and J. V. Sengers, *J. Chem. Phys.*, **90**, 763 (1990).
[51] R. F. Berg and M. R. Moldover, *Phys. Rev.*, **A42**, 7183 (1990).
[52] R. Krauss, J. Luettmer-Strathmann, J. V. Sengers and K. Stephan, *Int. J. Thermophys.*, **14**, 951 (1993).
[53] J. V. Sengers, *Transport Phenomena-1973*, ed. J. Kestin, AIP Conf. Proc. 11, p. 229 (1973).
[54] G. A. Olchowy and J. V. Sengers, *Phys. Rev. Lett.*, **61**, 15 (1988).

3.2 Thermal Conductivity

[1] U. V. Mardolcar and C. A. Nieto de Castro, *High Temp.-High Press.*, **24**, 551 (1992).
[2] J. W. Haarman, *Physica*, **52**, 603 (1971).
[3] M. J. Assael, C. A. Nieto de Castro, H. M. Roder and W. A. Wakeham, "*Measurement of Transport Properties of Fluids*", IUPAC: Blackwell Scientific Publications, Oxford, **7**, p. 189 (1990).
[4] R. Yamamoto, S. Matsuo and Y. Tanaka, *Int. J. Thermophys.*, **14**, 79 (1993).
[5] Y. Tanaka, M. Noguchi, H. Kubota and T. Makita, *J. Chem. Eng. Japan*, **12**, 171 (1979).
[6] Y. Tanaka, M. Nakata and T. Makita, *Int. J. Thermophys.*, **12**, 949 (1991).
[7] R. Mostert, Ph. D. Thesis, University of Amsterdam (1991).
[8] R. Mostert, H. R. van den Berg, P. S. van der Gulik and J. V. Sengers, *J. Chem. Phys.*, **92**, 5454 (1990).
[9] C. Zaihua, K. Tozaki and K. Nishikawa, *Jpn. J . Appl. Phys.,* **38**, L92 (1999).
[10] Y. Nagasaka, T. Hatakeyama, M. Okuda and A. Nagashima, *Rev. Sci. Instrum.*, **59**, 1156 (1988).
[11] K. Stephan, R. Krauss and A. Laesecke, *J. Phys. Chem. Ref. Data*, **16**, 993 (1987).
[12] J. Yata, private communication (1998).
[13] V. Vesovic, W. A. Wakeham, G. A. Olchowy, J. V. Sengers and J. T. R. Watson, *J. Phys. Chem. Ref. Data*, **19**, 763 (1990).
[14] Y. Tanaka and M. A. V. Mamaril, "*Physico-Chemical Properties for Chemical Engineering*", ed. Soc. Chem. Eng. Japan, **18**, p. 3 (1996).
[15] A. Michels, J. V. Sengers and P. S. van der Gulik, *Physica*, **28**, 1201 (1962).
[16] A. Michels, J. V. Sengers and P. S. van der Gulik, *Physica*, **28**, 1216 (1962).

[17] H. J. M. Hanley, J. V. Sengers and J. F. Ely, *Proceedings of 14th Int. Conf. on Thermal Conductivity,* ed. P. G. Klemens, T. K. Chu, Plenum Press, New York, p. 383 (1976).
[18] J. V. Sengers, *Int. J. Thermophys.*, **6**, 203 (1985).
[19] G. A. Olchowy and J. V. Sengers, *Int. J. Thermophys.*, **10**, 417 (1989).
[20] H. M. Roder, *Int. J. Thermophys.*, **6**, 119 (1985).
[21] H. M. Roder and C. A. Nieto de Castro, *High Temp.-High Press.*, **17**, 453 (1985).
[22] H. M. Roder and D. G. Friend, *Int. J. Thermophys.*, **6**, 607 (1985).
[23] A. Laesecke, R. A. Perkins and J. B. Howley, *Int. J. Refrig.*, **19**, 231 (1996).
[24] B. W. Tiesinga, E. P. Sakonidou, H. R. van den Berg, J. Luettmer-Strathmann and J. V. Sengers, *J. Chem. Phys.*, **101**, 6944 (1994).
[25] A. A. Clifford, P. Gray, A. I. Johns, A. C. Scott and J. T. R. Watson, *J. Chem. Soc. Faraday Trans. I*, **77**, 2679 (1981).
[26] A. C. Scott, A. I. Johns, J. T. R. Watson and A. A. Clifford, *J. Chem. Soc., Faraday Trans. I*, **79**, 733 (1983).
[27] A. I. Johns, S. Rashid, J. T. R. Watson and A. A .Clifford, *J. Chem. Soc., Faraday Trans. I*, **82**, 2235 (1986).
[28] A. I. Johns, S. Rashid, L. Rowan, J. T. R. Watson and A. A. Clifford, *Int. J. Thermophys.*, **9**, 3 (1988).
[29] R. A. Reid, J. M. Prausnitz and B. J. Poling, "*The Properties of Gases and Liquids*", McGraw-Hill, New York (1987).
[30] Y. Tanaka, S. Matsuo and S. Taya, *Int. J. Thermophys.*, **16**, 121 (1995).
[31] S. Matsuo, Y. Tanaka, N. Takada, H. Yamamoto and A. Sekiya, *J. Chem. Eng. Data*, **43**, 473 (1998).
[32] B. Le Neindre, Y. Garrabos and R. Tufeu, *Int. J. Thermophys.*, **12**, 307 (1991).
[33] P. Desmarest and R. Tufeu, *Int. J. Thermophys.*, **8**, 293 (1987).
[34] R. Tufeu and B. Le Neindre, *Int. J. Thermophys.*, **8**, 27 (1987).
[35] T. Makita, Y. Tanaka, Y. Morimoto, M. Noguchi and H. Kubota, *Int. J. Thermophys.*, **2**, 249 (1981).
[36] Y. Tanaka, H. Ueno, K. Arakawa, H. Kubota and T. Makita, *Proceedings of World Congress III of Chemical Engineering*, Tokyo, Japan, **2**, p. 79 (1986).
[37] N. Takada, S. Matsuo, Y. Tanaka and A. Sekiya, *J. Fluorine Chem.*, **91**, 81 (1998).
[38] J. Kestin, K. Knierim, E. A. Mason, B. Najafi, S. T. Ro and M. Waldman, *J. Phys. Chem. Ref. Data*, **13**, 229 (1984).
[39] E. Bich, J. Millat and E. Vogel, *J. Phys. Chem. Ref. Data*, **19**, 1289 (1990).
[40] H. J. M. Hanley, *J. Phys. Chem. Ref. Data*, **2**, 619 (1973).
[41] J. Assael, S. Mixafendi and W. A. Wakeham, *J. Phys. Chem. Ref. Data*, **15**, 1315 (1987).
[42] A. Boushehri, J. Bzowski, J. Kestin and E. A. Mason, *J. Phys. Chem. Ref. Data*, **16**, 445 (1987).
[43] H. J. M. Hanley and J. F. Ely, *J. Phys. Chem. Ref. Data*, **2**, 735 (1973).
[44] J. Millat and W. A. Wakeham, *J. Phys. Chem. Ref. Data*, **18**, 565 (1989).

[45] F. J. Uribe, E. A. Mason and J. Kestin, *J. Phys. Chem. Ref. Data*, **19**, 1123 (1990).
[46] M. J. Assael, J. Millat, V. Vesovic and W. A. Wakeham, *J. Phys. Chem. Ref. Data*, **19**, 1137 (1990).
[47] H. J. M. Hanley, W. M. Haynes and R. D. McCarty, *J. Phys. Chem. Ref. Data*, **6**, 597 (1977).
[48] K. Kodoya, N. Matsunaga and A. Nagashima, *J. Phys. Chem. Ref. Data*, **14**, 947 (1983).
[49] H. J. M. Hanley, R. D. McCarty and W. M. Haynes, *J. Phys. Chem. Ref. Data*, **3**, 979 (1974).
[50] B. A. Younglove and H. J. M. Hanley, *J Phys. Chem. Ref. Data*, **15**, 1323 (1986).
[51] H. J. M. Hanley and R. Prydz, *J. Phys. Chem. Ref. Data*, **1**, 1101 (1972).
[52] A. Laesecke, R. Krauss K. Stephan and W. Wagner, *J. Phys. Chem. Ref. Data*, **19**, 1089 (1990).
[53] N. Matsunaga and A. Nagashima, *J. Phys. Chem. Ref. Data*, **12**, 933 (1983).
[54] M. L. V. Ramires, C. A. Nieto de Castro, Y. Nagasaka, A. Nagashima, M. J. Assael and W. A. Wakeham, *J. Phys. Chem. Ref. Data*, **24**, 1377 (1995).
[55] J. V. Sengers, J. T. R. Watson, R. S. Basu, B. Kamgar-Parsi and R. C. Hendricks, *J. Phys. Chem. Ref. Data*, **13**, 893 (1984).
[56] J. V. Sengers and J. T. R. Watson, *J. Phys. Chem. Ref. Data*, **15**, 1291 (1986).
[57] K. Stephan and A. Laesecke, *J. Phys. Chem. Ref. Data*, **14**, 227 (1985).
[58] P. M. Holland, B. E. Eaton and H. J. M. Hanley, *J. Phys. Chem. Ref. Data*, **12**, 917 (1983).
[59] H. J. M. Hanley, K. E. Gubbins and S. Murad, *J. Phys. Chem. Ref. Data*, **6**, 1167 (1977).
[60] V. Vesovic, W. A. Wakeham, J. Luettmer-Strathmann, J. V. Sengers, J. Millat, E. Vogel and M. J. Assael, *Int. J. Thermophys.*, **15**, 33 (1994).
[61] P. M. Holland, H. J. M. Hanley, K. E. Gubbins and J. M. Haile, *J. Phys. Chem. Ref. Data*, **8**, 559 (1977).
[62] M. J. Assael, M. L. V. Ramires, C. A. Nieto de Castro and W. A. Wakeham, *J. Phys. Chem. Ref. Data*, **19**, 113 (1990).
[63] C. A. Nieto de Castro, S. F. Y. Li, A. Nagashima, R. D. Trengove and W. A. Wakeham, *J. Phys. Chem. Ref. Data*, **15**, 1073 (1986).
[64] R. Krauss and K. Stephan, *J. Phys Chem. Ref. Data*, **18**, 43 (1989).
[65] Y. Tanaka and T. Sotani, *Int. J. Thermophys.*, **17**, 293 (1996).
[66] R. Krauss, J. Luettmer-Strathmann, J. V. Sengers and K. Stephan, *Int. J. Thermophys.*, **14**, 951 (1993).
[67] R. Krauss, V. C. Weiss, T. A. Edison, J. V. Sengers and K. Stephan, *Int. J. Thermophys.*, **17**, 731 (1996).
[68] H. Lee, *Ind. Eng. Chem. Res.*, **29**, 488 (1990).

3.3 Binary Diffusion Coefficient

[1] K. K. Liong, P. A. Wells and N. R. Foster, *J. Supercrit. Fluids*, **4**, 91 (1991).
[2] G. Taylor, *Proc. Roy. Soc. London*, Ser. A., **219**, 186 (1953).
[3] G. Taylor, *Proc. Roy. Soc. London*, Ser. A., **223**, 446 (1954).
[4] G. Taylor, *Proc. Roy. Soc. London*, Ser. A., **225**, 473 (1954).
[5] R. Aris, *Proc. Roy. Soc. London*, Ser. A., **235**, 67 (1956).
[6] T. Funazukuri and N. Wakao, *Preprint for the AIChE 1993 Annual Meeting* (St. Louis) (1993).
[7] J. O. Hirshfelder, C. F. Curtiss and R. B. Bird, "*Molecular Theory of Gases and Liquids*", John Wiley & Sons, New York (1954).
[8] S. J. Bertucci and W. H. Flygare, *J. Chem. Phys*., **63**, 1 (1975).
[9] D. Chandler, *J. Chem. Phys*., **62**, 1358 (1975).
[10] R. A. Reid, J. M. Prausnitz and B. J. Poling, "*The Properties of Gases and Liquids*", McGraw-Hill, New York (1987).
[11] W. A. Wakeham, A. Nagashima and J. V. Sengers eds., "*Experimental Thermodynamics Vol. III: Measurements of the Transport Properties of Fluids*", Blackwell Scientific Publications, Oxford (1991).
[12] A. Alizadeh, C. A. Nieto de Castro and W. A. Wakeham, *Int. J. Thermophy*., **1**, 243 (1980).
[13] H. Nishiumi, M. Fujita and K. Ago, *Fluid Phase Equilibria*, **117**, 356 (1996).
[14] K. Ago and H. Nishiumi, *Ind. Eng. Chem. Res*., **37**, 1692 (1998).
[15] T. Funazukuri and N. Nishimoto, *Fluid Phase Equilibria*, **125**, 235 (1996).
[16] T. Funazukuri, S. Hachisu and N. Wakao, *Anal. Chem*., **61**, 118 (1989).
[17] T. Funazukuri, S. Hachisu and N. Wakao, *Ind. Eng. Chem. Res*., **30**, 1323 (1991).
[18] T. Funazukuri, Y. Ishiwata and N. Wakao, *AIChE J*., **38**, 1761 (1992).
[19] T. Funazukuri, C. Y. Kong and S. Kagei, *Int. J. Thermophys.,* **21**, 651 (2000).
[20] C. M. Silva and E. A. Macedo, *Ind. Eng. Chem. Res*., **37**, 1490 (1998).
[21] A. Akgerman, C. Erkey and M. Orejuela, *Ind. Eng. Chem. Res*., **35**, 911 (1996).
[22] K. K. Liong, P. A. Wells and N. R. Foster, *Ind. Eng. Chem. Res*., **30**, 1329 (1991).
[23] K. K. Liong, P. A. Wells and N. R. Foster, *Ind. Eng. Chem. Res*., **31**, 390 (1992).
[24] T. Wells, N. R. Foster and R. P. Chaplin, *Ind. Eng. Chem. Res*., **31**, 927 (1992).
[25] C. C. Lai and C. S. Tan, *Ind. Eng. Chem. Res*., **34**, 674 (1995).
[26] T. Funazukuri, C. Y. Kong and S. Kagei, *Ind. Eng. Chem. Res*., **39**, 4462 (2000).
[27] G. Knaff and E. U. Schlünder, *Chem. Eng. Process*., **21**,101 (1987).
[28] H. Higashi, Y. Iwai, Y. Takahashi, H. Uchida and Y. Arai, *Fluid Phase Equilibria*, **144**, 269 (1998).
[29] O. J. Catchpole and M. B. King, *Ind. Eng. Chem*., **33**, 1828 (1994).

[30] H. Higashi, Y. Iwai, Y. Nakamura, S. Yamamoto and Y. Arai, *Fluid Phase Equilibria*, **166**, 101 (1999).
[31] D. M. Lamb, S. T. Adamy, K. W. Woo and J. Jonas, *J. Phys. Chem.*, **93**, 5002 (1989).
[32] M. B. Iomtev and Y. V. Tsekhanskaya, *Russ. J. Phys. Chem.*, **38**, 485 (1964).
[33] Y. V. Tsekhanskaya, *Russ. J. Phys. Chem.*, **45**, 744 (1971).
[34] E. G. Vinkler and V. S. Morozov, *Russ. J. Phys. Chem.*, **49**, 1405 (1975).
[35] I. Swaid and G. M. Schneider, *Ber. Bunsenges. Phys. Chem.*, **83**, 969 (1979).
[36] R. Feist and G. M. Schneider, *Sep. Sci. & Technol.*, **17**, 261 (1982).
[37] H. H. Lauer, D. McManigill and R. D. Board, *Anal. Chem.*, **55**, 1370 (1983).
[38] P. G. Debenedetti and R. C. Reid, *AIChE J.*, **32**, 2034 (1986).
[39] P. R. Sassiat, P. Mourier, M.H. Caude and R.H. Rosset, *Anal. Chem.*, **59**, 1164 (1987).
[40] T. J. Bruno, *J. Res. Nat. Inst. Stand. Technol.*, **94**, 105 (1989).
[41] N. Dahmen, A. Dülberg and G. M. Schneider, *Ber Bunsenges. Phys. Chem.*, **94**, 384 (1990). Errata, **94**, 710 (1990).
[42] N. Dahmen, A. Kordikowski and G. M. Schneider, *J. Chromatography*, **505**, 169 (1990).
[43] C. Erkey, H. Gadalla and A. Akgerman, *J. Supercrit. Fluids*, **3**, 180 (1990).
[44] S. V. Olesik and J. L. Woodruff, *Anal. Chem.*, **63**, 670 (1991).
[45] T. J. Bruno, "*Supercritical Fluid Technology - Reviews in Modern Theory and Application*", ed. T. J. Bruno and J. F. Ely, pp. 293-324, CRC Press, Boca Raton (1991).
[46] S. Umezawa and A. Nagashima, *J. Supercrit. Fluids*, **5**, 242 (1992).
[47] J. L. Bueno, J. J. Suárez, J. Dizy and I. Medina, *J. Chem. Eng. Data*, **38**, 344 (1993).
[48] J. M. H. L. Sengers, U. K. Deiters, U. Klask, P. Swidersky and G. M. Schneider, *Int. J. Thermophys.*, **14**, 893 (1993).
[49] V. M. Shenai, B. L. Hamilton and M. A. Matthews, "*Supercritical Fluid Engineering Science: Fundamental and Applications*", ed. E. Kiran and J. F. Brennecke, ACS Symposium Series, **514**, p. 92 (1993).
[50] J. J. Suárez, J. L. Bueno and I. Medina, *Chem. Eng. Sci.*, **48**, 2419 (1993).
[51] T. J. Funazukuri, *J. Chem. Eng. Jpn.*, **29**, 191 (1996).
[52] D.Q. Tuan, M. E. Yener, J. A. Zollweg, P. Harriott and S. S. H. Rizvi, *Ind. Eng. Chem. Res.*, **38**, 554 (1999).
[53] A. Kopner, J. Hamm, J. Ellert, R. Feist and G. M. Schneider, *Chem. Eng. Sci.*, **42**, 2213 (1987).
[54] A. Eaton, D. B. Bukur and A. Akgerman, *J. Chem. Eng. Data*, **40**, 1293 (1995).
[55] S. R. Springston and M. Novotny, *Anal. Chem.*, **56**, 1762 (1984).
[56] J. H. Dymond, *Chem. Soc. Rev.*, **14**, 317 (1985).

[57] H. Liu, C. M. Silva and E. A. Macedo, *Ind. Eng. Chem. Res.*, **36**, 246 (1997).
[58] H. Liu and E. Ruckenstein, *Ind. Eng. Chem. Res.*, **36**, 5488 (1997).
[59] H. J. V. Tyrrell and K. R. Harris, "*Diffusion in Liquids*", Butterworth, London (1984).
[60] T. Funazukuri and N., Wakao, *Kagaku Kogaku Ronbunshu*, **21**, 204 (1995).
[61] M. A. Matthews and A. Akgerman, *AIChE J.*, **33**, 881 (1991).
[62] J. Dymond, *J. Chem. Phys.*, **60**, 969 (1974).
[63] K. R. Harris, *Physica*, **94A**, 448 (1978).
[64] K. R. Harris and N. J. Trappeniers, *Physica*, **104A**, 262 (1980).
[65] K. R. Harris, *J. Chem. Soc. Faraday Trans. I*, **78**, 2265 (1982).
[66] P. W. E. Peereboom, H. Luigjes and K. O. Prins, *Physica*, **156A**, 260 (1989).
[67] P. Etesse, J. A. Zega and R. Kobayashi, *J. Chem. Phys.*, **97**, 2022 (1992).
[68] A. Bondi, *J. Phys. Chem.*, **68**, 441 (1964).
[69] J. J. Erpenbeck and W. W. Wood, *Physical Review*, **A 43**, 4254 (1991).
[70] E. L. Cussler, "*Diffusion, Mass Transfer in Fluid Systems*", Cambridge University Press, New York, p. 195 (1984).
[71] S. T. Lee and S. V. Olesik, *Anal. Chem.*, **66**, 4498 (1994).
[72] I. Souvignet and S. V. Olesik, *Anal. Chem.*, **70**, 2783 (1998).
[73] T. Funazukuri and Y. Ishiwata, *Fluid Phase Equilibria*, **164**, 117 (1999).
[74] T. Funazukuri, N. Nishimoto and N. Wakao, *J. Chem. Eng. Data*, **39**, 911 (1994).
[75] T. Funazukuri, C. Y. Kong and S. Kagei, *Ind. Eng. Chem. Res.*, **39**, 835 (2000).
[76] J. T. Reaves, A. T. Griffith and C. B. Roberts, *J. Chem. Eng. Data*, **43**, 683 (1998).
[77] K. Ago, Y. Kataoka and H. Nishiumi, *JCPE Journal*, **11**, 77 (1999).
[78] K. Ago and H. Nishiumi, *J. Chem. Eng. Jpn.*, **32**, 563 (1999).
[79] D. K. Dysthe and B. Hafskjold, *Int. J. Thermophys.*, **16**, 1213 (1995).

4 Extraction and Separation Using Supercritical Fluids Using Supercritical Fluids

4.1 Principles for Supercritical Fluid Separation

4.1.1 Introduction

This section gives an introduction to the principles governing the use of a supercritical fluid (SCF) as a solvent for separation. The content is intended for explanation rather than review and references are cited only to major developments. The following three subsections (Sect. 4.1.2–4.1.4) are on basic aspects of phase equilibria and the subsequent five subsections (Sect. 4.1.5 – 4.1.9) on separation processes using a supercritical fluid as a separating agent.

4.1.2 Solubility of a Nonvolatile Compound in a Supercritical Fluid

The solvating power of a supercritical fluid is highly dependent on its temperature and pressure. Figure 4.1.1 shows that at low pressure the solubility of

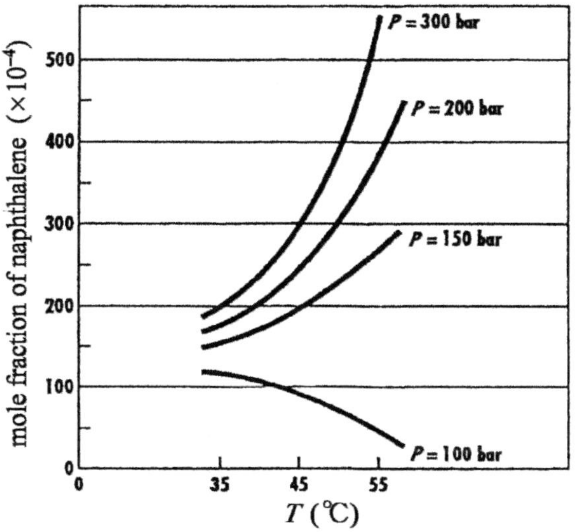

Fig. 4.1.1 Temperature dependence of solubility of naphthalene in supercritical CO_2 [12]. Reprinted from: L.T. Taylor: *Supercritical Fluid Extraction* (Wiley 1996) p. 5

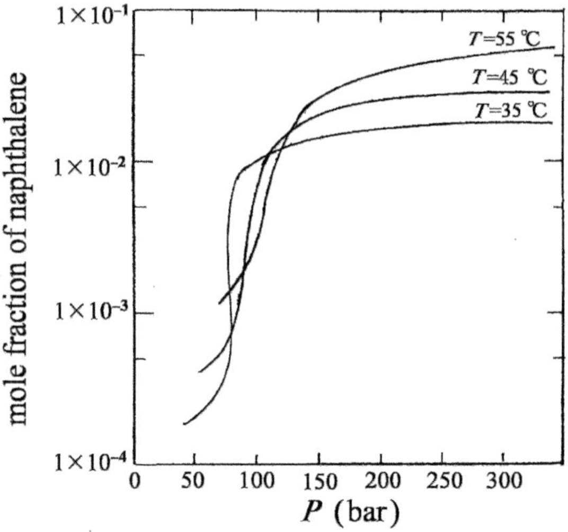

Fig. 4.1.2 Pressure dependence of solubility of naphthalene in supercritical CO_2

naphthalene in CO_2 surprisingly decreases with rising temperature; whereas at higher pressures it increases in a straightforward fashion. Figure 4.1.2 illustrates the change of the solubility of naphthalene with pressure under isothermal conditions. Beyond the critical pressure of CO_2 (73.8 bar) the solubilities of naphthalene in CO_2 are enhanced by one or two orders of magnitude. Between 7.5 and 13 MPa, reducing temperature also causes the solubility to be enhanced. This is called a *retrograde phenomenon* or a *temperature inversion effect* and is a manifestation of the complex relationship between the solubility and the temperature. When taking the density as the parameter in place of pressure, as shown in Fig. 4.1.3, the temperature dependence of the solubility becomes much simpler. From Figs. 4.1.2 and 4.1.3, the retrograde phenomenon comes about because the density of CO_2 increases dramatically with decrease in temperature close to its critical point. The density, not pressure, to a first approximation, determines the solvating power and solubility.

There are many empirical correlation methods of solubilities. Here the approaches recommended in the new book of Clifford [1] are described below. These approaches are to fit the data to simple equations, which can be used for design calculations. The *enhancement factor*, E, is defined as:

$$E = y_2 P / P_2^{\mathrm{S}}, \qquad (4.1.1)$$

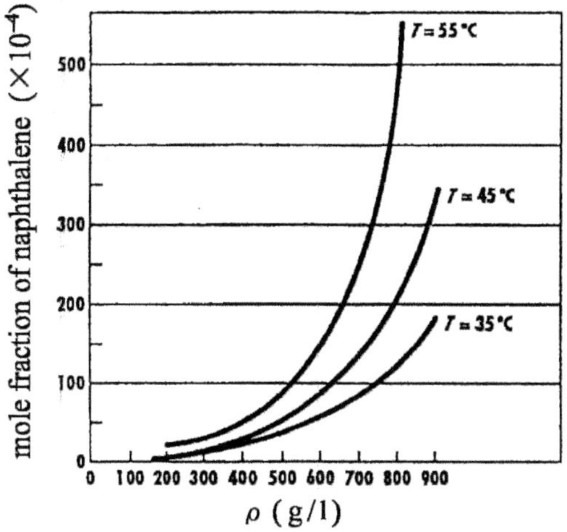

Fig. 4.1.3 Density dependence of solubility of naphthalene in supercritical CO_2 [12]

where y_2 is the solubility in mole fractions, P and P_2^S are pressure and vapor pressure of the solute respectively. Taking the natural logarithm of E, it is fitted to a linear equation of density:

$$\ln E = A + c\rho\,, \tag{4.1.2}$$

where A and c are constants. Bartle et al. [2] examined this approach for some 90 substances. To avoid extrapolation to zero density for the determination of the constant A, (4.1.2) was modified by introducing a reference density ρ_{ref} (= 700 kg/m^3):

$$\ln E = A + c(\rho - 700)\,. \tag{4.1.3}$$

Another problem is that vapor pressures are sometimes not available or not reliable for many compounds, particularly solid solutes because of experimental difficulties in measuring very low vapor pressures. Then (4.1.1) is substituted into (4.1.3):

$$\ln(y_2 P / P_{ref}) = A' + c(\rho - 700)\,, \tag{4.1.4}$$

where $A' = \ln(P/P_{ref}) + A$ and P_{ref} is a reference pressure (= 0.1 MPa). A' can be assumed to have the following equation based on similar behavior of the vapor pressure.

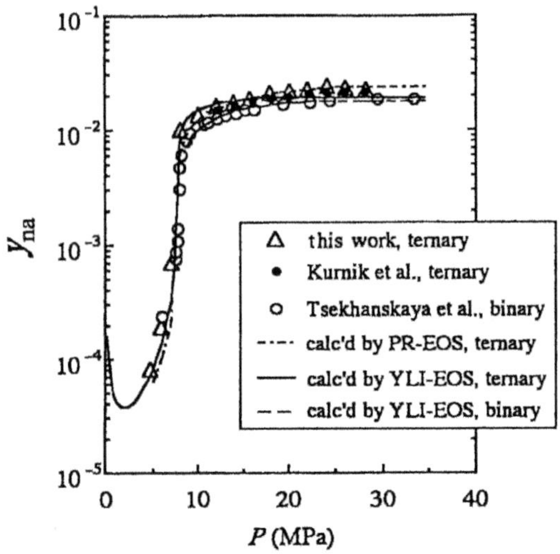

Fig. 4.1.4 Comparison of solubilities of naphthalene in CO_2 for CO_2 + naphthalene + phenanthrene and CO_2 + naphthalene systems [4]

$$A' = a + b/T, \tag{4.1.5}$$

where a and b are constants.

4.1.3 Solubility of a Solid Solute Mixture in a Supercritical Fluid

Supercritical fluid extraction processes for a multicomponent mixture of solutes are of practical interest, but the solubility data of mixed solutes are very scarce. The solubility isotherms of naphthalene + phenanthrene in CO_2 were measured by Kurnik and Reid [3] and Liu and Nagahama [4]. The data for naphthalene and phenanthrene are plotted as a function of pressure in Figs. 4.1.4 and 4.1.5, respectively. For comparison purposes, the solubilities of pure naphthalene and pure phenanthrene in CO_2 are also shown in the Figs. 4.1.4 [4] and 4.1.5 [4]. The solubilities of each solute in the ternary system (CO_2 (1) + naphthalene (2) + phenanthrene (3)) are higher by approximately 30% for naphthalene and 130% for phenanthrene than those in the binary system (CO_2 + naphthalene and CO_2 + phenanthrene) over the range of 10–30 MPa. The enhancement is due to the entrainer or cosolvent effect of the second solute component added to the first one. As a result of the solubility enhancement, the selectivity (= the ratio of solubility of each component in CO_2) calculated from the ternary data are plotted in Fig. 4.1.6 [4].

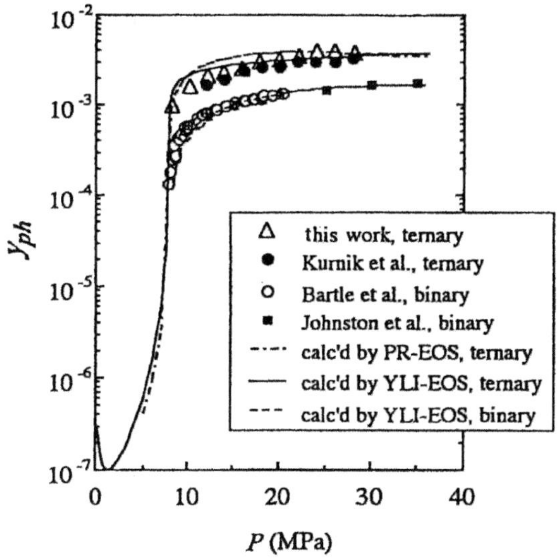

Fig. 4.1.5 Comparison of solubilities of phenanthrene in CO_2 for CO_2 + naphthalene + phenanthrene and CO_2 + phenanthrene systems [4]

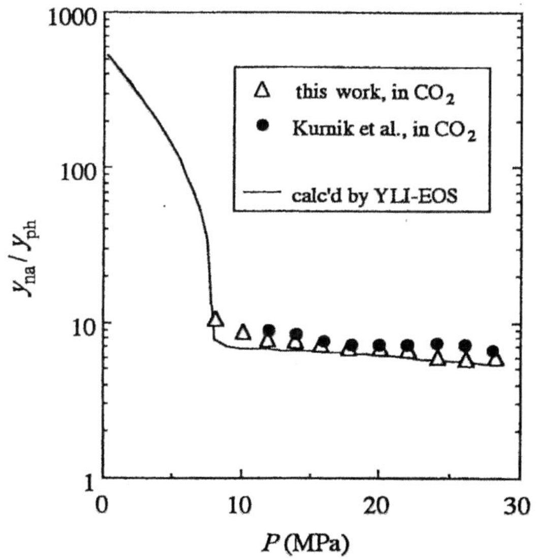

Fig. 4.1.6 Pressure dependence of selectivities of naphthalene to phenanthrene for the CO_2 + naphthalene + phenanthrene system at 308.2 K [4]. Reprinted from: G.-T. Liu, K. Nagahama: Solubility of organic solid mixture in supercritical fluids, *J. Supercritical Fluids* (Elsevier Science) **9**, 152 (1996)

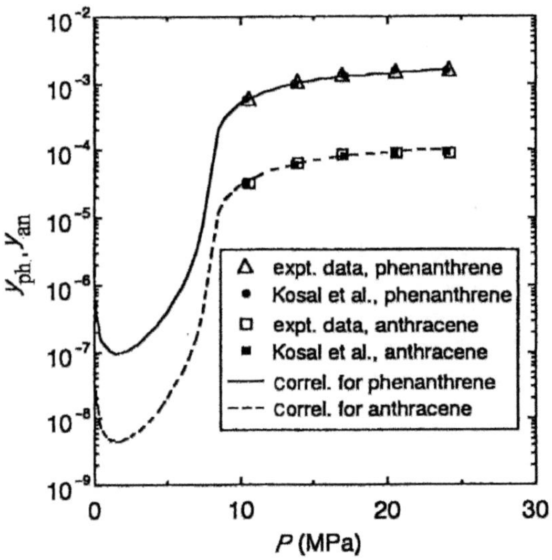

Fig. 4.1.7 Solubilities of phenanthrene and anthracene in CO_2 at 308.2 K (feed: a simple mixture of both pure solids) [5]

This figure shows that the selectivity of naphthalene to phenanthrene decreases drastically with increasing pressure. As the solid–liquid equilibrium of naphthalene + phenanthrene binary is a simple eutectic system with a eutectic temperature of 322.7 K, the solubility of those solutes in CO_2 is independent of the feed composition under fixed pressure and temperature. This is due to the fact that the number of Gibbs degrees of freedom for the solid–fluid phase equilibrium of the mixture containing CO_2 and eutectic binary is only two.

On the other hand, the solid–liquid equilibrium of the phenanthrene + anthracene binary is a solid solution system, that is, a single solid phase coexists with a liquid phase. The solubilities of the phenanthrene + anthracene mixture in CO_2 are dependent on the state of the feed mixture. If a simple mixture of two pure solids is used as a feed, the solubilities of each component are independent of feed composition at constant temperature and pressure as shown in Fig. 4.1.7 [5]. If a solid solution mixture prepared by melting, cooling and granulating is used as a feed solid, the mole fractions of each component in the fluid phase are dependent on the feed composition even at constant temperature and pressure as shown in Figs. 4.1.8 [5]. Figure 4.1.8 (b) is a solid–fluid phase equilibrium diagram with a solid solution as a feed. In this case, the number of Gibbs degree of freedom is three and the composition of solutes in CO_2 can vary in different feed compositions.

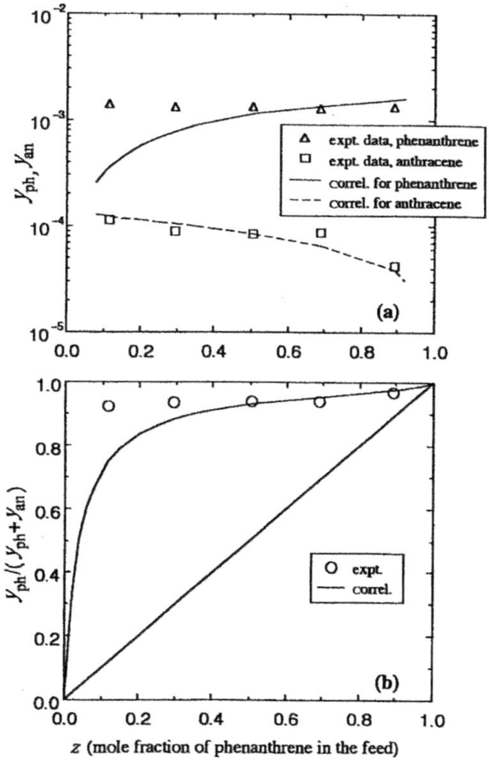

Fig. 4.1.8 Solubilities of phenanthrene and anthracene in CO_2 at 308.2 K and 17.0 MPa (feed: solid solution mixture): **(a)** dependence of solubilities on feed composition, **(b)** solid–fluid phase equilibrium relationship on CO_2, free basis [5]

In Figs. 4.1.4–4.1.7, the calculated curves are also plotted. The results were obtained by correlating the experimental data using the following equations [4, 5] (see (2.4.14)):

$$y_i = (P_i^S / P)(1/\phi_i) \cdot \exp[v_i^{sol}(P - P_i^S)/RT], \tag{4.1.6}$$

where y_i is the solubility of solute i in mole fraction, P_i^S is the vapor pressure of the pure solid component, v_i^{sol} is its molar volume, and P and T represent pressure and temperature. The fugacity coefficient ϕ_i was computed from an equation of state (EOS) by introducing adjustable binary interaction parameters between unlike components. Three types of EOS, the Soave–Redlich–Kwong (SRK) EOS, Peng–Robinson (PR) EOS and Yu–Lu–Iwai

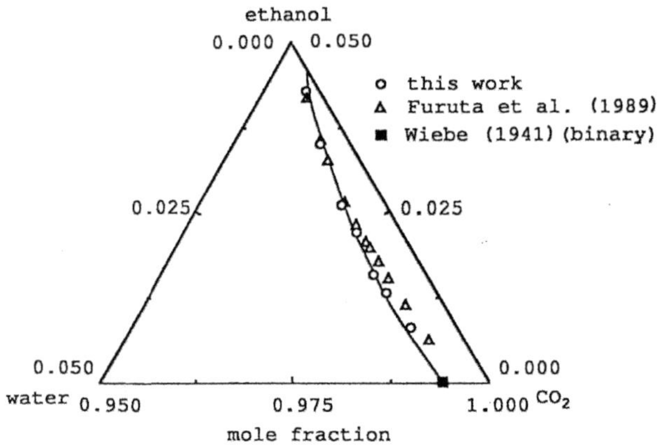

Fig. 4.1.9 Fluid phase composition in fluid–liquid equilibrium for the CO_2 + water + ethanol system at 333 K and 10 MPa [6]

EOS, were employed. The experimental solubilities are represented well by the above correlating method.

4.1.4 Fluid–Liquid Equilibria for a Ternary Mixture

Supercritical fluid extraction has been expected as an energy-saving process for the dehydration of alcohol. In order to develop the process, fluid–liquid equilibrium (FLE) data of supercritical fluid + alcohol + water mixtures are needed.

A lot of FLE data for the CO_2 + ethanol (EtOH) + water system have been reported in the literature. Figure 4.1.9 shows the fluid phase composition of the CO_2 + EtOH + water system at 333 K and 10 MPa [6] and Fig. 4.1.10 gives an X–Y diagram, where X and Y are the compositions of EtOH on the CO_2 free basis of the liquid and fluid phases. From Fig. 4.1.9 the mole fraction of EtOH in the fluid phase is less than 0.05 and that of water is less than 0.006. In Fig. 4.1.10 vapor–liquid equilibrium (VLE) data of the EtOH + water system at atmospheric pressure is also plotted, which means this system has an azeotropic point ($X = Y$) and this makes it impossible to dehydrate EtOH further by conventional distillation. On the other hand, the X–Y relationship of FLE data for CO_2 + EtOH + water at 333 K and 10 MPa indicates the possibility of complete dehydration of aqueous EtOH mixtures as shown in Fig. 4.1.10.

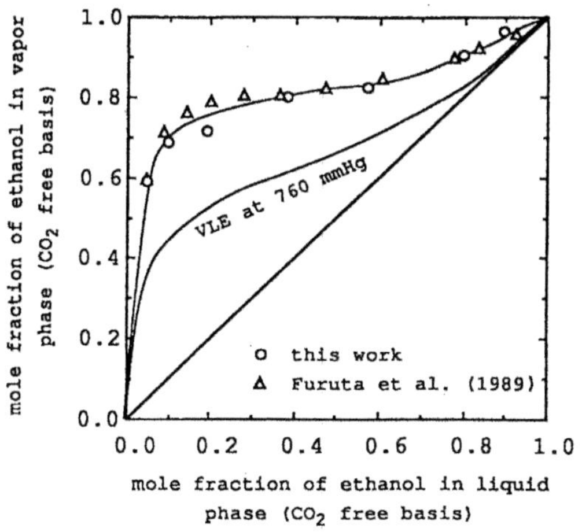

Fig. 4.1.10 Fluid–liquid equilibrium composition diagram for the CO_2 + water + ethanol system at 333 K and 10 MPa (CO_2 free basis) [6]

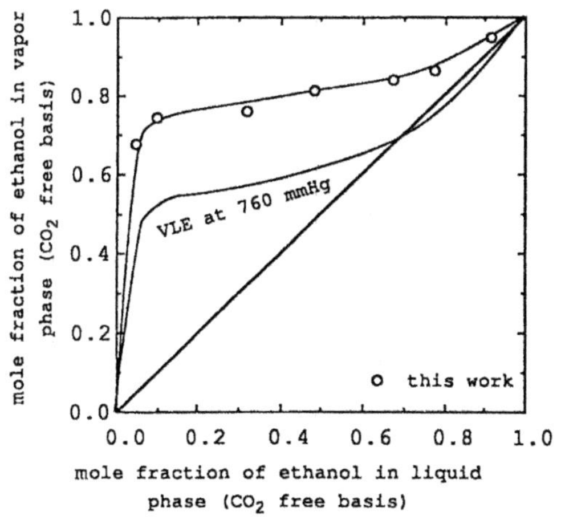

Fig. 4.1.11 Fluid–liquid equilibrium composition diagram for the CO_2 + water + 2-propanol system at 333 K and 10 MPa (CO_2 free basis) [6]

Similarly the 2-propanol + water system has an azeotropic point and it can be broken by introducing supercritical CO_2 into an aqueous 2-propanol solution. Figure 4.1.11 [6] shows the *X-Y* diagram for the CO_2 + 2-propanol + water system at 333 K and 10 MPa.

In general, EOS methods are most widely used to represent or calculate fluid–liquid equilibria. Especially simple cubic EOSs such as SRK-EOS and

PR-EOS may be successful in modeling the FLE. One of the significant questions faced when using cubic EOS, is whether the binary interaction parameter can be set equal to zero. For nonpolar mixtures containing hydrocarbons or nonpolar gases, cubic EOSs give good FLE prediction with the interaction parameter set to zero. Provided that the binary interaction parameter is fitted or adjusted to the data, the EOSs can handle polar mixtures such as CO_2 + alcohol + water. However, these EOSs were originally proposed for handling nonpolar substances and sometimes very large deviations between experimental and calculated FLE results may appear in the prediction of polar systems.

4.1.5 Adsorption in the Presence of a Supercritical Fluid

Studies on adsorption behavior under a supercritical fluid are needed to understand or develop supercritical fluid chromatography and adsorption/desorption separation with a supercritical fluid such as regeneration of loaded adsorbents and soil remediation. However, the number of adsorption equilibrium data and dynamic data under a supercritical fluid of practical interest are very scarce.

The adsorption equilibrium loadings of CO_2 on BPL activated carbon were measured by Tanida et al. [7] as shown in Figs. 4.1.12 and 4.1.13. They determined both apparent (Fig. 4.1.12) and true (Fig. 4.1.13) amounts of adsorbed CO_2 by use of a constant-volume apparatus. The former were determined directly from the mass balance between fed and remaining amounts of CO_2 in an equilibrium cell. The latter were corrected data by tak-

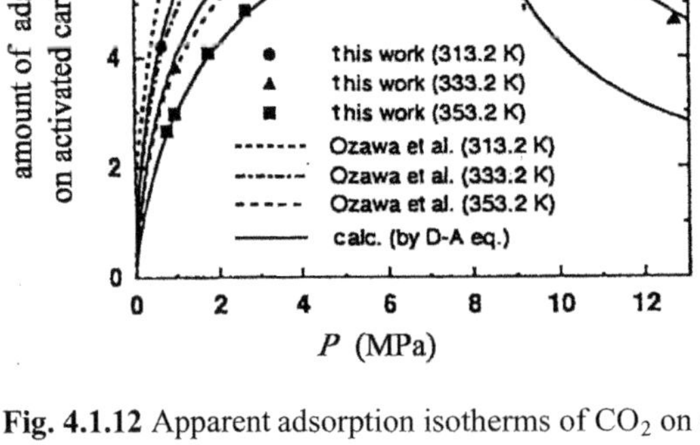

Fig. 4.1.12 Apparent adsorption isotherms of CO_2 on activated carbons [7]

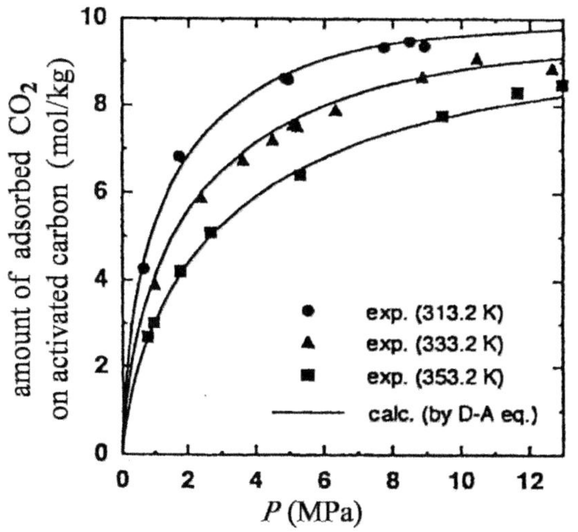

Fig. 4.1.13 True adsorption isotherms of CO_2 on activated carbons [7]

ing account of the density change of the adsorbed phase. From Fig. 4.1.12 curves of apparent adsorbed CO_2 at constant temperatures have maximum values and the maximum decreases with rising temperature. After passing the maximum, it looks like the behavior of liquid adsorption. However, the true amount of adsorbed CO_2 data, which were evaluated using the adsorbed phase density, have no irregular behavior and exhibit a general trend like gas adsorption under high pressure.

Some work on the adsorption equilibrium and dynamics of benzene on BPL activated carbon has been done [8, 9]. Equilibrium data are plotted in Fig. 4.1.14 along with the curves correlated by the Langmuir equation [9]. The equilibrium data were measured by time integration of the breakthrough curve using a fixed bed. The equilibrium loadings of benzene at constant pressure decrease with decreasing temperature and those at constant temperature increase with decreasing pressure. This behavior may be completely contrary to equilibrium relationships generally observed in gas adsorption at low pressure. It may be due to the following factors: first, the amount of CO_2 adsorbed increases with rising pressure as shown in Fig. 4.1.13 and also the adsorption comes about competitively between CO_2 and benzene; second, the solvating power of supercritical CO_2, that is the solubility of benzene, increases along with higher pressure.

For the design of adsorption processes the dynamic or rate data of adsorption under supercritical fluids are required. Porto et al. [9] studied the

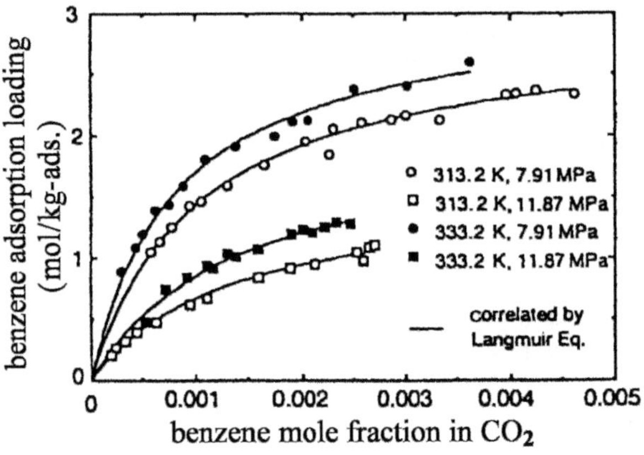

Fig. 4.1.14 Adsorption isotherms of benzene on activated carbons in the presence of CO_2 [9]

dynamics of adsorption of benzene dissolved in supercritical CO_2 on BPL activated carbon. They analyzed measured breakthrough curves and identified the controlling mass transfer process. Surface diffusion showed the main contribution to the diffusional flux in macropore and the effective surface diffusivities were in the order of 10^{-10} m^2/s. The orders of diffusivity were similar to those obtained from gas adsorption experiments at low pressure using activated carbon.

4.1.6 Supercritical Fluid Extraction

The extraction of valuable substances or the removal of useless materials using supercritical fluids is called supercritical fluid extraction (SFE). These processes have been operating on a commercial scale as well as an analytical scale for more than a decade. Large scale processes are the decaffeination of coffee beans and the production of hops extracts.

The SFE process basically consists of two steps: the extraction step and the separation step. In the extractor, the supercritical fluid (solvent) flows through the feed material and dissolves the soluble components (solutes). In the separator, the loaded solution (solvent + solutes) is expanded to low pressure and the extract is precipitated. Figure 4.1.15 shows typical SFE processes. Diagram (a), just described above, is the most popular scheme especially for extraction of natural products. In diagram (b), the precipitation is carried out by means of temperature change. In diagram (c), the separation takes place by

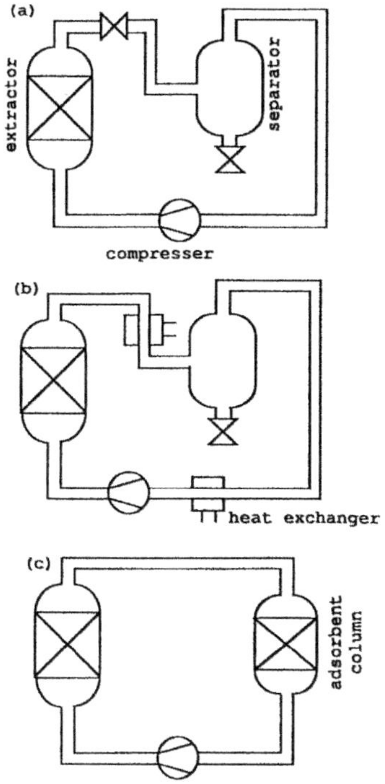

Fig. 4.1.15 Schematic diagram of typical supercritical fluid extraction processes

adsorbing the extract at constant pressure and temperature. In larger-scale plants, the unloaded solvent is often recycled to the extractor.

Extraction of the target component from solid materials such as coffee beans is influenced by the solubility, transportation by diffusion and the matrix of the solid materials. Extraction from solid materials is carried out by continuously contacting them with a constant flow rate of supercritical fluid solvent. Mostly the solid material forms a fixed bed, by which the solvent can extract the target compounds until they are depicted. Figure 4.1.16 [1] gives typical examples of extraction curves (extraction percent vs. time) which can be obtained in SFE experiments with a flow method as shown in Fig. 4.1.17. Curve (a) is obtained if the extraction process is controlled by the diffusion rate. Curve (b) is formed when the matrix effect is significant. When the solubility of the target compound is low, behavior like curve (c) is obtained. The gradient of the first part of the curves can be given by the equilibrium solubility, that is, the solubility may be determined from the gradients. But it is noted that the straight line can often be obtained by constant mass resistance. From

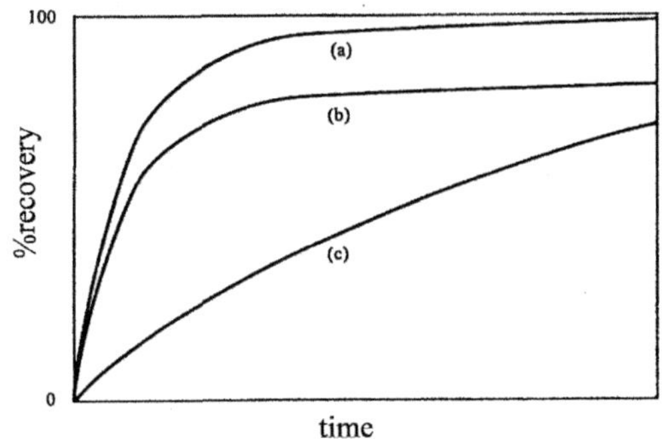

Fig. 4.1.16 Typical extraction curves of solid solutes: (**a**) diffusion-controlled, (**b**) significant matrix effect and (**c**) very low solubility of solute [1]

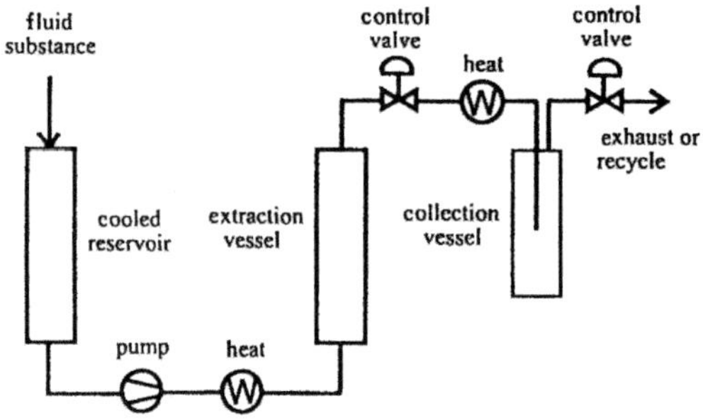

Fig. 4.1.17 Simple flow method for supercritical fluid extraction [1]

curve (a) in Fig. 4.1.16, which is a typical course of extraction from oil seeds, the extraction rate (extracted amount per unit time) is large and constant at the beginning of the period of extraction and rapidly declines during the second period. In cases of low initial concentration of extract in a solid substrate, such as the extraction of caffeine from coffee beans, the extraction rate is not high even at the beginning of the period and continuously declines over whole time course. As a result the extraction behavior becomes curve (c) in Fig. 4.1.16. The extraction curves may often have an initial region where the extraction rate increases with time [10]. This is caused by starting up the extraction with an extract free solvent. The extent of the initial region is dependent on the residence time of the solvent in the extractor. In the case of long residence time, such an initial region can extend over the whole extraction period.

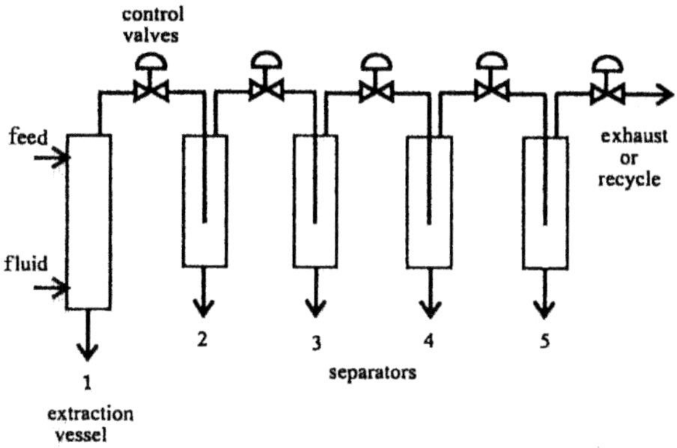

Fig. 4.1.18 Schematic diagram of polymer fractionation by isothermal decreasing pressure profiling [1]

4.1.7 Supercritical Fluid Fractionation of Polymers

Supercritical fluid fractionation of polymers can be carried out using isothermal decreasing pressure profiling or isothermal increasing pressure profiling [11]. The former is achieved by employing a multiple, sequential pressure reduction system as shown in Fig. 4.1.18 [1]. Using this system a synthetic oil or polymer can be fractionated into a homogeneous series of members continuously. A synthetic oil with a wide molecular weight distribution is fed at the top of the extractor and the supercritical fluid enters at the bottom. The extractor is operated under the temperature and pressure where most of the feed dissolves, leaving the heaviest fraction (fraction 1) that is eluted from the extractor. The extracted solution of lighter polymer and supercritical fluid is expanded through a control valve to precipitate a lighter fraction 2. The remaining solution is further reduced in pressure stagewise, which produces many fractions by molecular weight. After reaching the lowest pressure, the polymer-free fluid can be compressed again and recycled to the extractor.

Another polymer fractionation method is called isothermal increasing pressure profiling. The simple flow apparatus shown in Fig. 4.1.17 can be used. An amount of polymer or oil is charged to an extraction vessel and the extraction is carried out by increasing pressure stepwise. A low pressure is first used and the extraction is continued to the highest pressure until no more polymer fraction is collected in a cold trap vessel or until some maximum operating pressure is reached. Using the pressure increasing/collection cycle the charged polymer can be separated into several fractions based on molecular weight.

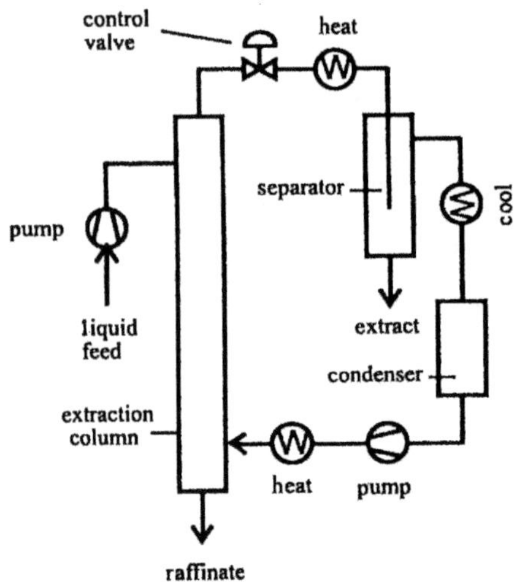

Fig. 4.1.19 Schematic diagram of countercurrent supercritical fluid extraction [1]

4.1.8 Multistaged Countercurrent Extraction

In order to improve the separation possible by simple batch supercritical extraction, multistaged countercurrent supercritical fluid extraction is applied. It looks like an intermediate between multistaged liquid–liquid extraction and fractional distillation with reflux. The advantages of the countercurrent process are firstly continuous and secondly more efficient than the batch extraction process. Batch extraction is a single-stage operation, whereas countercurrent extraction is a multistaged process in which a number of stages exist.

Figure 4.1.19 [1] illustrates countercurrent extraction using a supercritical fluid as a solvent. The main part is a vertical column involving packing or a tray. A liquid solution or solution containing solid solutes is fed at the top or the middle of the column. The extract phase is composed of the extracted solutes and a large amount of supercritical solvent emerges from the top of the column. The raffinate phase is composed of the remaining solution and a small amount of supercritical solvent and leaves the bottom of the column.

Clifford [1] divided the countercurrent processes into supercritical fluid extraction (top feed) and supercritical fractionation (middle feed) for convenience. In the latter case, refluxing occurs within the column by imposing a temperature gradient over the upper part of the column to precipitate the heavier components from the extracted phase. When the reflux is applied, the purity of the extract can be superior to that obtained without the reflux. However, a pure raffinate can be obtained without the reflux. This is because the lower part of the column corresponds to the stripping section and the upper part to the enrichment section, which are the terms used in distillation.

As mentioned before, multistaged countercurrent extraction using a supercritical solvent is similar to conventional distillation or liquid–liquid extraction. However, there are several limitations to the extent depending on the physical properties of supercritical solutions as follows:

(i) The density difference between both phases must be sufficient to avoid column flooding and this is critical in the stripping section.
(ii) The principles of supercritical fluid fractionation are different from those of distillation with liquid reflux.
(iii) The required height of the theoretical stages is inversely proportional on the flow rate.
(iv) Due to the very low solubility of solute in solvent, a high solvent to feed flow rate ratio is needed for reasonable production.
(v) Recovery or regeneration of spent solvent is much easier than that of liquid–liquid extraction.

4.1.9 Chromatographical Separation Using Supercritical Fluid

Chromatography has the highest separation efficiency compared with batch extraction and countercurrent extraction. For supercritical fluid chromatography (SFC), the mobile phase is a supercritical fluid and the stationary phase is a packed bed of solids or a capillary column wall coated with a thin layer.

The large diffusion coefficient of gases gives very narrow chromatographic peaks as seen in gas chromatography (GC). However, the diffusivity of liquids is many orders of magnitude smaller and gives wider chromatographic bands as in liquid chromatography (LC). The diffusion coefficient in a supercritical fluid is intermediate between gas and liquid data and also a supercritical fluid can dissolve many substances including low volatile compounds. Then SFC gives a useful option for the separation of a wide range of compounds. SFC can handle less volatile compounds which cannot be separated by GC. SFC is superior to LC in analytical speed and column efficiency.

The dissolving power of a supercritical fluid strongly depends on the density, which means the power can be adjusted by changing the temperature and pressure. The isothermal pressure programming (density programming) method can be used for SFC like temperature programming for GC and solvent programming for LC. The detailed theory and method for SFC are explained in Sect. 4.4.

4.1.10 Reduction of Organic Solvent Usage in Extraction

From an environmental point of view, today supercritical fluids are touted as a replacement for organic solvents. Liquid–liquid or liquid–solid extraction have very often been used and results in consuming large quantities of organic solvents. The US Environmental Protection Agency's program was designed to reduce industrial organic solvent usage in future. Its goal is a 33% reduction of 17 targeted chemicals by 1992 and a 50% reduction of these compounds by 1995, measured against the 1988 release inventory. Ten of the target chemicals are organic solvents: benzene, toluene, xylene, CCl_4, $CHCl_3$, CH_2Cl_2, tetrachloroethylene, trichloroethylene, trichloroethane, methyl ethyl ketone, methyl isobutyl ketone, which are commonly used for extraction in analytical and industrial scales.

Supercritical CO_2 offers good potential as a replacement solvent especially in analytical work and supercritical fluid extraction with CO_2 is a recent major topic [12] with over 300 references appearing in a year in *Chemical Abstracts*.

4.2 Process Development Studies for Separation of Chemicals

4.2.1 Introduction

In this section, we describe a number of process development studies with emphasis on the separation of chemicals, and highlight many facets of the development of supercritical fluid processes that need to be addressed, understood and solved for the evaluation of potential applications of supercritical fluid technology.

4.2.2 Separation of Organic + Water Liquid Mixtures

(a) Separation of Ethanol + Water

The oil shortages of the 1970s gave an impetus to the use of supercritical fluids for the separation of organic compounds from water. In particular, lots of workers published many papers and patents on the development of CO_2 separation of ethanol (EtOH) from aqueous solution produced by fermentation processes. McHugh and Krukonis [1] discussed the feasibility of CO_2 separation of EtOH from water. They showed that the distribution coefficient of

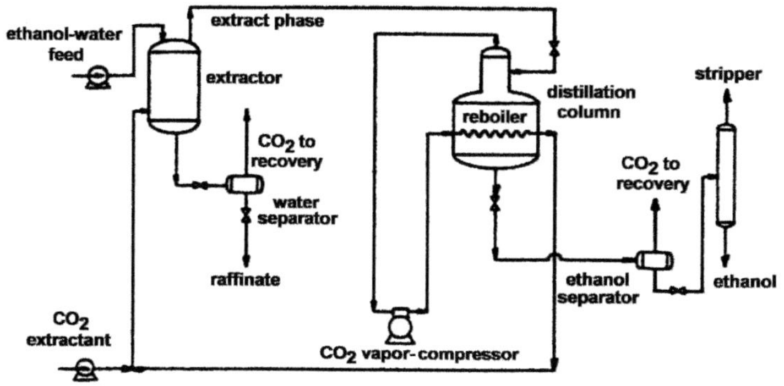

Fig. 4.2.1 Process diagram of dehydration of aqueous ethanol solution using CO_2 supercritical fluid extraction and distillation [1]. Reprinted from: M. McHugh, V. Kurukonis: *Supercritical Fluid Extraction*, 2nd ed. (Butterworth-Heinemann 1994) p. 170

EtOH in the range 0 wt% to 10 wt% EtOH averaged about 0.09 (weight basis) based on a number of distribution data sets over a wide range of extraction conditions, and also the selectivity of EtOH reached 81. Finally they concluded that due to the low distribution coefficients, the CO_2 process requires very high solvent-to-feed ratio and in order to achieve high recovery of EtOH from the feed stream, perfect separation of EtOH from the extract by use of distillation is necessary as shown in Fig. 4.2.1 [1], that is, it was needed to combine new supercritical CO_2 extraction with an old distillation process.

(b) Breaking Binary Azeotrope

Furuta et al. [2] and Suzuki et al. [3] reported that the CO_2 extraction of aqueous EtOH solution can break the azeotrope (90 mole% EtOH, at atmospheric conditions) based on the measured vapor–liquid equilibrium (VLE) data of the CO_2 + EtOH + water system. A diagram of the vapor composition under VLE conditions for the CO_2 + EtOH + water system measured by Suzuki et al. [3] at 333 K and 10 MPa is presented in Sect. 4.1 (see Fig. 4.1.9). The X–Y diagram for the system is also given in Fig. 4.1.10 [3] together with the VLE of the EtOH + water mixture at atmospheric pressure, where X and Y are liquid and vapor mole fractions on a CO_2 free basis. It is noted that it is possible to obtain pure EtOH using a multistaged extraction column since the vapor composition curve in Fig. 4.1.9 intersects the EtOH-CO_2 axis and also there is no azeotrope for the EtOH + water system as shown in Fig. 4.1.10. Before the measurements of Furuta et al. [2] and Suzuki et al. [3], most investigators concentrated on the VLE at 313 K and 10 MPa, at which condition there exists a plait (or critical) point for the CO_2 + EtOH + water system and also an azeotropic point ($X = Y$) for the EtOH + water system. When using hydrocarbons (ethane, propane and propylene) as a supercritical solvent to dehydrate aqueous EtOH solution, they can extract only EtOH by countercurrent extraction [4]. The existence of the plait point for the ternary mixture consisting of supercritical fluid, organic compound and water, makes it impossible to dehydrate aqueous solution further because at this point $X = Y$. Suzuki et al. [3] investigated the VLE data for the CO_2 + 2-propanol + water system and found that the azeotrope of the 2-propanol + water system was broken by adding CO_2 at 333 K and 10 MPa, as given in Fig. 4.1.11 [3]. Although at atmospheric conditions the azeotropic composition for the 2-propanol + water system locates at a water richer-composition than that for the EtOH + water system, as can be seen in Figs. 4.1.10 and 4.1.11, it is possible to remove water easily using supercritical CO_2 only.

(c) Energy-Saving Processes

In the past two decades a number of organic + water separations have been screened in feasibility tests. Some have received attention due to the large energy consumption for distillation separation. They include dioxane, acetone, formamide, *N,N*-dimethyl formamide and ethylene glycol. It is extremely difficult to extract these compounds from water with CO_2, because most of these organic compounds readily hydrogen bond to themselves and to water. Supercritical fluid extraction for separation of organics from water has the potential to reduce energy costs but that does not necessarily mean it will be less expensive overall. Sometimes capital and other operating costs can more than offset decreased energy costs.

4.2.3 Crystallization Separation of Aromatic Hydrocarbons

(a) Isomer Separation

Many kinds of aromatic isomers such as ortho-, meta- and para-xylenes, are used as raw materials in the chemical, polymer and pharmaceutical industries. The separation of the isomers is frequently very difficult due to the similarity in physical properties of the isomers. Since the species are isomers of each other, they have identical molecular weights and hence any separation process dependent on molecular-weight differences will fail.

In general distillation or crystallization is often employed for isomer separation in industry. Crystallization is dependent on the molecular shape differences, which bring about the difference in melting temperatures of each isomer. Para-

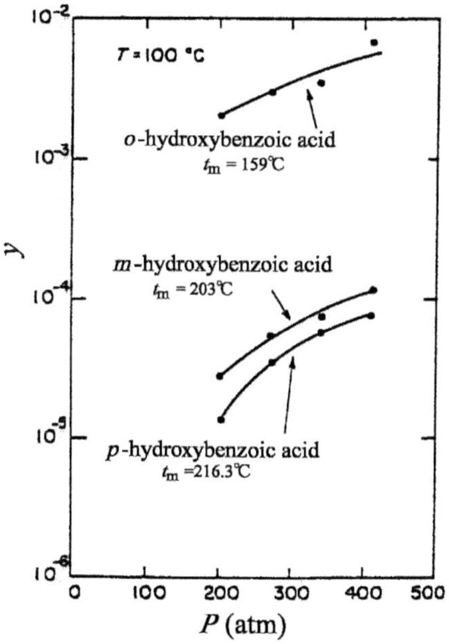

Fig. 4.2.2 Solubility of hydroxybenzoic acid isomers in CO_2 at 100 °C [1]

isomers can stack together readily into a crystal because of the symmetrical shape and result in a much higher melting point than the other isomers.

A lot of solubility data of aromatic isomers in supercritical fluids have been reported. Data for hydroxybenzoic acids are shown in Fig. 4.2.2 [1] where the melting temperatures of each isomer are included. The isomer that freezes most readily, *p*-hydroxybenzoic acid, exhibits the lowest solubility in CO_2 and its value is almost two orders of magnitude lower than the solubility of *o*-hydroxybenzoic acid (salycylic acid). In general there is not a large difference between the solubilities of meta- and para-isomers. The relationship between the melting point of the isomer and the solubility in a supercritical fluid is quite similar to that found with liquid organic solvents.

As a result the ortho-isomer can be easily separated from the other isomers using supercritical fluid extraction but separation between meta- and para-isomers is not easy.

(b) Retrograde Crystallization

The solubilities of substances in supercritical fluids are dependent on the pressure and temperature. Figure 4.2.3 shows schematically the solubility curves of individual aromatic hydrocarbons (A and B) for the A + B + supercritical

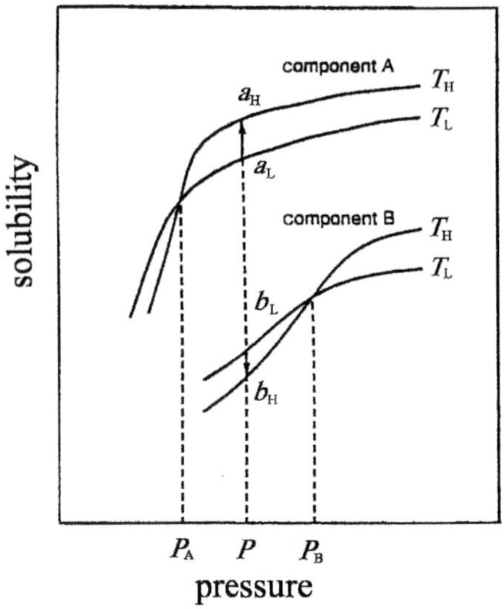

Fig. 4.2.3 Illustration of retrograde crystallization for separation of components A and B

fluid system at two different temperatures (T_H and T_L). In the pressure range less than P_A or P_B, the solubility of a substance at lower temperature (T_L) is bigger than that at higher temperature (T_H). This phenomenon is called *retrograde vaporization* because, upon reducing temperature, volatilization of the solid compound tends to occur. But at higher pressure beyond P_A or P_B, the solubility of each substance at higher temperature becomes bigger than that at lower temperature.

The intersection pressure of two isotherms, P_A or P_B, is called the cross-over pressure and let us suppose that P_B is higher than P_A. When a mixture of two solid substances (A and B) is dissolved into a supercritical fluid in an extraction chamber at pressure P ($P_A < P < P_B$) and at temperature T_L, the solubilities of component A and B are saturated and shown as a_L and b_L in Fig. 4.2.3. The supercritical fluid solution saturated with component A and B enters another high-pressure chamber (crystallizer) at temperature T_H at the same pressure as that of the extractor. Upon raising the temperature to T_H, the saturated solubilities of A and B change to a_H and b_H as indicated in Fig. 4.2.3. In this condition the supercritical solution is not saturated with component A and is supersaturated with component B. Therefore, the pure solid component B can deposit in the extractor and this process was named *retrograde crystallization* [5]. On the other hand, it is possible to operate reversibly the retrograde crystallization process. First the solid feed mixture is placed in the extraction chamber under T_H and P and extracted with CO_2. Second the

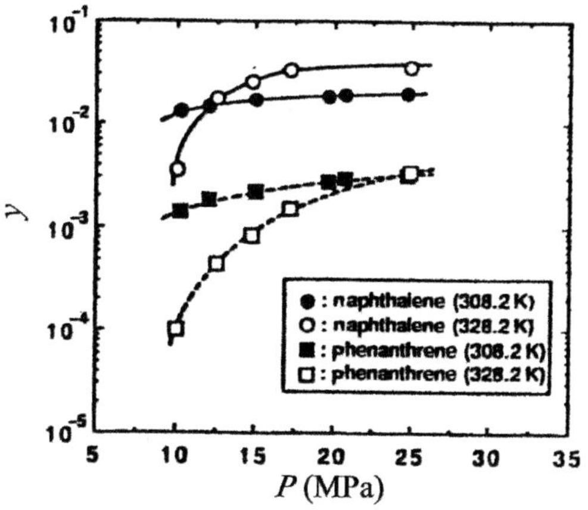

Fig. 4.2.4 Solubility isotherms for naphthalene and phenanthrene in CO_2 [6]

mixture solution is sent to the crystallizer whose temperature and pressure are kept at T_L and P, respectively. Lowering the temperature from T_H to T_L, the supercritical mixture is supersaturated with component A and a pure solid of A may precipitate in the crystallizer.

Sako et al. [6] reported the result of retrograde crystallization experiments for a naphthalene (A) + phenanthrene (B) mixture using supercritical CO_2. Temperatures of 308.2 K and 328.2 K were chosen as T_L and T_H respectively. The crossover pressure was 12 MPa for naphthalene and 23 MPa for phenanthrene as shown in Fig. 4.2.4 [6]. As mentioned above, the purification of each component can be made between 12 and 23 MPa. Under 14 MPa, they made two kinds of experiments: first raising the temperature to cause the deposition of pure phenanthrene and second lowering the temperature to precipitate pure naphthalene. The highest purification from an equi-composition mixture was 97.1 wt% for phenanthrene and 96.5wt% for naphthalene.

(c) Application of Rapid Expansion of Supercritical Solutions to Separation of Polycyclic Hydrocarbons

When a high-pressure supercritical solution containing solute and high-density supercritical solvent is passed through a nozzle or a capillary so that it is expanded within 10^{-6} seconds to low pressure, the solute solubility abruptly drops to $1/10^4$ or less in mole fraction and results in the deposition of solid solute. This process is called rapid expansion of supercritical solutions (RESS). The linear growth rate of naphthalene crystal by RESS reaches 0.02 to 23 m/s, greater by a factor of 10^5 or more than by conventional crystallization methods.

In order to apply the RESS process to the separation of aromatic hydrocarbons, Liu and Nagahama [7, 8] made experimental studies on two kinds of aromatic solid mixtures using CO_2 as a supercritical solvent. First the naphthalene + phenanthrene binary was studied as a eutectic solid mixture [7]. Fractional crystallization of this mixture by RESS and the enrichment of naphthalene were conducted. This was because the crystallization is a rate-governed process based on the balance between nucleation and growth rates. The solution saturated with a binary solute mixture passed rapidly through a nozzle with an orifice of 45 μm i.d. and was expanded into a crystallization cell at atmospheric pressure. Liu and Nagahama [7] used the enrichment factor β:

$$\beta = (x_{\mathrm{na}} / x_{\mathrm{ph}})/(y_{\mathrm{na}} / y_{\mathrm{ph}}), \tag{4.2.1}$$

where x is the mole fraction of precipitated crystal and y is that of the supercritical solution before expansion. If β is larger than unity, the enrichment of naphthalene is carried out. The values of β ranged from 1 to 3.2 and were dependent on operation factors, e.g., temperature of the crystallization cell and degree of supersaturation.

Employing an anthracene + phenanthrene mixture as solute and CO_2 as supercritical fluid, solubility as well as RESS experiments were conducted by Liu and Nagahama [8]. The solid−liquid equilibrium of this hydrocarbon mixture is a solid solution system, which means each component is miscible in the solid phase over the whole concentration range. The solubilities of phenanthrene and anthracene plotted against the phenanthrene composition of the feed are given in Fig. 4.1.8 [8]. Figure 4.1.8 (a) shows the dependence of solubilities on feed composition and a moderate effect of the solid solution feed composition on solubilities in supercritical CO_2. In Fig. 4.1.8 (b), the relationship between solid composition of feed and fluid composition of the CO_2-rich phase is plotted on a CO_2 free basis. From Fig. 4.1.8 (b), it is shown that a solid solution mixture such as phenanthrene + anthracene can easily be separated by supercritical fluid extraction.

From a thermodynamic point of view, the number of Gibbs degrees of freedom is three for the anthracene + naphthalene + CO_2 system and two for the naphthalene + phenanthrene + CO_2 system. This is because the former is a solid solution (one solid phase) mixture and the latter a eutectic (two solid phases) mixture. Therefore, the solubility of a binary solid solution system in CO_2 depends on the composition of the solid feed mixture even when fixing pressure and temperature. This is shown in Fig. 4.1.8. On the other hand, the solubility of a binary eutectic system in CO_2 is constant when fixing pressure and temperature.

4.2.4 Supercritical Fluid Adsorption Process

(a) Adsorption Separation of Dimethylnaphthalenes in a Supercritical Fluid

2,6-dimethylnaphthalene is an important raw material for the production of engineering plastics. It is purified from a mixture consisting of many structural isomers such as 2,7-dimethylnaphthalene. Iwai. et al. [9] tried to separate 2,6- and 2,7-dimethylnaphthalene through adsorption in a supercritical fluid mixture using zeolite as an adsorbent. They used a flow-type experimental apparatus and conducted experiments as follows: first, 2,6- and 2,7-dimethylnaphthalene are dissolved into supercritical CO_2. The supercritical fluid mixture saturated with the two structural isomers is forced to pass through the adsorption column packed with zeolite. The composition ratio of 2,6-/2,7-dimethylnaphthalene is 40:60 at the entrance of the adsorption bed. Figure 4.2.5 [9] shows an example of composition change at the exit of the adsorption column. The abscissa of this figure shows the total amount of CO_2 passing through the column. It is noted that in the initial stage nearly 100% of 2,6-dimethylnaphthalene could be obtained in the exit fluid, that is, 2,7-dimethylnaphthalene is fully adsorbed on the zeolite during this stage. After

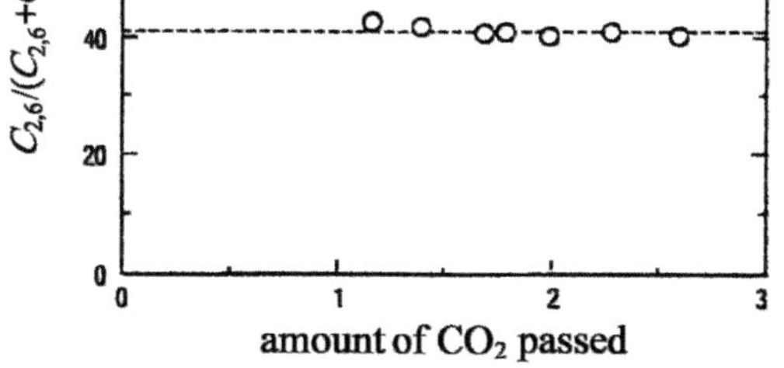

Fig. 4.2.5 Change of concentration of 2,6-dimethynaphthalene at column exit (308.2 K and 14.8 MPa) [9]

this stage the fluid composition at the column exit reaches the inlet composition as shown in Fig. 4.2.5. Testing various types of zeolites, Iwai et al. [9] concluded that Na-Y-type zeolite was the best for separation of 2,6- and 2,7-dimethylnaphthalenes.

(b) Regeneration of Activated Carbon

The regeneration of activated carbon with supercritical CO_2 was first demonstrated in the mid-1970s in both Japan and USA. In research sponsored by the US EPA, the feasibility of regenerating activated carbon was described at an ACS meeting in 1978 [10]. The advantages of the regeneration are reduced energy requirement and lower carbon loss compared with the conventional thermal regeneration process using steam or a high-temperature furnace. The organic materials adsorbed on activated carbon can be desorbed by supercritical CO_2, because many kinds of organic compounds can be dissolved by supercritical CO_2 and also the sites loaded with an organic component can be replaced by CO_2. The supercritical fluid regeneration of the loaded activated carbon involves three basic steps. The first is the adsorption step, where the solute or contaminant is transferred from the fluid phase into the adsorbent. The second is the desorption step. The pores of the carbon are flooded and replaced with supercritical fluid solvent, which dissolves and removes the adsorbed component. At the end of the second step, the activated carbon has been stripped of the adsorbed material, but the pores remain filled with solvent. The final step is the removal of the residual solvent, which can be taken away very easily by pressure reduction without heating or steaming.

Figure 4.2.6 [11] compares the breakthrough curves for the toluene vapor (adsorbate) / regenerated activated carbon system. The carbons were regenerated by a steam method and a supercritical CO_2 method operated at 313 K and 136 bar. It is seen that the adsorptive capacities of the carbon regenerated by steam dropped greatly after the first cycle and that there is a large difference in breakthrough time between the two methods. From the result shown in Fig. 4.2.6, it is concluded that the supercritical fluid regeneration method is superior to the steam method for adsorptive capacity.

(c) Recovery of Indole and Quinoline Using Supercritical Fluid Extraction and Adsorption

By combining the adsorption with supercritical fluid extraction processes, recovery of indole and quinoline from methylnaphthalene oil was examined using a semi-continuous supercritical fluid extraction apparatus [12]. First indole contained in naphthalene oil dissolved in supercritical CO_2 was trapped

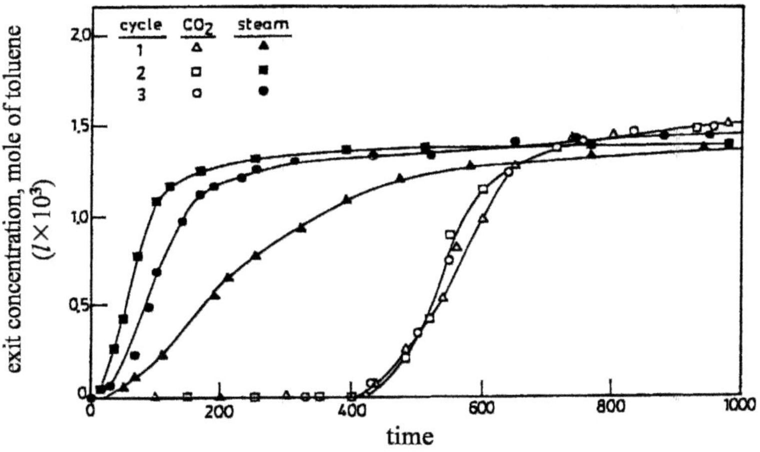

Fig. 4.2.6 Comparison of breakthrough curves for toluene using activated carbons regenerated by steam as well as supercritical CO_2 [11]. Reprinted from: C. H. Tan, D. C. Lion: Regeneration of activated carbon loaded with toluene by supercritical carbon dioxide, *Separation Sci. Technol.* (Marcel Dekker) **24**, 111 (1989).

on an anion exchange resin (Arnberlite IRA-904) and second the indole-free oil entered the bed of acidic adsorbent (10 wt% aluminum sulfate supported on silica gel).

After the ion exchange resin and the adsorbent were saturated with indole and quinoline, the feed of methylnaphthalene oil was stopped and adsorbed indole and quinoline were recovered by solvent extraction using methanol and tetrahydrofuran respectively. Figure 4.2.7 [12] shows a schematic process flow. Slightly acidic indole and basic quinoline can be selectively recovered by continuous supercritical CO_2 extraction with the two fix beds of adsorbents. The recovery of indole and quinoline can be done using methanol and tetrahydrofuran as entrainers.

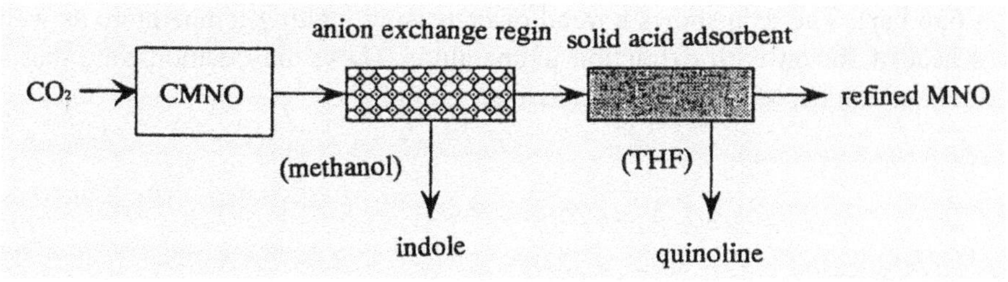

Fig. 4.2.7 Novel process for separation of indole and quinoline from crude naphthalene oil [12]

4.2.5 Polymer Processing

(a) Polymer Fractionation

Supercritical fluid fractionation of a synthetic oil or polymer is carried out in a multiple, sequential pressure reduction system. The solution of polymer and supercritical fluid which leaves the first vessel is expanded to the second one at some pressure, which causes the next highest molecular weight polymer in the solution to precipitate. The number of fractions is consistent with the number of pressure reduction stages, and the molecular weight range of the fractions is related to the pressure decrements per pressure reduction stage. This process is called isothermal decreasing pressure profiling fractionation [1] and is easily practiced on an industrial scale.

In order to operate polymer fractionation an appropriate supercritical fluid solvent must be chosen to dissolve or fractionate the polymer. The choice of fluid solvent for a given polymer fractionation is dependent on the interaction forces between the solvent–solvent, the solvent–polymer and the polymer–polymer pairs. CO_2 would not at all be considered as a suitable solvent. Ethylene, ethane, or propane are more advantageous for purifying, extracting, or fractionating the amino-terminated polymer.

Watkins et al. [13] demonstrate how linear low-density polyethylene (LLDPE) can be fractionated by molecular weight and backbone structure using a double fractionation method. The first step is to fractionate with respect to molecular weight and is carried out by the isothermally increasing pressure profile. The second step is to fractionate with respect to backbone structure using an isobarically increasing temperature profile. Table 4.2.1 [13] shows the results on the properties of fractions of LLDPE using an isothermally increasing pressure profile (from 250 to 675 bar) with propane at 130 °C. The parent polymer with a polydispersity of 5.0 was fractionated with a molecular weight range from 10 000 to 200 000 with polydispersity 1.2–1.6. Table 4.2.2 [13] shows the properties of each fraction which are obtained from the parent polymer by an isobarically increasing temperature profile (from 50 to 110 °C at 656 bar). The data shows a trend of increasing melting temperature as well as heat of fusion with extraction temperature. These may demonstrate that a fractionation based on structure may be accomplished.

Table 4.2.1 Properties of each fraction for linear low-density polyethylene obtained by an isothermally increasing pressure profile with propane at 130 °C [13]

fraction	*pressure (bar)*	*parent (wt%)*	M_w	M_w/M_n	ΔH^{fusion} *(J/g)*	T_{melt} *(°C)*
1	248	2.3				
2	296	3.7	10,700	1.2	104	115
3	379	6.3	16,200	1.3	108	116
4	421	8.8	29,100	1.2	114	121
5	455	11.2	32,300	1.5	106	123
6	483	12.4	50,100	1.3	107	123
7	507	11.5	61,800	1.3	110	123
8	527	11.0	79,600	1.3	114	123
9	545	11.2	108,000	1.4	124	121
10	558	10.1	155,700	1.4	115	121
11	600	8.6	177,300	1.3	132	123
12	655	2.9	273,000	1.6	146	126
		parent	160,000	5.0	113	123

Table 4.2.2 Properties of each fraction for linear low-density polyethylene obtained by an isobarically increasing temperature profiling at 65.6 MPa [13]

fraction	*temperature range (°C)*	*parent (wt%)*	*peak* T_{melt} *(°C)*	ΔH^{fusion} *(J/g)*	M_w	M_w/M_n
1	50–60	3.9	71.2	9.1	43,900	5.3
2	60–72	4.5	84.8	48.0	25,100	1.9
3	72–77	3.9	91.6	53.9	33,300	2.0
4	77–82	5.8	96.0	59.7	30,300	2.2
5	82–86	6.1	103.2	71.4	41,100	2.5
6	86–90	6.5	104.6	65.4	57,900	1.8
7	90–93	9.4	107.4	86.8	69,900	1.7
8	93–97	7.1	107.4	76.2	73,100	1.9
9	97–100	8.7	113.9	106.9	80,200	1.9
10	100–102	7.4	112.3	87.5	86,300	1.9
11	102–105	8.7	117.4	108.6	75,400	2.1
12	105–108	9.4	120.3	117.4	101,100	1.5
13	108–110	7.8	122.0	106.5	114,100	2.2
14	110+	10.7	125.6	115.9	221,600	3.0
		parent	123.1	113.3	106,000	5.0

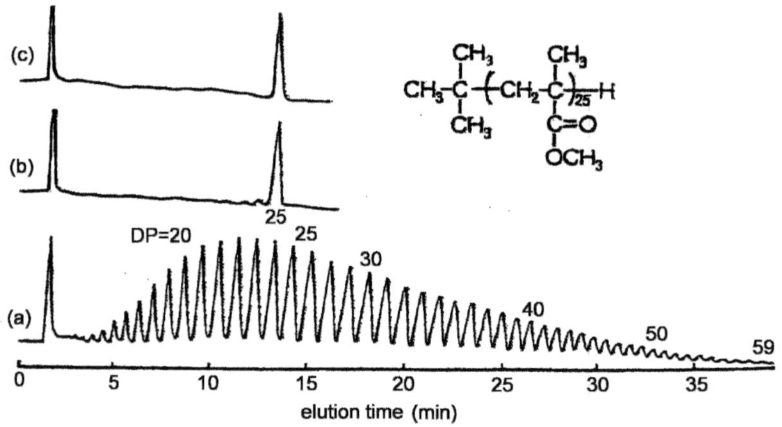

Fig. 4.2.8 Chromatograms of isotactic PMMA using supercritical fluid chromatography: (**a**) average degree of polymerization = 28.6, (**b**) 25mer with impurity, (**c**) purified 25mer by SFC [16]

(b) Oligomer Processing

Supercritical fluid extraction can be used for separation of low molecular weight oligomers from polymers in a wider range of molecular weights. Barry et al. [14] describes the removal of cyclic oligomers from polyoxyalkylene, the upgrading of a silicone polymer, designated OV-17, used as a gas chromatography stationary phase. The material tested, a polyphenylmethylsiloxane, is a very viscous polymer containing macrocyclic and low molecular weight oligomers that interfere with the performance of the material used as a chromatographic substrate. It is reported that supercritical CO_2 (140 bar and 35 °C) extracted 30% of the original polymer and the residual heavier portion yielded much improved characteristics in chromatography applications.

(c) Preparation of Monodisperse Polymer

The preparation of polymers having a uniform molecular weight is very important in polymer science or technology. Such monodisperse polymers are interesting for studying the structure and physical properties and very useful as standard reference materials. Supercritical fluid chromatography (SFC) is a very promising means for obtaining a uniform polymer. Ute et al. [15] reported that a uniform polymer (25mer isotactic polymethylmethacrylate (PMMA)) can be obtained easily from the parent polymer of an average degree of polymerization (DP) of 28.6 using SFC. Figure 4.2.8 [16] shows SFC curves of isotactic PMMA: (a) is the parent polymer with an average DP = 28.6, and (c) is the 25mer further purified by SFC of the crude fraction (b). As shown in Fig. 4.2.8, the parent polymer was completely separated into 10 to 59mers within 40 minutes. By repeating preparative separation of 19 to 29mers 49 times, a fraction of each component was obtained. They used a column (10 mm i.d.× 250 mm) packed with normal-phase silica gel and CO_2 + ethanol as a mobile phase. It was also possible to obtain uniform polymers such as 100mer isotactic PMMA, polystyrene, polyisocyanate, polyoxymethylene and poly(lactic acid).

4.2.6 Treatment of Various Waste Streams

Supercritical fluid extraction has been proposed as a clean-up technique for liquid or solid environmental samples. It includes major areas: extraction of organic contaminants from waste water, and remediation of contaminated soil. The main advantage of using supercritical extraction as a clean-up process is the nontoxicity of the supercritical solvent. The use of such a solvent eliminates concern over residual solvent concentration in a processed environmental matrix.

(a) Waste Water Treatment

Increasing amounts of waste water contaminated with toxic organic compounds are being produced every year. Lots of methods: distillation, incineration, adsorption and liquid extraction, are currently available for large-scale clean-up of contaminated water. Supercritical fluid extraction offers an attractive alternative to these conventional techniques. The broad range of extracted organic solute as well as the use of nontoxic solvents such as CO_2, has directed attention to this process as a viable method for cleaning up toxic waste present in industrial waste streams, surface water reservoirs and ground water reserves.

The extraction process for contaminated water is the same as that for conventional supercritical fluid extraction as mentioned before. In the field of waste water treatment using a supercritical fluid, most people have done fundamental studies of the distribution coefficient of contaminants between water and supercritical fluid phases.

Table 4.2.3 Soil characteristics and experimental distribution coefficient of phenol between CO_2 and soil phases [17]

material	surface area[a] (m^2/g)	% organic[b] (by weight)	% organic[c] (by weight)	K value[d]
sand	0.67	0.0	NA	~120[e]
Montmorillonite	256.0	0.0	NA	0.08 ± 0.01
soil 1	6.4	1.0	1.0	2.22 ± 0.19
soil 2	17.2 ± 0.1	1.4	1.4	1.52 ± 0.07
soil 3	15.4 ± 0.1	NA	0.7	1.61 ± 0.11
soil 4	67.3 ± 0.5	NA	2.1	0.78 ± 0.05
				0.73 ± 0.03
soil 5	52.7 ± 0.3	NA	1.5	0.81 ± 0.04
soil 6	1.1 ± 0.1	0.4	0.5	5.7 ± 0.6
				5.9 ± 1.8

NA: not available

[a] surface areas of the six soils were determined by N_2 BET
[b] as determined by wet chemistry
[c] as determined by TGA
[d] system conditions were 13.8 MPa, 310 K, and 100:1 ratio of solid to phenol mass
[e] the ratio of phenol in the CO_2 phase to the total phenol in the system was 0.995 ± 0.022; total number of samples was 25.

(b) Soil Remediation

Supercritical fluid extraction of organic compounds from soil and sediment is a promising technique for solid waste clean-up for remediation. So far many people have studied soil extraction using a supercritical fluid. Using supercritical CO_2 or ethylene, over 90% of chlorinated aromatic compounds such as PCBs and dioxins were able to be removed from model soil. The extraction results of a contaminant from soil are highly dependent on the characteristics of the soil. Table 4.2.3 [17] shows the distribution coefficients for phenol, where the coefficient is defined as the ratio of phenol contents in CO_2 to soil. The values distribute from 0.08 to 120. Extraction of sand results in perfect extraction of phenol to the CO_2 phase, showing sand has not adsorbed phenol. Extraction of clay (montnorillonite) yields the lowest distribution coefficient and took five days to attain equilibrium. Akgerman et al. [17] examined the resulting distribution coefficients of phenol with two parameters: surface area and content of natural organics in soil. With respect to the effect of organic content in natural soil on the coefficients, from Table 4.2.3, a larger percentage of natural organic in soil resulted in lower distribution coefficients of phenol. This shows that the natural organic materials can highly adsorb or trap phenol in the soil. Akgerman et al. investigated also the effect of the water content of soil on extraction of phenol. When water is added to the system, the extraction of phenol is better than for dry soil. They described the possible processes that could account for this behavior.

CF Systems Corporation developed a supercritical fluid extraction process to separate and recover oil from refinery sludges and to extract hazardous organic compounds from waste water, and sludge, sediment, and soil [17]. That process uses propane on contaminated soil and sludge and CO_2 on waste water. The solid feed is reduced in size and is slurried so that it can be pumped to the extractor. The CF Systems process was demonstrated on a pilot scale and was shown to be able to remove PCBs from sediments. A 300 barrel per day unit was installed in Texas, USA.

4.3 Extraction and Separation of Natural Products

4.3.1 Phase Equilibria for Components in Natural Materials

In supercritical extraction processes either from natural solid feed or liquid feed, knowledge of the solubility of the components is one of the most important factors for design and analysis. Experimental measurements of solubility have been done mostly by the static or dynamic methods. The data have been generally interpolated by using equations of state. Commercially important components in natural materials are lipids and essential oils.

Solubility data of lipid components such as fatty acid and glycerides available in the literature were listed by Maxwell [1]. Figures 4.3.1 and 4.3.2 show examples of the solubility data for fatty acids (myristic acid: $CH_3(CH_2)_{12}COOH$, palmitic acid: $CH_3(CH_2)_{14}COOH$, stearic acid: $CH_3(CH_2)_{16}COOH$) and higher alcohols (cetyl alcohol: $CH_3(CH_2)_{15}OH$, stearyl alcohol: $CH_3(CH_2)_{17}OH$, arachidyl alcohol: $CH_3(CH_2)_{19}OH$) [2, 3]. For both acids and alcohols, solubility is influenced by just one methyl group. Alcohol has higher solubility in supercritical CO_2 than acid for the same carbon number. The influence of cosolvent on the solubility was measured for acids (palmitic acid and stearic acid) and alcohols (cetyl alcohol and stearyl alcohol) in a supercritical CO_2 mixed with octane or ethanol [4]. The solubility of stearic acid is significantly increased by adding cosolvent, as presented in Sect. 2.4 (see Fig. 2.4.4). Ethanol is more effective on the solubilities of fatty acids than octane. On the other hand, the influence of -octane on the solubilities of higher alcohols is almost equal to that of ethanol.

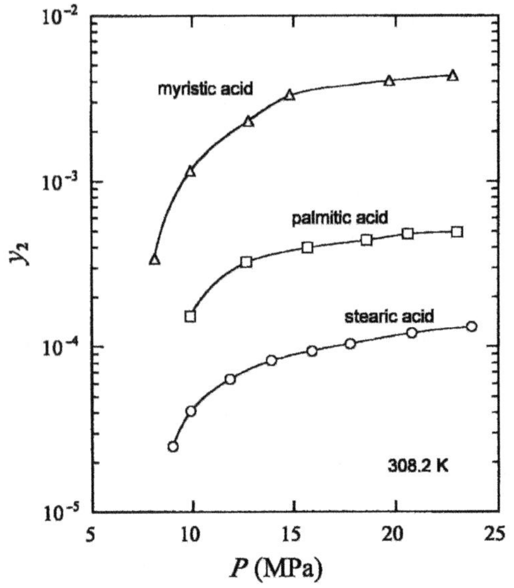

Fig. 4.3.1 Solubilities of fatty acids in supercritical CO_2

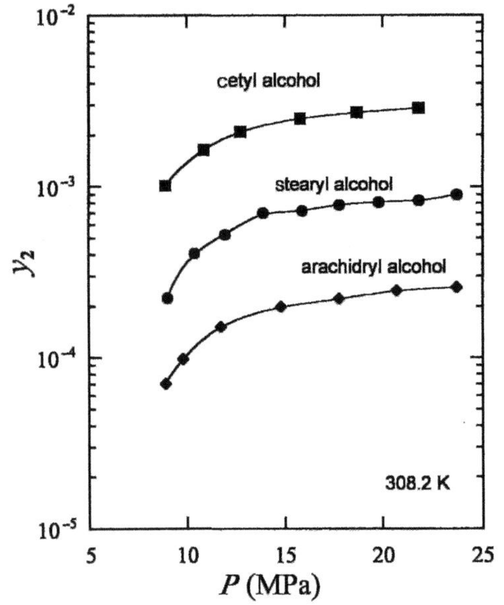

Fig. 4.3.2 Solubilities of higher alcohols in supercritical CO_2

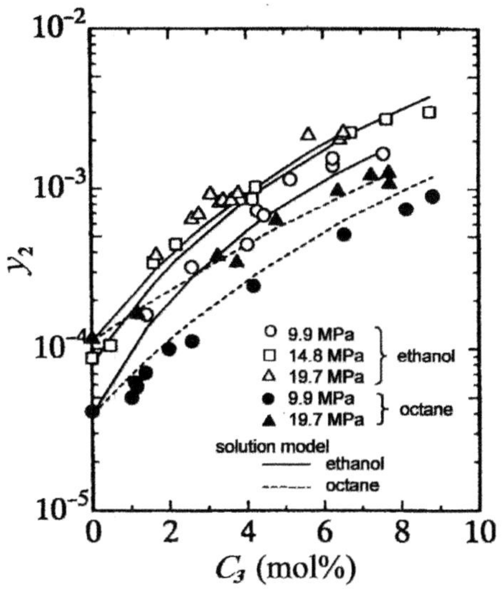

Fig. 4.3.3 Vapor–liquid equilibria for CO_2 (1) + limonene (2) and calculated results by the Peng–Robinson EOS [11]

These results imply that the polar–polar interaction is most influential on the solubilities of solid components in a supercritical fluid. The regular solution model coupled with the Flory–Huggins theory using the local composition of cosolvent around the solute molecule was applied to correlate the solubilities of solid components in supercritical CO_2 mixed with a cosolvent. The calculated results agreed reasonably with the experimental data as indicated in Fig. 2.4.4.

Solubility data of essential oil components were listed by Reverchon [5]. A main objective for the essential oil fractionation process is often deterpenation where hydrocarbon terpenes are separated from oxygenated aroma compounds. Most of the data available in the literature are binary phase equilibria of pure component + supercritical CO_2. Solubility data for limonene have been measured by several researchers [6–10], because limonene is a major component in most essential oils and separation between limonene and linalool is one of the most difficult systems in essential oils. Figure 4.3.3 shows recent work for limonene + CO_2 phase equilibria [11]. Vapor–liquid equilibria were correlated by the Peng–Robinson equation of state using conventional mixing rules with two interaction parameters. Equilibria for the linalool + CO_2 system, which is a major aroma component in orange oil, have been reported (see Fig. 2.4.1) [10, 12].

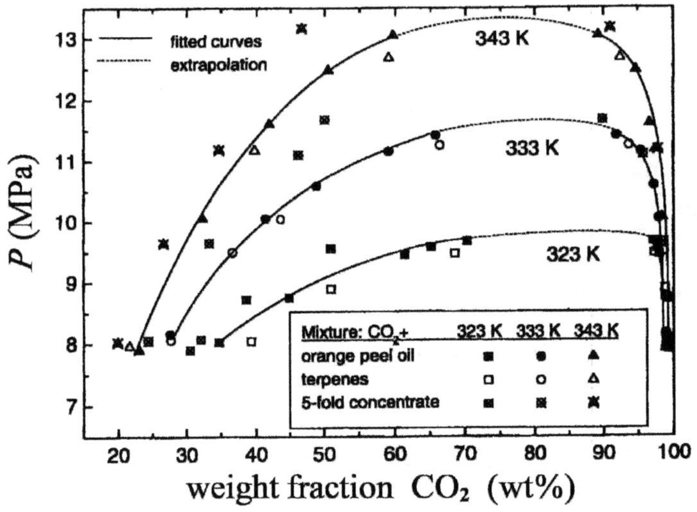

Fig. 4.3.4 Phase equilibria of citrus oil fractions [14]

Some authors have studied phase equilibria of essential oils as a mixture (e.g., lemon oil [13], orange oil [14]). Figure 4.3.4 shows phase equilibria of orange oil and its fractions. Mutual solubilities reveal the extent of the two-phase region with respect to pressure, which is important for the design of a countercurrent separation process. Orange oil consists of terpenes (around 98 wt%) and aroma components. Mutual solubilities at constant temperature and pressure increases for terpene fraction, whereas a five-fold aroma concentrate exhibits a much lower mutual solubility and increased critical points.

4.3.2 Solid Material Processing

Extraction of some compounds such as flavors and spices from natural solid materials is a mass transfer process, which is influenced by multicomponent solubility in the supercritical fluid phase, solute–solid interaction, complex structure of cellular matrices, and location of solute in the material. In the case of a plant leaf, waxes are located on the leaf surface, whereas essential oil compounds are usually located inside the cellular structure in vacuoles [5]. The extraction process from the porous material may consists of dissolving solutes into fluid phase in pores, diffusion within porous matrices, and mass transfer in an external fluid film around a solid particle as schematically described in Fig. 4.3.5. The extraction behavior significantly depends on the solute components and feed materials. Characterization of the extracts and the

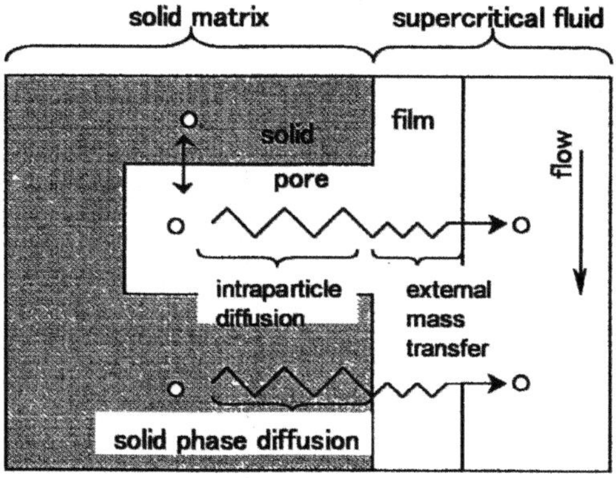

Fig. 4.3.5 Schematic drawing of the extraction process from natural materials

compositions has been reported [15]. Various mathematical models have been developed to analyze the extraction rate for various materials. For example, when wax and essential oil are extracted from a leaf, the concentration of wax in the supercritical fluid at the exit of the extractor is near the saturated solubility, whereas the concentration of essential oil is much smaller than the solubility. Vegetable oil from seeds is extracted easily from a partly crushed cell, but the extraction from a noncrushed cell is much slower due to mass transfer resistance.

(a) Diffusion Process

The diffusion process within a particle is described by Fick's law. For a porous particle, solutes diffuse in pore space by pore diffusion or surface diffusion. For a nonporous particle, solid diffusion may describe the mass transfer process. Mass transfer in a stagnant fluid layer around a particle can be expressed by the external fluid film mass transfer coefficient. The model can be simplified by imposing the linear driving-force approximation to represent the intraparticle diffusion and the external mass transfer.

(b) Local Dissolving Rate

When solutes exist in a particle as a solid or liquid phase entrapped in pores, solutes must dissolve into a supercritical fluid phase at the surface of the pore or liquid/supercritical fluid interface within a particle or at the outer surface of a nonporous particle. Two factors that may play an important role in the dis-

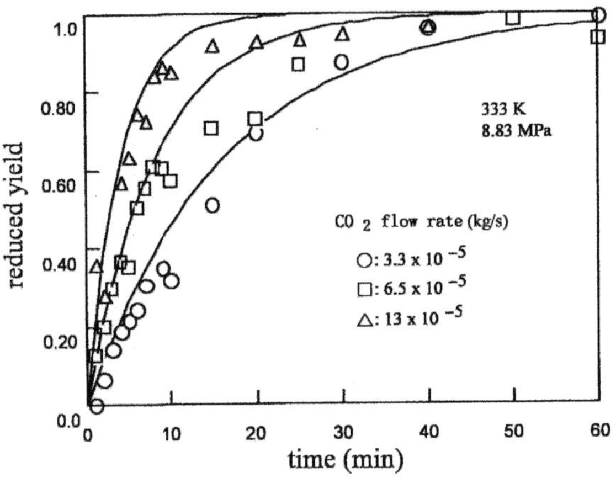

Fig. 4.3.6 Extraction curves at various flow rates for peppermint oil and modeling with linear adsorption equilibrium equation

solving process are the interaction between the solute and solid phase and the solubility of the solute in the supercritical fluid phase.

When the interaction between the solute and solid phase is influential, the dissolving process may be described by an adsorption/desorption mechanism. Recasens et al. [16] developed a mass transfer model with a Langmuir type adsorption/desorption rate. To describe caffeine extraction from soaked coffee beans, Peker et al. [17] used a model including solid/liquid adsorption equilibria in pores and liquid/supercritical fluid partition equilibria at the external surface of a particle. For cleanup of contaminated soil, extraction of organic materials was modeled by using Freundlich type adsorption equilibria.

As a simple equilibrium relationship, linear equilibria have been employed [18, 19]. For essential oil extraction from peppermint leaf, the extraction rate was much smaller than that expected from equilibrium solubility. Goto et al. [18] explained the behavior by assuming a linear adsorption equilibrium between essential oil and solid wax. Flow rate dependence of the extraction curves revealed that the extraction was not diffusion limited but equilibrium limited. As shown in Fig. 4.3.6, the extraction curves were simulated by the model.

The extraction process of solutes in natural materials may be influenced by both solute solubility in a supercritical fluid and solute–solid interaction (adsorption). In order to incorporate these effect, Goto et al. [20] used a BET type equilibrium equation where the solubility and adsorption equilibrium constant were involved. Extraction behaviors of both essential oil and cuticular

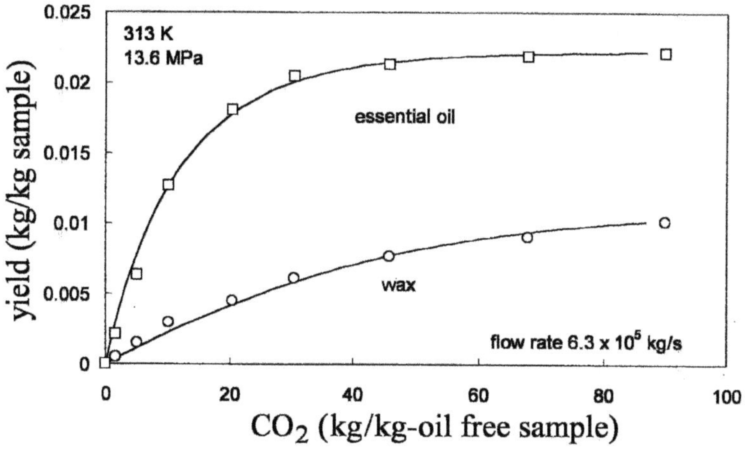

Fig. 4.3.7 Extraction curves for essential oil and cuticular wax from peppermint leaves and modeling with BET equilibrium equation

wax were simulated by the model as shown in Fig. 4.3.7 by using different adsorption equilibrium coefficients.

(c) Reactive Extraction

The extraction process from natural materials may involve a chemical reaction as well as physical phenomena. Reactive extraction of lignin and cellulose from biomass by using supercritical alcohol was modeled by Goto et al. [20–22]. The extraction data were analyzed by the model with a first-order parallel reaction for lignin and a first-order consecutive reaction for cellulose.

(d) Nonuniformity

Since natural materials often have a cellular structure, solutes trapped in the cell have to migrate through the cell wall. For some solutes, there is a considerable resistance to penetration of the cell wall. In the case of oil extraction from seeds, oil can be easily extracted from a crashed cell located in the outer region of a particle, whereas extraction of oil from the inner part consisting of uncrashed cells is restricted by the cell wall. This situation was modeled by a simple nonuniform model by Sovova [23]. A particle was assumed to be composed of an outer part where cells are partially broken and easily accessible and an inner part where cells are intact and protected by a wall. The extraction from the outer part was modeled by using external mass transfer coefficient

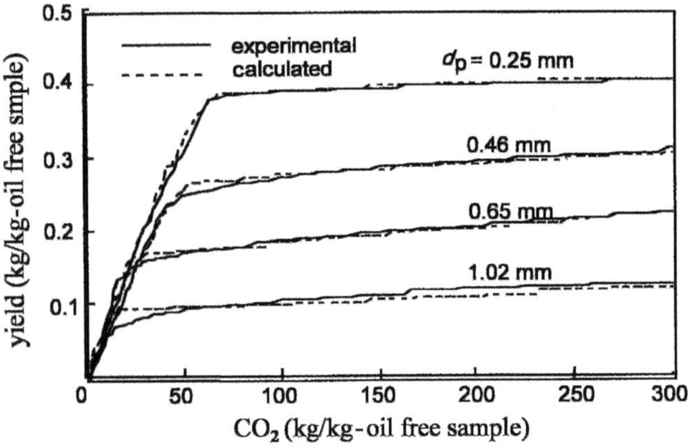

Fig. 4.3.8 Extraction curves at various particle sizes for tomato seed oil and calculation with nonuniform model [24,25]

with the driving force given by the discrepancy of fluid phase concentration from saturation concentration. The solid phase mass transfer coefficient was used to model extraction from the inner part. Vegetable oil extraction from seeds was successfully explained by the model for grape seed [24] and tomato seed [25] as shown in Fig. 4.3.8.

(e) Shrinking Core Model

When the dissolving process is irreversible or solutes exist in pores as a trapped liquid, the solubility of solute in the fluid phase is an important parameter in the rate equation. This situation can be described by a shrinking core model where a sharp boundary between the nonextracted core and the extracted outer shell region exists and the solute diffuses through the outer region. The boundary recedes toward the center of the particle with extraction. The shrinking core model was first applied to a supercritical fluid extraction process by King and Catchpole [26] and further developed by Goto et al. [27]. Although Roy et al. [28] have applied the model to ginger extraction, the model is more suitable to seed materials where the oil content is relatively high.

4.3.3 Liquid Material Processing

Separation of the liquid mixture by supercritical fluid extraction is a process involving partioning and mass transfer between the supercritical fluid phase and the liquid phase. Since natural materials consist of a number of components having similar properties, a simple extraction process cannot achieve sufficient separation. In this case, the following strategies may improve the separation: (a) countercurrent extraction process, and (b) adsorption/ desorption process.

(a) Countercurrent Extraction Process

A countercurrent contactor with a multistaged tower or packed bed tower may achieve higher separation. Since the supercritical fluid density changes significantly from gas-like to liquid-like by changing temperature or pressure, a supercritical fluid extraction tower can be regarded as a carrier gas distillation device or countercurrent solvent extractor.

A schematic drawing of a countercurrent extractor is shown in Fig. 4.3.9. In semi-batch operation mode, liquid feed is charged at the bottom and supercritical fluid flows from the bottom to the top. The composition of extracts changes with extraction time, where components with higher solubility are extracted earlier and those with lower solubility extracted later. More than two

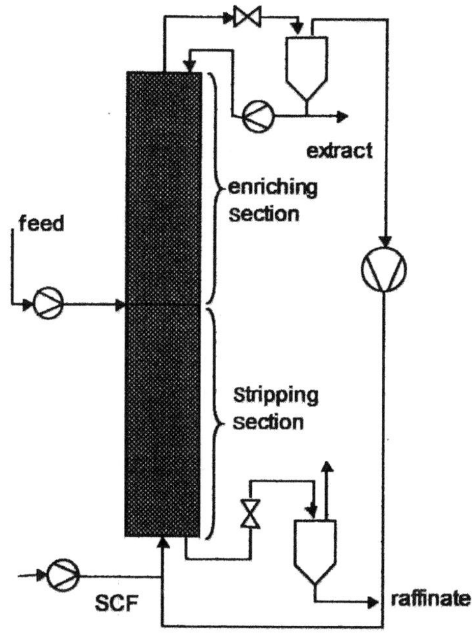

Fig. 4.3.9 Process scheme of a countercurrent supercritical fluid extractor

components can be separated into each fraction in series as a function of extraction time. The fractionation is enhanced by reflux where part of the extracted solutes is liquefied and flows down to give countercurrent contact with the supercritical fluid within the column. The reflux can be achieved by internal reflux induced by a temperature gradient along the column or external reflux from a separator set at lower pressure. In continuous operation mode, liquid feed is supplied continuously at the middle of the column and supercritical fluid is fed at the bottom. The upper part of the extraction column serves as an enriching section and the lower part as a stripping section. The feed material is basically fractionated into two fractions of extract and raffinate. The fraction with higher solubility (light components) can be obtained from the top and the fraction with lower solubility (heavy components) at the bottom. For the separation of more than two components into each fraction, several extractor units have to be combined. For the separation of n components $n-1$ extractors are necessary. The principle, design, and analysis of the countercurrent process are described by Brunner [29, 30].

The countercurrent extraction process has been applied to natural material processing such as citrus oils, unsaturated fatty acids, and squalene-tocopherol [30]. Fish oil processing has been extensively investigated for separating highly unsaturated fatty acids, EPA and DHA [31–33].

Another important application is citrus oil processing, which is one of the most important subjecs in the perfume and food industries. Citrus oil consists of terpenes, oxygenated aroma compounds, waxes and pigments. Small amounts of oxygenated compounds contribute to the specific flavor properties. Terpene content must be reduced to stabilize the products. The terpenes are conventionally removed by vacuum distillation or solvent extraction which may involve thermal degradation and organic solvent residue problems.

Since a simple extraction process does not achieve sufficient selectivity, two methods, the countercurrent extraction process and the adsorption/desorption process have been mainly investigated. Stahl et al. [6] proposed a continuous countercurrent extraction process for orange oils. With an axial temperature profile of 358 K in the middle, 348 K at the top, and 333 K at the bottom at a pressure of 8 MPa, they reduced the terpene content from 90 to 42% at the bottom. Perre et al. [34] described a pilot-scale apparatus.

Sato et al. [35–39] has developed the countercurrent extraction process for orange oil processing. They have used a 20 mm i.d., 2400 mm long column where the upper 3/4 part was packed with 3 mm Dixon packings. The extractor was used in both semi-batch mode and continuous mode of operation by using either cold-pressed orange oil or a model mixture composed of limonene, linalool, and citral (neral + geranial). In semi-batch mode, the operating condition used was 313–353 K with and without a temperature gradient at a

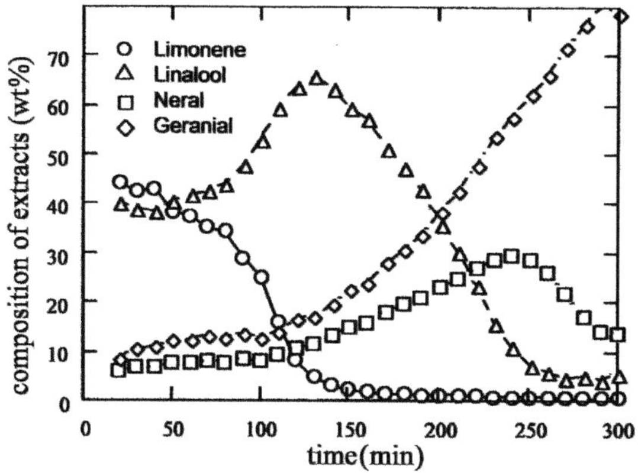

Fig. 4.3.10 Change in composition of extracts for semi-batch extraction of a model citrus oil mixture at 8.8 MPa and 313–333 K

pressure of 8.8 and 9.8 MPa. The temperature profile affected the separation behavior due to the internal reflux and countercurrent contact between the liquid phase and the supercritical fluid phase. Figure 4.3.10 shows the results for the semi-batch extraction of a model mixture operated with temperature gradient 313–333 K at 8.8 MPa. Limonene, linalool, neral and geranial were extracted sequentially. The separation selectivity increased by temperature gradient and the selectivity was 2.87 at the optimal condition of 313–333 K. The separation selectivity between limonene and linalool was defined by $(Y_{\text{limonene}} / X_{\text{limonene}})/(Y_{\text{linalool}} / X_{\text{linalool}})$. The internal reflux ratio was calculated by the measurement of the extraction rates at the top and bottom of the column, and was 7.6 at the above condition [36]. From the estimation of physical properties such as solubility, density, and viscosity in the column, they found that these properties were remarkably changed in temperature range 313–333 K.

In continuous operation, the terpene-rich fraction is recovered from the top of the column and oxygenated compounds are obtained from the bottom. Figure 4.3.11 shows the effect of the solvent to feed ratio (S/F ratio) on the extraction ratio of limonene at 333 K. The extraction ratio was defined by the ratio of the weight of limonene in extract to that in feed. The extraction ratio increased with the increase in the S/F ratio and pressure. The extraction ratio was larger for the raw orange oil than the model mixture. Figure 4.3.12 shows the effect of the S/F ratio on the separation selectivity at 333K. The selectivity also increased with the increase in the S/F ratio. When the model mixture was used as a feed, high selectivity up to 705 was observed. The selectivity of the raw orange oil was

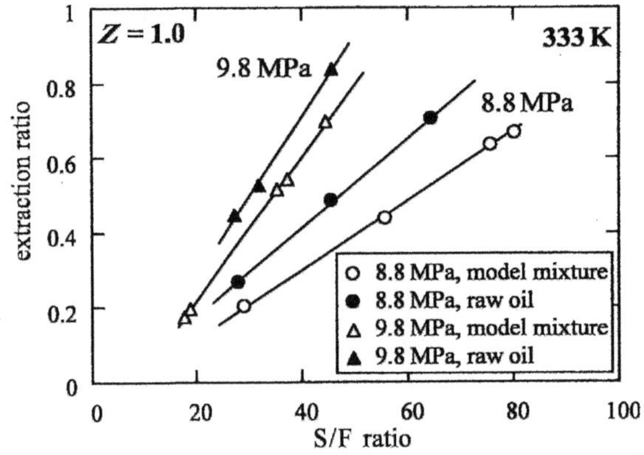

Fig. 4.3.11 Effect of solvent/feed (S/F) flow ratio on the extraction ratio of limonene for countercurrent extraction of an orange oil

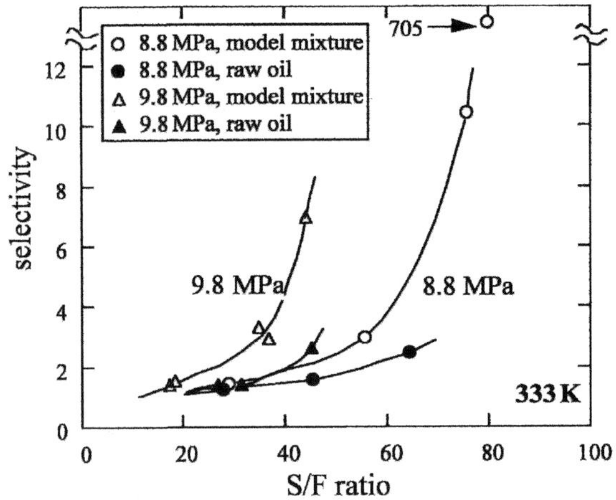

Fig. 4.3.12 Selectivity between limonene and linalool for countercurrent extraction of orange oil

lower than that of the model mixture due to the low stage efficiency induced by the high terpene content and interaction among the solutes.

For a system of the model mixture, the experimental HETS was calculated by using phase equilibria estimated by the Peng–Robinson equation of state with binary interaction parameters and the Soave–Redlich–Kwong equation of state without the parameters. Figure 4.3.13 shows the effect of the S/F ratio on the calculated HETS for the model mixture. The HETS decreased

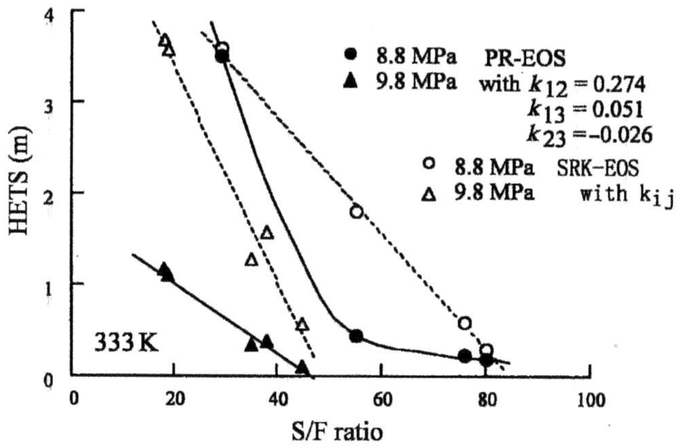

Fig. 4.3.13 HETS for countercurrent extraction of a model mixture

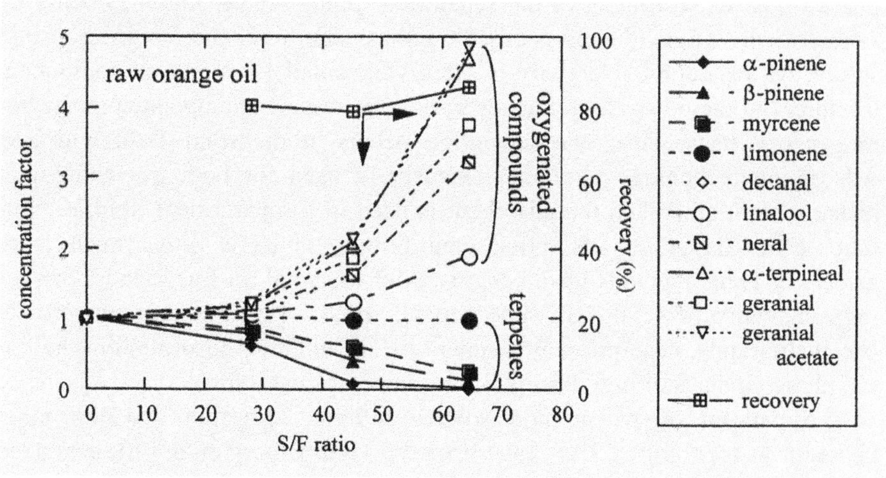

Fig. 4.3.14 Concentration factor of each component and recovery of oxygenated compounds in the raffinate at 333 K and 8.8 Mpa

from 4.0 to 0.2 m as the S/F ratio increased from 30 to 80 at 333 K, 8.8 MPa or from 20 to 45 at 333 K, 9.8 MPa. For raw orange oil processing, the HETS may be larger than that for model oil processing.

For raw orange oil processing, the effect of the S/F ratio on the concentration factor, defined by the concentration divided by that in the feed, was investigated for major constituents. Figure 4.3.14 shows the variation in concentration factor of each component and recovery of oxygenated compounds in raffinate at 333 K, 8.8 MPa. The solid symbols are terpenes, which are desired to be smaller than unity in raffinate, while the open symbols are oxygenated compounds, which are desired to be larger than unity in raffinate. Oxygenated compounds were concentrated with the increase in S/F ratio, without a decrease of the recovery yield. The deviations from unity for limonene and linalool were smaller than the other terpenes and oxygenated compounds. Therefore, when limonene and linalool were separated, the other components could be separated more selectively, that is, terpeneless oil could be obtained. The total concentration factor of oxygenated compounds in raffinate was 3.0 at the S/F ratio of 64 and 333 K, 8.8 MPa.

(b) Adsorption/Desorption Processes

Separation in the countercurrent extraction process is achieved based on phase equilibria. The operating condition is limited by the formation of a homogeneous phase, because two phases, the liquid phase and the supercritical fluid phase, must exist to achieve the separation. Since the separation process by supercritical fluid extraction is based on phase equilibria and the difference in solubility, separation selectivity is often very small for a natural component mixture. On the other hand, the adsorption process can be operated in the homogeneous phase since the separation occurs at the solid–fluid interface. Adsorption technology has been extensively used for both gas and liquid phase separation. When the adsorbent is used in a supercritical fluid, separation occurs based on adsorption equilibrium, which is often much more selective. Three methods have been used for essential oil fractionation by using an adsorbent in a supercritical fluid: preparative supercritical chromatography, desorption by a supercritical fluid after adsorption in the liquid phase, and adsorption/desorption in a supercritical fluid.

Separation of two or more solutes utilizing adsorption and desorption behavior in supercritical CO_2 have been reported. Bracey et al. [40] evaluated the separation process of aqueous fructose-glucose mixtures by adsorption at 8.3 MPa and desorption with supercritical CO_2 at 15.2 MPa containing a small amounts of ethanol. Iwai et al. [41] indicated the possibility of the adsorptive separation of an isomeric dimethylnaphthalene mixture in supercritical CO_2 by using zeolite. They measured the adsorption capacity for various types of zeolite. Lim and Rizvi [42] evaluated the adsorptive separation process for cholesterol reduction in anhydrous milk fat by supercritical CO_2 using magnesium silicate. Milk fat was passed through the adsorber at 24.1 MPa, where 83% of the cholesterol in the feed was adsorbed. Following its saturation with cholesterol, the adsorbent was regenerated with 10% ethanol in supercritical

CO_2 at 24.1 MPa, after which its adsorption capacity was close to that of the virgin adsorbent.

Adsorbents are also used for citrus oil processing. Yamauchi and Saito [43] fractionated lemon peel oil with a gradual increase in pressure (10–20 MPa at 313 K) by supercritical fluid chromatography. Four fractions were obtained, and were mainly composed of terpenes, oxygenated terpenes, oxygenated terpenes removed by ethanol addition, and high molecular-weight compounds. Barth et al. [44] and Chouchi et al. [45, 46] used supercritical CO_2 to desorb or extract the oxygenated aroma compounds after equilibrium saturation under ambient conditions in a batch operation. They obtained a high-quality essential oil containing less terpenes and less nonvolatiles by supercritical CO_2 desorption with increasing pressure. However, high pressure or cosolvent was required to regenerate the adsorber because the nonvolatiles such as waxes and pigments were more strongly adsorbed on silica gel than aroma compounds. These results suggested that aroma compounds could be more selectively adsorbed on silica gel than terpenes, relatively higher pressure is required to desorb them or regenerate the adsorbent, and the nonvolatiles must be removed to maintain the activity of the adsorbent.

Adsorption equilibria are important to design and develop the adsorption process. Many investigators have reported retrograde phenomena in the vicinity of the critical point where the dependency of the solubility and the adsorption equilibrium constant is different from the usual liquid- or gas-phase phenomena. Thus, it is necessary to know the temperature and pressure dependence of the adsorption equilibrium constant, K_i. The temperature dependence is given by [47, 48]

$$\left(\frac{\partial \ln K_2}{\partial T}\right)_{P,y_2} = \frac{\Delta H_2^{\text{trans}}}{RT^2} + \alpha^{\text{m}}, \tag{4.3.1}$$

where

$$\Delta H_2^{\text{trans}} = \left(h_2^{\text{IG}} - h_2^{\text{m}}\right) + \Delta H_2^{\text{ads}}, \tag{4.3.2}$$

$$\Delta H_2^{\text{ads}} = \left(h_2^{\text{s}} - h_2^{\text{IG}}\right), \tag{4.3.3}$$

α is the volume expansivity, h is the enthalpy, and ΔH_2^{ads} is the heat of adsorption of the solute. Superscripts IG, m, and s denote ideal gas, mobile phase, and stationary phase. Subscripts 1 and 2 denote supercritical fluid and solute. The temperature dependence of the adsorption equilibrium constant is determined by the relative values of $h_2^{\text{IG}} - h_2^{\text{m}}$ and $h_2^{\text{IG}} - h_2^{\text{s}}$. The adsorption equilibrium constant increases with temperature if $h_2^{\mathrm{IG}} - h_2^{\mathrm{m}} > h_2^{\mathrm{IG}} - h_2^{\mathrm{s}}$, which is satisfied in the retrograde region.

The pressure dependence of the adsorption equilibrium constant is given by

$$\left(\frac{\partial \ln K_2}{\partial P}\right)_{T,y_2} = \frac{v_2^{\mathrm{m}} - v_2^{\mathrm{s}}}{RT} - \beta^{\mathrm{m}}, \tag{4.3.4}$$

where v and β denote the molar volume and isothermal compressibility, respectively.

The adsorption equilibrium constants of limonene and linalool on silica gel in supercritical CO_2 measured by an impulse response technique are shown in Figure 4.3.15 [49]. Adsorption equilibrium constants were correlated linearly in a log-log plot as a function of the density of supercritical CO_2 independent of pressure and temperature. Adsorbed amounts decreased with the increase in the solvent density for both limonene and linalool. These results suggest the possibility of a process where oxygenated compounds are selectively adsorbed on the adsorbent at a lower pressure and then desorbed at a higher pressure. The adsorption isotherm was also measured for aroma and terpene fractions of orange oil by a step response method and correlated with a multicomponent Langmuir equation:

$$q_i = \frac{q_s K_i C_i}{1 + \Sigma K_i C_i}, \tag{4.3.5}$$

where adsorption equilibrium constants were correlated as a function of CO_2 density: $K_1 = 4.237 \times 10^8 \rho^{-2.874}$ for terpene, and $K_2 = 9.395 \times 10^{10} \rho^{-3.371}$ for aroma.

From the equilibrium relation it is evident that adsorption is favored at lower pressure and desorption is favored at higher pressure. The adsorption and desorption behavior of orange oil was measured at 313 K [50]. After the feed orange oil had breakthroughed the column at 8.8 MPa, pure supercritical CO_2 was passed through the column at 19.4 MPa to desorb the solutes. Figure 4.3.16 shows the desorbed amounts and the variation in concentration of solutes desorbed divided by that in the feed. Oxygenated aroma components were concentrated up to 50 times.

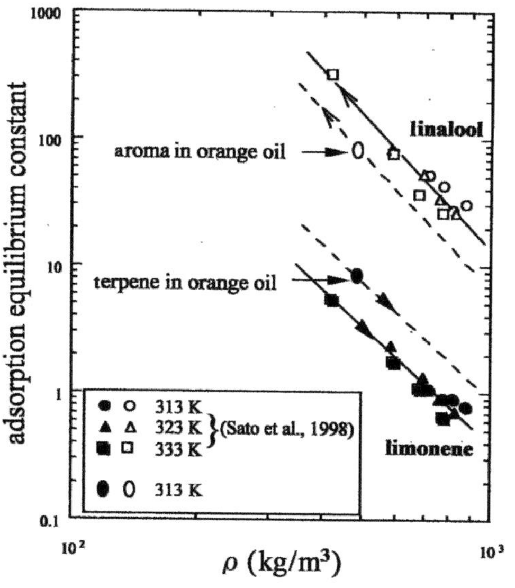

Fig. 4.3.15 Adsorption equilibrium constant of orange oil components on silica gel in supercritical CO_2 [49]

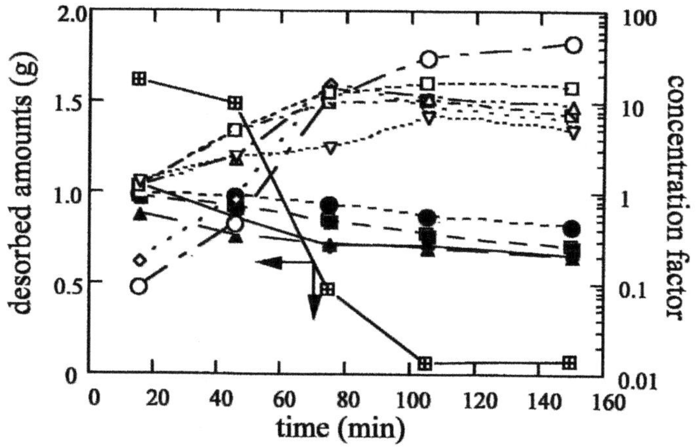

Fig. 4.3.16 Desorption curves from silica gel saturated by orange oil. Definitions of symbols are the same as in Fig. 4.3.14

4.3.4 Pressure Swing Adsorption

Pressure swing adsorption (PSA) is an important process for the separation of gas mixtures and has been commercialized for air drying, oxygen and nitrogen separation of air, hydrogen purification, and various other separations [51, 52], because of its low energy requirements and low capital investment costs. The PSA process is based on the regeneration of adsorber by the difference of adsorbed amounts as a function of pressure, since the adsorbed amounts decrease with decreasing pressure in gaseous systems. The process involves two fundamental steps: (1) adsorption at high pressure, where the preferentially adsorbed components are taken up from the feed gas; (2) desorption or regeneration step by pressure reduction, where the adsorbed components are removed from the adsorbent. Most commercial PSA processes are of multibed design to generate a near-continuous product stream. In the case of a two-bed process, one bed is in the adsorption step while the other one is simultaneously in the desorption step.

Sato et al. [53] applied pressure swing adsorption to a supercritical fluid for the fractionation of citrus oil. Figure 4.3.17 shows the schematic diagram of the experimental setup. A continuous cyclic operation between the adsorption step, where cold-pressed orange oil in supercritical CO_2 was continuously passed through the column at 8.8 MPa and 313 K, and the desorption step, where pure supercritical CO_2 was passed through the column at 19.4 MPa, including the rinse step, was demonstrated. All experiments were started with

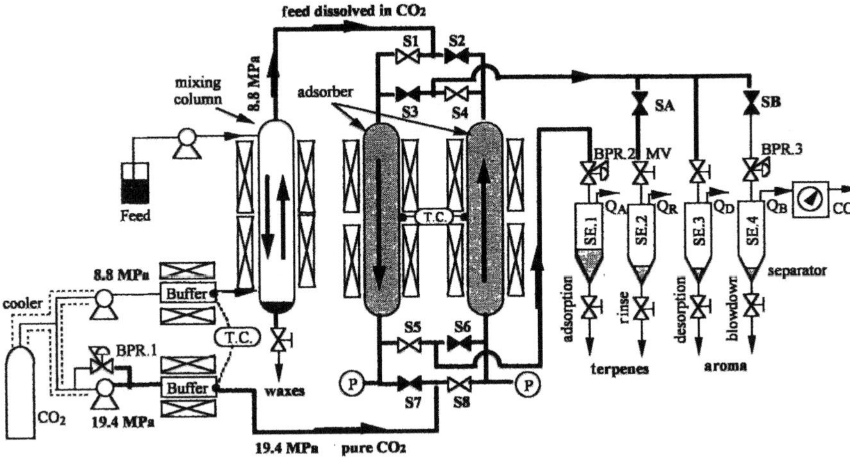

Fig. 4.3.17 Experimental setup of supercritical pressure swing adsorption for citrus oil processing

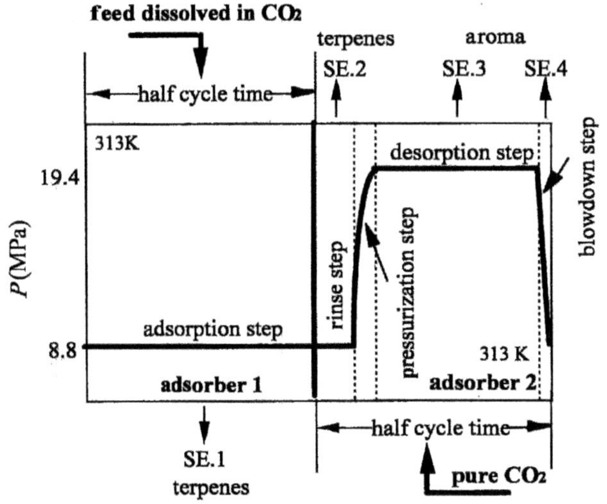

Fig. 4.3.18 Configuration of pressure swing adsorption for citrus oil processing in supercritical CO_2

clean beds. In general, 10 half-cycles were required to approach the cyclic steady state. The operation is shown in Figure 4.3.18.

The effect of desorption-to-adsorption CO_2 flow ratio Q_D/Q_A on the concentration factor and the recovery at a constant Q_A are shown in Figs 4.3.19 and 4.3.20, respectively. The concentration factor of 10 and the recovery of 65% were obtained at a Q_D/Q_A ratio of 2. An increase in the Q_D/Q_A ratio caused higher recovery in the desorption step. A mathematical model to simulate the pressure swing adsorption process has been developed. Model calculations agree roughly with the experimental results as shown in Figs. 4.3.19 and 4.3.20.

Figure 4.3.21 shows the gas chromatograms of the feed and the oil obtained from adsorption and desorption steps at a half-cycle time of 120 min. The chromatogram for the adsorption step shows that aroma compounds in orange oil are adsorbed on the silica gel in the adsorption step, therefore, terpenes make up the major portion of the oil obtained in the adsorption step. On the other hand, the chromatogram for the desorption step shows that aroma compounds adsorbed on the silica gel in the adsorption step are desorbed, so that aroma compounds make up the major portion of the product oil obtained in the desorption step.

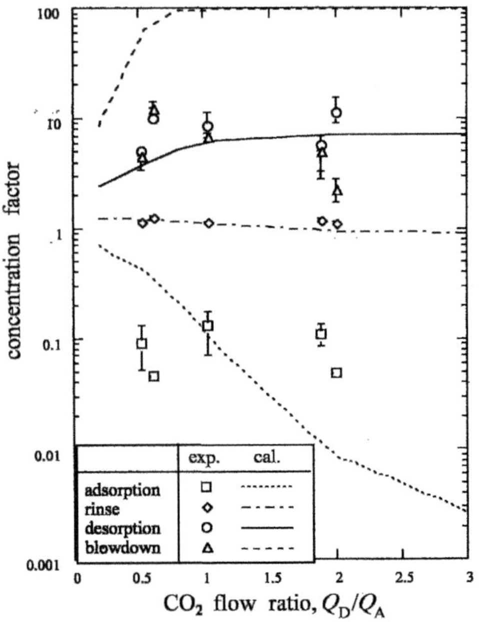

Fig. 4.3.19 Concentration factor for pressure swing adsorption of orange oil

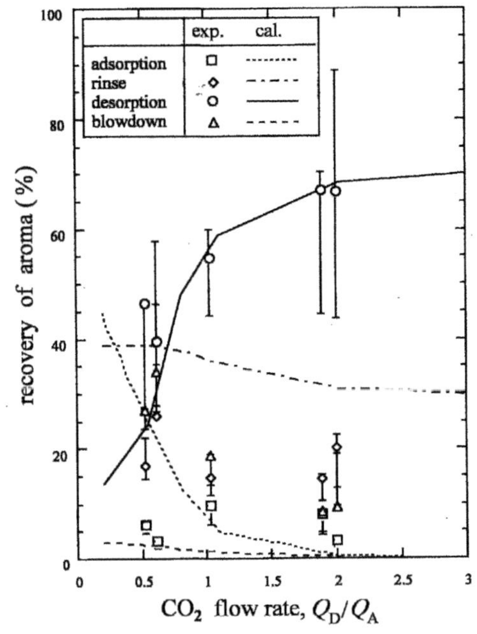

Fig. 4.3.20 Recovery of aroma for the pressure swing adsorption of orange oil

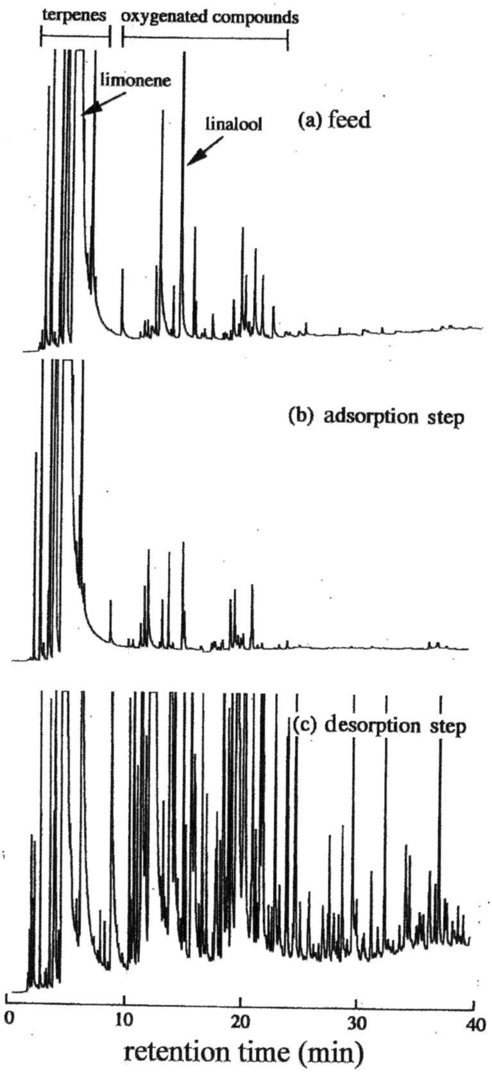

Fig. 4.3.21 Gas chromatograms of orange oil for (**a**) feed, (**b**) adsorption step, and (**c**) desorption step in pressure swing adsorption processing

4.4 Separation and Analysis by Supercritical Fluid Chromatography

Fluid Chromatography

4.4.1 Introduction

Supercritical fluid chromatography (SFC) is defined as a chromatographic technique in which the mobile phase is subjected to pressures and temperatures above the critical point. The subcritical region just below the critical point is often included. The mobile phase should have solvating power and thereby contribute to the chromatographic partition process. By varying the pressure or temperature it is possible to change the nature of the mobile phase from gas-like to liquid-1ike conditions. Therefore, SFC is generally considered as an intermediate technique between gas chromatography (GC) and liquid chromatography (LC).

Although chromatographic separation using a supercritical fluid as the mobile phase was first reported by Klesper et al. in 1962 [l], its growth was relatively slow until the early 1980s. In 1981, Novotny et al. introduced an open tubular column in SFC and demonstrated that high efficiency separation can be achieved [2]. In 1982, Gere et al. developed a packed column SFC instrument based on LC and demonstrated that rapid separation can be performed [3]. These events accelerated the investigation of SFC as well as the development of various associated techniques. SFC is presently recognized as a powerful analytical tool to supplement GC and LC, and is expected to be a separation and purification method in industry. Many companies provide SFC instruments, which are usually divided into two types, open tubular column SFC and packed column SFC.

4.4.2 Characteristics of Supercritical Fluid Chromatography

(a) Mobile Phase

Compared with a gas and a liquid, a supercritical fluid has favorable properties for use as the mobile phase in chromatography. A comparison of chromatographically important properties (density, viscosity and diffusivity) between gas, liquid and supercritical fluid is given in Appendix B [4]. Since a supercritical fluid has a liquid-like density and can dissolve many compounds, SFC can separate less volatile compounds which are not amenable to GC. Thermally labile compounds can be separated using a supercritical fluid with low critical temperature. More importantly, the solvent strength of a supercritical fluid can be widely altered by changing the pressure and temperature. This

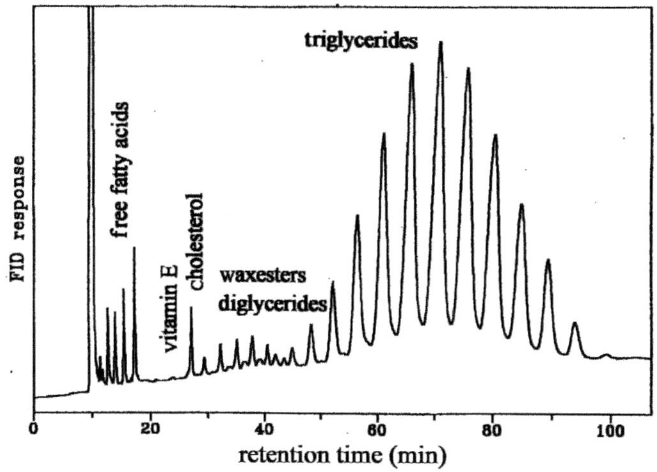

Fig. 4.4.1 Separation of sandeel oil. Conditions: DB-5 (20 m × 100 μm i.d., 0.4 μm d_f), 170 °C, CO_2, FID detection. Density was programmed at a rate of 0.001 g/cm^3·min from 0.300 g/cm^3 to 0.520 g/cm^3 [5]

implies that mixtures with a wide range of molecular weights can be separated without changing the chemical components of the mobile phase. SFC allows faster separation than LC, because the lower viscosity and higher diffusivity of supercritical fluids enables the rapid mass transfer in partition processes.

Although a large number of solvents can be used as the mobile phase, CO_2 is commonly used due to its favorable properties. Its relatively low critical temperature makes it possible to separate thermally labile compounds. It is available at high purity and low cost. An additional important feature of CO_2 is the compatibility with most detectors for GC and LC. CO_2 is relatively nonpolar and a good solvent for lipophilic compounds and hence finds wide applications for analysis of lipid-related compounds. However, for analytical purposes, the mobile phase is not always required to be a good solvent for solutes of interest, because separation is usually conducted under highly diluted conditions. CO_2 can dissolve various compounds with low to medium polarity such as foods, natural compounds and pharmaceuticals. Figure 4.4.1 shows one of most important applications of SFC, in which mixtures with varying molecular weight and polarity are separated in a single run [5]. Such separation is difficult to perform with GC and LC. In SFC, solutes with molecular masses of several thousands such as various oligomers can easily be separated and detected with a flame ionization detector (FID) (discussed later, see Fig. 4.4.9). Thus, SFC with a CO_2 mobile phase and FID is highly valuable for performing characterization of raw materials for both research and manufacturing purposes. Adding an appropriate organic solvent to CO_2 as a modifier (or an entrainer) largely improves the solubility problem, although the use of FID becomes impossible.

Separation of enantiomers has become increasingly important in the development of new pharmaceuticals and agricultural chemicals, due to the role of chirality in biological activity and toxicity. Many chiral stationary phases have been reported for use in GC and LC. The use of SFC attracts much attention in this field, because SFC often provides improved peak resolution and shortens the method development time when compared to LC. In Fig. 4.4.2 the complete separation of three peaks of enantiomers, which would require three different mobile phases in LC with the same column, are achieved in a single run using CO_2 + methanol eluent [6]. Furthermore, ease of mobile phase removal makes SFC attractive for preparative separation in analytical as well as production scales.

(b) Column Efficiency

Chromatographic separation efficiency is usually expressed in terms of plate height (H) or total number of theoretical plates (N). N can be calculated:

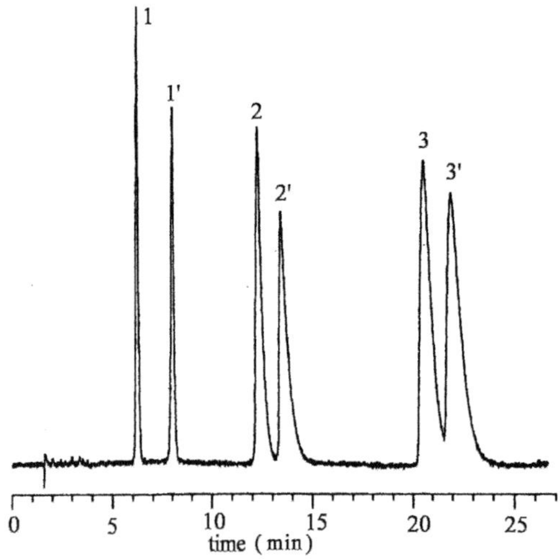

Fig. 4.4.2 Separation of *N*-(3,5-dinitrobenzoyl)-DL-valine methyl ester (1 and 1'), ancymidol (2 and 2') and proglumide (3 and 3'). Conditions: Cyclobond I 200 RN (25 cm × 4.6 mm i.d., 5μm), 30 °C, CO_2-methanol (90:10), 2.0 ml/min, 150 atm, UV detection [6]

$$N = 16\left(\frac{t_R}{W_b}\right)^2 = 5.54\left(\frac{t_R}{W_h}\right)^2, \tag{4.4.1}$$

where t_R is the retention time, and W_b and W_h are the peak width at the baseline and the half-height, respectively. When the column length is L, then H is N/L. The speed of analysis is usually related to the plate numbers per unit time. These are dependent on the column type and the property of the mobile phase, In GC, an open tubular column is preferred and in LC only a packed column is used, while both are used in SFC. Figure 4.4.3 shows a comparison of van Deemter curves (the relationship between plate height and the linear velocity of the mobile phase) for various chromatographic techniques under the conditions normally used, although a 50 μm i.d. open tubular column for GC (curve d) is just for comparison between GC and SFC using the same column [7].

The plate height for a typical GC column with 250 μm i.d. (curve **e**) is the highest among these columns. Due to its high permeability, however, a long column (more than 30 m) can be operated at the highest linear velocity, resulting in the highest total plate number (more than 100000) in a reasonable time. Since the diffusivity of a supercritical fluid is lower than that of a gas, an open tubular column with smaller i.d. (typically 50 μm and 10 m long) is used in

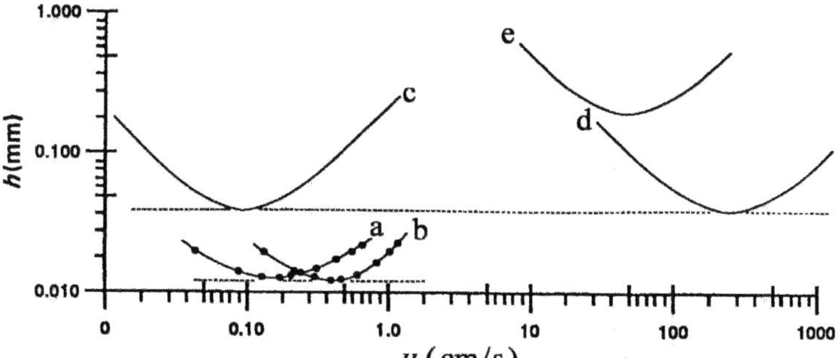

Fig. 4.4.3 Van Deemter curves for various column chromatographic techniques: (**a**) packed column LC, (**b**) packed column SFC, (**c**) open tubular column SFC, (**d, e**) open tubular column GC. Curves (**a**) and (**b**) are from experimental data, while curves (**c**)–(**e**) are calculated from the Golay equation. Conditions: (**a**) 10 cm × 4.6 mm i.d. packed column, C_{18} bonded silica particles (Hypersil), 5 μm, $k' = 2.85$, 30% H_2O in CH_3CN, 40 °C; (**b**) same as for (**a**) except CO_2 at 0.8 g/cm^3, $k' = 2.30$; (**c**) 50 um i.d. open tubular column, $k' = 2.30$, CO_2, 40 °C, 0.8 g/cm^3, $D_m = 8\times10^{-5}$ cm^2/s; (**d**) 50 μm i.d. open tubular column, $k' = 2.30$, D_m = 0.2 cm^2/s; (**e**) same as for (**d**) except 250 μm i.d. open tubular column [7]

SFC. If such a column is operated at the optimum linear velocity giving a minimum plate height, a total plate number comparable to that for GC is attainable (curve **c**). However, analysis speed is too slow at the optimum linear velocity, although operation at higher temperature and lower density can improve this situation to some extent. Therefore, a linear velocity more than ten times the optimum is usually used, resulting in a total plate number less than 50000. If the same column is used for LC (not shown here), the optimum linear velocity would be extremely slow, making the use of an open tubular column unrealistic. A further decrease in column dimension improves the analysis speed, but various technical problems will arises including sample injection techniques.

The plate height is the smallest for a packed column (curve a and b). When the same column is used, SFC can be operated at a linear velocity a few times faster than LC. The length of the packed column is usually limited to around 25 cm due to its lower permeability, resulting in a total plate number of around at 20 000. Therefore, packed-column SFC provides lower efficiency in terms of total plate number but faster analysis speed than open tubular column SFC.

(c) Effect of Pressure Drop

Besides separation efficiency, the effect of the pressure drop along the column must be considered. The pressure drop can usually be neglected in open tubular column SFC but is significantly large in packed-column SFC. Since a supercritical fluid is highly compressible, the pressure drop causes a density gradient along the column. In SFC, retention is strongly dependent on the density of the mobile phase . Retention in chromatography is usually measured as the capacity factor (k'):

$$k' = \frac{t_R - t_0}{t_0}, \tag{4.4.2}$$

where t_R is the retention time and t_0 is the time required by an unretained substance to elute from the column. $\log k'$ roughly changes in a linear fashion with density:

$$\log k' = a + b\rho, \tag{4.4.3}$$

where a and b are constants depending on the temperature and the compound type, and ρ is the density of the mobile phase. In addition, selectivity

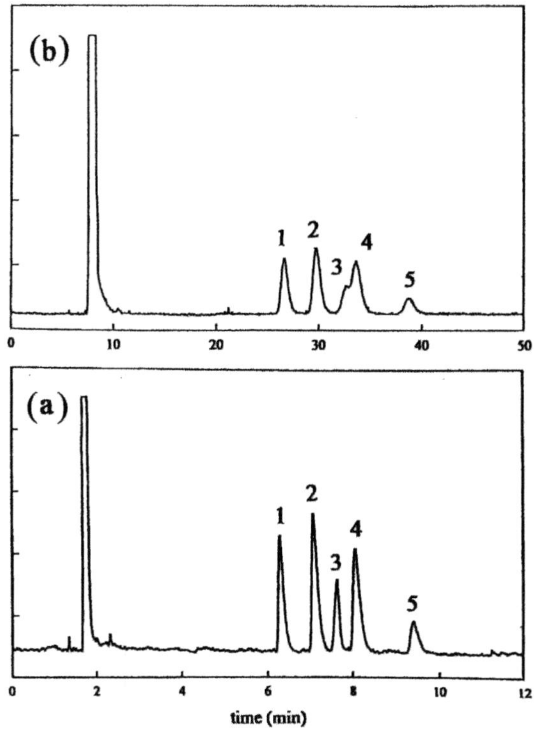

Fig. 4.4.4 Effect of flow rate on the separation of mixtures of liquid crystals used in an LCD display at an inlet pressure of 140 atm: (**a**) 600 ml/min. (**b**) 150 ml/min (gaseous flow at ambient conditions). Conditions: Zorbax ODS (25 cm × 4.6 mm i.d., 5 μm), 50 °C, CO_2, FID detection after splitting. (1) NC-Ph-cH-Et, (2) NC-Ph-cH-Pr, (3) MeO-Ph-cH-Pr, (4) NC-Ph-cH-Bu; (5) NC-Ph-cH-Pe [8]

often changes with pressure because the density change affects the solvation of solutes differently depending on the compound types. This implies that when a density gradient occurs, retention as well as selectivity may continuously change along the column. An example is shown in Fig. 4.4.4, where separations are performed under conditions of the same inlet pressure and different flow rate [8]. Since the inlet pressure, outlet pressure and flow rate are dependent upon each other, in this case the outlet pressure should be different. This behavior may be useful for a particular separation. However, the density gradient also changes the linear velocity of the mobile phase along the column. The excess pressure drop reduces the column efficiency because the linear velocity of the mobile phase at most positions in the column should be higher or lower than the optimum.

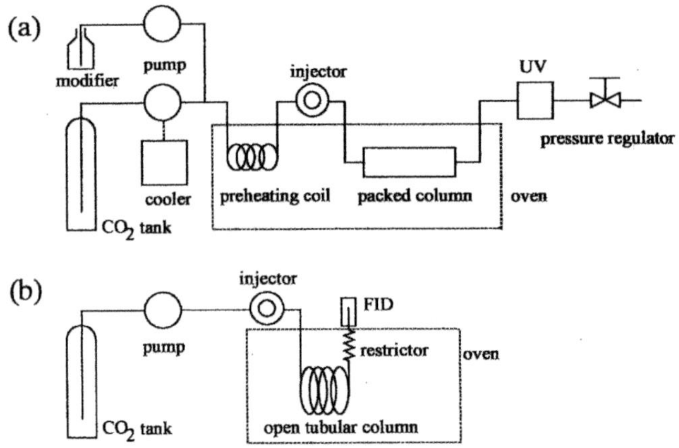

Fig. 4.4.5 Typical setup for SFC with (**a**) packed column and (**b**) open tubular column

4.4.3 Instrumentation

Figure 4.4.5 shows the typical setup for SFC with an open tubular column and a packed column. Packed-column SFC is based on LC, while open tubular column SFC is based on GC, but they have some unique aspects themselves. The most important aspects are the pressure-controlled pumping system which is essential for pressure programming and the flow or pressure restrictor at the end of column which is required to maintain supercritical pressure inside the column. Both types are commercially available, but they are often constructed from GC and LC parts by researchers.

(a) Column and Programming Method

In packed column SFC, an ordinary reciprocating LC pump is used with some modifications, which covers the required pressure range up to 500 atm. The pump head is cooled to effectively supply CO_2 as a liquid. The pump is usually operated at constant flow rate and the system pressure is kept constant or programmed with a pressure regulating valve at the column end. Preparation of mixed mobile phase as well as solvent composition programming can be performed in packed-column SFC as in LC. This setup can also be used for open tubular column SFC. However, the mobile phase must be split before the injector to get the homogeneity and stability of the modifier concentration at the slow flow rate required in open tubular column SFC.

In open tubular column SFC, a syringe pump is preferred due to its pulseless flow. In contrast with packed-column SFC, isobaric separation is rarely performed. In most cases, the pump is operated at pressure programming mode and the flow rate is controlled with a flow restrictor. This restrictor also plays an important role to interface the separation column under high pressure to the detectors working at lower pressure such as FID. This setup is also often used for SFC with a microbore packed column (less than 1 mm i.d.).

In addition to pressure and solvent composition programming, in SFC various methods are used to elute mixtures with a wide range of polarity and molecular weight in a single run. Temperature programming is always performed upward in GC, while in SFC both upward [9] and downward temperature programming [10] can be used depending on the conditions. The former is often used in combination with pressure programming to increase the density of the mobile phase.

(b) Detector

Success in a particular application often depends on the choice of the detector. Nearly all GC and LC detectors have been adopted in SFC. The most common detector for packed column SFC is an ultraviolet (UV) absorption detector with a high-pressure cell, because the entire system through the detector is under high pressure. Since unmodified CO_2 exhibits little background absorbance in the range down to 200 nm, a large number of compounds can be detected. UV is also often used in open tubular column SFC, but the cell volume must be reduced to minimize band broadening. A short piece of capillary tubing can be used as a cell. However, a capillary cell is very sensitive to refractive index changes of the mobile phase and pressure programming causes a baseline shift. Cooling the cell and keeping the cell pressure constant greatly reduces the baseline shift and allows detection at high sensitivity as shown in Fig. 4.4.6 [11].

The most popular detector for open tubular column SFC is FID. It provides a universal detection system for nearly all organic compounds. FID is also often used in packed-column SFC. However, the flow rate of a packed column is usually too high to sustain combustion of the FID flame, so that splitting of the effluent at the detector is needed. Using a microbore packed column may allows direct coupling to FID. Figure 4.4.7 is a unique application of packed-column SFC with FID, where type separation of fuel oil is performed [12]. This method is adopted as an American Society for Testing and Materials method (ASTMD5186).

Since an evaporative light scattering detector (ELSD) is a universal detector and compatible with the mixed mobile phase, it is a promising

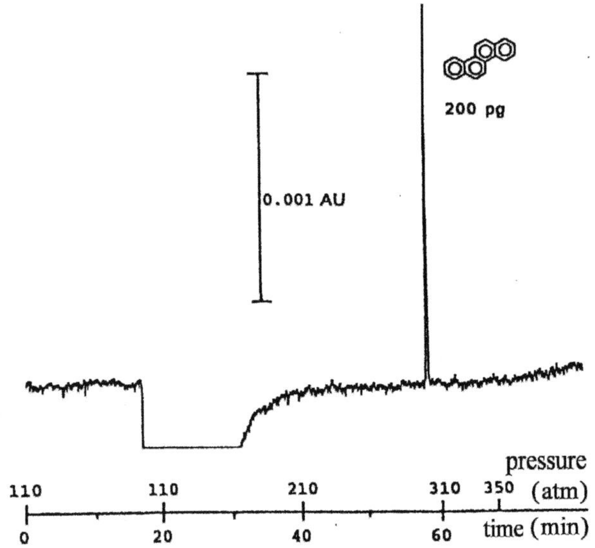

Fig. 4.4.6 Detection using a temperature- and pressure-controlled UV detector in open tubular column SFC. Conditions: SB-Biphenyl (10 m × 50 μm i.d., 0.25 μm d_f), CO_2, 90 °C oven temperature, 80 atm cell pressure, 18 °C cell temperature, UV 269 nm [11]

alternative detector when compounds without chromophore are separated by using a mixed mobile phase. Other common detectors include various selective detectors such as a thermionic detector (TID), a flame photometric detector (FPD), an electron capture detector (ECD) and a fluorescence detector (FD). On-line coupling to a mass spectrometer (MS) as well as Fourier transform infrared spectrometer (FT-IR) are also utilized.

(c) Restrictor

When a detector operating at lower pressure than the column is used as in the case of FID, a flow restrictor must be used to interface the high-pressure chromatographic system to the detector. Proper operation of the flow restrictor is thus crucial to the proper operation of the chromatographic system. Four types of restrictors shown in Fig. 4.4.8 have been used in SFC.

A linear restrictor with narrow-bore fused-silica tubing (5–25 μm i.d.), is successful for low-molecular weight volatile analytes, but often causes detector "spiking" and restrictor plugging when less volatile substances are chromatographed This type is mainly used for supercritical fluid extraction in

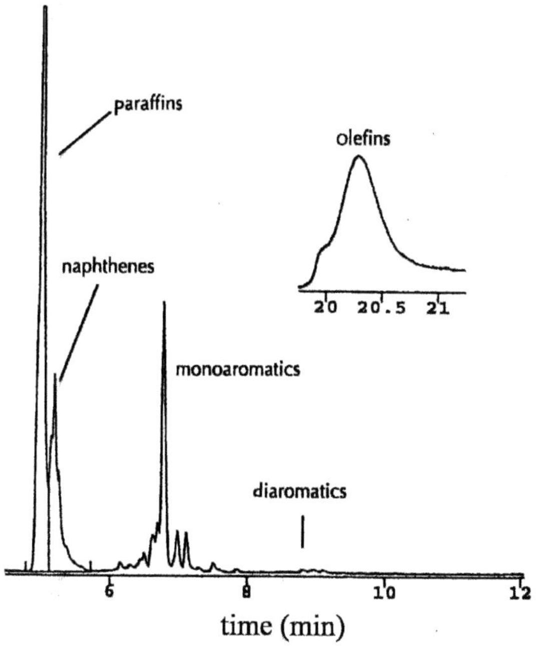

Fig. 4.4.7 Type separation of jet fuel oil by SFC. Four columns packed with silica gel (25 cm × 4.6 mm i.d., 5 μm) in series and a column packed with a sliver-loaded cation-exchanger (30 cm × 4.6 mm i.d., 5 μm) were coupled via a switching valve. Olefins were eluted by back-flushing the silver-loaded column and others were eluted through all columns. Conditions: CO_2, 148 atm, 29 °C, FID detection after splitting [12]

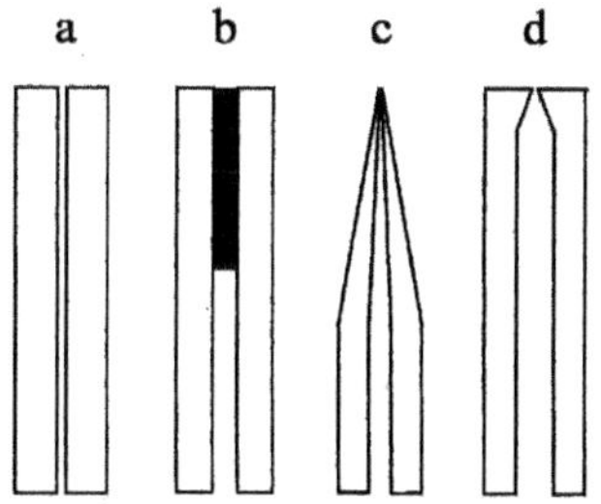

Fig. 4.4.8 Various restrictors used in SFC (**a**) linear (**b**) frit (**c**) tapered, (**d**) integral

combination with a solvent trapping method. A frit restrictor is a fused-silica tube (25–50 μm i.d) with a porous ceramic bed in its end. This type is the most resistant to plugging because decompression occurs through the multitude of paths in the porous ceramic bed. A tapered restrictor is produced from

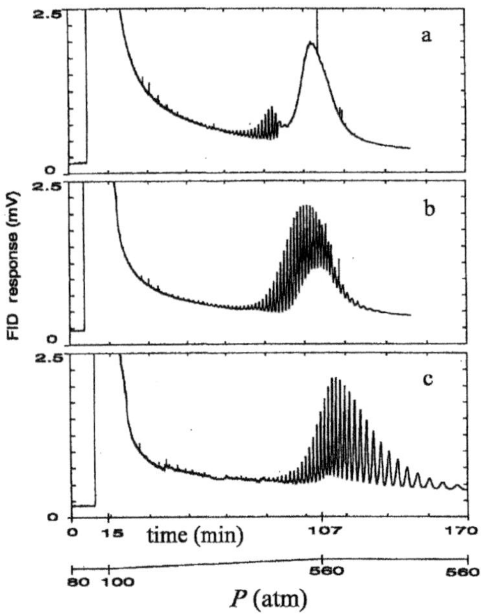

Fig. 4.4.9 Separation of the average molecular weight 2000 polyethylene glycol standard (PEG 2000) by using (**a**) a frit restrictor, (**b**) a tapered restrictor, and (**c**) an integral restrictor. Conditions: SB-Biphenyl-30 (10 m × 50 μm i.d., 0.25 μm d_f), 120 °C, CO_2, FID detection [13]

a fused-silica capillary (25–50 μm i.d.) by drawing it in a flame and cutting at the desired aperture. This type is mechanically very weak and is not easy to install into the detector. An integral restrictor is prepared by sealing the end of a fused-silica tube (25–50 μm i.d.) in a flame followed by abrasion of the tip until a small opening is achieved. This type is more robust than a tapered restrictor.

A frit restrictor is commercially available or attached to the commercial open tubular column. An integral restrictor is also commercially available, but often prepared by researchers themselves, as is the tapered restrictor. Although these three are satisfactory for usual applications, tapered and integral restrictors are superior to the frit restrictor for the separation of high molecular weight compounds as shown in Fig. 4.4.9 [13]. However, it must be kept in mind that the pressure dependency of the flow rate is different between tapered and integral restrictors as shown in Fig. 4.4.10 [13]. The linear velocity of the mobile phase rapidly increases with pressure for the tapered restrictor. Since an open tubular column is usually operated at considerably higher linear velocity than the optimum, column efficiency will be largely degraded at

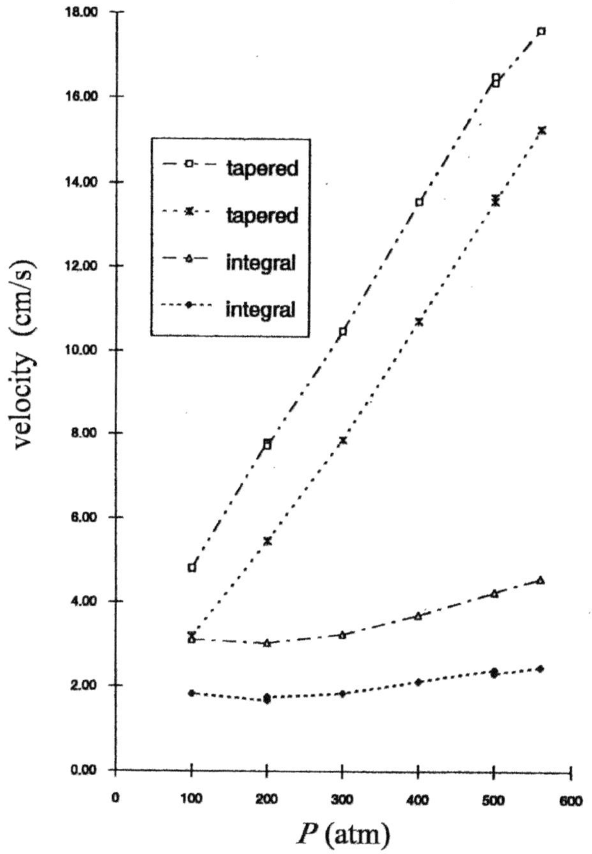

Fig. 4.4.10 Comparison of the linear velocity vs. pressure behavior of two tapered and two integral restrictors [13]

higher pressure when a tapered restrictor is used. On the other hand, the linear velocity is relatively constant during pressure programming when an integral restrictor is used.

(d) Sample Introduction

For packed-column SFC, a normal LC injection method using a high-pressure loop injector is used. Samples can be injected by dissolving in an appropriate solvent miscible with the mobile phase. However, the solvent strength of the sample solvent is usually stronger than that of the supercritical mobile phase, so that injection of solvent causes a transient increase in solvent strength of the

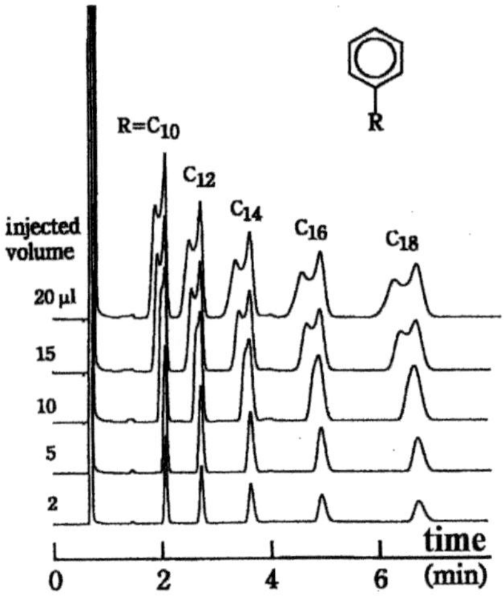

Fig. 4.4.11 Effect of sample volume on peak shape in packed column SFC. Conditions: Superpak Crest C18 (15 cm × 4.6 mm i.d., 5 µm), 60 °C, CO_2, 140 atm, 3 ml/min, UV 210 nm, 500 ppm alkylbenzenes in methylene chloride

mobile phase. The allowable injection volume without degradation of peak shape seems to be smaller in SFC than in LC. Figure 4.4.11 shows the effect of sample volume on the peak shape in packed-column SFC with a conventional analytical column, indicating that the sample volume should be less than 10 µl for this column [14], although this may be dependent on the properties of the sample solvent and the chromatographic conditions.

For open tubular column SFC, the situation is more severe, because of the small column dimensions. A commercially available injector with a sample volume of 0.06 µl is often used. However, injecting such a small volume makes it difficult to analyze trace components. A peak as low as 100 pg can be detected with FID. If it is desired to have a detection limit of 100 ppb in terms of solute concentration in the solution injected, the required injection volume would be 1 µl. Direct injection of this volume corresponding to the 0.5 m length of a 50 µm i.d column will spread solute over a meter or more of the column, resulting in the skewed or split peaks. So far various injection techniques allowing the injection of volumes of the order of a microliter have been developed; these are usually based on separating the solvent and solutes and then reconcentrating the solutes at the column inlet as a narrow band followed by pressure programming. Figure 4.4.12 shows an example with one of the injection methods, in which a joint with a small volume was placed ahead of the

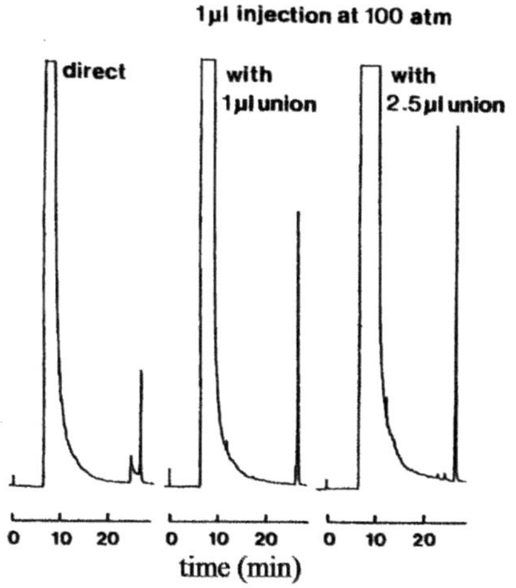

Fig. 4.4.12 Effect of chamber volume on peak shape in open tubular column SFC. Conditions: OV-17 (15 m × 100 µm, 0.25 um d_f), CO_2, FID detection. 1 µl injection of 100 ppm squalane in methylene chloride. Pressure was kept at 100 atm for 10 min and then programmed at a rate of 5 atm/min [15]

column to dilute the sample solvent with the mobile phase [15]. This dilution decreases the solvent strength and makes the solute refocusing effective.

4.4.4 Control of Retention and Selectivity

In GC, the solute vapor pressure determines the partition ratio. Changes in retention are made by adjusting the solute vapor pressure via the column temperature. In LC, retention is influenced by the solute–mobile phase interaction. This is usually adjusted by means of the chemical compositions of the mobile phase. In SFC, the density of the mobile phase is of primary importance as shown by (4.4.3). In addition, $\log k'$ also linearly changes with the reciprocal temperature, when the pressure is adjusted to afford the same density. However, when the temperature is changed at a given pressure, retention typically changes in a complex manner as shown in Fig. 4.4.13 [16]. This behavior can be explained as follows. At the lower temperature where the solute volatility is not a important factor for retention, solutes are partitioned primarily due to competition between the solute–mobile phase and the solute–stationary phase interaction as in LC. In this case, a loss in the mobile phase density caused by increasing temperature decreases the retention. At the higher temperature where the volatility contribution becomes more important, retention decreases with temperature as in GC.

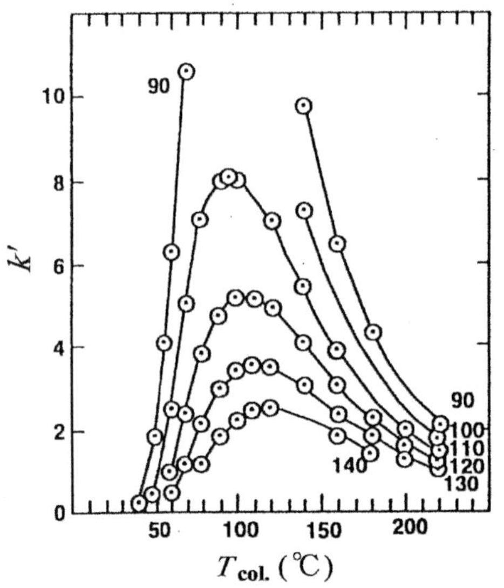

Fig. 4.4.13 Effect of temperature on retention of fluoranthene at various pressures: numbers next to curves indicate fixed pressures in atm. Conditions: methyl silicone (10 m × 50 μm i.d), CO_2 [16]

In SFC, both the stationary and mobile phase affect the retention and selectivity as in LC, but there are constraint on the mobile phase composition since the critical parameters are a function of the composition. Thus, the method used to optimize the separation is somewhat different from LC. Since retention is dependent on both temperature and density for a given mobile phase composition, there are various combinations of temperature and density capable of eluting a particular solute with a desired k' value. In addition, retention often changes in different way depending on the compound types. Thus, the pressure (or density) is first selected at a certain temperature and then the temperature is manipulated for tuning the selectivity. The effect of temperature on the selectivity is usually larger in SFC than in GC and LC [17], because the stationary phase swelling is strongly dependent on temperature under supercritical conditions and changes the property of the stationary phase.

4.4.5 Measurement of Solubility and Phase Behavior Using Supercritical Fluid Chromatography Techniques

Knowing the solubility in a supercritical fluid is very important in developing various supercritical fluid technologies. The most common method for determining solubilities is a gravimetric method. However, this method requires

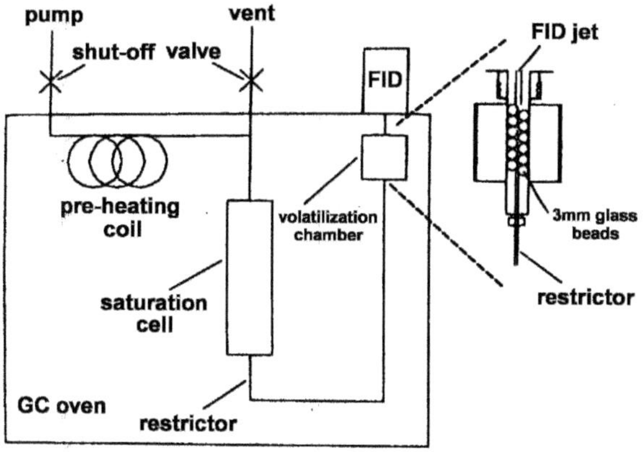

Fig. 4.4.14 Schematic diagram of the on-line FID solubility apparatus [18]

large amounts of analyte and lacks the necessary sensitivities for low-solubility compounds. In addition, such a batch process is time-consuming. Therefore, available data is usually limited to a narrow range. A saturation cell was coupled to FID as shown in Fig. 4.4.14 [18] and the solubilities of low- (as low as 1×10^{-8} in molar ratio) to moderate-solubility organics in CO_2 were determined over a wide range of temperature (40–200 °C) and pressure (150–400 atm) with day-to-day reproducibilities of less than 5% (relative standard deviation). This method considerably speeded up the solubility measurements, and data over a wide range were obtained. The solubilities determined by this method were in good agreement with published data from gravimetric or spectroscopic methods. The mixture critical curve is the locus of mixture critical points scanning the entire range of composition of the binary mixture. Experimental data is required for developing the equation of state to predict the phase behavior as well as for performing various supercritical fluid experiments. Although various binary mixtures containing CO_2 have been studied, only a few binary mixtures have been thoroughly studied. Most studies have been performed using a high-pressure view cell and either visual observation or light scattering detection. In this case, the experiments are generally performed under isothermal conditions with a known mixture composition and varying pressure. However, this method is time-consuming, making it difficult to obtain data over a wide range. A peak shape method for estimating the mixture critical curve for CO_2 + solvent mixtures was developed by using chromatographic equipment and FID [19]. Solvent is injected in CO_2 flowing in an open tubular column without stationary phase. If the resulting binary mixtures remains in one phase over all compositions, then the injected solvent

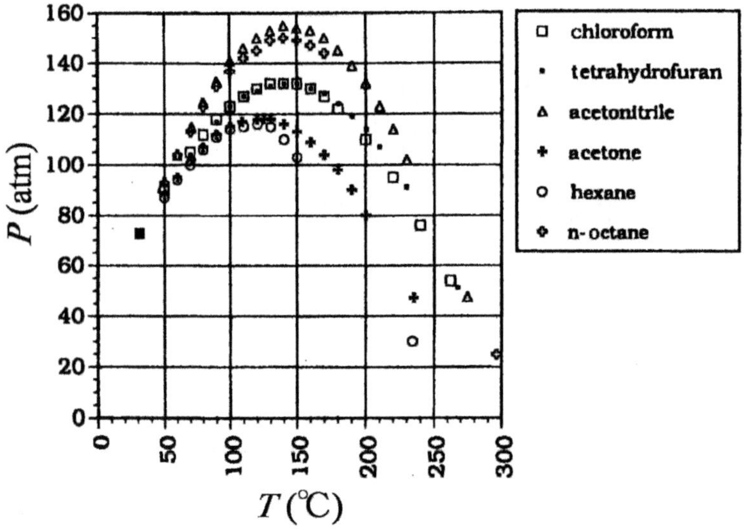

Fig. 4.4.15 Mixture critical curves for CO_2 combined with various solvents estimated using the peak-shape method. The last point of each curve is the critical point of the solvent [19]

plug mixes freely with the mobile phase and yields a peak as seen in chromatography. However, if a liquid phase forms, it dynamically coats the inner surface of the open tubular column, resulting in a rectangular peak shape. The results for several solvents (critical loci were reported for 13 solvents) are shown in Fig. 4.4.15 and the differences were within 3% compared with high-pressure view cell methods. Although this method estimates mixture critical pressures and temperatures but not the corresponding chemical compositions, it is a simple and rapid method.

References

4.1 Principles for Supercritical Fluid Separation

[1] T. Clifford, "*Fundamentals of Supercritical Fluids*", Oxford University Press, New York (1999).

[2] K. D. Bartle, A. A. Clifford, S. A. Jafar and G. F. Shilstone, *J. Phys. and Chem. Reference Data*, **20**, 713 (1991).

[3] R. T. Kurnik and R. C. Reid, *Fluid Phase Equilibria*, **8**, 93 (1982).

[4] G.-T. Liu and K. Nagahama, *J. Supercrit. Fluids*, **9**, 152 (1996).

[5] G.-T. Liu and K. Nagahama, *J. Chem. Eng. Jpn.*, **30**, 293 (1997).

[6] T. Suzuki, N. Tsuge and K. Nagahama, "*Solvent Extraction 1990*", ed. T. Sekine, Elsevier, Amsterdam, p. 1701 (1992).
[7] K. Tanida, Y. Sato and H. Masuoka, *Kagaku Kogaku Ronbunshu*, **22**, 385 (1996).
[8] H. Shojibara, Y, Sato, S. Takishima and H. Masuoka, *J. Chem. Eng. Jpn.*, **28**, 245 (1995).
[9] J. S. Porto, K. Tanida, Y. Sato, S. Takishima and H. Masuoka, *J. Chem. Eng. Jpn.*, **28**, 388 (1995).
[10] G. Brunner, "*Gas Extraction*", Springer, New York, p. 189 (1994).
[11] M. McHugh and V. Krukonis, "*Supercritical Fluid Extraction*", 2nd ed., Butterworth-Heinemann, Boston, p. 192 (1994).
[12] L. T. Taylor, *"Supercritical Fluid Extraction"*, John Wiley & Sons, New York, p. 5 (1996).

4.2 Process Development Studies for Separation of Chemicals

[1] M. McHugh and V. Kurukonis, *"Supercritical Fluid Extraction*", 2nd ed., Butherworth-Heinemann, Boston, p. 170 (1994).
[2] S. Furuta, N. Ikawa, R. Fukuzato and N. Imanishi, *Kagaku Kogaku Ronbunshu*, **15**, 519 (1990).
[3] T. Suzuki, N. Tsuge and K. Nagahama, *"Solvent Extraction 1990*", ed. T. Sekine, Elsevier, Amsterdam, p. 1701 (1992).
[4] H. Horizoe, T. Tanimoto, I. Yamamoto and Y. Kano, *Fluid Phase Equilibria*, **84**, 297 (1993).
[5] K. P. Johnston, S. E. Bany, N. K. Read and T. R. Holcomp, *Ind. Eng. Chem. Res.*, **26**, 2372 (1987).
[6] T. Sako and M. Sato, *J. Chem. Eng. Japan*, **23**, 770 (1990).
[7] G.-T. Liu and K. Nagahama, *Ind. Eng. Chem. Res.*, **35**, 4626 (1996).
[8] G.-T. Liu and K. Nagahama, *J. Chem. Eng. Japan*, **30**, 293 (1997).
[9] Y. Iwai, H. Uchida, Y. Mori, H. Higashi, T. Matsuki, T. Furuya, Y. Arai, K. Yamamoto and M. Mito, *Ind. Eng. Chem. Res.*, **33**, 2157 (1994).
[10] M. Mondell, R. P. de Filippi and V. J. Krukonis, Paper presented at the ACS Annual Meeting, Miami, FL (1978).
[11] C. H. Tan and D. C. Lion, *Separation Sci. Technol.*, **24**, 111 (1989).
[12] K. Sakanishi, H. Obata, I. Mochida and T. Sakaki, *J. Soc. Japan Energy*, **74(2)**, 109 (1995).
[13] M. McHugh and V. Kurukonis, *"Supercritical Fluid Extraction*", 2nd ed., Butherworth-Heinemann, Boston, p. 201 (1994).
[14] E. F. Barry, P. Ferioli and J. A. Hubball, *J. High Resolut. Chromatogr. Chromatogr. Commun.*, **6**, 172 (1983).
[15] K. Ute, N. Miyatake, T. Asada and K. Hatada, *Polym. Bull.*, **28**, 561 (1992).
[16] K. Ute, *"JASCO Report Special Issue*", ed. JASCO, Aug. 25, p. 54 (1997).
[17] A. Akgerman R. K. Roop, R. K. Hess and S.-D. Yeo, *"Supercritical Fluid Technology*", ed. T. J. Bruno and J. F. Ely, CRC Press, Florida, p. 479 (1991).

4.3 Extraction and Separation of Natural Products

[1] R. J. Maxwell, "*Supercritical Fluid Technology in Oil and Lipid Chemistry*", ed. J. W. King and G. R. List, AOCS Press, Champaign, p. 20 (1996).

[2] Y. Iwai, T. Fukuda, Y. Koga and Y. Arai, *J. Chem. Eng. Data*, **36**, 430 (1991).

[3] Y. Iwai, Y. Koga, H. Maruyama and Y. Arai, *J. Chem. Eng. Data*, **38**, 506 (1993).

[4] Y. Koga, Y. Iwai, Y. Hata, M. Yamamoto and Y. Arai, *Fluid Phase Equilibria*, **125**, 115 (1996).

[5] E. Reverchon, *J. Supercritical Fluids*, **10**, 1 (1997).

[6] E. Stahl, K. W. Quirin, D. Gerard, "*Dense Gases for Extraction and Refining*", Springer-Verlag, Berlin (1988).

[7] H. A. Matos, E. G. D. Azevedo, P. C. Simoes, M. T. Carronde and M. N. D. Ponte, *Fluid Phase Equilibria*, **52**, 357 (1989).

[8] Ph. Marteau, J. Obriot, R. Tufeu, *J. Supercritical Fluids*, **8**, 20 (1995).

[9] J. Suzuki and K. Nagahama, *Kagakukougaku Ronbunshu*, **22**, 155 (1996).

[10] F. Temelli, J. P. O'Connell, C. S. Chen and R. J. Braddock, *Ind. Eng. Chem. Res.*, **29**, 618 (1990).

[11] Y. Iwai, N. Hosotani, T. Morotomi, Y. Koga, and Y. Arai, *J. Chem. Eng. Data* **39**, 900 (1994).

[12] Y. Iwai, T. Morotomi, K. Sakamoto, Y. Koga and Y. Arai, *J. Chem. Eng. Data* **41**, 951 (1996).

[13] H. Kalra, S. Y.-K. Chung and C. Chen, *Fluid Phase Equilibria*, **36**, 263 (1987).

[14] G. Brunner, *J. Supercritical Fluids*, **13**, 283 (1998).

[15] E. Stahl and D. Gerard, *Perfumer and Flavorist,* **10**, 29 (1988).

[16] F. Recasens, B. J. McCoy and J. M. Smith, *AIChE J.*, **35**, 95 (1989).

[17] H. Peker, M. P. Srinivasan, J. M. Smith and B. J. McCoy, *AIChE J.*, **38**, 761 (1992).

[18] M. Goto, M. Sato and T. Hirose, *J. Chem. Eng. Japan*, **26**, 401 (1993).

[19] E. Reverchon and C. Marrone, *Chem. Eng. Sci.*, **52**, 3421 (1997).

[20] M. Goto, B. C. Roy, A. Kodama and T. Hirose, *J. Chem. Eng. Japan*, **31**, 171 (1998).

[21] M. Goto, O. Hortacsu and B. J. McCoy, *Ind. Eng. Chem. Res.*, **29**, 1091 (1990).

[22] M. Goto, J. M. Smith and B. J. McCoy, *Ind. Eng. Chem. Res.*, **29**, 282 (1990).

[23] H. Sovova, *Chem. Eng. Sci.*, **49**, 409 (1994).

[24] H. Sovova, J. Kucera and J. Jez, *Chem. Eng. Sci.*, **49**, 415 (1994).

[25] B. C. Roy, M. Goto, T. Hirose, O. Navaro and O. Hortacsu, *J. Chem. Eng. Japan*, **27**, 768 (1994).

[26] M. B. King and O. Catchpole, "*Extraction of Natural Products Using Near-Critical Solvents*", ed. M. B. King and T. R. Bott, Blackie Academic & Professional, Glasgow, p.184 (1993).

[27] M. Goto, B. C. Roy and T. Hirose, *J. Supercritical Fluids*, **9**, 128 (1996).

[28] B. C. Roy, M. Goto and T. Hirose, *Ind. Eng. Chem. Res.*, **35**, 607 (1996).
[29] G. Brunner, "*Gas Extraction*", Steinkopff, Darmstadt, Springer, New York (1997).
[30] G. Brunner, *J. Supercritical Fluids*, **13**, 283 (1998).
[31] W. B. Nilsson, E. J. Gauglitz Jr., J. K. Hudson, V. F. Stout and J. Spinelli, *J. Am. Oil Chem. Soc.*, **65**, 109 (1988).
[32] W. B. Nilsson, "*Supercritical Fluid Technology in Oil and Lipid Chemistry*", ed. J. W. King and G. R. List, AOCS Press, Champaign, p. 180 (1996).
[33] Y. Suzuki, M. Konno, K. Arai and S. Saito, *Kagakukougaku Ronbunshu*, **15**, 439 (1989).
[34] C. Perre, G. Delestre, L. Schrive and M. Carles, *Proceedings of 3rd International Symposium on Supercritical Fluids*, **2**, 465 (1994).
[35] M. Sato, M. Goto and T. Hirose, *Ind. Eng. Chem. Res.*, **34**, 3941 (1995).
[36] M. Sato, M. Goto and T. Hirose, *Ind. Eng. Chem. Res.*, **35**, 1906 (1996).
[37] M. Sato, M. Goto, A. Kodama and T. Hirose, "*Supercritical Fluids*", ed. M. A. Abraham and A. K. Sunol, ACS Symposium Series, **670**, p. 119 (1997).
[38] M. Goto, M. Sato, A. Kodama and T. Hirose, *Physica B*, **239**, 167 (1997).
[39] M. Sato, M. Goto, M. Kondo, A. Kodama and T. Hirose, *J. Supercritical Fluids* **13**, 311 (1998).
[40] W. Bracey, U. Akman and A. K. Sunol, *J. Supercritical Fluids* **4**, 60 (1991).
[41] Y. Iwai, H. Uchida, Y. Mori, H. Higashi, T. Matsuki, T. Furuya and Y. Arai, *Ind. Eng. Chem. Res.* **33**, 2157 (1994).
[42] S. Lim and S. S. H. Rizvi, *J. Food Science* **61**, 817 (1996).
[43] Y. Yamauchi and M. Saito, *J. Chromatography*, **505**, 237 (1990).
[44] D. Barth, D. Chouchi, G. D. Porta, E. Reverchon and M. J. Perrut, *J. Supercritical Fluids* **7**, 177 (1994).
[45] D. Chouchi, D. Barth, E. Reverchon and G. D. Porta, *J. Agriculture Food Chem.* **44**, 1110 (1996).
[46] D. Chouchi, D. Barth, E. Reverchon and G. D. Porta, *Ind. Eng. Chem. Res.* **34**, 4508 (1995).
[47] F. D. Kelly and E. H. Chimowitz, *AIChE J.*, **36**, 1163 (1990).
[48] A. Akgerman and G. Madras, "*Supercritical Fluids*", ed. E. Kiran and J. M. H. Levelt Sengers, Kluwer Academic Publishers, Dordrecht, p. 669 (1994).
[49] M. Sato, M. Goto, A. Kodama and T. Hirose, *Sep. Sci. Tech.*, **33**, 1283 (1998).
[50] M. Sato, M. Goto, A. Kodama, N. Tanoue and T. Hirose, "*High Pressure Chemical Engineering*", ed. Ph. Rudolf von Rohr, Ch. Trepp, Elsevier Science, Amsterdam, p. 303 (1996).
[51] D. M. Ruthven, S. Farooq and K. S. Knaebel, "*Pressure Swing Adsorption*", VCH, New York (1994).
[52] R. T. Yang, "*Gas Separation by Adsorption Processes*", Butterworth, Boston (1987).

[53] M. Sato, M. Goto, A. Kodama and T. Hirose, *Chem. Eng. Sci.*, **53**, 4095 (1998).

4.4 Separation and Analysis by Supercritical Fluid Chromatography

[1] E. Klesper, A. H. Corwin and D. A. Turner, *J. Org Chem.*, **27**, 700 (1962).
[2] M. Novotny, S. R. Springston, P. A. Peaden, J. C. Fjeldsted and M. L. Lee, *Anal. Chem.*, **53**, 407A (1981).
[3] D. R. Gere, R. Board and D. McManigill, *Anal. Chem.*, **54**, 736 (1982).
[4] M. L. Lee and K. E. Markides, *Science*, **235**, 1342 (1987).
[5] A. Staby, C. Borch-Jensen, S. Balchen and J. Mollerup, *Chromatographia*, **39**, 697 (1994).
[6] K. L. Williams, L. C. Sander and S. A. Wise, *J Chromatogr. A*, **746**, 91 (1996).
[7] L. G. Landall, *"Ultrahigh Resolution Chromatography"*, ed. S. Ahuja, American Chemical Society, ACS Symposium Series, **250**, Washington DC, p. 135 (1984).
[8] X. Lou, H. G. Janssen, H. Snijders and C. A. Cramers, *J. High Resolut. Chromatogr.*, **19**, 449 (1996).
[9] D. W. Later, E. R. Campbell and B. E. Richter, *J. High Resolut. Chromatogr. Chromatogr. Commun.*, **11**, 65 (1988).
[10] Y. Hirata. F. Nakata and S. Murata, *Chromatographia*, **23**, 663 (1987).
[11] Y. Hirata and S. Katoh, *J. Microcol. Sep.*, **4**, 503 (1992).
[12] M. P. Squicciarini, *J. Chromatogr. Sci.*, **34**, 7 (1996).
[13] J. D. Pinkston and R. T. Hentschel, *J. High Resolut. Chromatogr.*, **16**, 269 (1993).
[14] Y. Hirata, Y. Kawaguchi and K. Kitano, *Chromatographia*, **40**, 42 (1995).
[15] Y. Hirata. H. Koshiba and T. Maeda, *J. High Resolut. Chromatogr.*, **13**, 619 (1990).
[16] T. A Berger, *J. Chromatogr.*, **478**, 311 (1989).
[17] T. L. Chester and D. P. Innis, *J. Microcol. Sep.*, **5**, 127 (1993).
[18] D. J. Miller and S. B. Hawthorne, *Anal. Chem.*, **67**, 273 (1995).
[19] J. W. Ziegler, J. G. Dorsey, T. L. Chester and D. P. Inniss, *Anal. Chem.*, **67**, 456 (1995).

5 Material Processing Using Supercritical Fluids

5.1 Fractionation, Debindering and Drying

5.1.1 Fractionation of Polymers

Generally, synthetic polymers have a wide distribution in the molecular weights of components. Since the physical and mechanical properties of polymers are dependent on their molecular weights, the properties of synthetic polymers are distributed over a wide range. In order to use polymers as specialty materials such as hydraulic fluids, lubricants and surfactants, fractionation of polymers is necessary to obtain a polymer of a narrow molecular-weight distribution. However, for many polymers, conventional fractionation techniques such as molecular distillation, liquid solvent extraction and chromatographic fractionation are not practical.

One of the advantages of supercritical fluid solvents is the ability to change solubility power as a function of pressure and/or temperature. Therefore, supercritical fluid fractionation of polymers or synthetic oils is possible by using a multiple, sequential-pressure reduction system.

Figure 5.1.1 shows a schematic diagram of a continuous fractionation system employing a supercritical fluid to separate homogeneous-series members of a synthetic polymer, proposed by Krukonis [1]. While there are four extraction/separation vessels in this figure, the number of vessels is arbitrary. During the fractionation operation, polymer and supercritical fluid enter the primary extraction vessel. The supercritical fluid phase containing the dissolved lighter fractions of the polymer leaves the primary vessel from the top, while the heavy fraction leaves it from the bottom. Before the solution of supercritical fluid and the extracted lighter fractions enter the first separator, it is expanded to some lower pressure through the pressure reduction valve. The expansion causes the precipitation of a fraction having relatively high molecular weights, which is obtained as the first lighter fraction in the figure. The remaining stream is further reduced in pressure to fractionate the polymer. The molecular weight range of each fraction can be changed by adjusting the pressure in each vessel. After removing the lightest fraction, the almost solute-free gas is recompressed and recycled.

Krukonis [1] applied the above system to the fractionation of many polymers. An example of the results for poly(dimethyl siloxane) is listed in Table 5.1.1. The fractionation operation was carried out using supercritical

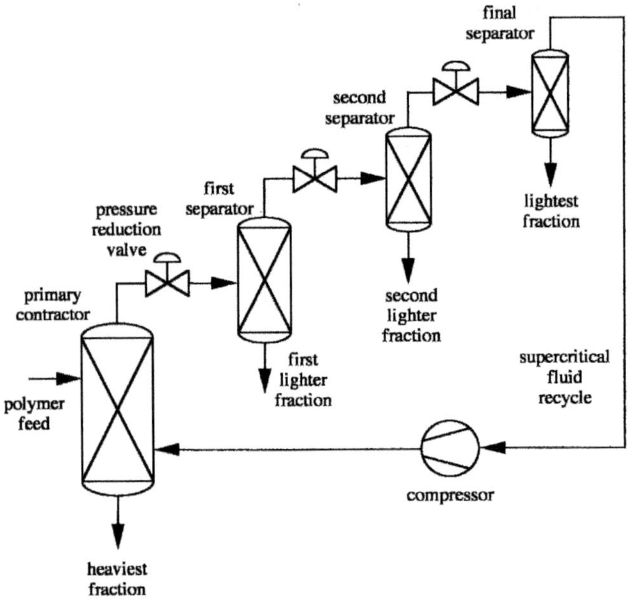

Fig. 5.1.1 Schematic diagram of multiple fractionation stage [1]

Table 5.1.1 Average molecular weights of parent poly(dimethyl siloxane) and its fractions separated by supercritical carbon dioxide [1]

fraction	number average molecular weight, $\overline{M}_n$	weight average molecular weight, $\overline{M}_w$	polydispersity index, $\overline{M}_w/\overline{M}_n$	wt.% of parent
parent	42,500	90,000	2.12	100.0
1	428	789	1.84	4.0
2	3,310	11,500	3.47	5.1
3	27,100	53,200	1.96	27.7
4	43,000	57,500	1.34	27.9
5	58,900	91,500	1.55	28.3
6	112,600	149,900	1.33	7.0

carbon dioxide at 80 °C and a pressure range from 1 800 to 6 500 psia. Six fractions were obtained from the parent polymer with a number-average molecular weight of 42 500. There are some more applications in literature [2–5].

5.1.2 Phase Separation of Polymer Solutions

In a solution polymerization process a polymer product should be separated from the solution. Conventional solvent-stripping techniques, such as steam stripping, however, are very energy-intensive. Another way to recover polymers from solutions is to make use of liquid–liquid separation of the polymer solution.

Figure 5.1.2 shows a schematic presentation of the phase behavior for a polymer solution [6]. As shown in Fig. 5.1.2 (a), a polymer–solvent mixture generally has LCST (lower critical solution temperature) and UCST (upper critical solution temperature) curves, although the UCST is not observed for many polymer solutions, since the solutions freeze before reaching it. The LCST behavior occurs because the free volume of solvents increases at a much faster rate than that of polymers. The chemical nature of the components, the molecular weight of the polymer and the critical properties of the solvent are closely related to the LCST [7]. The polymer solution located in a single liquid phase region can be split into a polymer-rich phase and a solvent-rich phase by increasing the temperature beyond the LCST isobarically. However, since the LCST usually exists near the critical temperature of the solvent for a mixture of a polymer and a good solvent, thermal degradation of the polymer can occur at such high temperatures.

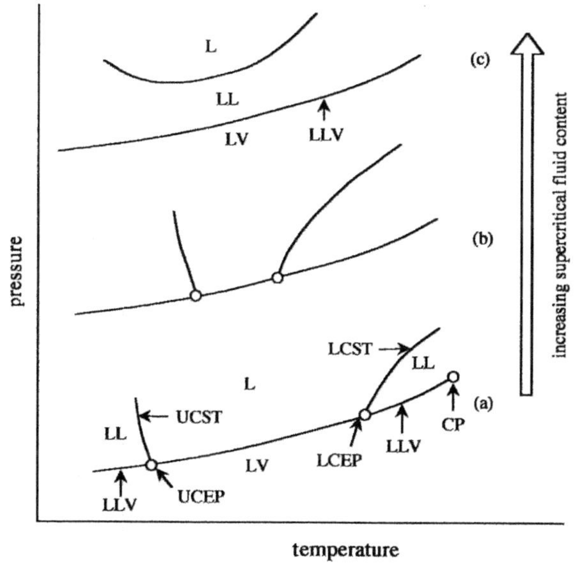

Fig. 5.1.2 Phase behavior of polymer solvent mixture [6]

Table 5.1.2 Effect of ethane addition on LCEP on the (polystyrene + toluene) system [10]

C_2H_6 conc. (wt. %)	LCEP temp. (°C)
0.0	284 [Ref. 11]
17.8	122
22.0	53
24.9	UCST and LCST merge

Irani et al. [8] proposed a process that lowers the LCST of a polymer solution by adding a supercritical anti-solvent to the solution. The added supercritical solvent lowers the critical temperature of the mixed solvent, and reduces the attractive interaction between the polymer and the solvent. Unlike a liquid additive, the density of a supercritical additive can be varied widely, which has a very large effect on the free volume of the solvent. Pressure is a key parameter in this method, because the solubility of the supercritical additive and the density of the mixed solvent are controlled with it. As a result, the LCST curve shifts to lower temperatures, as shown in Fig. 5.1.2 (b), and at a high enough pressure the LCST and UCST curves may merge into a single curve (Fig. 5.1.2 (c)).

This SCF-additive process is closely related to the phase equilibria for supercritical fluid + liquid solvent + polymer systems. McHugh and coworkers have measured them for several ternary systems [9,10]. Table 5.1.2 shows the shift of the LCEP (lower critical end point) temperature with the addition of ethane to the polystyrene + toluene mixture [10]. Without any ethane, the LCEP for the polystyrene + toluene system occurs at 284 °C [11]. With the addition of 22.0 wt% ethane the LCEP is lowered to almost ambient temperature. Finally, the LCST and UCST curves merge if the ethane concentration increases to 24.9 wt%. McHugh and Guckes reported the phase behavior of mixtures consisting of random poly(ethylene-co-propylene), a mixed organic solvent (mainly hexane), and a supercritical fluid additive [9]. Methane, ethylene, propylene or carbon dioxide was used as the supercritical fluid additive. The amount and type of supercritical fluid additive had a dramatic effect on the shape and location of the phase boundary curves of the polymer solution. They found that the LCEP temperatures of the polymer solutions consisting of random poly(ethylene-co-propylene) and various pure or mixed solvents can be represented by a single line as a function of the critical temperature of the pure or mixed solvents, where the critical temperatures of mixed solvents were estimated by using Kay's rule. The lower the critical temperature of the solvents, the lower the LCEP temperature.

Bungert et al. [12] examined supercritical anti-solvent fractionation. They measured the phase equilibria for the monodisperse and bidisperse (polystyrene + cyclohexane + carbon dioxide) system. Adding a small amount of carbon dioxide shifted the LCST to dramatically lower temperatures. Pressure had a pronounced effect on the partitioning of polystyrenes of different molecular weights in the two liquid phases. Martin et al. [13] measured the phase behavior for the (polypropylene + pentane + carbon dioxide) system.

5.1.3 Devolatilization of Polymers

The removal of volatile substances, such as unreacted monomers, low-molecular-weight oligomers and solvents, from polymers, which is called devolatilization, is one of the most important operations in polymer manufacture from the standpoints of product quality and influence on the environment and human bodies. The volatile impurities are usually removed by vacuum or inert gas stripping of polymer melts or particles. However, since the diffusion coefficients of the impurities in polymers are very low at low temperatures, the operating temperature is usually raised to accelerate the diffusion of the impurities in polymers. Hence, the devolatilization process consumes a great quantity of energy. Furthermore, degradation of the polymer can occur at high temperatures. The supercritical fluid extraction (SFE) technique using carbon dioxide as a solvent, which is typically carried out near room temperature, is therefore attractive for the removal of impurities in such thermally labile polymers.

The key to the supercritical fluid extraction of impurities from polymers is phase equilibria and diffusion coefficients in the system composed of a supercritical fluid, low-molecular-weight volatile components and a polymer. The high solubility of volatile components in dense supercritical fluids may make it possible to operate near room temperature. Although the diffusion coefficients of impurities in polymers are very low at low temperatures, the plasticizing effect of supercritical fluids may increase the diffusion coefficients; namely, supercritical fluids dissolved in polymers plasticize them.

Copelin [14] applied supercritical fluid extraction to the removal of cyclic ethers from homopolymers and copolymers of tetrahydrofuran (THF) and alkylene oxide (AO). In the production of THF-AO homopolymers and copolymers to prepare polyurethane, oligomeric cyclic ethers (OCE) are also produced and can comprise 7–15 % of the polymer. These cyclic ethers are undesirable because they tend to degrade the polyurethane's properties. The experimental conditions and results for the extraction of OCE from the polymers are listed in Table 5.1.3. OCE could be extracted by bubbling super-

Table 5.1.3 Experimental conditions and results for the extraction of cyclic ether oligomers (CEO) [14]

polymer	SCF	T (K)	P (MPa)	initial conc. of CEO (wt.%)	final conc. of CEO (wt.%)
THF/PO copolymer	ethylene	318	6.9	8	2
THF homopolymer	propylene	278	10.3	1	0.2
EO homopolymer	ethylene	293	6.9	8	4.6

critical ethylene or propylene through the polymers at rather mild conditions. The molecular weight of OCE was less than about 500.

Barry et al. [15] described the upgrading of polyphenylmethylsiloxane (OV-17) used as a stationary phase in gas chromatography. This polymer contains low-molecular-weight oligomers that reduce the performance in a temperature-programmed analysis, because they bleed from the chromatographic column at high temperatures. They reported that 30% of the original polymer was extracted with supercritical carbon dioxide at 35 °C and 140 atm. The residual polymer of heavier molecular weight had improved characteristics as a chromatographic stationary phase compared to that of the original polymer.

Yamamoto and Nakae [16] reported the extraction of methylene chloride (T_b = 40 °C) from granular polycarbonate (PC) having diameters from 0.59 to 1.17 mm. Supercritical carbon dioxide was continuously passed through a fixed bed of 400 mg of PC that involves 1.82 wt% of methylene chloride. The concentration of methylene chloride in the polymer was determined with a gas chromatograph before and after extraction. As a result of experiments at temperatures of 40, 50 and 70 °C and pressures of 10, 15 and 20 MPa, it was confirmed that almost perfect removal of methylene chloride was possible for 3 h even below the glass transition temperature (T_g = 155 °C) of PC. In addition, Fick's model was employed to yield the diffusion coefficient of the removal process. The obtained diffusion coefficients increased with increasing temperature and pressure.

Sasaki et al. [17–21] studied the extraction of low-molecular-weight impurities from polymers for model systems of benzene + poly(vinyl acetate) (PVAc), toluene + PVAc and benzene + polystyrene (PS) with supercritical carbon dioxide. A polymer film of known impurity concentration was placed in a high-pressure vessel and supercritical carbon dioxide flowed continuously through the vessel for a certain time at constant temperature, constant pressure and constant flow rate. The concentration of extracted impurity in the outlet flow was measured quantitatively and continuously by using a UV spectrophotometer. The final concentration of remaining impurity in the polymer after

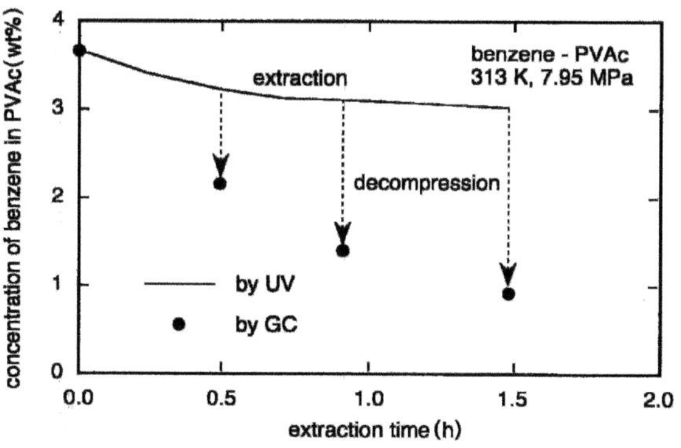

Fig. 5.1.3 Results of the supercritical fluid extraction of benzene from PVAc [19]

decompression was also determined with a gas chromatograph. An example of the experimental results is shown in Fig. 5.1.3, where PVAc films with a thickness of 2.3 mm and an initial benzene concentration of 3.65 wt% were used [19]. In the extraction process at 7.95 MPa and 313 K, the concentration of benzene remaining in PVAc, which was determined based on UV analysis and is denoted by a solid line, decreased with extraction time. However, after the decompression the concentration of benzene in PVAc, which was measured with a gas chromatograph and is shown with circles, was much reduced compared to the concentration just before decompression. This is because carbon dioxide is accompanied by benzene when carbon dioxide is quickly released from the polymer. Thus they explained two different mechanisms of impurity removal, namely "removal by extraction" and "removal by decompression". The amount of removal by decompression increased with extraction time. This may be due to the increase in the amount of dissolved carbon dioxide in the polymer.

For the amount of removed benzene, they applied Fick's diffusion equation in two different ways. In a simple way they considered the diffusion of benzene in a polymer solution saturated with carbon dioxide [19]. They assumed that the concentration of dissolved carbon dioxide, thickness of PVAc film and the diffusion coefficient of benzene in the solution of PVAc and carbon dioxide were constant during the extraction process, and that the concentration of benzene on the surface of the film was zero. As shown in Fig. 5.1.3, this simple model can correlate the experimental result of removal by extraction. A model with a different value of diffusion coefficient can also correlate total removal (extraction and decompression), although this apparent diffusion coefficient had little physical meaning. The diffusion coefficients determined by fitting to the experimental removal rate are listed in Table 5.1.4 as well as

Table 5.1.4 Diffusion coefficients of benzene under different conditions

system	T (K)	P (MPa)	D (m^2/s)	Ref.	remarks
benzene in CO_2	313	7.95	5×10^{-8}	[22]	infinite dilution
benzene in PVAc + CO_2	313	7.95	7.34×10^{-12}	[19]	removal by extraction
benzene in PVAc + CO_2	313	7.95	1.06×10^{-10}	[19]	removal by extraction and decompression
benzene in PVAc	313	0.1	4.8×10^{-17}	[23]	infinite dilution
benzene in PVAc	313	0.1	1.7×10^{-14}	[23]	10 wt.% benzene

literature values of diffusion coefficients of benzene in supercritical carbon dioxide [22] and that in pure PVAc [23]. It can be seen that the determined diffusion coefficient of benzene in PVAc in the presence of supercritical carbon dioxide lies between those in supercritical carbon dioxide and in pure PVAc. PVAc can be plasticized by the dissolution of carbon dioxide, since the solubility of carbon dioxide in PVAc is about 30 wt% at 313 K and 8.0 MPa [24]. This simple model could represent the experimental extraction rate for each polymer sample of the same film thickness fairly well. However, the determined diffusion coefficients were dependent on the thickness of the polymer films.

Sasaki et al. proposed another model where the mass transfer of benzene and carbon dioxide in PVAc was strictly treated [21]. In this model it was assumed that benzene and carbon dioxide in PVAc diffused in the opposite direction during extraction. The diffusion coefficient of benzene in the polymer, D_B, was assumed to be a function of the local concentration of carbon dioxide dissolved, as follows;

$$D_B = D_{0B} \exp(\alpha_B \cdot u_C), \tag{5.1.1}$$

where u_C is the concentration of carbon dioxide in PVAc [kg-CO_2/kg-PVAc] and D_{0B} and α_B are adjustable parameters. The diffusion coefficient of carbon dioxide is also dependent on its local concentration and was expressed according to the experimental results for the carbon dioxide + PVAc system. The model also considered the volume change of the polymer due to the dissolution of carbon dioxide based on the dilation experiments for the system. A moving coordinate system proposed by Schoeber [25] was used to represent the expansion of a polymer film as a function of local concentration of carbon dioxide. Furthermore, the partition coefficient of benzene between the supercritical fluid phase and polymer phase was introduced using experimental data measured with gas–liquid partition chromatography and a piezoelectric sorption method by Wang et al. [26]. Thus the model solved the diffusion equations for benzene and carbon dioxide in the polymer, phase equilibria on the surface of the polymer film, swelling of the polymer film, and material balance for benzene simultaneously. The two adjustable parameters, D_{0B} and α_B, in (5.1.1) were determined by fitting the calculated extraction rate to that of all experiments carried out for the system at different polymer thicknesses, different initial concentrations of benzene, and different extraction times. The experimental results could be represented successfully by this strict model. The calculated concentration profiles of benzene and carbon dioxide explained the acceleration of the diffusion rate of benzene near the polymer surface. As mentioned by this model, the plasticizing effect of dissolved supercritical fluid plays a very important role in the extraction of volatile impurities from polymers.

5.1.4 Debindering of Molded Fine Ceramics

The injection molding technique of fine ceramics has come to be recognized as one of the most efficient methods for the mass production of complex shapes. In this technique amounts as large as 40 vol% of organic compounds are usually added to the ceramic powder as a binder to provide green strength and plasticity, and these compounds must be removed prior to sintering. Thermal decomposition, which is currently used as a debindering process, requires the most time in the processes composing the entire manufacturing procedure, in order to avoid defects such as expansion and cracking in green bodies, particularly for large size parts. This is why alternative techniques are needed to produce defect-free green bodies and to reduce the debindering process time.

Yokawa and Nakajima [27–29] applied the supercritical fluid extraction technique to the debindering process of molded fine ceramics. In their studies they examined a mixed binder that consisted of low-molecular-weight binder components (LMBC) and high-molecular weight binder components (HMBC). As a result, they suggested the following principle for binder combination: (1) select an LMBC as a main component; (2) add a certain amount of HMBC as a subcomponent; (3) add a lubricant or plasticizer for better fluidity; and (4) adjust the content of the HMBC to control the extraction yield of the LMBC. They found that SCF debindering reduced the debindering process time to a great extent. Furthermore, they pointed out many advantages in supercritical fluid debindering such as small thermal deformation of products, relatively uniform binder distribution in the debindered product, recycling of removed binders and no requirement of exhaust gas treatment.

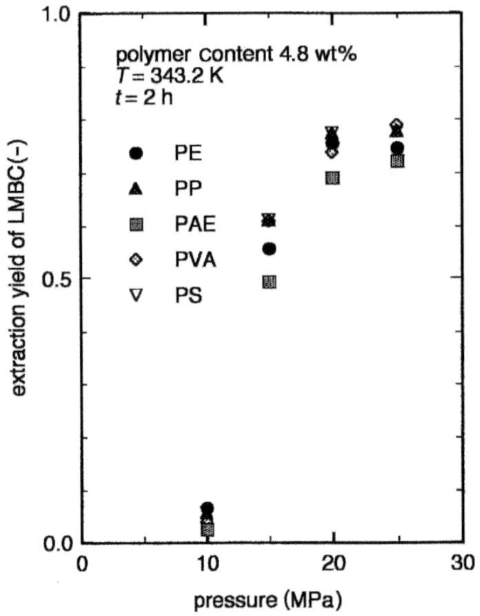

Fig. 5.1.4 Effect of pressure on the supercritical fluid debindering of molded alumina [30]

Takishima et al. [30, 31] investigated the combination of binder components for supercritical fluid debindering. Among several waxes they selected a paraffin wax that has a mean melting point of 320 K as a main binder component, because it could be easily extracted from molded alumina by supercritical carbon dioxide. Other paraffin waxes of higher melting points, polyethylene wax, microcrystalline wax and bees wax showed lower extraction yield. Five thermoplastic resins, polyethylene (PE), polypropylene (PP), acrylic ester copolymer (PAE), poly(vinyl alcohol) (PVA) and polystyrene (PS), were also examined in the extraction experiments from injection-molded rectangular parallelepiped samples with a dimension of 3.7 × 5.0 × 65.6 mm. The samples consisted of 87.0 wt% of alumina fine powders, 2.2 wt% of stearic acid, the paraffin wax and one of the thermoplastic resins. The content of the thermoplastic resins was varied from 0.9 to 4.8 wt%, and that of the paraffin wax was adjusted according to the thermoplastic resin content.

The effects of extraction pressure and temperature on the extraction yield of LMBC (the paraffin wax and stearic acid) are shown in Figs. 5.1.4 and 5.1.5, respectively, for the samples containing 4.8 wt% of the thermoplastic resin [30]. While the extraction yields are very low at pressures below 10 MPa, they dramatically increase with a pressure of around 15 MPa. More than 70% of the initial LMBC can be extracted for 2 hours above 20 MPa. Temperature has little effect on the extraction yield as shown in Fig. 5.1.5.

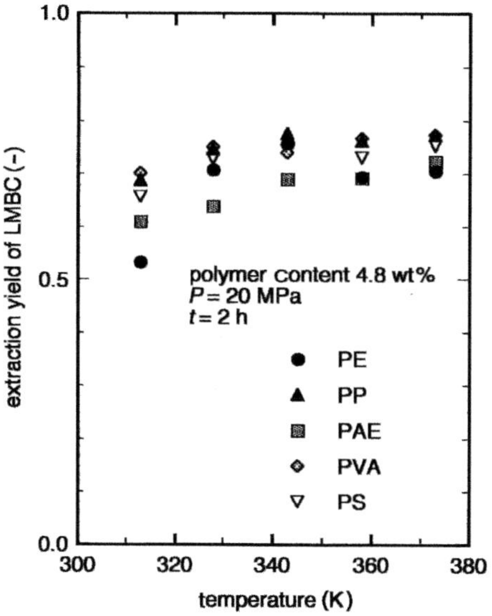

Fig. 5.1.5 Effect of temperature on the supercritical fluid debindering of molded alumina [30]

The extraction yields increased with the reduction of the thermoplastic resin content; around 90% of the LMBC were extracted for the samples containing 0.9 wt% of the thermoplastic resins for 2 hours.

Table 5.1.5 lists the condition of debindered samples. All samples containing less than 1.5 wt% of the thermoplastic resins showed bubbles and/or cracks after extraction. Among the five thermoplastic resins examined, PE and PP were better than PAE, PVA and PS on the soundness of debindered samples. The samples containing more than 2.8 wt% of PE or PP showed no defect after debindering at pressures above 20 MPa and temperatures from 313 to 373 K. The

Table 5.1.5 Conditions of debindered samples [30] ($P = 20$ MPa, $T = 343$ K, $t = 2$h)

polymer	content of polymer (wt. %)			
	0.9	1.5	2.8	4.8
PE	*	**	****	****
PP	*	*	****	****
PAE	**	**	**	***
PVA	*	*	***	***
PS	*	*	*	**

**** : no defects and high handling performance
*** : no defects but low handling performance
** : some small defects
* : large defects or destroyed

Table 5.1.6 Comparison of bending strength for sintered samples [30]

debindering method+	binder combination++	debindering time	bending strength (GPa) minimum	maximum	average	Weibull coefficient (-)
SFE	for SFE	4	0.36	0.43	0.39	23.3
TD	for SFE	24	0.30	0.41	0.37	10.2
TD	for TD	60	0.31	0.49	0.42	8.3

\+ SFE : supercritical fluid extraction at 313 K and 20 MPa
TD : thermal decomposition

++ for SFE : 6.0 wt. % of paraffin wax, 2.2 wt. % of stearic acid and 4.8 wt. % of PE
for TD : totally 13 wt. % of PE, PP and acrylic resin

reason, as explained by the authors, is that PE and PP have similar chemical structures to that of the paraffin wax and lower melting point temperatures and glass transition temperatures than the others. These thermoplastic resins may be mixed with the paraffin wax more homogeneously and are more flexible. While the debindered green bodies contain the thermoplastic resin, this is necessary in order to keep the green bodies stiff. The green bodies can be sintered without removing the thermoplastic resin.

The bending strength of sintered bodies is listed in Table 5.1.6 [30]. In this table values for the samples debindered by the conventional thermal decomposition technique are also added. The bending strength of the samples debindered with supercritical carbon dioxide was almost the same as that of those debindered by thermal decomposition. In addition, a considerably higher value of the Weibull coefficient for the samples debindered with supercritical carbon dioxide indicates that the distribution of the bending strength is narrower than those of the other samples. It should be noticed in this table that the supercritical extraction technique reduced the debindering time to less than one tenth of the conventional method.

Takishima et al. [32] further proposed a supercritical fluid debindering process of molded ceramics where activated carbon is used to recover binder components from the supercritical fluid phase. In this method ceramic green samples are placed on the upper side of the extraction vessel and granular activated carbon is on the bottom. Under the extraction condition, the density of supercritical carbon dioxide containing extracted LMBC around the green bodies is larger than that of almost pure carbon dioxide around the activated carbon. Therefore natural convection occurs in the vessel where the dense supercritical carbon dioxide containing LMBC goes down and the less dense supercritical carbon dioxide goes up. Thus, the extracted LMBC is removed from the supercritical fluid phase by adsorbing on the activated carbon without mixing the phase mechanically or circulating the phase through an additional separator. The use of activated carbon in the high-pressure vessel had the same effect on the debindering rate as flowing pure carbon dioxide at a high linear velocity of about 1 cm/s around the green bodies.

Since major mass transfer resistance exists in the diffusion process of LMBC in ceramic green bodies, the extraction rate of the LMBC can be represented with Fick's diffusion equation. Nishikawa et al. [29] evaluated the value of the effective diffusion coefficient of LMBC in the green bodies. The value was in the range from 2.3×10^{-10} to 3.5×10^{-10} m^2/s for green bodies containing different amounts of LMBC at temperatures of 333−353 K and a pressure of 19.6 MPa. Takishima et al. [32] also reported the values of 2.2×10^{-10} m^2/s for the sample listed in Table 5.1.6. Chartier et al. [33] showed the dependence of temperature and pressure on the diffusion coefficient of a paraffin in molded alumina and the values were from 1.2×10^{-10} to 3.7×10^{-10} m^2/s. In these cases, the apparent diffusion coefficients of LMBC in green bodies increase with increasing extraction temperature and pressure, which may be due to the reduction of viscosity of the green bodies.

The major problem in conventional thermal decomposition debindering consists of the very long process time. Thermal decomposition of binder components generates gaseous hydrocarbons, which cause a very large volume change in green bodies at low pressures. In order to avoid defects due to gasification, the heating rate should be very low so that the decomposed components do not create bubbles but diffuse out from the green bodies. On the other hand, in the supercritical fluid debindering technique the binder components can be extracted without a large change in volume because of high pressure conditions. The supercritical fluid debindering technique is, hence, the best way to remove binders in green bodies quickly without causing defects. This technique is being examined for various ceramic and metal green bodies [34, 36] and is expected to be commercialized in the near future.

5.1.5 Drying of Alcogels

The sol-gel process for the production of ceramic materials is widely known. In this process gels in aqueous alcohol solutions, which are called *hydrogels* or *alcogels* according to the solvent, can be obtained through the hydrolysis and condensation of metal alkoxides or salts. The gels are extremely porous and have large surface areas and pore volumes. In order to use this material with its unique structure in a dried state, it is required to remove the liquid solvent inside the gels and to replace it with a dry gas. When the solvents are removed by simple evaporation, however, the formation of a liquid−vapor interface within the gel results in surface tension causing considerable shrinkage and

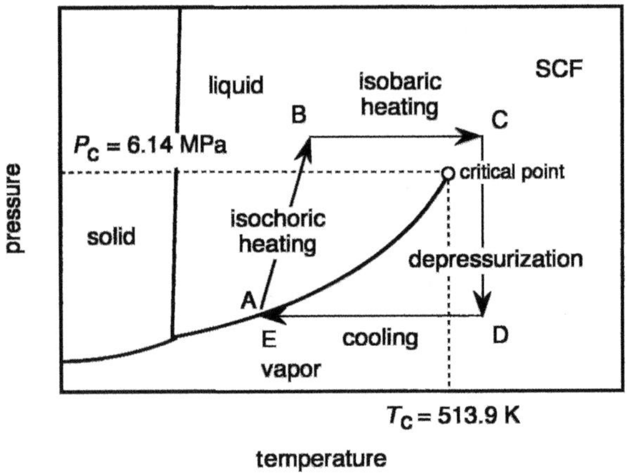

Fig. 5.1.6 Pressure–temperature diagram for ethanol with the passage of supercritical ethanol drying of alcogel

collapse. The resultant hardened glassy gels are called *xerogels*. Therefore, it is necessary to eliminate the internal tensile forces in the pores of a gel by preventing the liquid–vapor interface in the drying process.

This is possible by drying a gel in a supercritical condition of solvents. Kistler [37] first described the supercritical drying of a gel. In his method, silica gels were made by hydrolyzing water glass with hydrochloric acid. The gel was then washed with water to remove the chloride and the water inside the gel was replaced by ethanol. A lower alcohol such as methanol or ethanol is used as a solvent to be removed, because its critical temperatures and critical pressures are lower than those of water. Drying of the alcogel was carried out in an autoclave containing an excess amount of ethanol. The temperature was raised until the pressure exceeded the critical pressure of ethanol, and then the vapors in the autoclave were slowly vented at temperatures higher than the critical temperature.

This process can be explained with the P–T diagram of ethanol shown in Fig. 5.1.6. First the gel is immersed in an excess amount of liquid ethanol (point A in the figure), while there may be a saturated vapor phase above the liquid phase. As the temperature increases, the vapor phase soon disappears and the pressure of the liquid increases isochorically, if the overall density of ethanol is higher than the critical density. After the pressure exceeds the critical value (B), the liquid is heated isobarically to a temperature slightly above the critical value (C) while releasing some liquid. Under this condition the liquid in the autoclave has been transformed into a supercritical fluid without liquid–vapor phases being present at any time. The fluid in the autoclave is then slowly vented, maintaining the temperature, and is transformed into a gas as the pressure decreases (D). Now the dispersion medium of the gel is a low-pressure gas. The gaseous ethanol may be replaced with air before cooling the gel to room temperature (E).

In this whole process from A to E in Fig. 5.1.6 the dispersion medium of the gel changes continuously from a liquid to a supercritical fluid and to a gas without two-phase coexistence. There is no liquid–vapor meniscus, and therefore no surface tension or capillary forces developed within the gel. The elimination of surface tension allows the gel to be dried without shrinkage and the gel matrix to remain nearly intact. The resultant gel consists of a dry network of solid containing large amounts of homogeneously dispersed air within the structure, and is called an *aerogel*. Aerogels have higher surface areas, lower bulk densities, larger pore sizes, and greater pore volumes than simply dried xerogels. In addition to the unique structure, the microscopically dispersed air in gels offers many unusual properties, for example, extremely low thermal conductivities and diffraction index, and fascinating acoustic properties. Not only are they used as insulators, adsorbents, catalytic supports, catalysts and ceramic precursors, but special applications of aerogels as Cerenkov detectors [38] and transparent insulated windows [39] have receiving much attraction. Hundreds of papers have been published on the preparation, structural analysis, properties and applications of aerogels. There are good reviews by Teichner et al. [40], Livage and Lemerle [41], and Gesser and Goswami [42].

Recently, Poco et al. [43] developed a rapid supercritical extraction (RSCE) process that can be used to prepare near-net shape aerogels very quickly. The RSCE process is very similar to injection molding commonly used in the processing of plastics. In this process, a liquid solution containing precursor chemicals is injected directly into a two-piece, sealed mold, having connecting tubing. The mold is rapidly heated in a vacuum chamber and the precursor chemicals react to form a gel within the mold. As heating continues, the liquid trapped within the gel expands to create a hydrostatic pressure within the mold, while some liquid leaks out through the tubing. After a short time, both the temperature and pressure within the mold exceed the critical point of the liquid and it becomes a supercritical fluid. The fluid is removed from the delicate gel network through an opening in the mold by rapid decompression. After the mold is sufficiently cooled, it is opened and the finished aerogel part is removed.

The entire process, i.e., filling the mold, gelation, heating to the critical point of the entrained liquid, decompression of the supercritical fluid, cooling the mold and disassembly, is typically accomplished in less than 3 h, which is about 30 times faster than existing supercritical drying processes. This is because the gel forms during the heating process and the liquids can be rapidly purged from

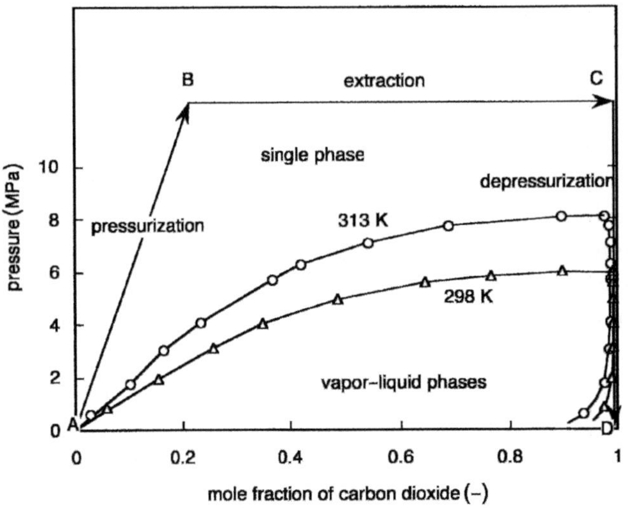

Fig. 5.1.7 Vapor–liquid equilibria for a carbon dioxide methanol system [45] with the passage of liquid or supercritical carbon dioxide drying of alcogel

the confined gel without cracking due to hydrostatic pressure. This process can produce aerogels having precise sizes and shapes maintaining high surface tolerances, because the mold totally defines them. The silica aerogels produced by the RSCE process exhibit higher strength, lower surface area, larger particles, and larger necks between particles than those produced by conventional SCE using the same gel precursors. This is probably the result of accelerated reactions due to gelation at higher temperatures.

Another useful process of the supercritical drying of alcogel is to use liquid or supercritical carbon dioxide [44]. Carbon dioxide and lower alcohols are completely miscible at temperatures around the critical temperature of carbon dioxide and at pressures slightly above its critical pressure. Under these conditions alcohols within the gel can be replaced with carbon dioxide without vapor–liquid coexistence. Figure 5.1.7 shows the passage of the drying process using supercritical carbon dioxide as well as vapor-liquid equilibria of a carbon dioxide + methanol mixture at 25 and 40 °C [45]. First an alcogel is placed in an autoclave and the autoclave is filled with methanol (point A in Fig. 5.1.7). Next liquid carbon dioxide is introduced in the autoclave to the desired operating pressure (B). The composition of carbon dioxide in the liquid phase increases almost linearly with pressure. Then methanol in the autoclave is replaced with carbon dioxide by pumping carbon dioxide into the autoclave and releasing the carbon dioxide methanol mixture from the autoclave at the operating pressure and an operating temperature slightly below or above the critical temperature of carbon dioxide. The operating pressure must be higher than the maximum pressure of the vapor–liquid coexistence curve (vapor pressure of carbon dioxide or critical pressure of carbon dioxide + methanol mixture, for operating temperatures below or above the critical temperature of carbon dioxide, respectively). After the replacement is finished (C), carbon dioxide is released from the autoclave to atmospheric pressure (D) at temperatures above its critical temperature. This process also eliminates the vapor–liquid interface within the gel.

Van Bommel and de Haan [46] discussed the economics of a large-scale drying process that uses supercritical carbon dioxide to obtain monolithic aerogel plates. The process uses five parallel extractors with an internal diameter of 1 m, height of 3 m and an inner volume of about 2 m^3. For the most economical design they chose a drying temperature of about 35 °C and a drying pressure of about 80 bar. The carbon dioxide + ethanol mixture from the extractors is separated at 20 °C by pressure reduction to 50 bar, and gaseous carbon dioxide is liquefied and recycled to the extractors. The costs of equipment, engineering, production, maintenance, labor and overheads were estimated and compared for aerogel plates with thicknesses of 1, 2 and 3 cm. The total cost for drying a 1 cm thick silica aerogel plate with a total production of 3.36×10^6 kg/year was around 1.2 US\$/m^2 and it increased exponentially with increasing plate thickness. This sharp increase in the production cost is mainly due to the sharp decrease in production, since a much longer drying time is needed for a thicker plate.

While hundreds of papers have been published on the preparation, structure and properties of aerogels as mentioned above, some studies by Japanese researchers are reviewed below.

Tajiri et al. [47] examined the effects of supercritical drying media on the structure and properties of silica aerogels. Silica alcogel samples were prepared from tetramethylorthosilicate and were supercritically dried with methanol, ethanol, 2-propernol and carbon dioxide. The density, transmittance, specific surface area and pore size of the resultant aerogels were investigated. The surface of the aerogels prepared by supercritical drying of alcohol is covered by the corresponding alkyl groups with almost the same molar concentration regardless of the kind of alcohol used. The alkyl group with larger volume makes the pore size of the network smaller, and it makes the skeletal density of the aerogel smaller because the molar density of alkyl groups is smaller than that of silica. Thus the aerogel prepared by supercritical drying of 2-propanol showed excellent transparency.

Mizushima and Hori [48] reported the preparation of a palladium-supported alumina aerogel as a catalyst for methane combustion. Aluminum sec-butoxide, ethyl acetoacetate and water were mixed to yield an alumina sol in ethanol solvent. Palladium chloride was added to the hydrolyzed alumina sol along with pyridine, making a palladium chloride and pyridine complex in the sol. The sol was maintained at 60 °C for a week to promote the gelation and aging process. The alumina wet gel containing palladium (1 wt% as metal) was dried by two different supercritical drying methods. In one method the gel was dried under the supercritical condition of ethanol in an autoclave at 270 °C and 26.5 MPa. In the other, ethanol within the gel was extracted by supercritical carbon dioxide in an extractor at 80 °C and 15.7 MPa. The dried Pd-supported alumina gels were fired at 1200 °C for 5 h. The fired aerogels were then broken into 2–3 mm grains, packed into a quartz tube, and placed in a fixed-bed reactor. A gas mixture consisting of 1 vol% of methane and 99 vol% of air was flowed into the quartz tube at various temperatures. The combustion ratio of methane was determined with a gas chromatography. As a result, the Pd-supported alumina aerogel prepared in the autoclave (supercritical ethanol condition) exhibited greater combustion activity; it showed 50–60 °C lower temperature in perfect combustion than the aerogel dried in the extractor. The fired aerogels dried in the autoclave exhibited a much larger specific surface area than that dried in the extractor. And palladium oxide existed in the palladium particles on the former aerogel, while it did not exist in the latter. They concluded that, for the aerogel dried in the extractor, the sintering of the alumina support caused sintering and vaporization of palladium particles and deterioration of the activity. They also reported a similar investigation for a palladium-supported alumina-silica aerogel [49].

Goto et al. [50] examined the structure and catalytic activity of nickel oxide (NiO)/alumina aerogels dried with supercritical carbon dioxide. The mixture of nickel acetate, water and methanol was dropped into butanol solution of 10 wt% aluminum sec-butoxide under agitation at room temperature. A gel was produced immediately as a precipitate. The gel was washed by methanol and methanol within the gel was removed by supercritical drying using carbon dioxide at 333 K and 20 MPa. The gel was then calcinated at 573 K under nitrogen flow for 3 h. The ground nickel oxide/alumina gel was packed as a catalyst in a 10 mm i.d. U-tube reactor. From the results of partial oxidation of iso-butylene, the nickel oxide/alumina aerogel catalyst was highly selective to methylacrolein over acetone at Ni/Al ratios around 0.4, whereas xerogel catalyst was not selective. With the aid of observations using TEM they concluded that smaller particles of nickel oxide are highly divided in the aerogel. This may be due to the highly porous structure of the aerogel, where coaggregation of nickel particles was restricted during drying and calcination.

In the application of silica aerogels the resistance to moisture is a significant problem, since adsorption of water on the surface of aerogels results in unfavorable changes in their structure and properties. In particular, aerogels produced by supercritical carbon dioxide drying contain a large number of hydrophilic silanol groups on the surface. Yokogawa et al. [51, 52] produced a hydrophobic silica aerogel by the modification of the aerogel surface. Hexamethyldisilazane or methyltrimethoxysilane were used for the replacement of silanol groups with trimethylsilil groups before or in the supercritical carbon dioxide drying process. The weight, size, transmittance of visual light and thermal conductivity of the modified aerogels were maintained during the moisture-resistance test for one week with a temperature of 60 °C and relative humidity of 90 %, while those of unmodified aerogels changed significantly. Takishima et al. [53] also examined the modification of aerogel surface by using methyltrimethoxysilane, dimethyldimethoxysilane or trimethylmethoxysilane. Solubilities of these compounds in supercritical carbon dioxide were measured prior to the modification experiments to determine the operating conditions. The effect of the modification was evaluated by measuring the amount of water adsorbed on the aerogels, and the aerogel modified with methyltrimethoxysilane exhibited the lowest water adsorption.

5.2 Crystallization Using a Supercritical Fluid

There are two methods to form solids with supercritical fluids: the rapid expansion of supercritical solutions (RESS) method and the gas (supercritical) antisolvent (GAS or SAS) method. RESS can be used to form solids of compounds that are quite soluble in the supercritical fluid, whereas GAS is used for sparingly soluble substances.

5.2.1 Rapid Expansion of Supercritical Solutions (RESS)

(a) Principle of RESS

High-pressure supercritical fluids (SCFs) can dissolve even nonvolatile solid compounds at room temperature and their solubilities can be significantly controlled through the adjustment of temperature as well as pressure.

When a solid solute compound is dissolved in high-density SCF, it forms an SCF solution consisting of SCF and a small amount of solute component. Next, the SCF solution is passed through a nozzle or a capillary tube so that it is very rapidly expanded to some low pressure, e.g. atmospheric pressure. The rapid expansion brings about a very large reduction of the solubility and gives the solute an extremely high supersaturated condition to result in abrupt sepa-

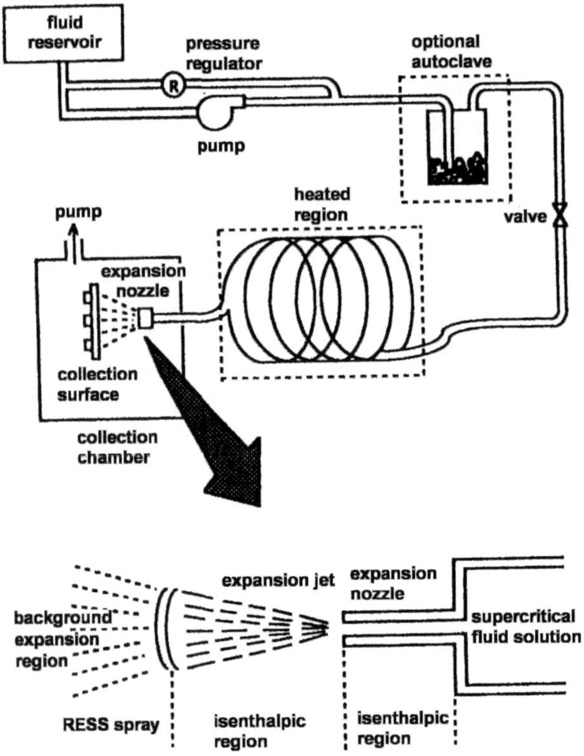

Fig. 5.2.1 Principle of the RESS method [2]

rating out as solids The very high degree of supersaturation causes rapid nucleation as well as growth of the solute. This is the principle of RESS. In general, this high degree of supersaturation can be reached within a very short period of 10^{-5}–10^{-6} s and the physical fluctuation caused by rapid expansion, such as temperature and pressure, propagates at the velocity of sound. These two factors have a great influence on the properties of the formed solid.

The principle was first proposed by Krukonis [1] as a technology for grinding solid materials into fine particles using SCF. Later, Matson et al. developed and named the RESS technique in the late 1980s [2,3]. Figure 5.2.1 illustrates the schematic apparatus consisting of an SCF reservoir, an autoclave for forming an SCF solution and an expansion nozzle for forming a solid phase.

Table 5.2.1 Growth rate of naphthalene crystals from supercritical solutions [4]

crystallization method	growth rate (m/s)
S C G	$3\times10^{-9}-9\times10^{-8}$
S E S S	$8\times10^{-9}-4\times10^{-7}$
R E S S	$2\times10^{-2}-2.3\times10$

(b) Formation of Solid by RESS

Solid formation results from two mechanisms: nucleation and crystal growth. The rates of the two mechanisms depend mainly on the concentration of solutes or degree of supersaturation. From general crystal growth theory, when the degree of supersaturation increases the particle size reduces because nucleation is dominant over crystal growth; when the degree of supersaturation lowers, the particle size increases because crystal growth is accelerated.

The crystal morphology obtained by RESS is dependent particularly upon the expansion conditions such as the type of expansion nozzle, temperature and pressure. The shape of the solid may be spheres, fibers and/or thin films.

Table 5.2.1 gives a comparison of the growth rate of naphthalene particles produced by various crystallization methods using supercritical fluids [4]. The SCG (single crystal growth) method is used to grow a single crystal placed in the supercritical solution dissolving naphthalene. The SESS (slow expansion of supercritical solutions) method is an RESS method in which the expansion speed is considerably lowered. Table 5.2.1 indicates that the growth rates with the SCG and SESS methods are $10^{-7}-10^{-10}$ m/s, comparable to conventional crystal growth rates in aqueous solutions. In contrast, the growth rate with RESS is 0.02–23 m/s, which is greater by a factor of 10^5 or more than with other crystallization methods.

(c) Hydrodynamic Behavior of a Fluid during the RESS Operation

The RESS process is characterized by the occurrence of a very high degree of supersaturation in less than 10^{-5} s, causing nuclei to be generated and fine crystals to grow in an extremely short period of time.

In RESS processes, spray crystallization is generally employed as shown in Fig. 5.2.1, that is, a supercritical solution is sprayed through a nozzle or a capillary in the form of tiny droplets into a temperature controlled chamber.

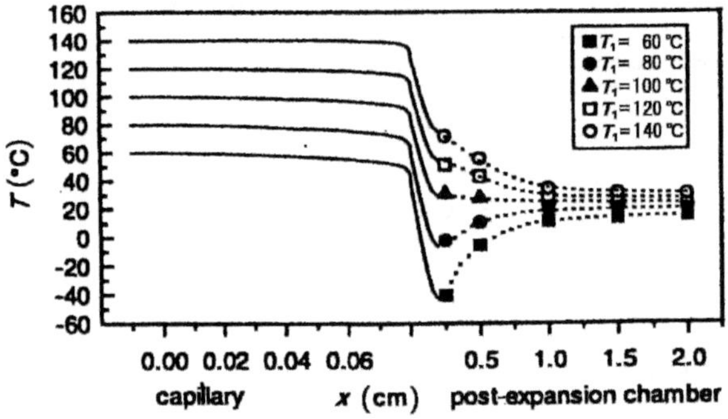

Fig. 5.2.2 Measured and calculated temperature profiles along the expansion of CO_2 (T_1 = temperature before expansion, nozzle inlet pressure = 200 bar) [6]. Reprinted from: E. Reverchon, P. Pallado: Hydrodynamic modeling of the RESS process, *J. Supercritical Fluids* (Elsevier Science) **9**, 216 (1996)

The product crystallizes in solid particles as a result of the abrupt reduction of solute solubility and hence of the high degree of supersaturation

According to the analysis by Tom et al. [5], the RESS (spray) method is affected by the following three factors: temperature drop, expansion of supercritical solvent and supersaturation of the solute, which are caused by rapid depressurization. They also pointed out that RESS depends on the structure (orifice or capillary) and size of expansion nozzles. The fluid behavior in the nozzle and the free-jet region after expansion can be modeled and analyzed as a contracted flow involving frictional pressure loss. It is carried out by solving simultaneous equations for continuity of material, momentum and energy. The calculated results by Tom et al. [5] show that pressure and temperature drops already occur at the inlet of the capillary, that is, nucleation and crystal growth is gradually in progress in a capillary nozzle. On the contrary, when using an orifice nozzle, nucleation and its growth occur rather in the free-jet region formed in the downstream than in the orifice because of the difference in size and/or structure of nozzle used. The phase change of supercritical fluid during expansion is of importance because it may be the main factor in changing the morphology of the formed solids.

Reverchon et al. [6] measured the flow rate at the nozzle outlet and temperature profiles in the free jet region using CO_2 and trifluoromethane. Figure 5.2.2 compares the experimental and calculated temperature profiles for pure CO_2 by changing the fluid temperature before entering the orifice. The dimension of the nozzle is 0.8 mm long with a diameter of 40 μm. The x-axis represents two distances: one is that from the nozzle inlet in the orifice and the other is that from the nozzle outlet in the post-expansion chamber. The calculated lines in Fig. 5.2.2 are consistent with experimental observations.

(d) Applications of RESS

To date, RESS concerns mostly the production of solid particles which have been dissolved in a supercritical fluid. RESS application to organic substances was an early focus in pharmaceuticals. The RESS process can yield to powders made of small size particles with controlled distribution and is also suitable for therapeutic purposes with no residual trace of organic solvent. The low solubilities of pharmaceutical compounds narrow the application of RESS in its field.

Applications of RESS to polyaromatic compounds such as naphthalene, phenanthrene, anthracene and benzoic acid have been reported by many workers. Organic polymers are particularly interesting solutes for use in the RESS process because they can form fibers or thin films. Upon decompression of the supercritical solution, the polymer precipitates to form either a solid or a liquid, and polymer fibers form when the polymer solute precipitates as a liquid phase. Fiber diameters are typically in the 1 − 5 μm range.

Various types of supercritical solvents like CO_2, nitrous oxide, trifluoromethane, ethylene and pentane are used for organic solutes and a few percent of polar solvent was added for enhancement of solubilities in supercritical solvents.

The spray technique is widely used for elaboration of inorganic materials. There are few examples of the production of inorganic powders by CO_2 RESS. The limitation comes from the very low solubilities of inorganic compounds. Supercritical water can dissolve some oxides but shows a too weak solvent power towards nitrides or sulfides of transition metals.

(e) Crystallization of Polyaromatic Compounds by RESS

For a better understanding the RESS process, we will give an example: Nagahama and Liu's work [7, 8] on the crystallization of pure naphthalene, phenanthrene and anthracene as well as their mixture using CO_2 as a solvent.

Apparatus

Figure 5.2.3 shows the experimental apparatus [7]. It consists of four sections: a liquid CO_2 pressurizing pump (P), a solute dissolving and extracting unit (E), an extract crystallizing unit (S) and a CO_2 flow rate measuring unit (G). Figure 5.2.4 shows the detail of the expansion nozzle placed in the crystallization unit. At the exit of the nozzle the orifice, with a length of 280 μm and 35−90 μm i.d., is equipped for producing a large pressure drop or rapid expansion of supercritical solutions.

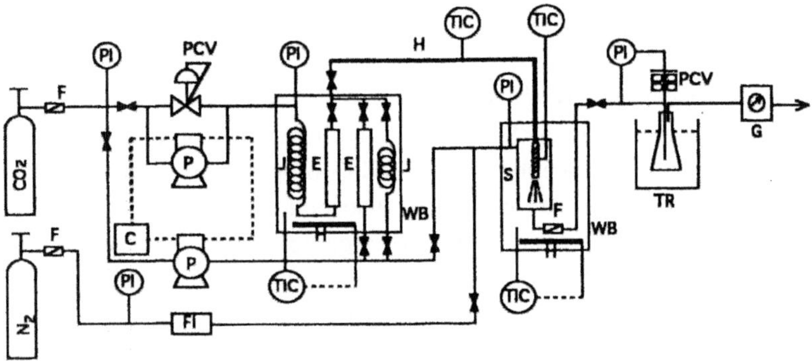

Fig. 5.2.3 Experimental setup for RESS: C, cooler; E, equilibrium cell; F, filter; F1, high-pressure flow meter; G, gas meter; H, heater; J, preheater coil; P, HPLC pump; PCV, back pressure regulator; P1, pressure gauge; S, crystallization cell; TIC, temperature controller; TR, trap; WB, constant temperature bath [7]

The RESS experiment was made by flowing supercritical CO_2 to dissolve the solute compound to saturation and rapidly expanding the solution of CO_2 + solute into the crystallization cell through the nozzle. The formed particles are collected into a sample bottle or on a glass plate. In order to change the degree of supersaturation, Nagahama and Liu [8] varied the nozzle temperature and added pure CO_2 to the solution before expansion. The morphology of solid particles was observed via optical microscopy and the crystal characteristics via X-ray diffraction.

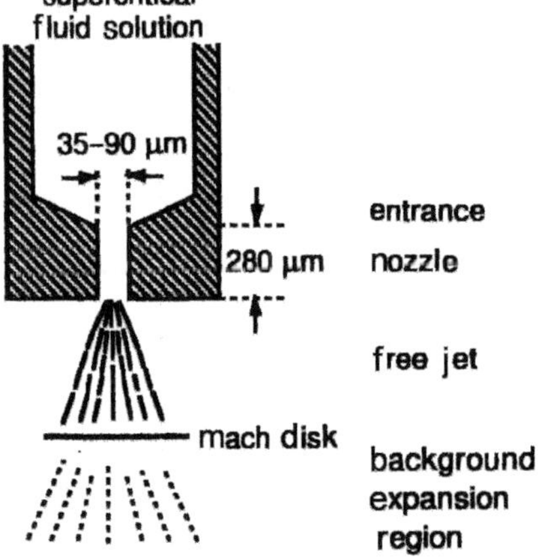

Fig 5.2.4 Expansion nozzle [7]

Crystallization of Pure Solute

Table 5.2.2 gives the experimental RESS results for pure naphthalene by varying the operating conditions [7]. Figure 5.2.5 shows the change of morphology when increasing the nozzle temperature from 87 to 125 °C. The increase in the nozzle temperature corresponds to a decrease in the degree of supersaturation and the average diameter of the crystal was raised as shown in Table 5.2.2 and Fig. 5.2.5. Figure 5.2.6 shows the change of morphology for pure anthracene when diluting the solution by adding fresh CO_2 after extraction [8]. The degree of dilution R is the ratio of the flow rates of the added fresh CO_2 to the saturated solution. As shown in Fig. 5.2.6, increasing the value of R from 0 to 2 causes the particle size to be increased significantly. This may be due to the decrease of solute concentration, i.e. depression of supersaturation in the down stream.

Table 5.2.2 RESS experiment on pure naphthalene using CO_2 [7]

run no.	P_{ext} (MPa)	T_{noz} (℃)	orifice diameter (μm)	P_{cr} (MPa)	T_{cr} (℃)	y_{na} (x10^4)	particle size (μm)	maximum particle size (μm)
			increasing pre-expansion temperature					
P-1	20.40	87	40	2.50	45	167	5–8	25
P-2	20.31	105	40	2.50	45	167	8–10	27
P-3	20.40	125	40	2.50	45	167	15–25	42
			decreasing expansion chamber temperature					
P-3	20.40	125	40	2.50	45	167	15–25	42
P-4	20.36	125	40	2.50	30	167	8–12	80
P-5	20.34	125	40	2.55	15	167	5–10	20
			decreasing naphthalene concentration					
P-6	20.49	125	90	2.50	45	167	8–13	40
P-7	15.02	125	90	2.50	45	145	10–17	45
P-8	10.89	125	75	2.50	45	113	15–25	40
			decreasing expansion chamber pressure					
P-9	20.10	86	75	4.36	45	167	3–7	15
P-1	20.40	87	40	2.50	45	167	5–8	25
P-10	20.39	87	40	0.10	45	167	5–8	23
			increasing orifice diameter					
P-3	20.40	125	40	2.50	45	167	15–25	42
P-11	20.49	125	75	2.55	45	167	20–25	40
P-6	20.49	125	90	2.50	45	167	8–14	40

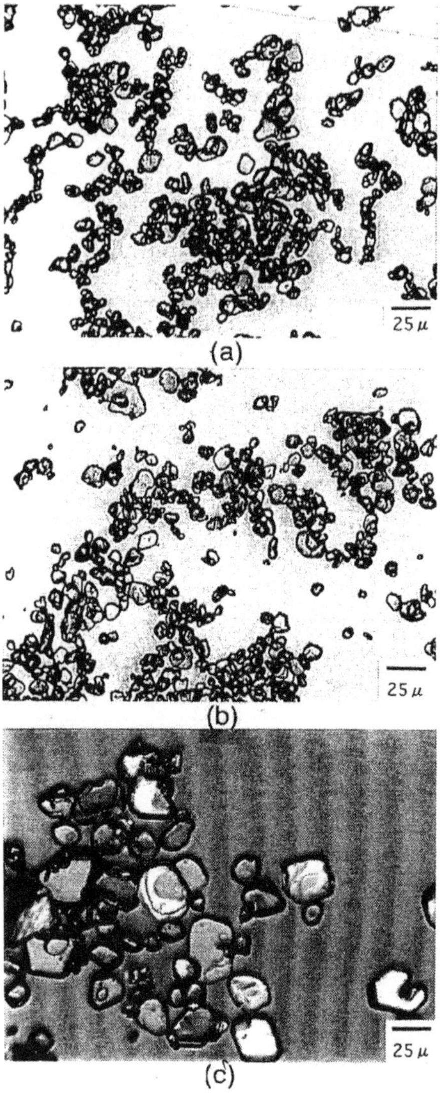

Fig. 5.2.5 Naphthalene crystals obtained via RESS: (**a**) T_{noz} = 87 °C, (**b**) T_{noz} = 105 °C, (**c**) T_{noz} = 125 °C [7]

Fig. 5.2.6 Influence of degree of dilution R on forming anthracene crystals: (**a**) $R = 0$, (**b**) $R = 2$ [8]

Crystallization of Naphthalene + Phenanthrene (Eutectic) Mixture

For the purpose of developing a new separating method, RESS crystallization experiments were performed for the naphthalene + phenanthrene system using the same apparatus as shown in Fig. 5.2.3 [7]. The morphology as well as the composition of the collected solid particles were analyzed by optical microscopy and gas chromatography. The maximum purity of naphthalene in the formed solid mixture reached 0.96 mole fraction, which is 1.08 times higher than the saturated solubility. Figure 5.2.7 shows solid particles produced by widely varying the feed concentration of the solute mixture to the nozzle. Pure

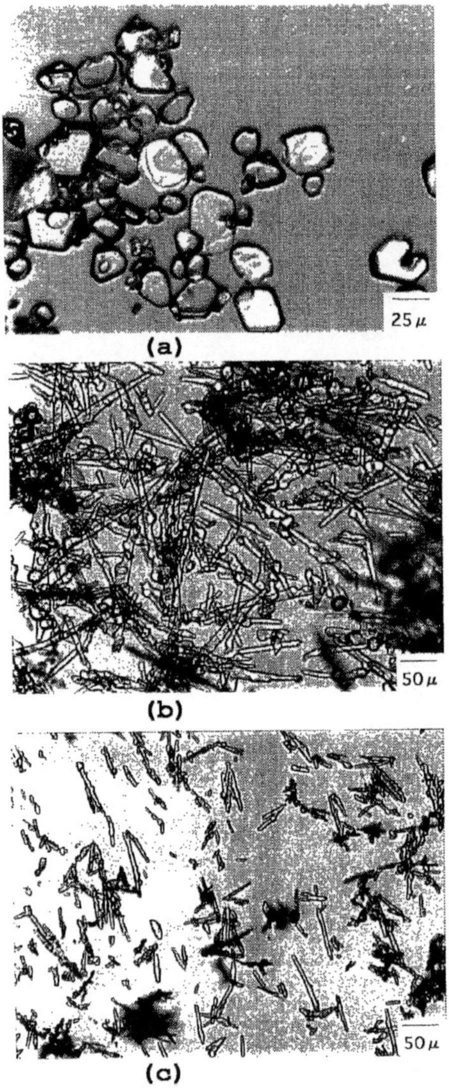

Fig. 5.2.7 Composition dependence of crystal morphology for a naphthalene + phenanthrene system: (**a**) pure naphthalene, (**b**) naphthalene mole fraction = 0.195 and (**c**) pure phenanthrene [7]

naphthalene yields relatively large flat plates and pure phenanthrene forms as small needle-like particles. The solid particles produced from the mixture seem to be a mixture of each pure crystal because the solid–liquid equilibrium of this mixture is eutectic. This was also confirmed by an analysis of the X-ray diffraction pattern for the solid mixture.

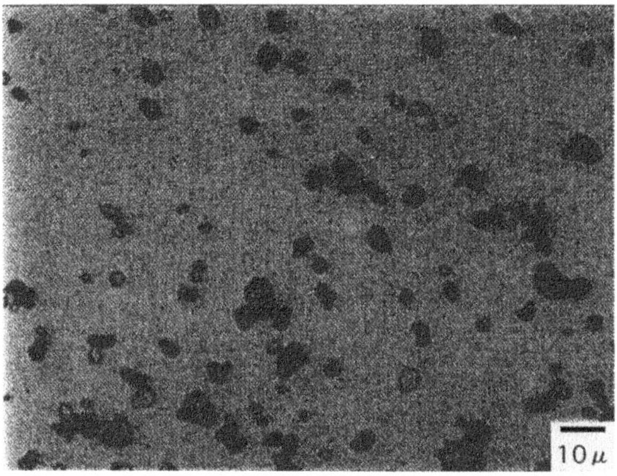

Fig. 5.2.8 Crystals of anthracene + phenanthrene mixture obtained via RESS [8]

Crystallization of Anthracene + Phenanthrene (Solid Solution) Mixture

A mixture of anthracene + phenanthrene forms a solid solution in solid-liquid equilibrium. The solubility and RESS behavior for this solid solution were investigated by Liu and Nagahama [8]. As the number of Gibbs degrees of freedom in the solid-fluid equilibrium for CO_2 + anthracene + phenanthrene is equal to three, the solubility of this mixture can be determined when given not only the temperature and pressure, but also a fixed solid composition. This means the solubilities of anthracene + phenanthrene in CO_2 change depending on the solid composition.

Figure 5.2.8 is a photograph of crystals formed by RESS. Comparing Fig. 5.2.8 with Figs. 5.2.6 (pure anthracene) and 5.2.7 (c) (pure phenanthrene), the crystal shapes and sizes in Fig. 5.2.8 are completely different from those in Figs. 5.2.6 and 5.2.7 (c). This suggests that the crystals of Fig. 5.2.8 may not be a mixture of pure solids but may be a solid solution. Also X-ray diffraction observation gave further support for this fact.

(f) Advantages and Disadvantages of RESS

One advantage of RESS is the relatively narrow distribution of particle size with small average value (about micrometers). Furthermore, it is very rapid and dose not require a vacuum. From a process management standpoint, crystallization and drying occur simultaneously and leave products free from solvent. And also RESS reduces the waste volumes required for filtration and washing by simplifying solid separation [9].

The disadvantage of RESS is that it can only be applied to materials which are soluble in SCF and since the solubility of most materials in SCF is low, a large amount of fluid is needed to prepare a small amount of solute particles. Therefore, RESS is a comparatively costly technique in terms of fluid usage and can only be used in processes involving high value added products.

5.2.2 Gas Anti-Solvent (GAS)

(a) Features of the GAS method

A gaseous and supercritical fluid can be used in the crystallization as an anti-solvent, which is called the *gas anti-solvent* (GAS) method. When the solubility decreases by dissolving the anti-solvent (gaseous or supercritical state) in the solutions, the solute crystallizes. The GAS method is usually applicable at lower temperature and lower pressure than with the RESS method to various kinds of compounds by choosing an appropriate liquid solvent. The pioneering work for the GAS method was carried out by Gallagher et al. [10]. They applied the GAS method using carbon dioxide, chlorodifluoromethane and dichlorodifluoromethane to manipulate the crystal size distribution for explosive nitroguanidine, naphthalene and others, and the effect of various expansion paths on particle size was investigated. As an anti-solvent, CO_2 is mostly used; for example, the crystallization of β-carotene and acetaminophen were carried out in toluene and n-butanol solutions by Chang and Randolph [11], and the change of partial molar volume for each component in the liquid phase and the dependence of the solubility of β-carotene on CO_2 pressure were shown. The GAS method is applicable to the separation of materials which is difficult to separate by conventional methods. Shishikura [12] reported the separation of carboxylic acids in acetone solutions. Citric acid was successfully separated from other organic acids, including the main by-product of citric acid fermentation, and it was shown that for efficient separation the supersaturation degree of the solution and the solubility of the solute are important. Recently Catchpole et al. [13] also achieved the separation of lecithin from the soya oil and coriander triglycerides from essential oil using the GAS method with CO_2. Tai and Cheng [14] applied the GAS method to the crystallization of organic compounds such as acetaminophen and sucrose, and inorganic compounds such as ammonium chloride. It was reported that the growth process of crystals in the GAS method is similar to those in conventional crystallization except for millimeter-sized inorganic crystals.

The features of the GAS method can be written as follows, when carbon dioxide is used as an anti-solvent.

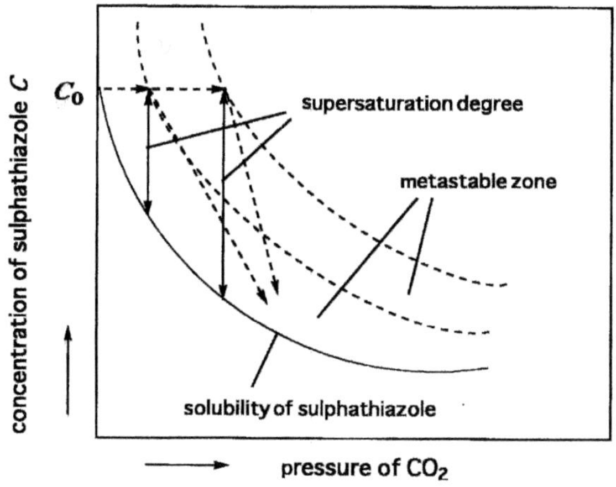

Fig. 5.2.9 Metastable zone and supersaturation degree in crystallization with the GAS method

(i) The crystallization rate can be controlled by pressurization of the anti-solvent at low temperature and low pressure.

(ii) Separation of the anti-solvent after crystallization is easy compared to conventional methods.

(iii) The toxicity of CO_2 is very low in comparison with other common organic solvents.

(b) Pressurization Conditions in the GAS Method

In the GAS method, control of the pressurization pattern is essential, because the pressurization pattern determines the changing figure of the supersaturation degree which is the driving force of nucleation and crystal growth in solutions. Gallagher et al. [10] indicated the relationship between the pressurization pattern and crystal size. They observed that the crystal size of nitroguanidine changes with the pressurization rate (expansion path), and with high pressurization rate, fine particles were obtained. This is due to the high nucleation rate. The pressurization pattern can be compared with the cooling pattern in conventional cooling crystallization. The dependence of the supersaturation degree on the pressurization rate can be explained using the concept of the metastable zone width [15]. In Fig. 5.2.9, it is shown that slow pressurization accords to the slow increase of supersaturation and the metastable zone width is narrow. In this case the nucleation rate may be slow and the crystal size becomes large.

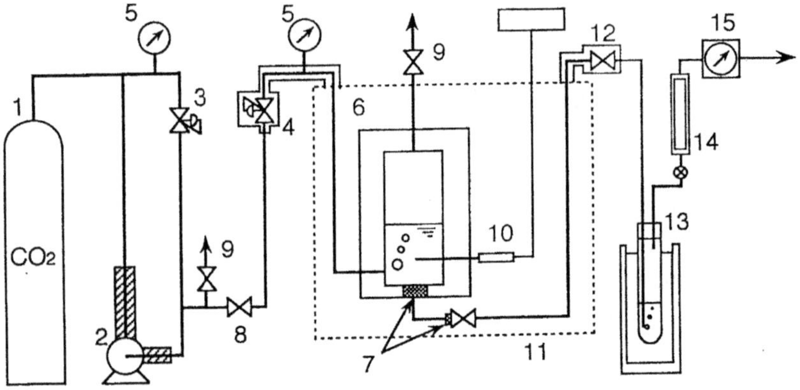

Fig. 5.2.10 Schematic diagram of experimental apparatus; 1, CO_2 cylinder; 2, pump; 3, back pressure regulator ; 4, pressure regulator; 5, pressure gauge; 6, cell; 7, filter; 8, stop valve; 9, relief valve; 10, thermocouple; 11, thermostated water bath; 12, metering valve; 13, cold trap; 14, flow meter; 15, dry gas meter

As mentioned above, to develop the GAS method in practical use, the dependence of the crystallization behavior on the pressurization pattern is very important. In the following section, an example of the detailed relationship between the pressurization pattern and the crystallization behavior of sulphathiazole, and the morphology and crystal size on these pressurization conditions is shown [16]. Furthermore, as the sulphathiazole has some polymorphs, the crystallization behavior of the polymorphs is also described.

(c) Crystallization of Sulphathiazole by the GAS method

Experimental Method and Apparatus

In this section, the crystallization of the sulphathiazole ($C_9H_9N_3O_2S_2$) (SUT), a pharmaceutical, from ethanol by GAS using CO_2 is described. The apparatus for the GAS method is presented schematically in Fig. 5.2.10. The high-pressure cell (volume 145 cm^3) equipped with a tempered glass window is used as a crystallizer. In the cell (6), the ethanol solutions saturated with SUT (0.0143 mol/l) at the set temperature (298.2 K) are fed and the cell is dipped in the thermostatic water bath (11). Carbon dioxide is pumped into the cell controlling the pressure in the cell by the regulator (4), and the crystallization behavior is observed through the window of the cell and the expansion degree of the solution ($\Delta V / V_0$) is measured with the cathetometer. After crystallization the crystals are filtrated and subjected to microscopic observation and the crystal size, morphology and the amount of the crystals are measured.

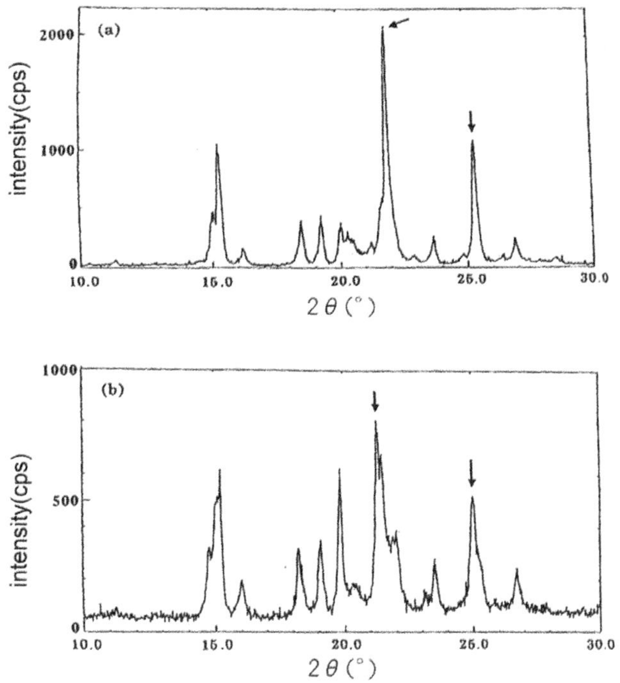

Fig. 5.2.11 X-ray diffraction pattern of the crystals: (**a**) material (**b**) with stepwise pressurization

It is known that SUT has some polymorphs. Higuchi et al. [17] showed type I and II X-ray diffraction patterns of the crystals. With the X-ray diffraction pattern the composition of the SUT polymorphs in the crystals precipitated by the GAS method was determined [18]. In Fig. 5.2.11 (a) the X-ray diffraction pattern of the material (special grade Tokyo Kasei) is shown. From the pattern the material appeared to belong to type I in reference [17].

Pressurization Pattern

It is thought that the pressurization rate is an important factor in the GAS method, because it is the increase rate of the supersaturation. Therefore, to examine the pressurization rate two extreme conditions are selected, i.e. "stepwise pressurization" and "rapid pressurization" (Fig. 5.2.12). In addition to these methods, a third pressurization pattern is devised here. If the nucleation and growth process of crystals can be separated in the crystallization, this should be advantageous for the control of the crystal size. From this point of

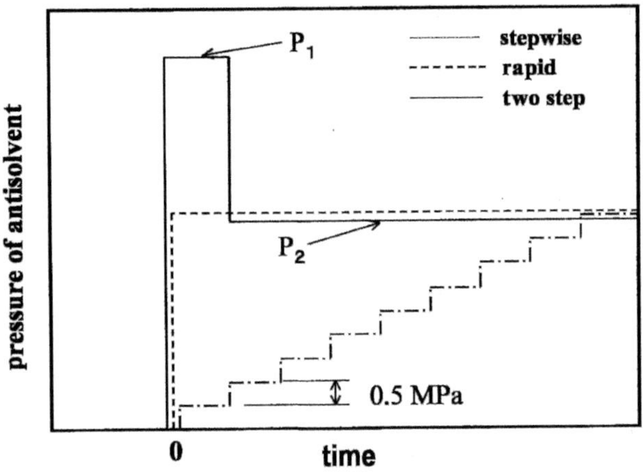

Fig. 5.2.12 Pressurization patterns of stepwise, rapid and two-step pressurization

view, a “two-step pressurization” method is adopted, i.e., first a very large pressure is instantly imposed to accelerate strikingly the nucleation, and as the second step the pressure is reduced to the second value at a “decompression point” to control the growth rate of the crystals (Fig. 5.2.12).

Stepwise Pressurization

With the stepwise pressurization of carbon dioxide, the solution expands gradually and reaches equilibrium values. In this method, carbon dioxide is added stepwise by 0.5 MPa increments after the solution expansion has stopped. The dependence of the expansion ratio of the liquid phase to the initial volume, $\Delta V / V_0$, on time at the maximum pressure of 5.5 MPa is shown in Fig. 5.2.13 The maximum expansion ratio of the volume is 1.01 and the nucleation is observed at the expansion ratio of about 0.55, when the pressure was 5.0 MPa. In this case, exceptionally large pillar-like crystals (2–6 mm in the longitudinal direction) are obtained (Fig. 5.2.14 (a)). This may be due to the slow increase of supersaturation, which causes a narrow metastable zone width and a low nucleation rate (Fig. 5.2.9).

The X-ray diffraction pattern of this crystal (Fig. 5.2.11 (b)) indicates that this pattern is different from that of type II in reference [17] and similar to that of type I. However, strictly speaking they are different from type I. On this account, we denote this polymorph as type III.

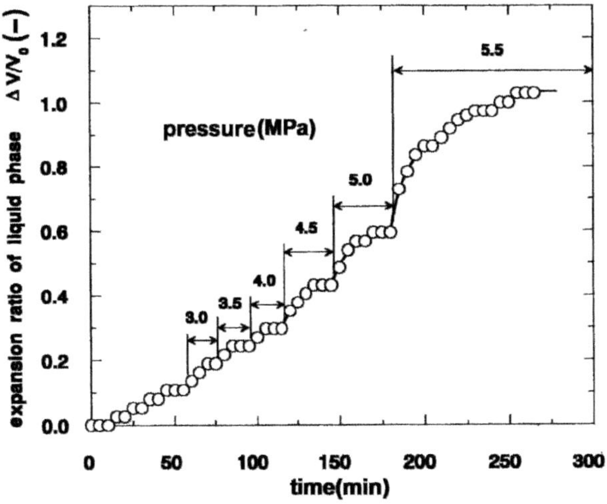

Fig. 5.2.13 Dependence of expansion ratio of solutions on time in the case of stepwise pressurization

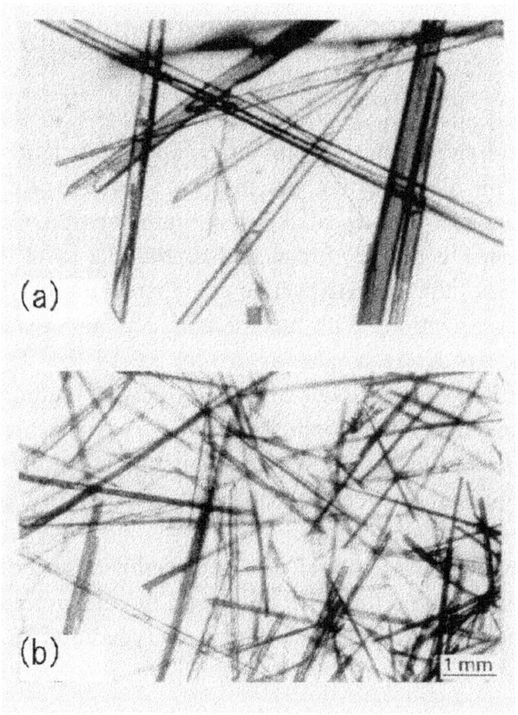

Fig. 5.2.14 Crystals precipitated in (**a**) stepwise and (**b**) rapid pressurization methods

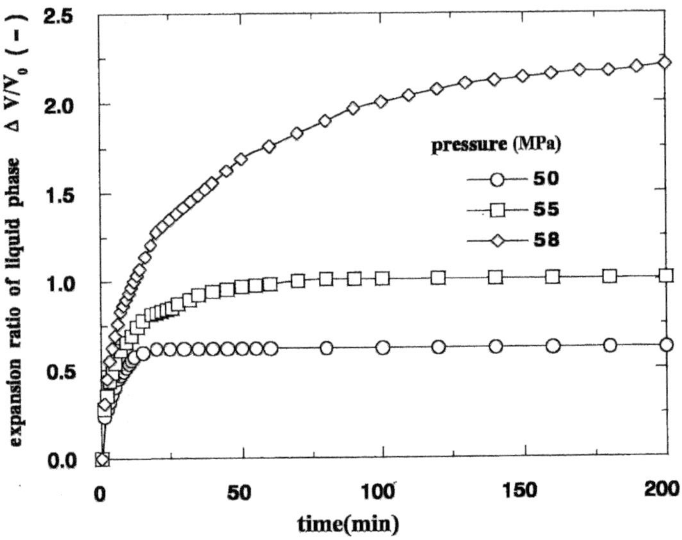

Fig. 5.2.15 Dependence of expansion ratio of solutions on time in the case of rapid pressurization

Rapid Pressurization

In the rapid pressurization method, the maximum pressure is applied instantly. Therefore, it is expected that the nucleation rate is increased and small crystals precipitate. The results of the expansion ratio at the maximum pressures of 5.0, 5.5 and 5.8 MPa are shown in Fig. 5.2.15. It can be seen that the increasing rate of the expansion ratio, i.e., the increasing rate of supersaturation, is much larger than in stepwise pressurization. With this method, crystals of size 0.5−3.0 mm are obtained at each pressure as shown in Fig. 5.2.14 (b). The crystals are only about a third or half of that obtained by the stepwise method, which are larger than the size expected. The morphology of the crystals is also not so different from the crystals obtained in stepwise pressurization. Furthermore, a similar crystal size is obtained between the different pressures in the same manner as the case of stepwise pressurization.

The X-ray diffraction pattern of the crystals obtained in this case indicates that the crystals are a mixture of type I and III. However, the discrimination of the morphology difference seems to be difficult between type I and III.

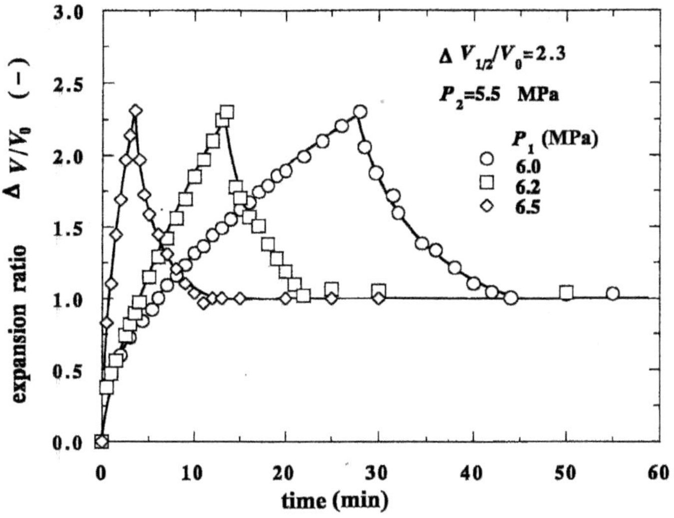

Fig. 5.2.16 Dependence of expansion ratio of solutions on time in the case of two-step pressurization (key parameter is the first pressure, P_1)

Two-Step Pressurization

In this method, first a very large pressure is instantly imposed, and secondly the pressure is changed to the lower value at a "decompression point" (Fig. 5.2.12). As the operational parameters, the first pressure (P_1), the second pressure (P_2) and the expansion ratio of the solution volume ($\Delta V_{1/2} / V_0$) at a decompression point from P_1 to P_2 are chosen. In the second, the pressure is controlled first to be slightly lower than P_2 for a short while to shorten the time to attain the equilibrium state. In Fig. 5.2.16, the behavior of the volume change corresponding to the pressurization pattern is shown. The volume expands first strikingly and with decompression it decreases to the stationary value.

The expansion rate of the solution increases remarkably with the first pressure and, during the expansion of solutions by imposing the first pressure, nucleation is observed. In this method a much smaller size (from a few tens to several hundred micrometers) of crystals than those obtained by the stepwise and rapid pressurization precipitate as shown in Fig. 5.2.17 (P_1 = 6.5 (a), 6.2 (b) and 6.0 MPa (c)). The crystallization of such smaller crystals indicates that the nucleation rate is particularly increased with large supersaturation and violent bubbling and that the growth of the crystals is suppressed compared with the stepwise and rapid pressurization methods. This change of crystal size may be due to the supersaturation degree and bubbling period, which are effective

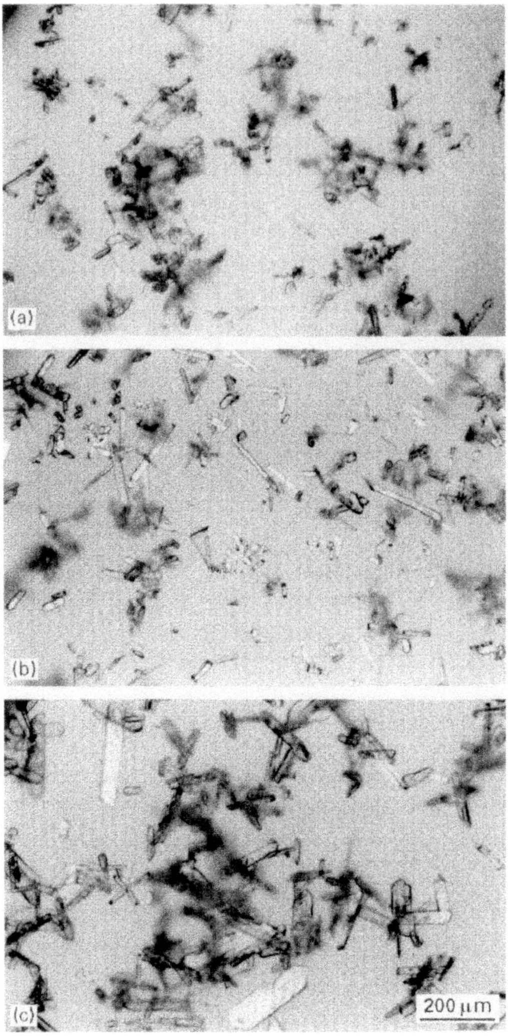

Fig. 5.2.17 Crystals obtained by two-step pressurization method: (**a**) P_1 = 6.5 MPa; (**b**) P_1 = 6.5 MPa; (**c**) P_1 = 6.0 MPa (P_2 = 5.5 MPa; $\Delta V_{1/2} / V_0$ = 2.3)

for the nucleation. On the other hand, the effect of the second pressure appears to be not so distinctive.

From the X-ray diffraction pattern, it is observed that in this method mixtures of type I and type III polymorphs precipitate in a similar manner with rapid pressurization. However, the ratio of type I is increased compared to that in the rapid pressurization method. These results suggest that under large supersaturation, nucleation of type I preferentially occurs.

Comparison between Pressurization Patterns

The crystal size and polymorphous crystallization behavior depend on the pressurization patterns. The difference of the crystal size corresponds to the pressurization rate and the metastable zone width. However, the crystal size is not so different between the stepwise pressurization and rapid pressurization method and the difference is rather small in comparison with the expectation. On the other hand, by two-step pressurization crystals with a wide size range are produced. The results demonstrate the capability of the control of the crystal size with this method. It also appears that the crystallization behavior of type I and type III polymorphs of sulphathiazole depends on the pressurization patterns. These results suggest the usefulness of the GAS method not only for crystal size control but also for the polymorphic control.

5.3 Processing of Polymeric Materials Using Supercritical Fluids

The application of supercritical fluid technology to high-performance polymers is becoming increasingly important as such polymers become more prevalent in industry. Above all, recent interest has been expressed in foaming of polymers and crystallization of plastics induced by SCF. Since the production of chlorofluorocarbons, which have been used as physical blowing agents for industrial foaming processes, were banned in 1997 according to the Montreal Protocol, CO_2 has attracted much attention as a new environmentally benign alternative blowing agent.

Microcellular plastic (MP) is a newly invented polymeric foam material that has several superior mechanical and thermal properties compared with conventional foam plastics. This material has cell size from 0.1 to 10 μm and cell number density from 10^9 to 10^{15} cells/cm^3 [1]. In the production of an MP, supercritical CO_2 and N_2 are used as physical blowing agents. As other applications, supercritical fluids can induce crystallization of amorphous polymers [2]. Since the crystallized polymers exhibit high strength and durability in harsh conditions, a variety of applications as replacements for metal parts are expected.

In the processing of these polymeric materials, the solubility and diffusivity of an SCF in the polymer are important properties. The investigations of polymeric materials processing using SCFs and of related properties are explained in this section.

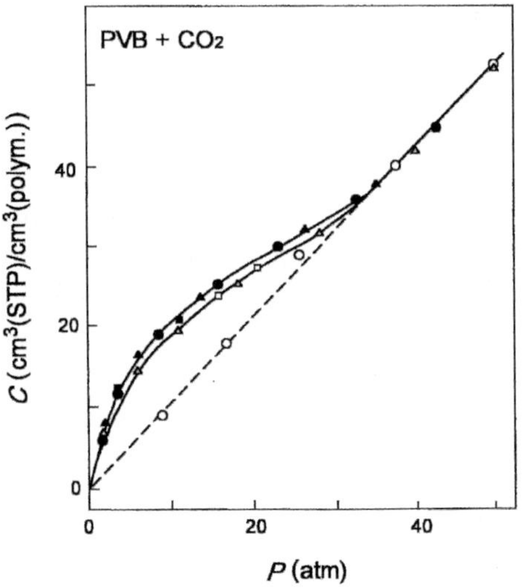

Fig. 5.3.1 Comparison of the first sorption isotherm (○) measured for CO_2 in PVB which had not been exposed to high-pressure CO_2, with the first desorption isotherm (●) and subsequent sorption and desorption isotherms (△, □; ▲, ■) at 25 °C

5.3.1 Solubility and Diffusivity of Gases in Polymers

The solubility and diffusivity of gases in molten polymers have been reported before 1970 by several authors [3–9]. However, the solubility and diffusivity of CO_2 and N_2 are lacking. Since sorbed CO_2 induces a large volume change of polymer, the measurements of solubility and diffusivity are difficult.

Kamiya and coworkers [10–12] actively reported gas sorption and dilation of polymers. They have investigated partial molar volumes of gas in polymer, the Flory–Huggins interaction parameter χ, and the isothermal glass transition. Sorption isotherms of the glassy poly(vinyl benzoate) (PVB) + CO_2 system are shown in Fig. 5.3.1. They found "conditioning effects" in the system as described below. In the first CO_2 sorption run on a PVB film, which had never been exposed to high-pressure CO_2, the sorption isotherms obtained below the glass transition temperature T_g were almost linear. At pressures below the glass transition pressure, these isotherms lay below all desorption and subsequent sorption isotherms. In the case of N_2 and Ar, however, no such effects were found. They also developed the modified dual mode sorption model [13] consisting of the Langmuir adsorption and the Flory–Huggins terms.

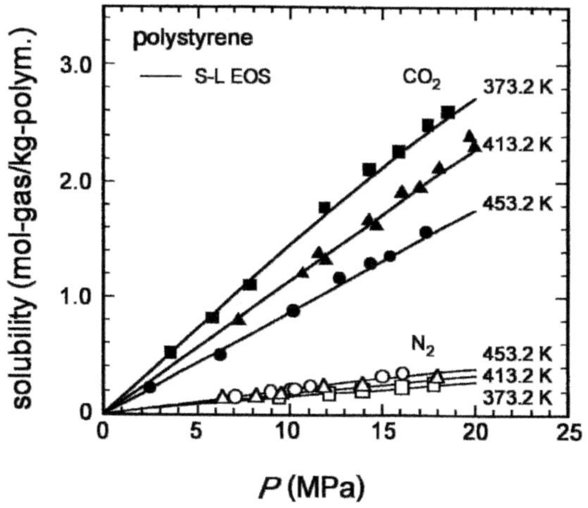

Fig. 5.3.2 Solubility of CO_2 and N_2 in polystyrene

Sato et al. [14,15] measured the solubility of CO_2 and N_2 in polystyrene (PS), polypropylene (PP), and high-density polyethylene (HDPE). The volume of the polymers under high temperatures and high pressures, which are required to determine the solubility accurately, was obtained by *PVT* measurements [16]. Change in the polymer volume due to dissolution of gases was predicted using the Sanchez and Lacombe equation of state (S–L EOS) [17,18], because measurement of the swollen molten polymer volume turned out to be difficult. The diffusion coefficients of gases in polymers were determined by measuring pressure vs. time curves during gas dissolution.

The measured solubilities of CO_2 and N_2 in molten PS are shown in Fig. 5.3.2. The solubilities of both the gases increased almost linearly with pressure, and those of carbon dioxide decreased with an increase in temperature. This tendency has been generally observed in many gas + polymer and solute + solvent systems. The solubility of nitrogen above 373.2 K, however, increased with increasing temperature. This tendency is called *reverse solubility* and has been observed for gases of low critical temperatures such as He, H_2, N_2 and O_2. The solubilities of CO_2 in the three polymers at 453.2 K are compared in Fig. 5.3.3, along with S-L EOS calculations. CO_2 exhibited the highest solubility in polypropylene (PP) and the lowest solubility in polystyrene (PS). The solubility of CO_2 in PP and high-density polyethylene (HDPE) were about 2 and 1.7 times greater than in PS, respectively.

Henry's constants, K_P [kg·MPa/cm^3(STP)] for nitrogen in PP, polyethylene (PE) and PS are compared with literature values (Durrill and Griskey [8] and Lundberg et al. [7]) in Fig. 5.3.4. The data for PP and PE are very similar,

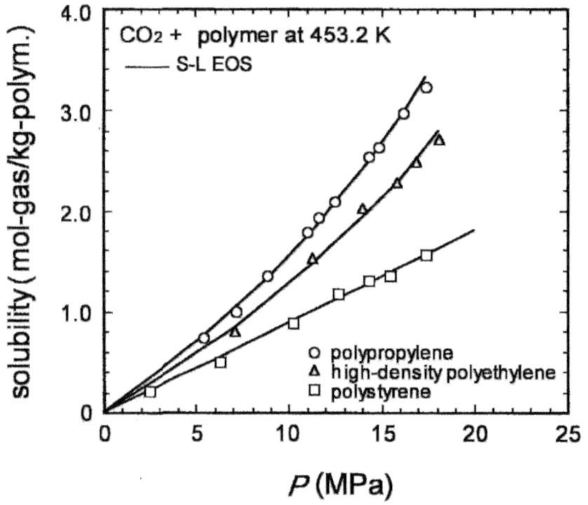

Fig. 5.3.3 Solubility of CO_2 in polypropylene, high-density polyethylene, and polystyrene at 453.2 K

which may be due to the similar structure of the two polymers. Zhong and Masuoka [19] reported a similar result in the correlation of K_P for organic vapors in PP and PE. The $1/K_P$ for PS exhibited a distinct minimum at a temperature of about 350 K. Burgess et al. [20] reported a similar solubility minimum for a series of noble gases in poly(methyl acrylate) at its glass transition temperature. The temperature of minimal solubility obtained by Sato et al. is about 10 K lower than its glass transition temperature of PS (372.6 K).

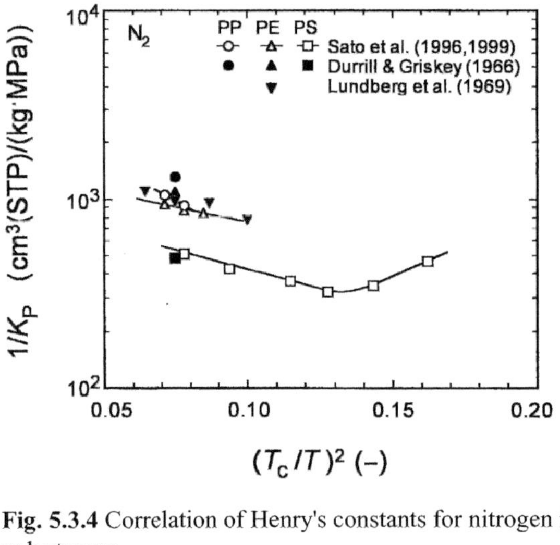

Fig. 5.3.4 Correlation of Henry's constants for nitrogen in polypropylene, polyethylene, and polystyrene

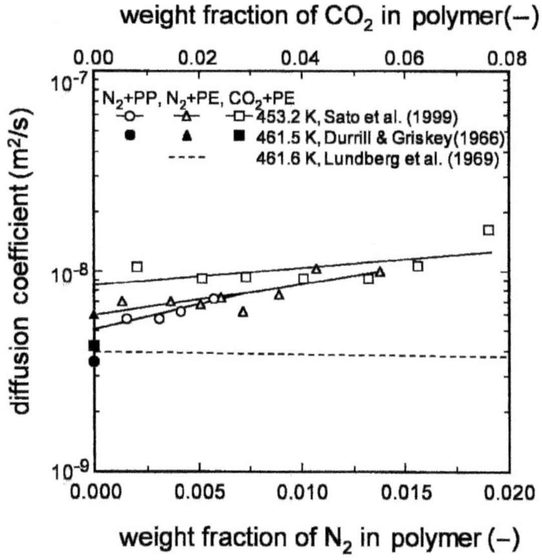

Fig. 5.3.5 Diffusion coefficients of N_2 + polypropylene, N_2 + high-density polyethylene, CO_2 + high-density polyethylene

Burgess et al. explained that the minimum solubility was caused by a hole-filling contribution to gas sorption. However, the hole-filling contribution probably varies depending on the gas, due to differences in gas–polymer interactions and the molecular sizes of the gas and the holes.

Diffusion coefficients of the N_2 + PP, N_2 + HDPE, and CO_2 + HDPE systems at 453.2 K [15] are shown in Fig. 5.3.5. The diffusion coefficients in these molten polymers exhibited a weak concentration dependence, whereas those in glassy polymers showed a large dependence as is well known. These data had the same order of magnitude of 10^{-9} m^2/s.

Sato et al. [21] investigated the solubility of fluorocarbons (HFC-134a, HFC-152a) and chlorofluorocarbon (HCFC-142b), which are used as alternative blowing agents, in PS at temperatures from 348.2 to 473.2 K and pressures up to 3.2 MPa. The solubilities of the three blowing agents increased proportionally with pressure and decreased with temperature, as shown in many polymer and condensable gas systems. In Fig. 5.3.6, the solubility of the three blowing agents, CFCs [22] and CO_2 [14] in PS at 373 K were compared. The solubility of the blowing agents in PS is of the order of CFC-11 > HCFC-142b > CFC-12 ≥ HFC-152a > HFC-134a > CO_2. The order of solubility is similar to the order of the boiling point temperature of the blowing agents, except for CFC-12.

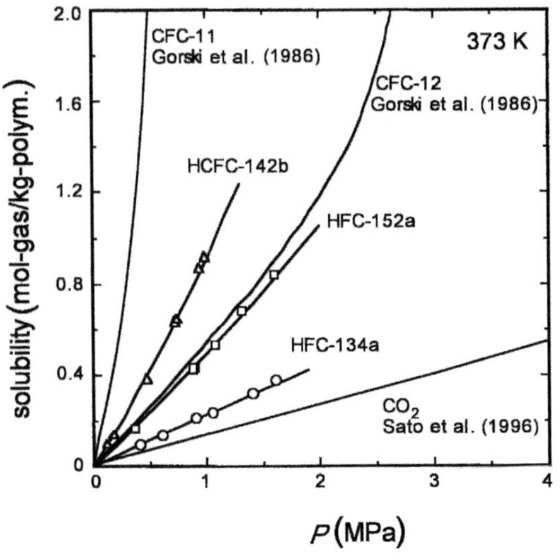

Fig. 5.3.6 Comparison of solubilities of various gases in polystyrene at 373 K.

5.3.2 Crystallization of Plastics Induced by Supercritical Fluid

Chiou et al. [23] observed glass transition temperature reductions due to gas sorption with the DSC method. Chiou et al. [24] and Mizoguchi et al. [25] found that amorphous polymers crystallized in high-pressure CO_2. These phenomena are similar to solvent-induced crystallization due to plasticization of the polymer, as is well known.

Mizoguchi et al. [25] studied that CO_2 induced crystallization of poly(ethylene terephthalate) (PET). They measured the change of the intensities of some bands of the infrared spectra and density of PET films treated by thermal and CO_2 induced crystallization. In the thermal crystallization at 85 °C (10 K above T_g), the crystallization half-life, $t_{1/2}$, is 36 days. On the other hand, in the case of exposure to 50 atm CO_2 at 85 °C, PET began to crystallize within several minutes and $t_{1/2}$ was about 10 min, which was close to that of thermal crystallization at 120 °C as shown in Fig. 5.3.7. Namely, the crystallization rate was greatly increased by the sorption of CO_2.

Mizumoto et al. [26] studied CO_2-induced stereocomplex crystallization of *isotactic*- and *syndiotactic*-poly(methyl methacrylate) mixtures at a temperature of 30–70 °C and pressure up to 20 MPa. The melting point temperature and heat of fusion are shown in Fig. 5.3.8. The melting point temperature of the polymer rose as the CO_2 treatment temperature and pressure increased.

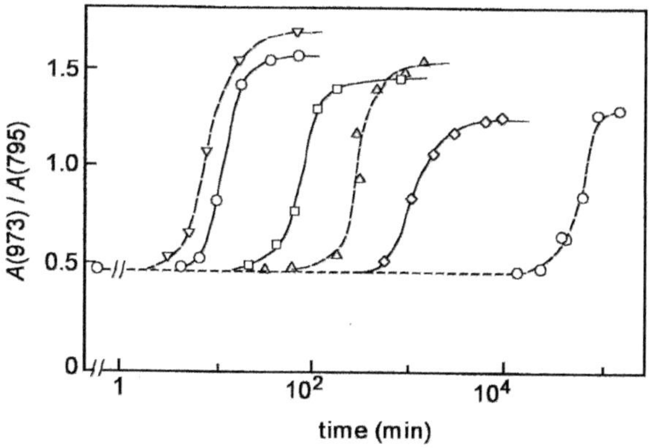

Fig. 5.3.7 Crystallization curves obtained by IR absorption for PET. Relative absorbance vs. $\log t$: (-------) thermal crystallization at 85 (○), 100 (△), and 120°C (▽); (——) crystallization by exposure to 50 atm CO_2 at 35 (◇), 65 (□), and 85°C (○)

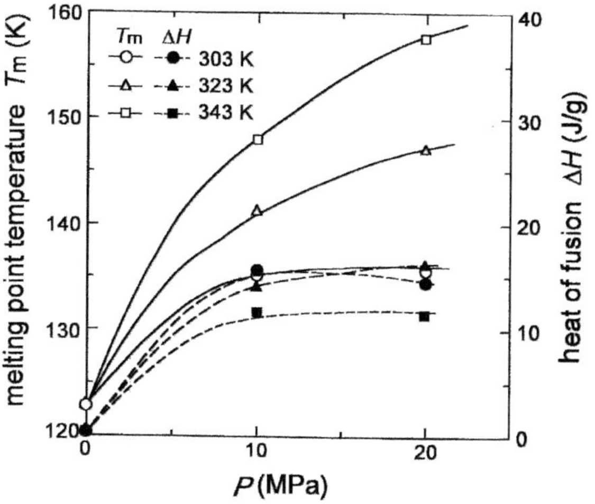

Fig. 5.3.8 Pressure dependence of melting point temperature and heat of fusion for stereocomplex poly(methyl methacrylate)

On the other hand, the heat of fusion of the polymer did not exhibit a large temperature and pressure dependence. They also investigated foaming of the mixture in the expectation that high crystallinity caused high cell density and small cell size in the foam.

Takada et al. [27] investigated the effect of CO_2 and N_2 on the crystallization of high-impact polypropylene from the viewpoint of melting tem-

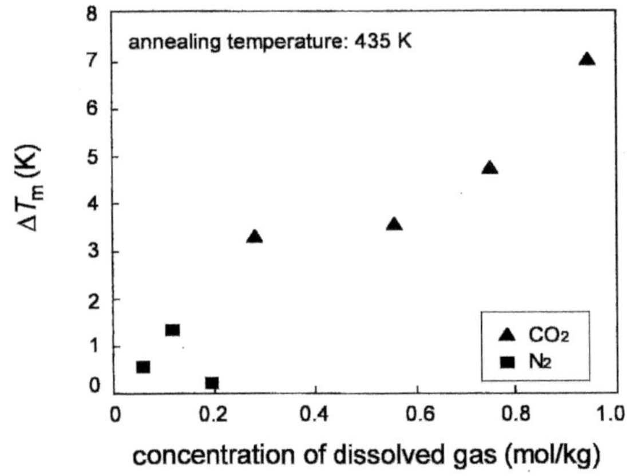

Fig. 5.3.9 Effect of the concentration of dissolved gas on ΔT_{m}

perature, i.e., lamella thickness. They crystallized the molten polypropylene by cooling under high-pressure CO_2. In the melting point temperature measurement at atmospheric pressure, the results showed that the melting temperature increased with increasing dissolved CO_2 concentration as shown in Fig. 5.3.9. ΔT_{m} is calculated by $\Delta T_{\mathrm{m}} = T_{\mathrm{m}} - T_{\mathrm{ma}}$, where T_{ma} is the melting temperature of the polymer re-crystallized under atmospheric air pressure. The melting point temperature of the polymer re-crystallized in high-pressure N_2 did not change significantly. They concluded that the dissolved CO_2 in the polymer would reduce the interchain interaction and increase the mobility of the polymer chains. As a result, the increased mobility enhanced the lamella thickness during annealing the polymer at high pressure.

5.3.3 Batchwise Production of Microcellular Thermoplastics

Polymeric foams are widely used as heat insulators, food trays, support materials, and so on, because of their many advantageous characteristics such as low thermal conductivity, light weight, and high impact strength.

Shimbo et al. [28] investigated the mechanical properties and viscoelastic behavior of microcellular CPET (a polyethylene terephtalate resin containing a polyolefin nucleating agent) foamed with CO_2. In order to investigate the effect of cell size, microcellular CPET samples having a nearly constant cell density of 10^9 cells per cubic centimeter, and cell diameters, D, in the range of 3–10 μm were produced. The relationship between ultimate strength, specific strength and average cell size is shown in Fig. 5.3.10 where the ultimate and specific strength at $D = 0$ denote that of unfoamed CPET. The specific

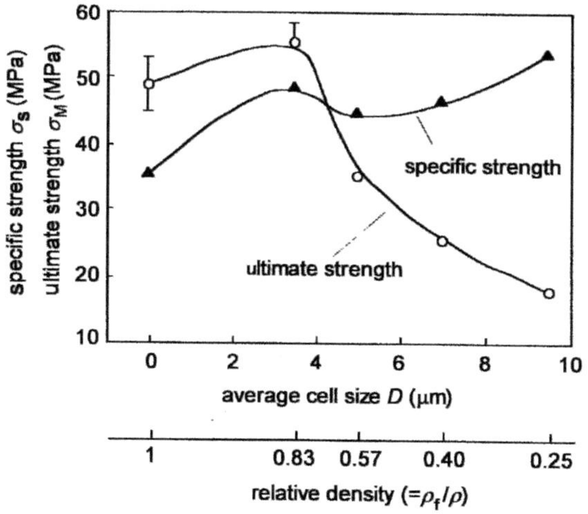

Fig. 5.3.10 Relationship between ultimate strength, specific strength and average cell size for CPET.

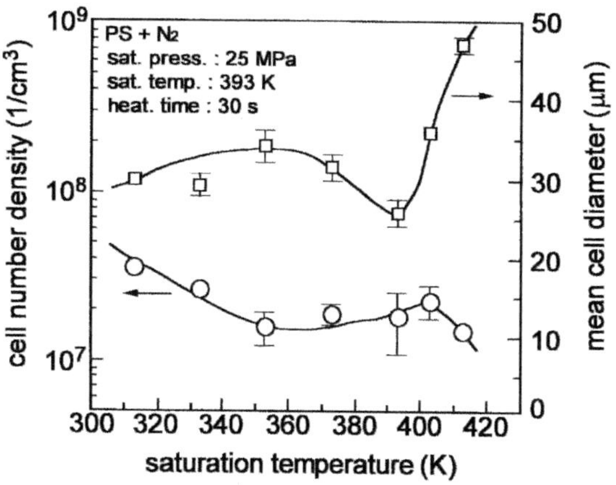

Fig. 5.3.11 Pressure dependence on cell number density and average cell diameter for N_2 + PS system

strength increased with increasing cell diameter and that of D = 9.5 μm was 40% higher than that of unfoamed CPET.

Sumarno et al. [29–31] investigated PS foaming by using N_2 and CO_2 at temperatures above its T_g. They observed a minimum of the cell number density and maximum mean cell diameter at 350 K for the PS/N_2 system shown in

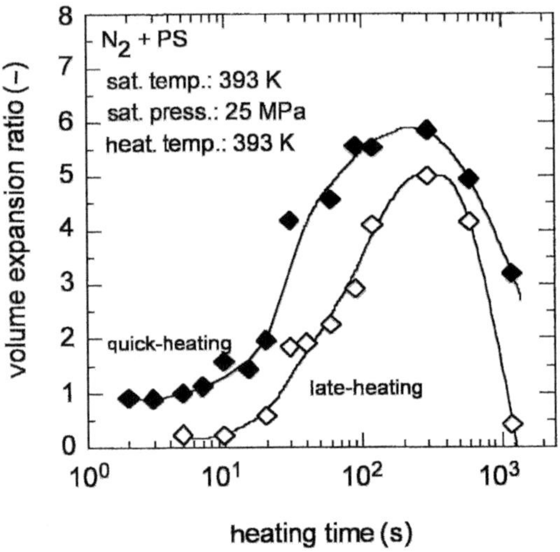

Fig. 5.3.12 Effect of the heating time on volume expansion ratio

Fig. 5.3.11. As described above, the solubility of N_2 exhibited a minimum at 350 K; solubility was important for the structure of the cell. They proposed two heating methods to foam the PS. One is "late heating", in which the sample is thermally insulated from the wall of the autoclave with a Teflon insulator. The samples were taken out of the autoclave immediately after decompression. After a time interval (generally 3 minutes from the end of the decompression), they were heated in a silicon oil bath for a certain time. On the other hand, in the "quick heating method", the samples were allowed in the autoclave after decompression for the desired heating time to heat the samples in terms of the heat capacity of the autoclave. In the latter method, instead of the Teflon insulator, the samples were placed on a brass tray in the autoclave. After the heating process the samples were immersed in an ice-water bath to freeze the internal structure. The degree of supersaturation of the polymer solution by quick heating should be higher than that by late heating, since the heating process by quick heating started at saturation concentration. The difference in volume expansion ratios between the two heating methods is shown in Fig. 5.3.12. Both volume expansion ratios increased with heating time and then indicated a maximum around 300 s. In the quick heating method, dissolved gas was used effectively.

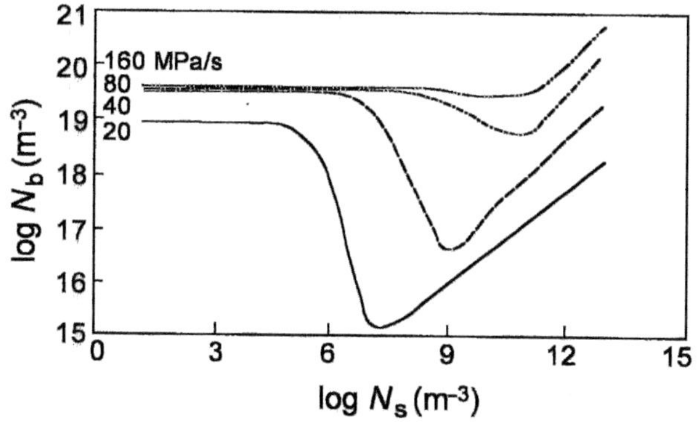

Fig. 5.3.13 Change of N_b against N_s at different pressure release rates

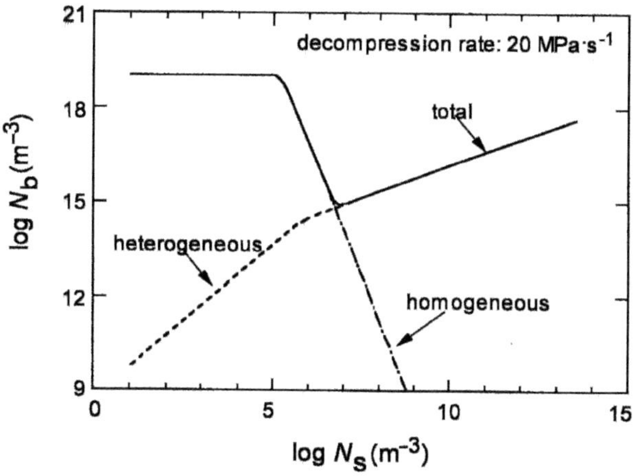

Fig. 5.3.14 Number density of homogeneously and heterogeneously nucleated bubbles

Inamori et al. [32, 33] investigated numerical simulations for polymeric foaming on a polyethylene/N_2 physical foaming process where homogeneous and heterogeneous nucleations were simultaneously induced by decompression. A simulation result of the number density of bubbles, N_b vs. the number density of heterogeneously nucleated bubbles N_s at various decompression rates is shown in Fig. 5.3.13. The number density of bubbles shows a minimum at slow decompression. The number density of homogeneously and heterogeneously nucleated bubbles at a decompression rate of 20 MPa/s is shown in Fig. 5.3.14 to explain the minimum. The number of heterogeneously nucleated bubbles increases with increasing the number of heterogeneous nuclea-

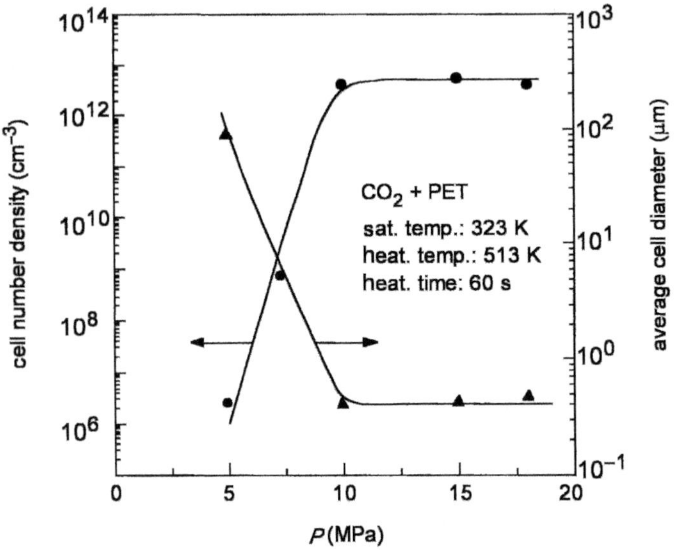

Fig. 5.3.15 Pressure dependence on cell number density and average cell diameter for CO_2 + PET system

tion site, whereas the number of homogeneously nucleated bubbles decreases. In heterogeneous nucleation, the nucleation of lower supersaturated gas can occur, because the free energy change of nucleation is small. In the slow decompression condition, heterogeneous nucleation can occur at the initial step of the decompression, and the dissolved gas is consumed for heterogeneous nucleation. Hence, the number of homogeneously nucleated bubbles decreases. In rapid decompression, homogeneous and heterogeneous nucleation occur simultaneously. Hence, the minimum is not observed at rapid decompression (160 MPa/s).

Chikatsune et al. [34, 35] studied the batch-wise production of PET foaming by CO_2 and N_2 as a blowing agent. Crystallization of PET induced by CO_2 was observed above 5 MPa, while the crystallization was not observed in the case of N_2. The pressure dependence of the number density and average cell diameter of PET foamed by CO_2 is shown in Fig. 5.3.15. The crystallinity of PET increased with increasing pressure and had a constant value at pressures higher than 7 MPa, which is not shown in the figure. The number density of the cell was more than 10^{12} cm^{-3} above 10 MPa. Homogeneous nucleation probably dominated the foaming at lower pressures whereas heterogeneous nucleation caused by the crystalline dominated the foaming at high pressures. This implies that the crystalline plays an important role in the structure of the foam.

5.4 Hydrothermal Synthesis of Metal Oxide Fine Particles at Supercritical Conditions

5.4.1 Principle of Hydrothermal Synthesis in Supercritical Water

When a metal salt solution is heated up, metal (hydro)oxides are formed due to the shift of the reaction equilibrium. At higher temperatures, the equilibrium shifts further toward metal oxide formation,

$$M^{x+} + xOH^- = M(OH)_x, \tag{5.4.1}$$
$$M(OH)_x = MO_{x/2} + (x/2)H_2O. \tag{5.4.2}$$

The method of producing metal (hydro)oxides from aqueous solutions just by elevating the temperature is hydrothermal synthesis. Ordinarily, this method is operated at around 100–200 °C [1–6]. Recently, a new approach has been proposed for hydrothermal crystallization using supercritical conditions [7–13].

At supercritical conditions, the reaction rate and reaction equilibrium changes greatly, due to the drastic change of dielectric constant at around the critical point. Some models are proposed for describing the equilibrium constant at sub- and supercritical conditions In a recent study, it has been demonstrated that the following equation involving the Born term for the dielectric constant effect can be used to describe successfully the metal oxide solubility in sub- and supercritical water [14]:

$$\ln K = \ln K_0 + \frac{\Delta H_0}{R}\left(\frac{1}{T} - \frac{1}{T_0}\right) - \frac{\omega}{RT}\left(\frac{1}{\varepsilon} - \frac{1}{\varepsilon_0}\right) - \frac{2Z_A Z_B}{RT}\alpha\sqrt{I}, \tag{5.4.3}$$

where H_0 is the enthalpy of reaction at the reference state, I is the ionic strength, K_0 is the equilibrium constant at the reference state, R is the gas constant, Z_i is the charge of component i, α and ω are constants determined by the system, ε is the dielectric constant of the fluid and ε_0 is that at a reference state. This equation clearly indicates that the reaction equilibrium will change greatly, due to the great change of dielectric constant with slight changes of temperature and pressure around the critical point. The foregoing suggests that the atmosphere of hydrothermal synthesis (reaction rate, nucleation rate, solubility of metal (hydro)oxides, soluble chemical species distributions, etc.) significantly varies around the critical point. Another specific feature of supercritical water for hydrothermal synthesis is the phase behavior around the critical point. A supercritical aqueous solution is miscible with oxygen or hydrogen gas to form a homogeneous oxidizing or reducing reaction atmosphere.

In this section, first the specific features of hydrothermal synthesis of metal oxides at supercritical conditions will be discussed and then various applications of this method will be introduced.

5.4.2 Key Features of Hydrothermal Synthesis in Supercritical Water

(a) Applicability for Various Systems

For hydrothermal synthesis at supercritical conditions, the flow apparatus schematically shown in Fig. 5.4.1 is employed. The fed metal salt solution is mixed with supercritical water fed from another line to heat up rapidly to the reaction temperature and start the reaction. At the exit of the reactor, to quench the reaction, cooling water is directly fed to the line, and a cooling water jacket is used. By this rapid heating method, the effect of the heating-up period on the reaction can be eliminated and thus the specific features of hydrothermal synthesis at the supercritical condition can be clearly discussed. The reaction temperature is in the range 300–400 °C, pressure from 25 to 40 MPa and the residence time of the solution in the reactor is from 0.7 sec to 90 min.

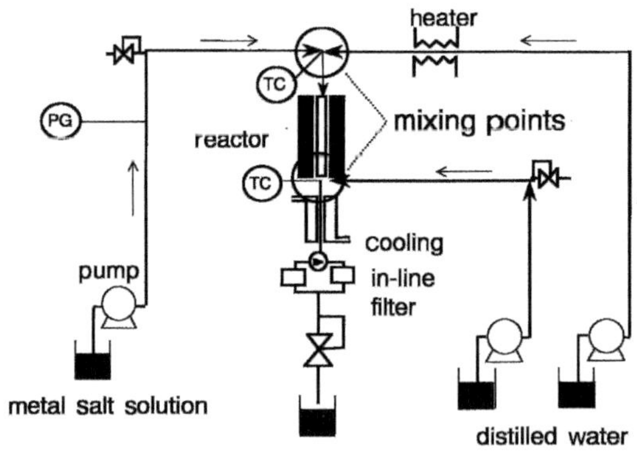

Fig. 5.4.1 Apparatus for supercritical hydrothemal synthesis [7]

Table 5.4.1 Conditions and obtained particles

starting materials	products	particle size (nm)	morphology
$Al(NO_3)_3$	AlOOH	80 – 1,000	hexagonal plate rhombic needle-like
$Fe(NO_3)_3$	α-Fe_2O_3	– 50	spherical
$Fe_2(SO_4)_3$	α-Fe_2O_3	– 50	spherical
$FeCl_2$	α-Fe_2O_3	– 50	spherical
$Fe(NH_4)2H(C_6H_5O_7)_2$	Fe_3O_4	– 50	spherical
$Co(NO_3)_2$	Co_3O_4	– 100	octahedral
$Ni(NO_3)_2$	NiO	– 200	octahedral
$ZrOCl_2$	ZrO_2(cubic)	– 10	spherical
$Ti(SO_4)_2$	TiO_2	– 20	spherical
$TiCl_4$	TiO_2 (anatase)	– 20	spherical
$Ce(NO_3)_3$	CeO_2	20 – 300	octahedral
$Fe(NO_3)_3$, $Ba(NO_3)_2$	$BaO \cdot 6Fe_2O_3$	50 – 1,000	hexagonal plate
$Al(NO_3)_3$, $Y(NO_3)_3$, $TbCl_3$	$Al_5(Y+Tb)_3O_{12}$	20 – 600	dodecahedral

Table 5.4.1 summarizes the metal salt used, the experimental conditions and results [7, 8]. Figures. 5.4.2–5.4.6 are scanning electron microscope (SEM) photographs of the produced metal oxide fine particles. The metals used here are Fe, Al, Ni, Co, Ti, Zr and the anion was NO_3, Cl, SO_4, CO_3, and ammonium citrate. As shown here, this method can be applied for various systems.

(b) Production of Ultra-Fine Particles

As shown in Table 5.4.1 and Figs. 5.4.2–5.4.6, fine particles are obtained in most cases. A series of experiments with CeO_2 formation has been conducted for elucidating the mechanism of fine particle formation [9]. Above the critical temperature (374 °C), conversion of cerium nitrate at the shortest reaction time (0.4–0.6 s) is nearly 100%, while for the lower temperature reactions, more than several seconds are required for 100% conversion. This is probably because of the drastic reduction of dielectric constant above the critical points, which favors the polarity reducing reactions of hydrothermal synthesis (Reactions 5.4.1 and 5.4.2). This implies the hydrolysis reaction is completed and a great amount of monomers are produced within an extremely short time at a mixing tea. Regarding particle growth, particles became larger with increasing residence time at subcritical conditions even after the 100% conversion level,

Fig. 5.4.2 CoO_2 particles obtained by hydrothermal synthesis at the supercritical condition (0.0094M $Co(NO_3)_2$, 673 K, 35MPa) [7]

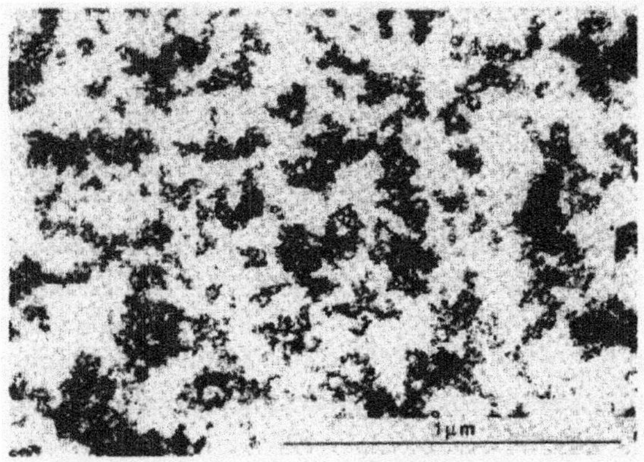

Fig. 5.4.3 ZrO particles obtained by hydrothermal synthesis at the supercritical condition (0.02M $ZrCl_2O$, 673–763 K, 30 MPa) [7]

whereas particles grow little at supercritical conditions. It seems that the precursor, which is soluble in subcritical water but insoluble in room temperature water, is gradually included in the crystals, showing the crystal growth in subcritical water.

Although the detailed mechanism has not been clarified yet, the formation of ultra-fine particles in supercritical water seems to be attributed to the extremely high hydrolysis reaction rate and the low solvent power of supercritical water that most likely leads to a high degree of supersaturation. The mechanism is shown in Fig. 5.4.7.

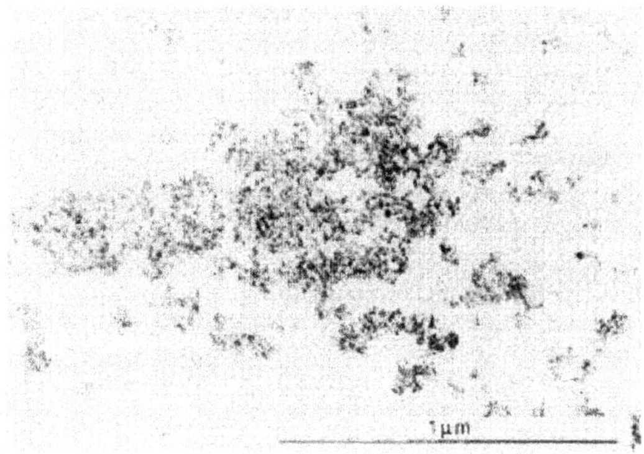

Fig. 5.4.4 TiO_2 particles obtained by hydrothermal synthesis at the supercritical condition (0.038 M $TiCl_4$ at 673–723 K and 30 MPa) [7]

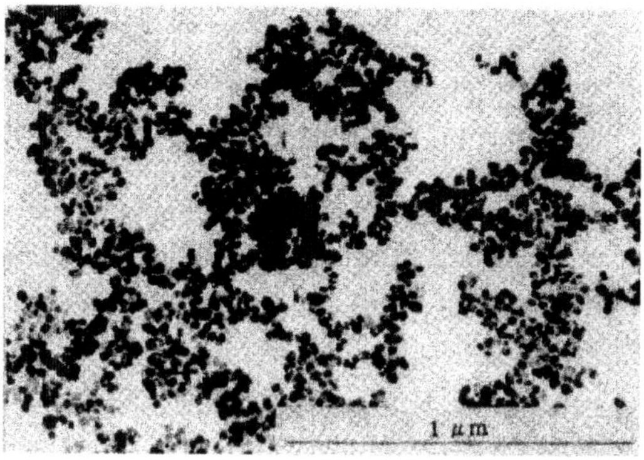

Fig. 5.4.5 Fe_2O_3 particles obtained by hydrothermal synthesis at the supercritical condition (0.016 M $Fe(NO_3)_3$ at 673 K and 35 MPa) [7]

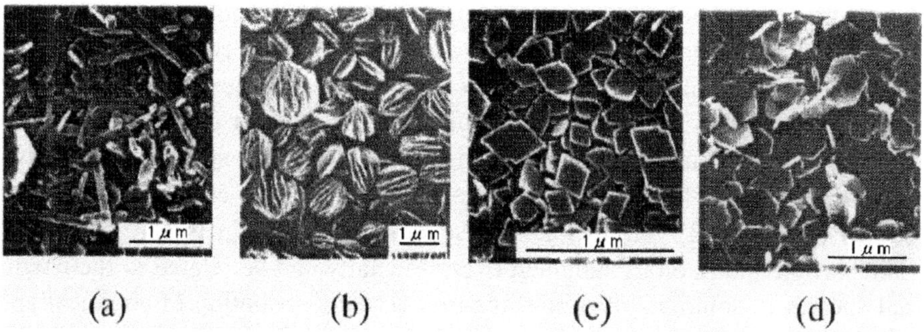

Fig. 5.4.6 AlOOH particles obtained by hydrothermal synthesis around the critical point [8]

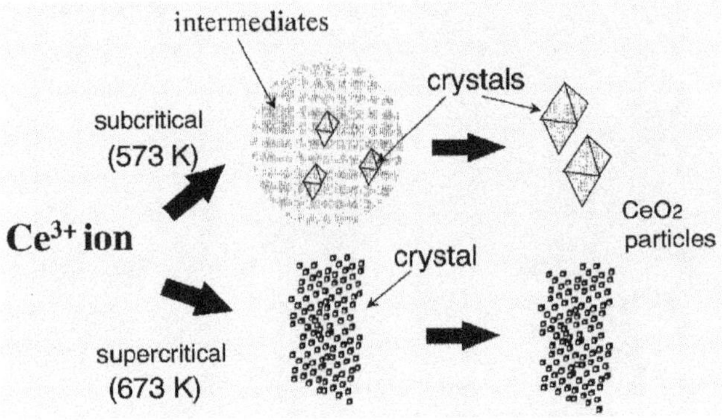

Fig. 5.4.7 Mechanism of ultra-fine particle formation in supercritical water [13]

(c) Control of Oxidizing and Reducing Atmosphere

When ferrous ammonium citrate is used as a feed, magnetite particles are formed. In this case, Fe^{3+} is reduced to Fe^{2+}. This is probably because of the reduction of Fe^{3+} by CO that is formed by the thermal decomposition of ammonium citrate. Since CO is miscible with SCW, a homogeneous reaction atmosphere is generated, which allows the production of magnetite particles in a single phase. This implies that by introducing O_2 or H_2, the reducing and oxidizing atmosphere can be controlled.

(d) Control of Morphology

The morphology of AlOOH particles changes greatly with temperature and pressure around the critical point. The morphology of AlOOH particles is rhombic plate in a lower concentration or at moderate pH, and hexagonal plate or sword-like particles at the higher pH or in higher concentration. Rhombic plate particles are more preferably formed at supercritical conditions.

In hydrothermal synthesis, the crystal habit is generated by the difference of the surface potential energy of each face against chemical species in the solution. Thus in this analysis, the chemical species distribution ($Al(OH)_x^{3-x}$, OH, H, and NO_3) and pH at the reaction condition were estimated using (5.4.3). The results are schematically shown in Fig. 5.4.8 [10]. The particle morphology obtained experimentally can be related to the chemical species in solution. By using this model, the distribution of chemical species for the AlOOH system (Al^{3+}, $Al(OH)^{2+}$, $Al(OH)_2^+$, $Al(OH)_3$, $Al(OH)_4^-$, NO_3^-) in subcritical (350 °C, 30 MPa) and supercritical water (400 °C, 30 MPa) is calculated. Particle morphology seems to be determined by selec-

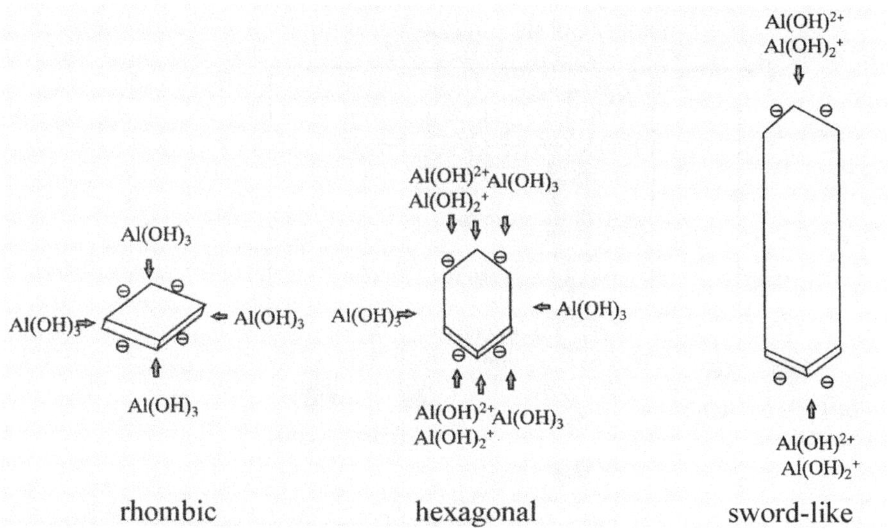

Fig. 5.4.8 Morphology change of AlOOH particles with variation of chemical species distribution [10]

tive adsorption of positively charged species, $Al(OH)_2^+$, on the negatively charged surface of the AlOOH crystal [10].

5.4.3 Applications of Hydrothermal Synthesis in Supercritical Water

(a) Method

The apparatus shown in Fig. 5.4.9 is used for the production of $BaO_6Fe_2O_3$. A barium and iron (III) nitrate aqueous solution is fed to a reactor, mixed first with potassium hydroxide aqueous solution fed from another line and then with high-temperature water at the mixing point, MP2. After this preheating, the solution is mixed at MP3 with supercritical water at 450–500 °C, which rapidly heats the mixture up to the reaction temperature. Similar apparatus is used also for $YOAl_2O_3$ or $LiCoO_2$ production. For $YOAl_2O_3$ production, a mixed aqueous solution of $Y(NO_3)_3$ and $Al(NO_3)_3$ is used as a starting solution, and for $LiCoO_2$ production, a H_2O_2 aqueous solution after heating up to supercritical state is mixed with a starting solution of $LiNO_3$ and $Co(NO_3)_2$, for the oxidation of Co (from Co^{2+} to Co^{3+}). H_2O_2 is decomposed into oxygen gas and water in a preheating region, but it forms a homogeneous phase at supercritical conditions.

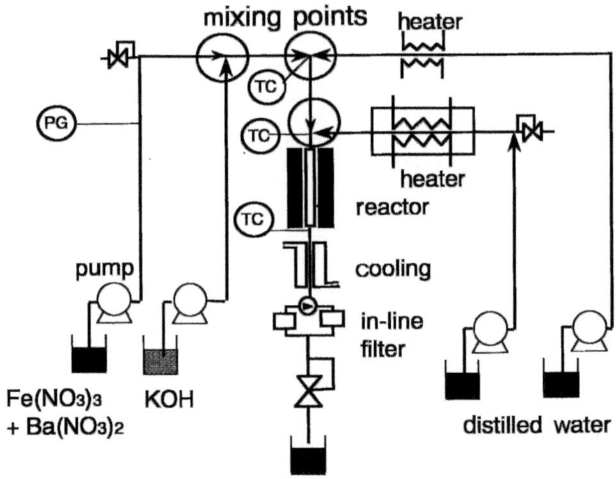

Fig. 5.4.9 Apparatus for continuous production of barium hexaferrite particles [11]

(b) Rapid and Continuous Production of Barium Hexaferrite Fine Particles

A perpendicular magnetizing recording method is used for producing high-density recording media. Barium hexaferrite ($BaO{\cdot}6Fe_2O_3$) is of hexagonal plate and the magnetization takes place perpendicular to the plate plane. However, the control of morphology is difficult task by the solid state method, which is energy consuming and takes a long time.

It was reported that barium hexaferrite hexagonal plate particles could be produced with a reaction time of approximately 1 minute, by the above method. Figure 5.4.10 shows a SEM photo of the obtained particles. The $BaO{\cdot}6Fe_2O_3$ particles obtained have magnetic properties comparable with commercial barium hexaferrite, while those obtained at subcritical conditions have poorer magnetic properties. This might be because the products obtained in subcritical conditions have more hydroxyl functional groups on the surface of the particles than those produced in supercritical water, according to thermogravimetric analysis. This suggests that *in situ* heat treatment has been achieved in hydrothermal synthesis at supercritical conditions.

(c) Production of Phosphor (YAG:Tb) Fine Particles

Rapid and continuous production of phosphor (YAG:Tb) fine particles has been successfully achieved by the proposed method [12]. At temperatures over

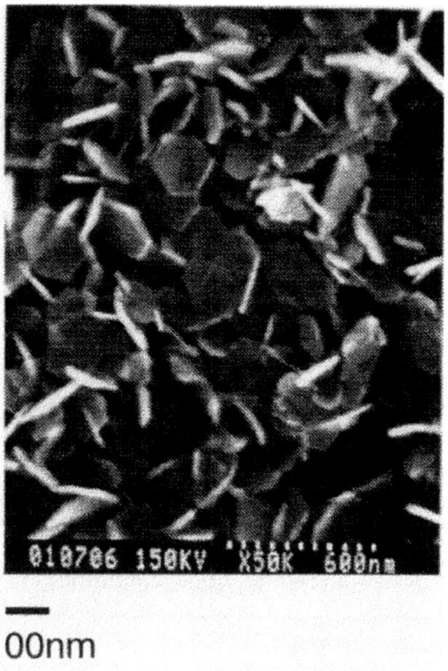

Fig. 5.4.10 Barium hexaferrite particles obtained at supercritical conditions

the critical point, continuous and YAG:Tb fine particles can be obtained in a reaction time less than 60 s. Emission spectra with UV excitation of as-received particles agree with typical spectra pattern of YAG:Tb, even without any heat treatment.

(d) Production of $LiCoO_2$ Fine Particles

Continuous production of $LiCoO_2$ lithium ion battery cathode materials has also been successfully achieved by the proposed method [13]. Figure 5.4.11 shows the XRD patterns for the particles obtained. As shown here, the main products obtained at subcritical conditions were Co_3O_4, which is a mixture of divalent and trivalent Co ions. Thus under these conditions, sufficient oxidation had not been achieved. However, $LiCoO_2$ can be produced in a single phase at supercritical conditions. This is probably because supercritical water and oxygen gas form a homogeneous phase at these conditions. The obtained particles were single crystals and this led to the high stability for the charge–discharge cycles.

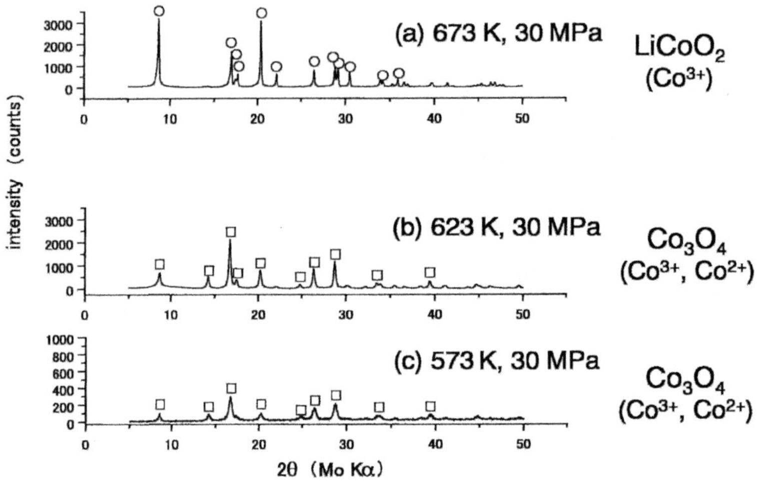

Fig. 5.4.11 XRD paterns of the particle obtained in sub- and supercritical conditions [13]

(e) Remarks

The specific features of hydrothermal synthesis at supercritical conditions are summarized as follows:

(i) Ultra-fine particles can be produced, due to the rapid hydrolysis reaction and the lower solubility of metal oxides in supercritical conditions compared with subcritical conditions.
(ii) The morphology of the produced particles can be controlled.
(iii) O_2, CO or H_2 gases can be used for controlling the oxidizing or reducing atmosphere. This method has been applied for the production of magnetic materials, $BaO_6Fe_2O_3$, phosphor, YO Al_2O_3, and lithium ion battery cathode materials.

References

5.1 Fractionation, Debindering and Drying

[1] V. J. Krukonis, in "*Supercritical Fluid Extraction. Principles and Practice*", ed. M. A. McHugh and V. J. Krukonis, Butterworths, Boston (1986).

[2] S. K. Kumar, S. P. Chhabria, R. C. Reid and U. W. Suter, *Macromolecules*, **20**, 2550 (1987).
[3] I. Yilgor, J. E. McGrath and V. J. Krukonis, *J. Polym. Bull.*, **12**, 491 (1984).
[4] J. J. Watkins, V. J. Krukonis, P. D. Condo, D. Pradhan and P. Ehrlich, *J. Supercrit. Fluids*, **4**, 24 (1991).
[5] M. Daneshvar and E. Gulari, *J. Supercrit. Fluids*, **5**, 143 (1992).
[6] M. A. McHugh and V. J. Krukonis, "*Supercritical Fluid Extraction. Principles and Practice*", Butterworths, Boston (1986).
[7] B. Allen and C. H. Baker, *Polymer*, **6**, 181 (1965).
[8] C. A. Irani, C. Cosewith and S. S. Kasegrande, U.S. Patent 4,319,021 (1982).
[9] M. A. McHugh and T. L. Guckes, *Macromolecules*, **18**, 674 (1985).
[10] A. J. Seckner, A. K. McClellan and M. A. McHugh, *AIChE J.*, **34**, 9 (1988).
[11] S. Saeki, N. Kuwahara, S. Konno and M. Kaneko, *Macromolecules*, **6**, 246 (1973).
[12] B. Bungert, G. Sadowski and W. Arlt, *Fluid Phase Equilibria*, **139**, 349 (1997).
[13] T. M. Martin, A. A. Lateef and C. B. Roberts, *Fluid Phase Equilibria*, **154**, 241 (1999).
[14] H. B. Copelin, US Patent 4,306,058 (1981).
[15] E. F. Barry, P. Ferioli and J. A. Hubball, *J. High Resolut. Chromatogr. Chromatogra. Commun.*, **6**, 172 (1983).
[16] K. Yamamoto and M. Nakae, *Kagaku Kogaku Ronbunshu*, **15**, 673 (1989).
[17] M. Sasaki, S. Takishima and H. Masuoka, *Proceedings of 1st International Symposium on Supercritical Fluids*, Nice, p. 711 (1988).
[18] M. Sasaki, S. Takishima and H. Masuoka, *Sekiyu Gakkaishi*, **32**, 67 (1989).
[19] M. Sasaki, S. Takishima and H. Masuoka, *Sekiyu Gakkaishi*, **33**, 304 (1990).
[20] M. Sasaki, S. Takishima and H. Masuoka, *Kagaku Kogaku Ronbunshu*, **16**, 1020 (1990).
[21] M. Sasaki, S. Takishima and H. Masuoka, *Kagaku Kogaku Ronbunshu*, **16**, 1029 (1990).
[22] J. Swaid and G. M. Schneider, *Ber. Bunsenges. Phys. Chem.*, **83**, 969 (1979).
[23] R. J. Kokes and F. A. Long, *J. Am. Chem. Soc.*, **75**, 6142 (1953).
[24] S. Takishima, K. Nakamura, M. Sasaki and H. Masuoka, *Sekiyu Gakkaishi*, **33**, 332 (1990).
[25] W. J. A. H. Schoeber, Ph. D. thesis, Eindhoven Technical University (1976).
[26] N.-H. Wang, K. Hattori, S. Takishima and H. Masuoka, *Kagaku Kogaku Ronbunshu*, **17**, 1138 (1991).
[27] N. Nakajima, S. Yasuhara and M. Ishihara, US Patent 4,731,208 (1988).
[28] S. Yokawa and N. Nakajima, *Kagaku Kogaku*, **52**, 513 (1988).

[29] E. Nishikawa, N. Wakao, N. Nakajima, R. Sato and S. Yokawa, *Kagaku Kogaku Ronbunshu*, **16**, 647 (1990).
[30] S. Takishima, H. Matsumoto, H. Masuoka, Y. Mukai and T. Sakai, *Kagaku Kogaku Ronbunshu*, **17**, 243 (1991).
[31] S. Takishima, H. Nagasaki, H. Masuoka, Y. Mukai and T. Sakai, "*Solvent Extraction 1990*", ed. T. Sekine, Elsevier, Amsterdam, p. 1719 (1992).
[32] S. Takishima, H. Matsumoto, H. Nagasaki, H. Masuoka, Y. Mukai and T. Sakai, *Kagaku Kogaku Ronbunshu*, **17**, 716 (1991).
[33] T. Chartier, M. Ferrato and J. F. Baumard, *J. Am. Ceram. Soc.*, **78**, 1787 (1995).
[34] A. Miyamoto, H. Nishio and T. Miyashita, *J. Australas. Ceram. Soc.*, **26**, 127 (1990).
[35] Y. Yoshikawa and M. Yoda, *Nippon Seramikkusu Kyokai Gakujutsu Ronbunshi*, **98**, 765 (1990).
[36] L. A. Merz, *54th Annu. Tech. Conf. Soc. Plast. Eng.*, **1**, 612 (1996).
[37] S. S. Kistler, *Nature*, **127**, 741 (1931).
[38] M. Cantin, M. Casse, L. Koch, R. Jouan, P. Mestreau, D. Roussell, F. Bonnin, J. Moutel and S. J. Teichner, *Nucl. Instrum. Methods*, **118**, 177 (1974).
[39] M. Rubin and C. M. Lampert, *Sol. Energy Mater.*, **1**, 393 (1983).
[40] S. J. Teichner, G. A. Nicolaon, M. A. Vicarini and G. E. E. Gardes, *Adv. Colloid Interface Sci.*, **5**, 245 (1976).
[41] A. Livage and J. Lemerle, *Annu. Rev. Mater. Sci.*, **12**, 103 (1982).
[42] H. D. Gesser and P. C. Goswami, *Chem. Rev.*, **89**, 765 (1989).[43] J. F. Poco, P. R. Coronado, R. W. Pekala and L. W. Hrubesh, *Mat. Res. Soc. Symp. Proc.*, **431**, 297 (1996).
[44] P. H. Tewari, A. J. Hunt and K. D. Lofftus, *Mater. Lett.*, **3**, 363 (1985).
[45] K. Ohgaki and T. Katayama, *J. Chem. Eng. Data*, **21**, 53 (1976).
[46] M. J. van Bommel and A. B. de Haan, *J. Non-Cryst. Solids*, **186**, 78 (1995).
[47] K. Tajiri, K. Igarashi and T. Nishio, *J. Non-Cryst. Solids*, **186**, 83 (1995).
[48] Y. Mizushima and M. Hori, *J. Mater. Res.*, **10**, 1424 (1995).
[49] Y. Mizushima and M. Hori, *J. Mater. Sci.*, **30**, 1551 (1955).
[50] M. Goto, Y. Machino and T. Hirose, *Microporous Mater.*, **7**, 41 (1996).
[51] H. Yokogawa, M. Yokoyama, H. Masuoka, S. Takishima and T. Sumiyoshi, *Proceedings of the 16th Japan Symposium on Thermodynamic Properties*, Hiroshima, p. 261 (1995).
[52] H. Yokogawa and M. Yokoyama, *J. Non-Cryst. Solids*, **186**, 23 (1995).
[53] S. Takishima, Y. Suzuki, Y. Konishi, T. Ohmori, Y. Sato and H. Masuoka, *Proceedings of the 4th International Symposium on Supercritical Fluids*, Vol. B, Sendai, p. 267 (1997).

5.2 Crystallization Using a Supercritical Fluid

[1] V. J. Krukonis, *Paper 140f, presented at AIChE Ann. Meeting*, San Francisco, Nov. (1984).
[2] D. W. Matson, J. L. Fulton, R. C. Petersen and R. D. Smith, *Adv. Cer. Mat.*, **1 (3)**, 242 (1986).
[3] D. W. Matson, J. L. Fulton, R. C. Petersen and R. D. Smith, *Ind. Eng. Chem. Res.*, **26**, 228 (1987).
[4] C. Y. Tai and C.-S. Cheng, *Trans. Inst. Chem. Eng.*, Part A, **75** (Feb.), 228 (1997).
[5] J. W. Tom and P. G. Debenedetti, *J. Supercritical Fluids*, **7**, 9 (1994).
[6] E. Reverchon and P. Pallado, *J. Supercritical Fluids*, **9**, 216 (1996).
[7] G.-T. Liu and K. Nagahama, *Ind. Eng. Chem. Res.*, **35**, 4626 (1996).
[8] G.-T. Liu and K. Nagahama, *J. Chem. Eng. Japan.* **30**, 293 (1997).
[9] P. Subra and P. Jestin, *Powder Technol*, **2**, 103 (1999).
[10] P. M. Gallapher, M. P. Coffey, V. J. Krukonis and N. Kulastis, ACS Symposium Series, **406**, 334 (1989).
[11] C. J.Chang and A. D. Randolph, *AIChE J.*, **36**, 939 (1990).
[12] A. Shishikura, H. Takahashi, S. Hirohama and K. Arai, *J. Supercri. Fluids*, **5**, 303 (1992).
[13] O. J. Catchpole, S. Hochmann and S. R. J. Anderson, *"High Pressure Chemical Engineering"*, ed. Ph. Rudolf von Rohr and Ch. Trepp, p. 309 (1996).
[14] C. Y. Tai and C.-S. Cheng, *J. Crystal Growth*, **183**, 622 (1998).
[15] J. W. Mullin, *"Crystallization"*, Butterworth-Heinemann Ltd., Oxford (1993).
[16] M. Kitamura, M. Yamamoto, Y. Yoshinaga and H. Masuoka, *J. Crystal Growth* **178**, 378 (1997).
[17] W. I. Higuchi, P. D. Bernardo and S. C. Mehtha, *J. Pharm. Sci.* **56**, 200 (1967).
[18] M. Kitamura and H. Masuoka, *Proceedings of the 14th International Symposium on Industrial Crystallization*, p.166 (1999).

5.3 Processing of Polymeric Materials Using Supercritical Fluids

[1] D. F. Baldwin, D. E. Tate, C. B. Park, S. W. Cha and N. P. Suh, *J. Jpn. Soc. Polym. Processing (SEIKEI-KAKOU)*, **6**, 187 & 245 (1994).
[2] J. R. Lee and R. G. Kander, "High-temperature Properties and Applications of Polymeric Materials", ACS Symposium Series, 603, ed. M. R. Tant, J. W. Connell and H. L. N. McManus, Chapter 8, American Chemical Society (1995).
[3] D. M. Newitt and K. E. Weale, J. Chem. Soc. (London), 1541 (1948).
[4] J. L. Lundberg, M. B. Wilk and M. J. Huyett, J. Appl. Phys., 31, 1131 (1960).
[5] J. L. Lundberg, M. B. Wilk and M. J. Huyett, J. Polym. Sci., 57, 275 (1962).

[6] J. L. Lundberg, M. B. Wilk and M. J. Huyett, Ind. Eng. Chem. Fundam., 2, 37 (1963).
[7] J. L. Lundberg, E. J. Mooney and C. E. Rogers, J. Polym. Sci., Part A-2, 7, 947 (1969).
[8] P. L. Durrill and R. G. Griskey, AIChE J., 12, 1147 (1966).
[9] P. L. Durrill and R. G. Griskey, AIChE J., 15, 106 (1969).
[10] Y. Kamiya, J. National Institute of Materials and Chemical Research, 5, 121 (1997).
[11] Y. Kamiya, Bulletin of Industrial Products Research Institute, 111, 35 (1988).
[12] Y. Kamiya, Y. Naito, K. Mizoguchi and K. Terada, Kobunshi Ronbunshu, 53, 765 (1996).
[13] Y. Kamiya, T. Hirose, K. Mizoguchi and Y. Naito, J. Polym. Sci. Part B: Polym. Phys., 24, 1525 (1986).
[14] Y. Sato, M. Yurugi, K. Fujiwara, S. Takishima and H. Masuoka, Fluid Phase Equilibria, 125, 129 (1996).
[15] Y. Sato, K. Fujiwara, T. Takikawa, Sumarno, S. Takishima and H. Masuoka, Fluid Phase Equilibria, 162, 261 (1999).
[16] Y. Sato, Y. Yamasaki, S. Takishima and H. Masuoka: J. Appl. Polym. Sci., 66, 141 (1997).
[17] I. C. Sanchez and R. H. Lacombe, J. Phys. Chem., 80, 2352 (1976).
[18] I. C. Sanchez and R. H. Lacombe, Macromolecules, 11, 1145 (1978).
[19] C. Zhong and H. Masuoka, Ind. Eng. Chem. Res., 36, 2509 (1997).
[20] W. H. Burgess, H. B. Hopfenberg and V. T. Stannett, J. Macromol. Sci. B5, 23 (1971).
[21] Y. Sato, T. Iketani, S. Takishima and H. Masuoka, Polym. Eng. Sci., 40, 1369 (2000).
[22] R. A. Gorski, R. B. Ramsey and T. Dishart, J. of Cellular Plastics, 22, 21 (1986).
[23] J. S. Chiou, J. W. Barlow and D. R. Paul, J. Appl. Polym. Sci., 30, 2633 (1985).
[24] J. S. Chiou, J. W. Barlow and D. R. Paul, J. Appl. Polym. Sci., 30, 3911 (1985).
[25] K. Mizoguchi, T. Hirose, Y. Naito and Y. Kamiya, Polymer, 28, 1298 (1987).
[26] T. Mizumoto, N. Sugimura, M. Moritani, Y. Sato and H. Masuoka, "Macromolecules 1999", Bath, UK (1999).
[27] M. Takada, S. Takeda, M. Tanigaki and M. Ohshima, *The Polymer Processing Society 15th Annual Meeting*, S. Hertogenbosch, The Netherlands (1999).
[28] M. Shimbo, D. F. Baldwin and N. P. Suh, *J. Jpn. Soc. Polym. Processing (SEIKEI-KAKOU)*, **6**, 863 (1994).
[29] Sumarno, K. Hayashi, Y. Sato, S. Takishima and H. Masuoka, *Proceedings of the 5th Meeting on Supercritical Fluids Materials and Natural Products Processing, Tome 1: Materials*, Nice, France, pp. 33–38 (1998).
[30] Sumarno, Y. Sato, S. Takishima and H. Masuoka, J. Appl. Polym. Sci., 77, 2383 (2000).

[31] Sumarno, T. Sunada, Y. Sato, S. Takishima and H. Masuoka, Polym. Eng. Sci., 40, 1510 (2000).
[32] K. Inamori, K. Hayama, M. Takeda, M. Ohshima and M. Tanigaki, J. Jpn. Soc. Polym. Processing (SEIKEI-KAKOU), 11. 195 (1999).
[33] K. Inamori, K. Hayama, M. Ohshima and M. Tanigaki, J. Jpn. Soc. Polym. Processing (SEIKEI-KAKOU), 11. 471 (1999).
[34] T. Chikatsune, Y. Sato, S. Takishima, H. Masuoka, H. Yamamoto and M. Takasugi, Society of Chem. Eng. Japan Tokushima Meeting, Tokushima, Japan (1999).
[35] T. Chikatsune, K. Hayashi, Sumarno, Y. Sato, S. Takishima, H. Masuoka, H. Yamamoto and M. Takasugi, "*Symposium on Polymer Processing 1998*, Osaka, Japan (1998).

5.4 Hydrothermal Synthesis of Metal Oxide Fine Particles at Supercritical Conditions

[1] K. Ohno and T. Abe, *J. Electrochem. Soc.*, **133(3)**, 638 (1986).
[2] W. F. van der Weg, J. A. Popma, A. T. Vink, *J. Appl. Phys.*, **57(12)**, 5450 (1985).
[3] K. M. Kinsman and J. McKittrick, *J. Am. Ceram. Soc.*, **77(11)**, 2866 (1994).
[4] R. P. Rao, *J. Electrochem. Soc.*, **143(1)**, 189 (1996).
[5] O. A. Lopez, J. McKittrick and L. E. Shea, *J. Luminescence*, **71**, 1 (1997).
[6] T. Takamori and L. D. David, *J. Ceram. Soc. Bull.*, **65(9)**, 1282 (1986).
[7] T. Adschiri, K. Kanazawa and K. Arai, *J. Amer. Ceram. Soc.*, **75(4)**, 1019 (1992).
[8] T. Adschiri, K. Kanazawa and K. Arai, *J. Amer. Ceram. Soc.*, **75(9)**, 2615 (1992).
[9] Y. Hakuta, T. Adschiri and K. Arai, *J. Mat. Sci. Let.*, **17**, 1211 (1998).
[10] Y. Hakuta, T. Adschiri, H. Hirakoso and K. Arai, *Fluid Phase Equilibria*, **158–160**, 733 (1999).
[11] Y. Hakuta, T. Adschiri, T. Suzuki, T. Chida, K. Seino and K. Arai, *J. Amer. Ceram. Soc.*, **81(9)**, 2461 (1998).
[12] Y. Hakuta, K. Seino, H. Ura, T. Adschiri, H. Takizawa, K. Arai, *J. Mat. Chem.*, **9(10)**, 2671 (1999).
[13] T. Adschiri, Y. Hakuta and K. Arai, Ind. Eng. Chem. Res. **39**, 4901 (2000).
[14] K. Sue, Y. Hakuta, R. L. Smith, T. Adschiri and K. Arai, *J. Chem. Eng. Data*, **44**, 1422 (1999).

6 Reactions in Supercritical Fluids

6.1 Reactions and Reaction Theory in Supercritical Fluids

in Supercritical Fluids

6.1.1 Introduction

In the development of technology for the chemical industry, solvent selection is of great importance, but as solvents are generally chosen from those that are liquid at atmospheric pressure, the selection of solvents has been very limited. Of particular interest in recent years have been the attempts at replacing liquid organic solvents with supercritical fluids (SCFs), since the physical properties of these solvents can be broadly controlled with temperature and pressure.

Beginning in the 1980s, attention focused on the controllability of the dissolution strength and phase equilibrium of SCFs, and researchers have worked on their development as new extraction and separation solvents [1, 2]. In particular, supercritical carbon dioxide ($scCO_2$) is nontoxic, odorless, and inexpensive, and also has a critical temperature that is near room temperature. It is therefore capable of separation operations at low temperatures and can be used in the separation of substances that are thermally labile. It also promises high selectivity in separation, and there are few problems with post-separation residual solvent. For these reasons, development has proceeded on $scCO_2$ extraction processes chiefly in the areas of food and pharmaceuticals, as is discussed in Chap. 4.

The controllability of physical characteristics of SCFs is not limited to extraction and separation solvents, but also offers promise for reaction solvents, and recently research and development are proceeding in these applications [3–5]. This section will explain the characteristics of reactions in SCFs. First an explanation is given of the physical characteristics and solvent characteristics of SCFs, and this is followed by an explanation of the principal and specific features of the reactions in SCFs.

6.1.2 Characteristics of Reactions in Supercritical Fluids

(a) Supercritical Fluids

Whether a fluid is in a gaseous phase or liquid phase depends on the kinetic energy of its molecules and the energy of the intermolecular forces. When the intermolecular forces dominate to a greater extent the molecular kinetic energy, a fluid will be a liquid with a certain measure of order, but when kinetic energy is dominant, it will be a gas with a random structure. Kinetic energy grows as temperature rises, and the shorter the distance between molecules (i.e., the higher the concentration), the more dominant the intermolecular forces will become. Thus, in general, a liquid will become a gas when its temperature is raised, and a gas will liquefy when it is compressed. However, above a certain temperature, molecular kinetic energy is greater no matter what the distance between the molecules, and the substance will become a noncondensable fluid. This temperature, which is called the critical temperature, is unique to each substance, and a fluid at or above its critical temperature is an SCF.

Because the state of molecules near the critical temperature is characterized by the balance between the formation of system order by intermolecular force and dissipation by molecular kinetic energy, on a microscopic level this is a state in which the molecules are changing places very actively while maintaining a certain extent of order, and rather small changes in temperature and pressure will bring about major changes in average fluid density and cluster structure.

All physical properties of a fluid depend on its average structure and dynamical structure. Important properties for a reaction solvent are density, viscosity, diffusivity, thermal conductivity, and various solvent properties such as dielectric constant, ionic product, and hydrogen bonding. Because the density of an SCF is between that of a gas and a liquid, the averages of the above mentioned properties are also usually between those of the gas and liquid. This fact is sometimes very important in considering the characteristics of the reaction in SCFs. Some examples of solvent properties are shown in Appendices A and B.

As already mentioned, the properties of a fluid depend on its average structure and dynamical structure. Because these structures change greatly near the critical point, various physical characteristics also undergo great changes in conjunction with them. Graphical representations of the temperature and/or pressure dependence of the relevant properties of three important SCFs, i.e. water, carbon dioxide and methanol, will be presented in Sect. 6.2. Here only the $P\rho T$ relation and the dielectric constant dependence on temperature and pressure are presented in Figs. 6.1.1 [6] and 6.1.2 [7], respectively. These figures show that near the critical temperature, the physical properties change greatly. The characteristics of the reaction solvent change greatly in accordance with the change of physical properties. As a result, reactions differ greatly with the change of temperature and pressure of the SCFs, as will be exemplified in Sects. 6.3 and 6.4.

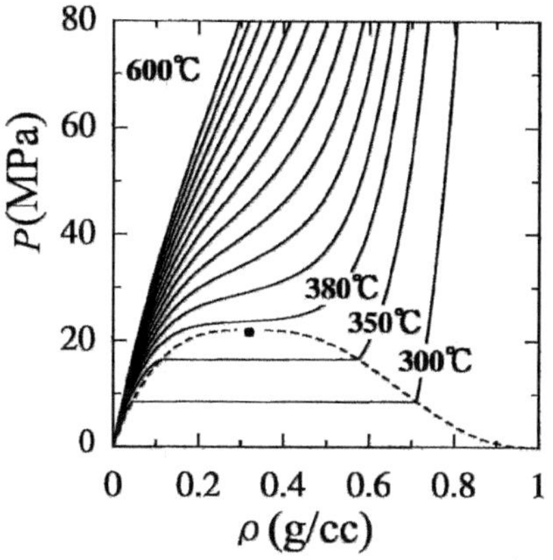

Fig. 6.1.1 $P\rho T$ relationship of water [6]

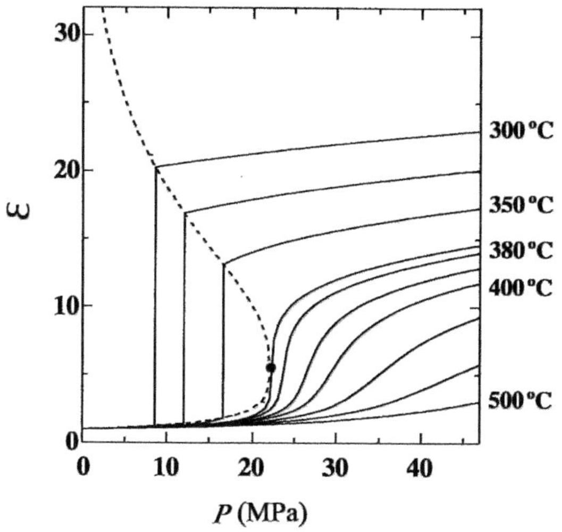

Fig. 6.1.2 Water dielectric constant dependence on temperature and pressure [7]

(b) Phase Behavior

Actual fluids are multicomponent systems, so to determine the states of such systems necessitates considering also the energy of interaction between molecules of different kinds. When an SCF is involved, the interaction between molecules of different kinds also changes greatly near the critical point, and as a result the phase behavior also changes greatly.

The phase behavior of a two-component systems is summarized into six types, as is explained in Sect. 2.2 (see Fig. 2.2.3). Here, we will explain one example which is relevant to supercritical water (SCW) reactions. The phase behavior of a (water + gas) system and a (water + hydrocarbon) system near the critical points is shown in Figs. 6.1.3 [8] and 6.1.4 [9], respectively. In the higher-temperature regions (right-hand side), components mix uniformly at arbitrary compositions. At room temperature, forcing a gas into water will produce bubbles, but SCW, which is a high-density vapor, will completely mix with a gas. On the other hand, water and hydrocarbons in the low-temperature range will refuse to mix and undergo phase separation, but when water is supercritical they mix uniformly. This mutual miscibility with hydrocarbons is understandable when considering that the dielectric constant of supercritical water is about 2–10 and is approximately equivalent to that of a polar organic solvent as will be explained in Sect. 6.2 in more detail. The homogeneous mixing of solvent and reactant is very important for a reaction to proceed without any mass transport limitation. This merit is indeed adopted in various SCF reactions such as supercritical water oxidation (SCWO).

6.1.3 Reaction Mechanism and Reaction Rate in Supercritical Fluids

(a) Determining Factors of the Reaction Rate

Important reaction classes from mechanistic viewpoints are bimolecular and unimolecular reactions. The active species through which the reaction proceeds are produced through the encounter of reactant molecules in the case of bimolecular reactions. The diffusion process is a prerequisite for the encounter. If the diffusion does not proceed rapidly enough, the overall rate is determined by the diffusion, but not by the reaction itself. For the formation of activated species from which the unimolecular reaction proceeds, the intermolecular diffusion to allow for the encounter between the reactant molecule and third-body molecules, and a certain kind of friction within the solvent cage which induces intramolecular energy transfer, are both important. These physical processes are incorporated in the relevant Kramer's turnover theory. However, the Kramer's turnover is studied in detail only in a limited number of unimolecular reactions, specifically in photochemical reactions, and thus the theory will not be explained in this section.

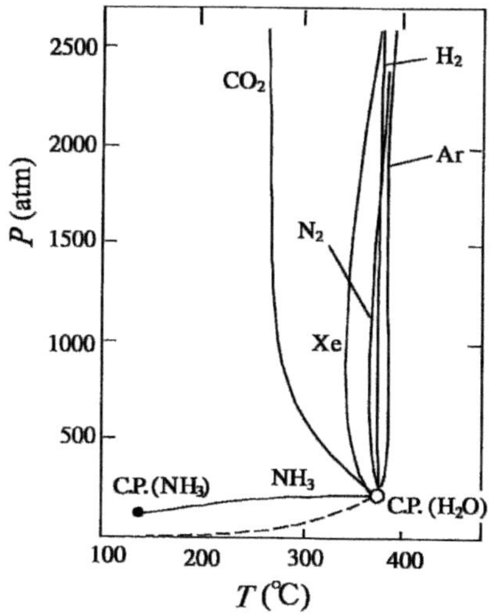

Fig. 6.1.3 Phase behavior of water + gas systems [8]

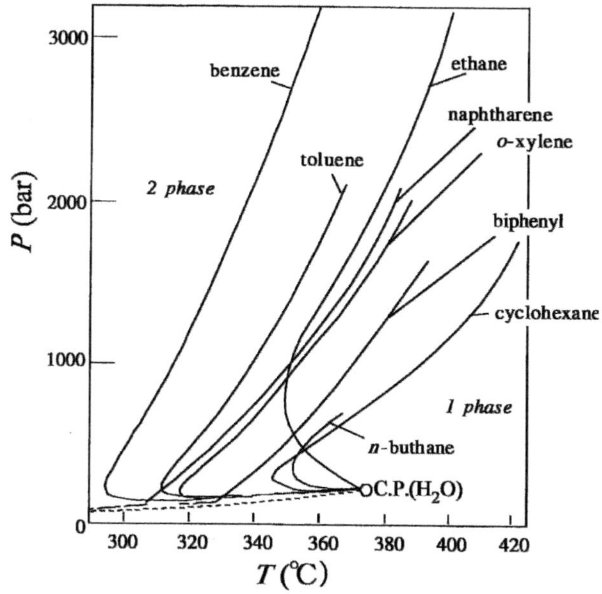

Fig. 6.1.4 Phase behavior of water + hydrocarbon systems [9]

(b) Diffusion Limited Reactions

In this section, the rate of diffusion will be briefly described. Consider that molecule B is transported through diffusion to a distance r from molecule A. In the case of steady-state diffusion, the diffusion flux J_B is described as

$$J_B = -4\pi r^2 (D_A + D_B) \frac{d[B]_r}{dr}, \tag{6.1.1}$$

due to Fick's law. The sign of J_B is taken negative because the flux is directed towards the origin from infinity. Here D_A and D_B are the diffusion constants of the individual molecules in the solvent. $[B]_r$ is the concentration of B molecules at r. Integrating the above equation while noting that the concentration of B at infinity is denoted as $[B]$,

$$[B]_r = \frac{J_B}{4\pi r(D_A + D_B)} + [B] \tag{6.1.2}$$

is obtained. If the chemical reaction between A and B proceeds at infinite rate when $r = R$, the steady-state diffusion flux is derived as

$$J_B = -4\pi (D_A + D_B) R[B]. \tag{6.1.3}$$

Thus the equivalent reaction rate constant k_D for the diffusion process is

$$k_D = 4\pi (D_A + D_B) R. \tag{6.1.4}$$

The diffusion constants are frequently estimated from the viscosity data of the solvent using the Stokes–Einstein law. For a spherical molecule of radius r,

$$D = \frac{k_B T}{6\pi r \eta(T)}, \tag{6.1.5}$$

where k_B is the Boltzmann constant, and $\eta(T)$ is the viscosity coefficient at absolute temperature T. When the encounter radius R is equated to $r_A + r_B$, and $r_A = r_B$ is assumed, the diffusion limited rate constant is

$$k_D = \frac{8 k_B T}{3\eta(T)}. \tag{6.1.6}$$

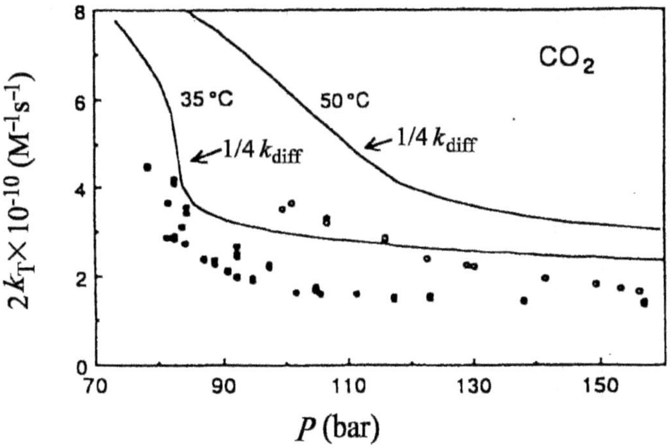

Fig. 6.1.5 Recombination rate constants of benzyl radicals and diffusion rate constants divided by 4 in CO_2 solvent [10]

If the bimolecular reaction rates exceed the diffusion rates, the overall reaction rates are determined by the diffusion rates. Though the viscosity coefficient of SCFs is usually much smaller than those of ordinary liquid solvents, some reactions such as radical–radical recombination reactions, which proceed without any appreciable activation energy, are indeed diffusion-limited, and the overall rates are specifically dependent on the density change of the SCFs.

Reactions such as triplet–triplet annihilation and radical recombination reactions are frequently limited by diffusion process. Figure 6.1.5 shows the comparison of the diffusion rate constants divided by 4 for the correction of the spin statistics of radical recombination reactions and the measured recombination rate constants of benzyl radicals at two temperatures in $scCO_2$ [10]. The diffusion rate constants are obtained using (6.1.6) based on the measured viscosity data. They increase very rapidly when approaching the critical pressure. Though the recombination rate constants are slightly below 1/4 of k_D, the relatively good agreement between them shows that the benzyl radical recombination reaction occurs near the spin-statistically allowed diffusion-controlled limit at all solvent densities studied.

The encounter of the radicals is suppressed by the diffusion resistance. On the contrary, the mutual diffusion of fragments such as those produced through photodissociation is frequently suppressed by the surrounding solvents, which is called the cage effect. The cage effect is important in the understanding of reaction progress in liquids. SCFs have a different cage effect from liquids, which will be discussed in Sect. 6.4.2 with various examples.

(c) Reaction Rate in Terms of Transition State Theory

In most reactions other than radical–radical reactions, the reaction itself has a relatively large activation energy, and thus is not diffusion-limited. In such reactions, the rate constant can be described in terms of a transition state theory. The difference in potential between the transition state and the reactants is the activation energy. When there is a solvent, the states (chemical potential) of the reactants and the transition state change, and as a result the activation energy will also change. Thus we can easily understand that the reaction rate changes by the solvent.

How can we evaluate the change of the chemical potential of a given substance according to the solvent? For a reaction, A + B = P, in general, the reaction rate can be expressed as

$$\text{rate} = k[\mathrm{A}][\mathrm{B}]. \tag{6.1.7}$$

According to Laidler [11], the transition state theory expresses the reaction rate constant as

$$\ln k = \ln\left(\kappa \frac{k_{\mathrm{B}}T}{h} K_0^{\,*}\right) + \ln \frac{f_{\mathrm{A}} f_{\mathrm{B}}}{f_{\mathrm{X}}}, \tag{6.1.8}$$

where f_{A} and f_{B} are the fugacity coefficients for components A and B, h is Planck's constant, and $K_0^{\,*}$ is an equilibrium constant for A, B and the transition state. The first term is for the reaction rate of the gas phase and can be evaluated. The second term is the difference from the intrinsic rate, and is estimated taking into account the effect of the dielectric constant ε of the medium and other nonelectrostatic effects, as will be shown below. The fugacity coefficient can be split into several fugacity coefficients related to the dielectric constant of the solvent, ionic strength, and so on.

One of the main factors for the solvent effect is the electrostatic interaction. For the case of dipolar substances, Kirkwood's expression for the free energy of transfer of a strong dipole moment μ from a vacuum to a medium of dielectric constant ε can be employed.

$$RT \ln f = -\frac{C\mu_{\mathrm{A}}^{\,2}}{a^3}\left(\frac{\varepsilon - 1}{2\varepsilon + 1}\right), \tag{6.1.9}$$

where a is the molecular radius and C is a constant. Amis et al. [12] proposed a similar equation with a different function of dielectric constant:

$$RT \ln f = -\frac{C' \mu_A^{\ 2}}{a^3}\left(1 - \frac{1}{\varepsilon}\right), \tag{6.1.10}$$

where C' is a constant. At a higher dielectric constant ($\varepsilon > 4$), namely at $P >$ 200 bar at 400 °C, the Kirkwood term has a similar function to the Amis although C' must be $C/2$.
For the ion–ion reaction in the same way, we obtain

$$RT \ln f = -\frac{C'' z^2 e^2}{r}\left(1 - \frac{1}{\varepsilon}\right), \tag{6.1.11}$$

where C'' is a constant, r is the radius of the ion and z is the ionic charge. As shown in Fig. 6.1.2, the dielectric constant will change greatly near the critical point due to slight changes in temperature and pressure, showing that the fugacity coefficient, namely the reaction rate, will change greatly.

Another solvent property that can influence reactions involving ions or polar molecules is the ionic strength I,

$$I = \frac{1}{2}\sum_i z_i^2 C_i\ , \tag{6.1.12}$$

where C_i is the concentration of ion i. The Debye–Hückel expression [13] can be used to quantify the influence of ionic strength on a chemical reaction involving ions at low concentrations.

$$\ln f_i = -Q z_i^2 \sqrt{I} + b_i I\ , \tag{6.1.13}$$

where

$$Q = \frac{N_A^{\ 2} e^3 (2\pi)^{3/2}}{(\varepsilon k_B T)^{3/2} (1000)^{1/2}}\ . \tag{6.1.14}$$

Here, N_A is the Avogardo number, and the term $b_i I$ in (6.1.13) is a correction factor.

Thus, based on the above information on the solvent effect, the reaction rate can be expressed as

$$\ln k = \ln k_0$$
$$+\left[\frac{C'}{RT}\left(\frac{\mu_A^2}{a_A^3}+\frac{\mu_B^2}{a_B^3}-\frac{\mu_{\neq}^2}{a_{\neq}^3}\right)+\frac{C''e^2}{RT}\left(\frac{z_A^2}{r_A}+\frac{z_B^2}{r_B}-\frac{(z_A+z_B)^2}{r_{\neq}}\right)\right]\left(\frac{1}{\varepsilon}-1\right)$$
$$+2Qz_A z_B\sqrt{I}+(b_A+b_B-b_{\neq})I\,, \qquad (6.1.15)$$

where the first term of the right-hand side is for the gas phase reaction rate, the second term is the electrostatic solvent effect, the third term is for the effect of ionic strength and the last term is the other solvent effects including the effect of nonelectrostatic interactions and rearrangement of the solvent.

Recently, some researchers have suggested that the local dielectric constant around the substance is different from the bulk one, around the critical point. One means of probing solvent effects on chemical reaction is the use of the spectral shift (see Sect. 1.2). Because of solute–solvent interactions, a solute's electronic energy levels can be affected by the solvent. Since the ground and the excited state have a unique electron distribution individually, the electrostatic solvent effect on these states is different. Thus the absorption or emission spectrum deduced from the band gap is shifted with the dielectric constant of the field. The observed spectrum shift cannot be explained by the bulk dielectric constant and a local dielectric constant due to the solvation has been suggested. Thus around the critical point, it is essential to know how to estimate the local dielectric constant. Some examples of the local density augmentation will be shown in Sects. 6.2–6.4.

One of the explanations of the pressure effect in reaction rate theory derives the following equation, which is sometimes adopted in order to indicate the remarkable features of the supercritical solvent:

$$\left(\frac{\partial \ln k_x}{\partial P}\right)_T = -\frac{\Delta v^{\neq}}{RT}\,, \qquad (6.1.16)$$

where $\Delta v^{\neq}$ is the activation volume, and is the difference between the partial molar volumes, $\bar{v}$, of the activated complex and the reactants,

$$\Delta v^{\neq} = v^{\neq} - a\bar{v}_A - b\bar{v}_B\,. \qquad (6.1.17)$$

The activation volume in a supercritical phase near the critical point, unlike that in a liquid, can change markedly with pressure, since the dielectric constant or other fluid properties changes greatly around the critical point and also the local dielectric constant changes more significantly around the critical

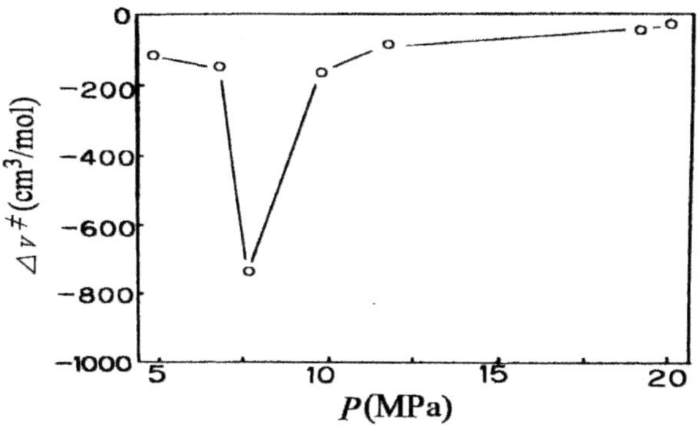

Fig. 6.1.6 Activation volume of Diels–Alder reaction between isoprene and methyl acrylate in CO_2 solvent [14]

point. Diels–Alder reactions frequently show a large activation volume change near the critical points of solvents, and one such example for the reaction between isoprene and methyl acrylate [14] is shown in Fig. 6.1.6. Some other related reactions will be described in Sect. 6.4. At the same time, it is worth noting that if the activation volume analysis is conducted on the apparent reaction rate constant, the evaluated constant includes the effects of pressure on diffusion, electrostatic interactions, the compressibility of the fluid phase, and phase behavior.

The foregoing suggests that the reaction rate changes greatly near the critical point, and that it is possible to control it by manipulating pressure and temperature. Some examples of the kinetic features of SCF reactions will be shown in the following sections.

6.2 Characteristics of Supercritical Fluids as Reaction Solvents

as Reaction Solvents

6.2.1 Introduction

The nature of a supercritical solvent and how the reaction progress is affected by the medium have been discussed in Sect. 6.1. Most chemical substances have their own critical points and can be used as reaction media. However, from the practical viewpoint, in particular for realizing environment-friendly reaction processes, the most important solvents are carbon dioxide and water. They are nontoxic and easily available. A variety of solvent properties are realized by these two solvents. Methanol may be added to the above two solvents as a third. The vapor pressure and the critical point of each fluid are shown in Fig. 6.2.1.

In this section, the solvent properties and their influence on some reactions will be presented. First, the features of supercritical fluids (SCFs) as reaction media from the scientific and technological viewpoints are summarized.

Scientific Viewpoints

(i) Solvent properties such as density, viscosity, diffusivity and polarity are usually between the corresponding gas and liquid. The range of these properties is shown in Appendix B. High mass transfer is realized by the low viscosity and high diffusivity of SCFs compared to liquids. High heat transfer is also achieved by the large thermal conductivity of SCFs compared to liquids.

(ii) The above solvent properties are changeable over a wide range by manipulating the temperature and pressure or by the addition of a small amount of cosolvent. The most appropriate conditions may be selected by changing the temperature and pressure or selecting the cosolvent, using only one solvent.

(iii) SCFs can dissolve inorganic gases or moreover can sometimes make a homogeneous phase with various species.

(iv) A relatively weak molecular association compared with that in the liquid phase is very desirable in some organic reactions.

(v) Clusters, in which a reactant is surrounded by a lot of supercritical solvent molecules, are frequently formed near the critical point, which accelerate the reaction rate and change the reaction selectivity in a very specific manner.

Technological Viewpoints

(vi) Integration of a reaction with a separation process is realized using SCFs. As a simple example, an actually solvent-free reaction process is realized in the case of reactions in supercritical carbon dioxide (scCO_2), because CO_2 goes away after reducing the pressure.

Because the above three SCFs have quite different properties from each other, they are applied to different reactions, in particular, in practical usage. The applicability of the three SCFs to chemical reactions are illustrated in Fig. 6.2.2. Typical reactions are as follows. First, scCO_2 with a low critical temperature of 30.9 °C is used for organic synthesis and organic material processing. It offers an organic solvent-free reaction process and can give abundant and safe

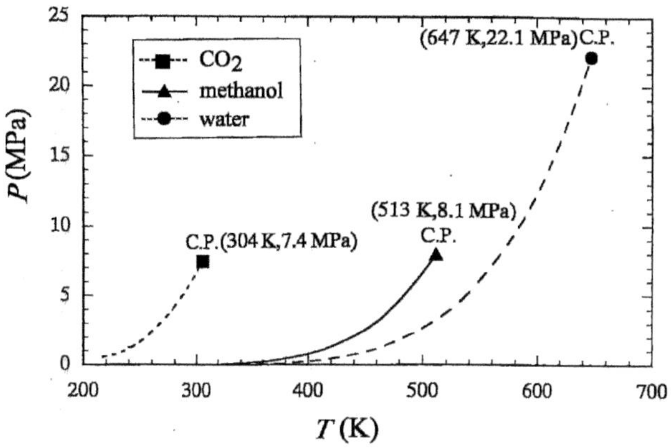

Fig. 6.2.1 Critical points and vapor pressures of CO_2, water and methanol

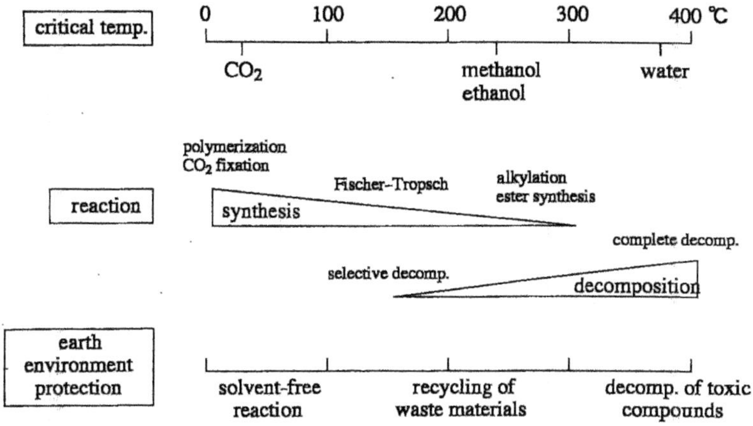

Fig. 6.2.2 Applicability of SCFs to chemical reactions

carbon resources. Secondly, supercritical water (SCW) with a high critical temperature (374.1 °C) and pressure (22.12 MPa) is suitable for waste treatment. At the same time, a lot of interesting reactions are found to proceed in SCW. The third, supercritical methanol ($scCH_3OH$) with intermediate critical temperature (239.4 °C) and pressure (8.09 MPa) between carbon dioxide and water, is effective for chemical recycling of some condensation polymers such as polyethylene terephthalate.

6.2.2 Supercritical Carbon Dioxide

$ScCO_2$ reactions have been studied both from the scientific and technological viewpoints. Firstly, the critical temperature of CO_2 is near ambient conditions and thus research as a typical SCF is easily carried out. Secondly, CO_2 is an environment-friendly solvent which may be a potential alternative of organic solvents. At the same time, in some reactions, CO_2 can be used as both a reaction medium and reactant in reactions such as hydrogenation of $scCO_2$ to formic acid and the efficient production of carbonates from $scCO_2$.

The dependence of the pressure and temperature on the density is presented in Fig. 6.2.3. A large number of experimental data of phase equilibrium and solibility for mixtures including $scCO_2$ and organic solutes are published. Solvatochromic studies (see Sect. 1.2) have found that the dipolarity and polarizability of $scCO_2$, as measured by the Kamlet–Taft π^* parameter, is slightly lower (0.1 at *ca.* 16 MPa and 313.2 K) than that of alkanes (0.0). Similarly the polarity, measured on the E_T (30) scale, is 31–33, comparable to the value of 31 for alkanes. $ScCO_2$ can dissolve a wide range of organic compounds including fluoropolymers which are soluble only in chlorofluorocarbons [1]. Permanent gases such as H_2 are sparingly soluble in liquid solvents, but they are highly miscible with $scCO_2$. For example, the concentration of H_2 in a supercritical mixture of 8.6 MPa of H_2 and 12.2 MPa of CO_2 at 323.2 K is 3.2 M, while the concentration of H_2 in tetrahydrofurane (THF) under the same pressure is merely 0.4 M [2]. For many organic compounds, the solubility limit can be raised by adding a small amount of a volatile cosolvent, usually a polar or protic compound such as small alcohols, acetone or amines. Water as a cosolvent of $scCO_2$ dramatically changes the performance of the solvent.

In conclusion, $scCO_2$ is an attractive solvent for organic synthesis (homogeneous reactions and catalytic reactions) and organic material processing. This is because the solvent property of $scCO_2$ is similar to that of a liquid nonpolar solvent such as hexane. It can dissolve nonpolar or slightly polar compounds with relatively small molecular weight. Unique phenomena such as the dramatic increase in the reaction selectivity is also frequently observed in the close vicinity of the critical point.

6.2.3 Supercritical Water

Water has its critical point at a very high temperature (647.3 K) and pressure (22.12 MPa), as shown in the $P\rho T$ diagram in Fig. 6.1.1. Sometimes, sub- and supercritical water are very corrosive. These facts have caused a delay in research using SCW until now. However, the properties of sub- and supercritical water are very specific and attractive, and nowadays much research is being done.

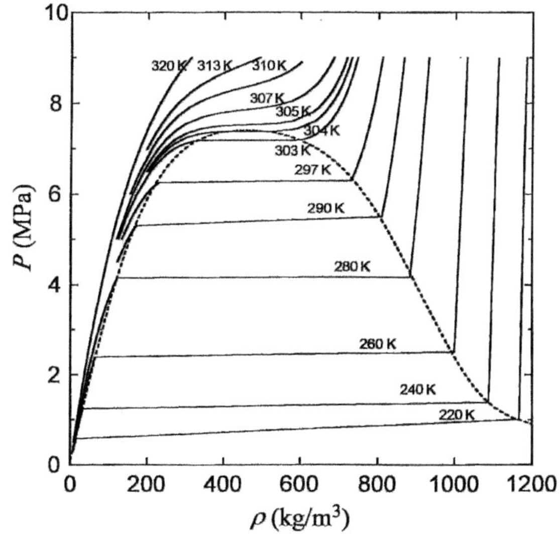

Fig. 6.2.3 $P\rho T$ relationship of CO_2

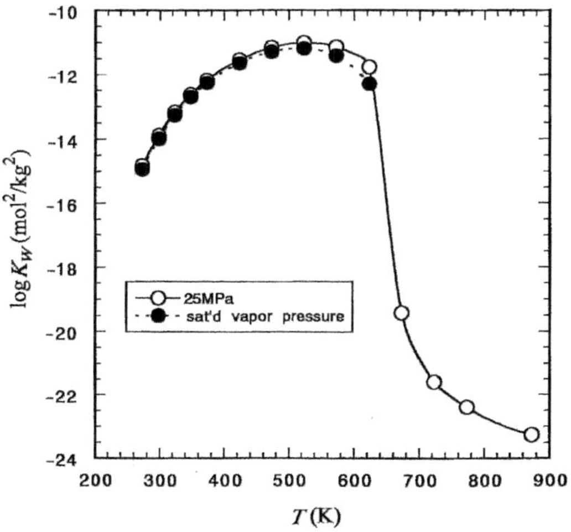

Fig. 6.2.4 Ion product of water (K_W) from liquid to supercritical state

According to the study of Uematsu and Frank [3], the static dielectric constant ε of water changes from 78.5 at ambient temperature and pressure to 5.9 at 673.2 K, 30 MPa, as already shown in Fig. 6.1.2. For the ion product of water (K_W) Marshall and Franck [4] reported, as shown in Fig. 6.2.4, that the value of K_W at a constant pressure of 25 MPa first increases with temperature until K_W reaches maximum (10^{-11}) at 523.2 K. K_W then decreases to 10^{-19} at 673.2 K and 10^{-22} at 773.2 K. The changes in these physical properties of water lead to the complete miscibility of nonpolar organics under supercritical conditions. This phase behavior was already described in Figs. 6.1.3 and 6.1.4. Due to the large change of values of ε and K_W for water with temperature and sometimes with pressure, the character of the reaction may change from that of ionic reactions to that of radical reactions with temperature and pressure changes. Now, various kinds of reactions are studied, among which are rearrangement, elimination, hydrogenation/dehydrogenation (related to the water–gas shift reaction), C–C bond formation, hydration/dehydration, partial oxidation, and complete oxidation. For practical applications, much attention is paid to complete oxidation and depolymerization of some plastics through hydrolysis. These two types of reactions will be discussed in more detail in Sect. 6.3.

6.2.4 Supercritical Methanol

The third, $scCH_3OH$ with intermediate critical temperature (512.6 K) and pressure (8.09 MPa) between carbon dioxide and water, is used for both decomposition and synthesis of organic compounds. The solvent property of $scCH_3OH$ is not so different from that of SCW, but $scCH_3OH$ can make the reaction conditions much milder than SCW by its lower critical temperature and pressure. The alcoholysis is effective for chemical recycling of some condensation polymers such as polyethylene terephthalate. Furthermore the high reactivity is attractive to build up an environment-friendly alkylation process without any acid. Some specific solvent properties of methanol will be explained below.

Figure 6.2.5 shows the temperature and pressure dependence of the Kamlet–Taft scale of solvent property π^* of CH_3OH at 313–453 K [5]. The solvent polarity decreases dramatically with increase in the temperature. For example, at a temperature higher than 453 K the π^* of methanol is getting close to that for nonassociated solvents such as diethyl ether and diisopropyl ether at ambient conditions. The winding chains of sequentially hydrogen-bonded methanol molecules are considerably broken at this temperature. This result is supported by NMR [6], and Raman spectroscopic studies [7]. It is reported that around 70% of the total hydrogen bonding between methanol molecules is broken at the critical point, and more than 90% disappears at the supercritical condition of 573 K and 10 MPa.

The main application of $scCH_3OH$ is solvolysis to decompose organic compounds such as waste plastics and biomass. For example, polyethylene terephthalate (PET) and polyehtylene naphthalate (PEN) are easily depoly-

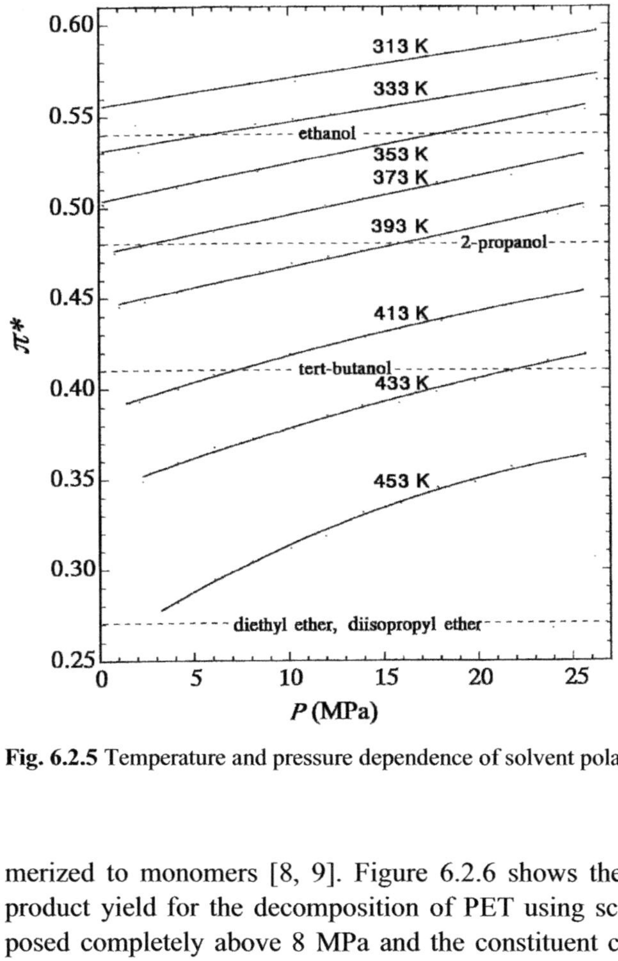

Fig. 6.2.5 Temperature and pressure dependence of solvent polarity π^* of $scCH_3OH$

merized to monomers [8, 9]. Figure 6.2.6 shows the pressure effect on the product yield for the decomposition of PET using $scCH_3OH$. It was decomposed completely above 8 MPa and the constituent components were recovered as either monomers and 1:1 oligomers.

A new alkylation technique using supercritical alcohol has been developed recently. $ScCH_3OH$ is used not only as a reaction solvent but also as an alkylation reagent. Methanol is a preferable alkylation alternative for conventional hazardous reagents such as dimethylsulfate, methyl halide and toluene. A typical example is the alkylation of hydroquinone to produce methylhydroquinone shown in Fig. 6.2.7 [10]. Compared with conventional Friedel–Crafts alkylation, the new alkylation using $scCH_3OH$ does not require an acid catalyst. It is remarkable that no polyalkylated compounds were detected using $scCH_3OH$, while the Friedel–Crafts reaction gives a complex product mixture of mono- and polyalkylated compounds. As shown in Fig. 6.2.7, moreover, etherification of phenol also occurs easily by $scCH_3OH$ without catalyst.

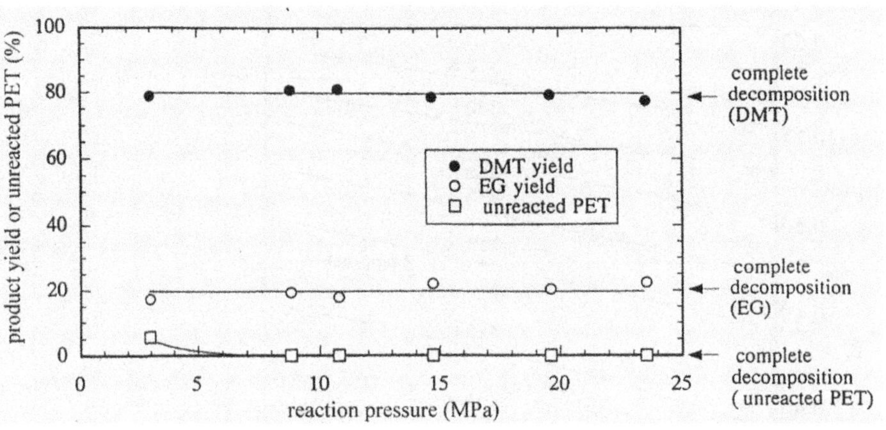

Fig. 6.2.6 Pressure effect on product yields for PET decomposition with $scCH_3OH$ at 573 K and 30 min in reaction time

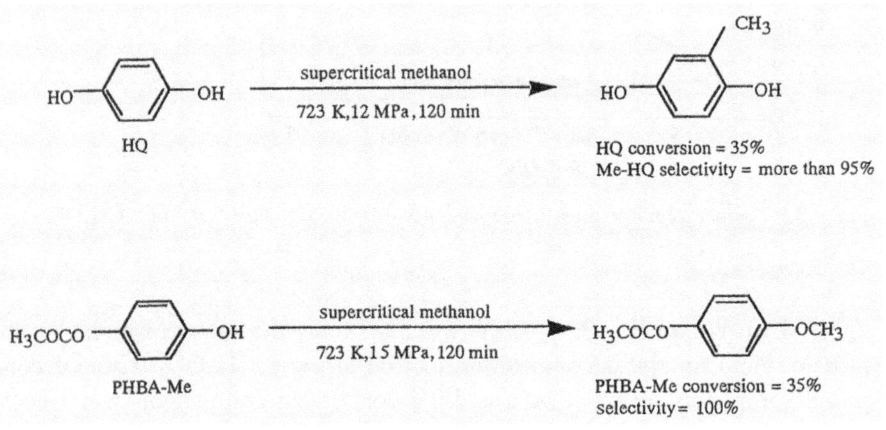

Fig. 6.2.7 Chemical reactions with $scCH_3OH$

6.3 Oxidation and Hydrolysis Reactions in Supercritical Water

in Supercritical Water

6.3.1 Oxidation and Related Reactions

Supercritical water oxidation (SCWO) is a technique for the treatment of hazardous materials in water at temperatures and pressures above the critical point of water. High destruction efficiencies of this process have been established for a wide variety of organic materials, and most of the organic compounds have been found to be oxidized within reasonable residence times. The complete oxidation of organic compounds in supercritical water has received much attention as a promising technology for waste treatment, not only because of the higher reaction rates compared to the conventional "wet oxidation" process, but also because of its unique physical and chemical properties as a solvent. In this section, we discuss the fundamental chemistry of SCWO from the engineering point of view by reviewing the previous studies on the reaction kinetics and mechanism in the oxidation of inorganic and organic compounds at supercritical conditions.

(a) Characteristics of Supercritical Water Oxidation

One advantage of SCWO for the treatment of waste is that the decomposition rates are extremely large for most organic compounds. The processing time for the complete conversion is on the order of seconds or minutes except for stable compounds such as ammonia and acids, which need higher temperatures and longer processing times. This phenomenon is related to the unique physical properties of SCW, some of which were already explained in Sect. 6.2.3. Due to the small values of dielectric constant and ion product for water at supercritical conditions, ionic reactions are suppressed, and the solvent medium favors radical reactions.

Besides the high temperature, which promotes faster rates for each elementary reaction, the advantage of the SCWO technique is that reactions of organic compounds with oxygen take place in a single phase. Conventional wet oxidation, which is operated below the critical temperature and pressure of water, involves more than two phases, where the mass transport at the phase boundary could be rate-limiting. In SCW, on the contrary, organics and water are miscible, and the solubility of inorganic gases such as oxygen and air in water is also very high. Therefore, there are no interphase transport limitations in SCWO. Higher concentrations of reactants in water are also expected, which may lead to the enhancement of the overall efficiency of the degradation process.

(b) Global Reaction Rates in Supercritical Water Oxidation

The global rates of the oxidation reactions give useful and important information for process design, and help the better understanding of global reaction mechanisms. To obtain accurate global kinetics, a large amount of experimental data under different reaction conditions should be accumulated and analyzed statistically.

There have been many studies on the global reaction rates of SCWO for various compounds, most of which adopted the following equation based on the global rate law,

$$d[S]/dt = -k[S]^a[O_2]^b, \tag{6.3.1}$$

where [S] denotes the concentration of compound to be oxidized. The global rate constant k has an Arrhenius-type temperature dependence as follows,

$$k = A\exp(-E_a / RT), \tag{6.3.2}$$

where A, E_a, R and T denote the pre-exponential factor, activation energy, gas constant and temperature, respectively. The reported values of reaction orders and kinetic parameters for several compounds are summarized in Table 6.3.1. The reaction order for most substrates is close to unity, whereas that for oxygen varies from zero to 0.5. The reaction order with respect to oxygen in wet oxidation is almost unity, which suggests that the dependence of the decomposition rate on the oxygen concentration is smaller at supercritical conditions. In the following sections, the previous studies of the global kinetics for several model compounds are introduced.

Table 6.3.1 Reported values of reaction orders and kinetic parameters in the SCWO of several compounds

compounds	reaction orders		$\log A$ [1]	activation energy (kJ/mol)	ref.
	fuel	oxygen			
hydrogen	1.10 ± 0.25	0.02 ± 0.29	24.4 ± 4.9	390 ± 60	[2]
carbon monoxide	0.96 ± 0.30	0.34 ± 0.24	8.5 ± 3.3	134 ± 32	[1]
methane	0.99 ± 0.08	0.66 ± 0.14	11.4 ± 1.1	42.8 ± 4.3	[3]
methane	1.84 ± 0.084	-0.06 ± 0.065	17.1 ± 0.49	60 ± 1.4	[4]
methanol	0.89 ± 0.69	0.12 ± 0.66	28.8 ± 10.5	107 ± 30	[5]
acetic acid	0.72 ± 0.15	0.27 ± 0.15	9.9 ± 1.7	168 ± 21	3)
phenol [2]	0.85 ± 0.04	0.50±0.15	2.34 ± 0.28	51.8 ± 4.2	4)
phenol	1.041 ± 0.002	0.381 ± 0.003	8.69 ± 0.006	124.776 ± 0.093	[9]

1) The reaction order of water (0.42 ±0.05) is also taken into consideration.
2) The unit of A changes according to the reaction orders of fuel and oxygen.
3) J. C. Meyer, P. A. Marrone and J. W. Tester, *AIChE J.*, 41(9), 2108 (1995).
4) S. Gopalan and P. E. Savage, *AIChE J.*, 41(8), 1864 (1995).

Carbon Monoxide and Hydrogen

Carbon monoxide and hydrogen are the compounds that have been selected as a model compound for the study of kinetics and mechanism in SCWO. The SCWO of these compounds have been investigated because they are the simplest of the reactive compounds, and its oxidation in the gas phase has already been extensively investigated.

Information about the SCWO of carbon monoxide also helps understanding the SCWO of larger organic compounds, as carbon monoxide is frequently an important intermediate in the complete oxidation of those carbon compounds. Holgate et al. [1] reported that, in the SCWO of carbon monoxide, there are two global reaction pathways in the formation of carbon dioxide; one is a direct oxidation (reaction 6.3.3) and the other is a water–gas shift reaction (reaction 6.3.4).

$$CO + (1/2)O_2 = CO_2 \qquad \text{(direct oxidation)} \qquad (6.3.3)$$
$$CO + H_2O = CO_2 + H_2 \qquad \text{(shift reaction)} \qquad (6.3.4)$$

They concluded that these two kinetic pathways are separable, and that the rate of the direct oxidation rises more rapidly and its relative contribution increases with increase of temperature, as shown in Fig. 6.3.1 [1].

The kinetic study of hydrogen oxidation is important, not only because hydrogen could be formed as an intermediate product in a reaction such as the water–gas shift reaction, but because hydrogen is a good candidate for an auxiliary fuel in SCWO processes. Holgate and Tester [2] determined the kinetics of hydrogen oxidation in SCW, where the reaction rate is proportional to the hydrogen concentration and independent of the oxygen concentration. They also observed the apparent induction time in the hydrogen oxidation reaction, as shown in Fig. 6.3.2, though the physical reason for the induction time was not specified.

Methane and Methanol

The kinetics of the oxidation of methane have been examined by Webley and Tester [3], and Steeper et al. [4]. As shown in Table 6.3.1, the global rate expressions obtained in their work do not necessarily agree well, mainly because of the difference in the reaction conditions such as temperature, pressure, and methane concentration. Steeper et al. [4] reported the first Raman spectroscopic measurement of reaction products, which suggested the capability of *in situ* analysis of the reaction intermediate. This novel technique provided a methodological improvement in the analysis of the reaction progress in SCWO.

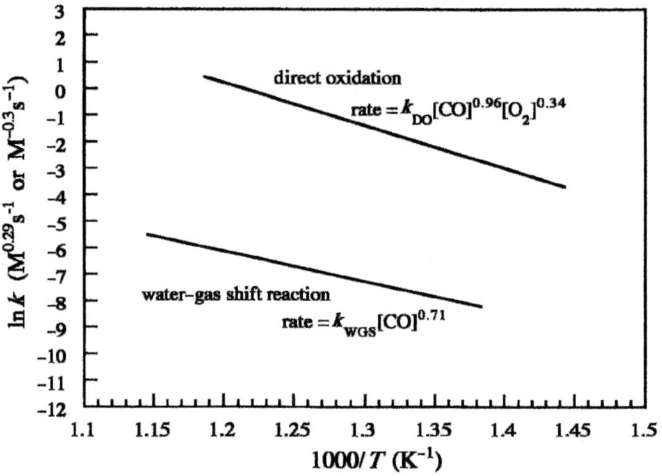

Fig. 6.3.1 Arrhenius plots for direct oxidation and water-gas shift reaction in the SCWO of carbon monoxide [1] Reprinted from: Carbon monoxide oxidation in supercritical water: H. R. Holgate, P. A. Webley, J. W. Tester and R. K. Helling, *Energy and Fuels* (American Chemical Society) **6**, 586 (1992).

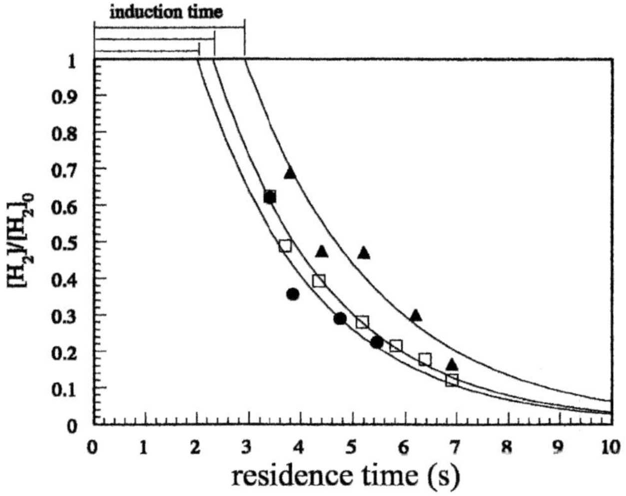

Fig. 6.3.2 Normalized hydrogen decay profiles in the SCWO of hydrogen at 550 °C [2]. $[H_2O]_0 = 4.25\times10^{-3}$ mol/cm^3. ▲: $[H_2]_0 = 3.06\times10^{-6}$, $[O_2]_0 = 1.55\times10^{-6}$ mol/cm^3; □: $[H_2]_0 = 2.06\times10^{-6}$, $[O_2]_0 = 1.04\times10^{-6}$ mol/cm^3; ●: $[H_2]_0 = 1.065\times10^{-6}$, $[O_2]_0 = 0.54\times10^{-6}$ mol/cm^3. Curves are exponential fits to data

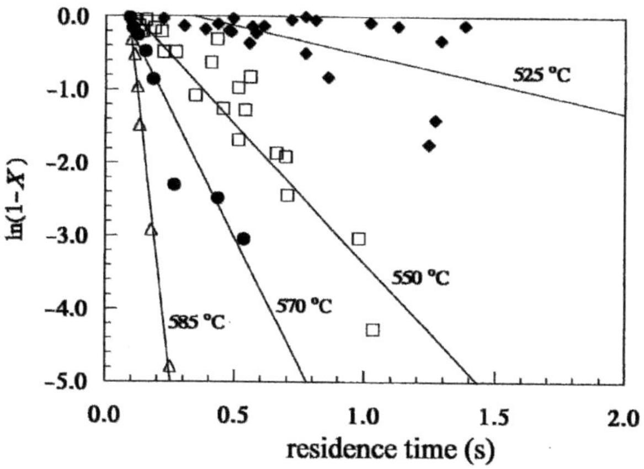

Fig. 6.3.3 First-order plot for methanol disappearance kinetics [6]. The lines represent fits of the experimental data to $\ln(1-X) = -k(t-t_{\text{ind}})$ at 525 °C (◆), 550 °C (□), 570 °C (●), and 585 °C (△)

For the SCWO of methanol, Tester et al. [5] reported the global rate expression, where the reaction rate is proportional to the methanol concentration and independent of the oxygen concentration. Brock et al. [6] employed a simple reaction model, in which methanol is oxidized consecutively to carbon dioxide *via* formaldehyde and carbon monoxide, and clarified the dependence of the pseudo-first-order rate constant of each step on the reaction temperature. Brock et al. confirmed that this reaction has an induction period whose length decreases as the temperature increases, as shown in Fig. 6.3.3. A similar trend has been reported by Rice et al. [7], where the oxidation rate of methanol as well as the subsequent production and destruction of the primary product, formaldehyde, were investigated using *in situ* Raman spectroscopic analysis. These experimental results suggest that the apparent "time lag" in the beginning of the reaction is not due to a physical reason such as mixing time, but is due to the chemical kinetics in the radical chain reaction.

Phenol

Since phenol is a ubiquitous pollutant in industrial waste, several investigations on the kinetics of phenol oxidation in SCW have been published recently [8–11]. But the global kinetics obtained in these studies do not necessarily agree well, especially in terms of the dependence on the water concentration. The discrepancies of the rate expressions may be caused by the difference in the reaction regimes, such as temperature, pressure, and reactant concentrations. When dealing with a relatively large compound as a reactant, the reaction mechanism becomes complicated, thus, the number of intermediate products and species taken into account should also increase. Therefore, in the case of such a large molecule as phenol, the destruction rate of the reactant is more likely affected by these intermediate species than the case of smaller molecules. This can lead to the variation of the reaction kinetics between different reaction regimes. The discrepancies in the previous rate expressions imply that there might be a methodological limitation in the application of global kinetic analysis for the data taken at a wide range of reaction conditions with a single rate expression, especially when the reactant has a large molecular weight and a complicated structure.

(c) Reaction Mechanism and Elementary Reaction Modeling

It is recognized that SCWO is a radical-chain reaction and consists of many elementary reactions, as in combustion chemistry. In order to understand the complete mechanism, it is necessary to build a reaction model in which all relevant elementary reactions are involved. Several researchers have attempted to develop a detailed radical mechanism in SCWO using elementary reaction models. There are some difficulties, however, in the application of this methodology for systems involving larger molecules, because of the complexity of the reaction mechanism. Therefore, successful models have been provided only for relatively simple molecules, such as CO, H_2, methane, and methanol. Tester and coworkers [12] examined the modeling for CO and H_2, where the experimental data were fairly well reproduced. Brock and Savage [13] proposed a single reaction mechanism for H_2 and C_1 compounds that involves 22 species and 148 reversible reactions. The kinetic parameters used in these models were derived from the values in combustion chemistry with some modifications based on theoretical reasons. The resulting models reproduce the experimental data taken at higher temperature well, though the accuracy of the model prediction decreases as the temperature becomes closer to the critical temperature. Table 6.3.2 shows the comparison of the experimental reaction orders with the predicted orders for several compounds. It was found from the sensitivity analysis that there are fewer than 20 elementary reactions which control the SCWO kinetics for the disappearance of H_2, CO, methane, and methanol. In the SCWO of methanol, Brock et al. [6] proposed a reaction network, as shown in Fig. 6.3.4, where OH and HO_2 radicals play important roles for the radical-chain mechanism.

Table 6.3.2 Comparison of the experimental reaction orders with the reaction orders predicted by the model in the SCWO of several compounds [13]

compounds	T (K)	reaction conditions: initial concentration (mol/cm³) $[H_2O]_0$	$[fuel]_0$	$[O_2]_0$	results: experiments fuel	experiments oxygen	model fuel	model oxygen
hydrogen	823	4.28×10^{-3}	1.50×10^{-6}	1.00×10^{-6}	1.10 ± 0.25	0.02 ± 0.29	0.86 ± 0.13	0.00 ± 0.10
carbon monoxide	770	4.93×10^{-3}	1.50×10^{-6}	1.50×10^{-6}	0.96 ± 0.30	0.34 ± 0.24	0.86 ± 0.04	0.02 ± 0.05
methane	873	3.85×10^{-3}	3.00×10^{-6}	4.00×10^{-6}	0.99 ± 0.08	0.66 ± 0.14	1.09 ± 0.10	0.34 ± 0.14
methanol	766	5.00×10^{-3}	3.00×10^{-6}	3.00×10^{-6}	0.89 ± 0.69	0.12 ± 0.66	1.23 ± 0.11	0.00 ± 0.05

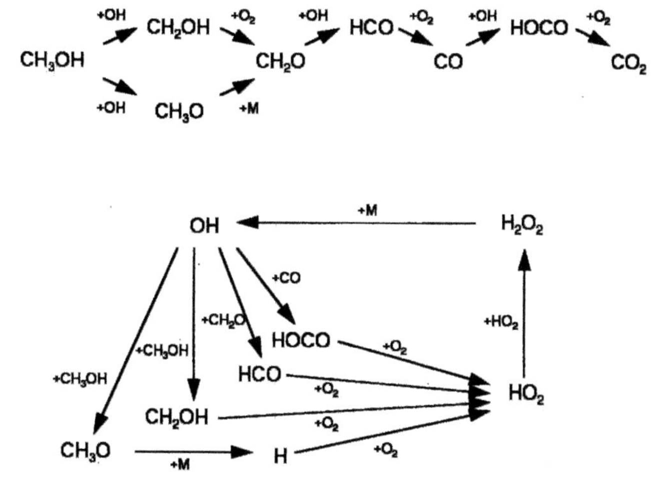

Fig. 6.3.4 Radical reaction network in the SCWO of methanol [6]

These fundamental models for the analysis of simple molecules, however, have not been easily extended to the oxidation of more complicated compounds. An elementary reaction model was proposed for the phenol oxidation in SCW [14], though it could not satisfactorily predict the kinetics of phenol SCWO. Boock and Klein proposed detailed mechanisms for the oxidation of C_1–C_3 alcohols and acetic acid [15], and mixtures [16] in high-temperature water. In this work, they adopted a mechanism-based but philosophically different way of modeling called the "lumping strategy", where each of the reactions is classified into one of eight reaction categories as shown in Fig. 6.3.5. Basically, the elementary reactions in the same reaction family are assumed to have a single pre-exponential factor, because these reactions are considered to go through a similar transition state, and the Polanyi structure/reactivity relationships are

initiation			
1	hydrogen abstraction	$RH + O_2$	$\rightarrow R\cdot + HO_2\cdot$
propagation			
2	oxygen addition	$R\cdot + O_2$	$\rightarrow RO_2\cdot$
3	H-abstraction	$RO_2\cdot + RH$	$\rightarrow ROOH$
4	isomerization	$RO_2\cdot$	$\rightarrow HOOR'\cdot$
5	β-scission	$HOOR'\cdot$	$\rightarrow HO\cdot + O{=}R''$
chain transfer			
6	disproportionation	$2RO_2\cdot$	$\rightarrow O_2 + 2RO\cdot$
branching			
7	decomposition	$ROOH$	$\rightarrow RO\cdot + HO\cdot$
termination			
8	radical recombination	$RO_2\cdot + RO_2\cdot$	$\rightarrow$ products

Fig. 6.3.5 Reaction families in the lumping strategy for supercritical water oxidation [15]

used for the estimation of the activation energies. The resulting models provide good fits to the experimental data, which suggests the validity of this methodology as a conventional technique for the modeling of the reaction mechanism.

(d) Technical Applications of Supercritical Water Oxidation to the Decomposition of Hazardous Materials

The decomposition of dioxins (molecular structures are shown in Fig. 6.3.6) and PCBs is an example of a practical application of SCWO to the treatment of hazardous chemicals. These chlorinated aromatic compounds are known to be toxic and decomposition-resistant organic compounds. Now, much attention is being given to SCWO as a reliable and economical decomposition technique. The experimental results given in Table 6.3.3 [17] show that this method can decompose dioxins very efficiently. The kinetics and reaction mechanism can be discussed by consulting the reactions of simple model compounds such as phenol.

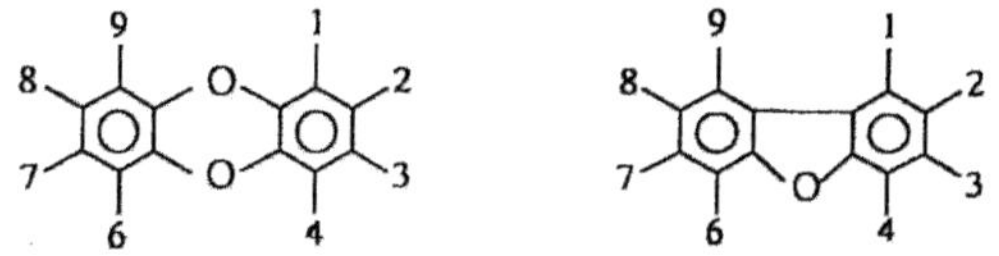

polychlorinated dibenzo-p-dioxin (PCDD)

polychlorinated dibenzofuran (PCDF)

Fig. 6.3.6 Molecular structures of dioxins

Table 6.3.3 Decomposition of dioxins with supercritical water + oxidizer

	charged fly ash	R-1	R-2
charged fly ash (g)	–	2.00	2.00
charged water (g)	–	7.33	7.54
oxidizer		0.1 MPa air	0.5MPa O_2
T_4CDD (ng/g)	2.0	1.2	0.27
P_5CDD (ng/g)	5.9	0.25	0.79
H_6CDD (ng/g)	12.1	0.12	0.10
H_7CDD (ng/g)	34	0.035	0.15
O_8CDD (ng/g)	38	0.085	0.04
total PCDD (ng/g)	92	1.69	1.41
decomp. yield (%)	–	98.2	98.5
T_4CDF (ng/g)	5.5	2.64	0.79
P_5CDF (ng/g)	8.7	0.29	0.30
H_6CDF (ng/g)	2.8	0.20	0.28
H_7CDF (ng/g)	38	0	0.05
O_8CDF (ng/g)	37	0.015	0
total PCDF (ng/g)	92	3.14	1.42
decomp. yield (%)	–	96.6	98.5
sum of PCDD + PCDF (ng/g)	184	4.83	2.83
total decomp. yield (%)	–	97.4	98.5

conditions: temperature, 673 K; pressure, 30 MPa; reaction time 30 min.

total decomposition yield (%) = {1–(residue of PCDD+PCDF)/(charged PCDD+PCDF)} x 100
abbreviations

T_4CDD = tetrachlorodibenzo-*p*-dioxin
P_5CDD = pentachlorodibenzo-*p*-dioxin
H_6CDD = hexachlorodibenzo-*p*-dioxin
H_7CDD = heptachlorodibenzo-*p*-dioxin
O_8CDD = octachlorodibenzo-*p*-dioxin

T_4CDF = tetrachlorodibenzofuran
P_5CDF = pentachlorodibenzofuran
H_6CDF = hexachlorobenzofuran
H_7CDF = heptachlorodibenzofuran
O_8CDF = octachlorodibenzofuran

(e) Future Scope of Supercritical Water Oxidation

As mentioned in the previous sections, SCWO is a promising technique for waste treatment, and the applicability of this technique to industrial processes has already been proved for some specified cases. In contrast to these application-oriented studies, however, the fundamentals of the kinetics in SCWO have yet to be completely clarified. For instance, the specific reason for the induction period observed in the SCWO of H_2 and methanol has not been given, and the effect of water density on the reaction rate is another kinetic issue yet to be fully discussed. Also, more investigations on heterogeneous reactions, such as wall effects, catalysis, and corrosion of the reactor, will be required for the design and development of SCWO processes. In order to establish this technique as a novel and conventional method for waste treatment, much more intensive efforts should be put into fundamental studies of the chemistry in SCF, as well as the accumulation of reaction engineering data for a variety of chemical compounds.

6.3.2 Hydrolysis Reactions

(a) Introduction

One reaction of great importance in high-temperature water or SCW is noncatalytic hydrolysis. Townsend et al. [18] conducted a basic study on the decomposition of coal-related model compounds and reported that in high-temperature water, hydrolysis takes place without any acid or base catalyst. Harrell et al. [19] elucidated the mechanism of noncatalytic hydrolysis of benzonitrile in high-temperature water.

Various applications of noncatalytic hydrolysis are expected, one of which is the hydrolysis of condensation polymers to monomers. Nowadays, significant waste problems exist in the world and thus the development of new chemical recycling processes is needed. Judging from the above studies, there appears to be considerable opportunity for recovering monomers from the condensation polymers with hydrothermal techniques. Recently, Adschiri et al. have proposed new processes of polymer hydrolysis to recover monomers using hydrolysis in SCW [20–24]. This section first explains the mechanisms of hydrolysis in high-temperature water, and then highlights the key features of hydrolysis of cellulose, polyethylene terephthalate (PET) and bisphenol A (BPA) tar processes that recover the monomers in SCW.

(b) Mechanism of Hydrolysis in High-Temperature Water

What is the reason that hydrolysis takes place in high temperature water without a catalyst? One possible explanation may be the increased concentration of $[H^+]$ and $[OH^-]$ in high-temperature water, as suggested by the increase in the ion product of water. This change of the concentration of $[H^+]$ and $[OH^-]$ influences the acid $[H^+]$ or base $[OH^-]$ catalyzed hydrolysis.

Recently, Harrell et al. [19] studied the mechanism of non-catalytic hydrolysis of benzonitrile in high-temperature water in detail, focusing the point of the study on the determination of the attacking nucleophile (H_2O, H^+, or OH^-) and the substrate (RC = N, RC = NH_3^- or RC = NH^+). The results of the Hammett plot analysis for the rate dependence on substitute suggested that nucleophilic attack takes place at the nitrile carbon (RC* = N) on the rate determining step. The pressure dependence on the hydrolysis rate in SCW was evaluated and the results were compared with that of the ion product of water, $K_W^{1/2} = [H^+] = [OH^-]$. The acid or base catalyzed hydrolysis rate should be linearly proportional to the concentration of $[H^+]$ or $[OH^-]$. Solvent isotope kinetic analysis has also been introduced for determining the attacking nucleophile. Their results clearly indicated that the predominant nucleophile in high-temperature water is not H^+ nor OH^-, but the H_2O molecule. In addition to the above study, the effect of $[H^+]$ and $[OH^-]$ on the reaction rate was evaluated in high-temperature water by adding acid or base catalysts, in order to confirm that the $[H^+]$ or $[OH^-]$ concentration in high-temperature water was too low to promote catalytic hydrolysis. Their study implied that hydrolysis proceeds through the S_N2 reaction of the unprotonated nitrile as the substrate with H_2O as the nucleophile. Their elucidated dipole–dipole reaction mechanism is shown in Fig. 6.3.7.

Generally, the electrostatic solvent effect on a dipole–dipole reaction can be described by the Kirkwood equation [25] as explained in Sect. 6.1. The relevant rate constant derived from (6.1.8) and (6.1.9) is,

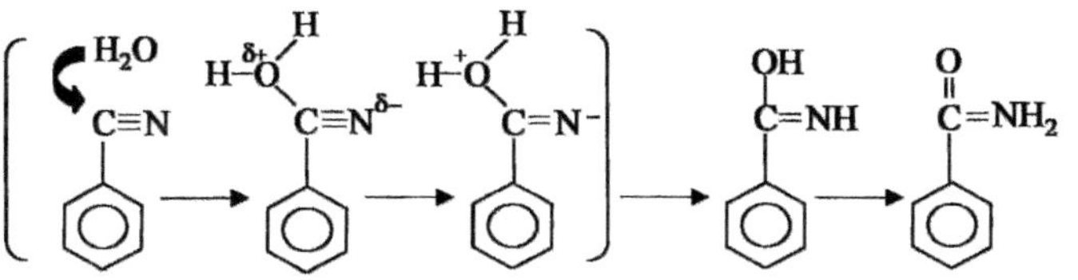

Fig. 6.3.7 Proposed nucleophilic attack of water during the hydrolysis of benzonitrile [19]

$$\ln k = \ln k_0 + \frac{C}{RT}\left(\frac{\mu^{\neq}}{(a^{\neq})^3} - \sum_i \frac{\mu_i}{a_i^3}\right)\frac{\varepsilon - 1}{2\varepsilon + 1}. \tag{6.3.5}$$

Although the evaluation of local dielectric constant and local solution structure around the solutes is needed for the detailed description of hydrolysis in SCW, this equation implies that reactions forming more polar transition state species than the reactants are promoted in a higher electric field. For most cases reported, the hydrolysis rate increases with increasing pressure (dielectric constant) in SCW, which can be explained by the transition state being more polar than the substrate and the water molecule as shown in Fig. 6.3.7.

(c) Cellulose Hydrolysis to Chemicals

Biomass is a major component of municipal and agricultural wastes and cellulose is the main constituent of biomass. Adschiri et al. have proposed a method for converting cellulose to chemicals via hydrolysis in high-temperature water. Below are concepts that have been found in that group of studies [22–24].

In their work, they employed a flow type apparatus as shown in Fig. 6.3.8. A microreactor was used that had a volume ranging from 0.03 to 5.27 cm^3. A cellulose–water slurry was mixed with preheated water and thus at the mixing point, the slurry was rapidly heated to its reaction temperature. At the end of the reactor, cooling water was injected directly into the reaction stream to terminate the reaction. This rapid heating and quick quenching method allows accurate determination of extremely short residence times.

Figures 6.3.9–6.3.11 show HPLC (high-pressure liquid chromatography) chromatograms of the products obtained at almost 100% conversion level at 320 °C, 350 °C and 400 °C, respectively. As shown here, most of the products at 320 °C were glucose pyrolysis products, namely erythrose, glycolaldehyde, glyceraldehyde, dihydroxyacetone, pyruvaldehyde, 1,6-anhydroglucose, 5-hydroxymethyl-2-furfural (5-HMF) and acids. At 350 °C, the main products were also glucose pyrolysis products. However, at 400 °C, that is in SCW, cellulose disappeared at 100% conversion level in 0.05 s and most of the products obtained were hydrolysis products, including glucose, fructose and oligomers of the glucose (cellobiose, cellotriose, cellotetraose cellopentaose, and cellohexaose). These results clearly demonstrate that hydrolysis of cellulose in SCW is a promising technology for converting cellulose to hydrolysis products with high selectivity.

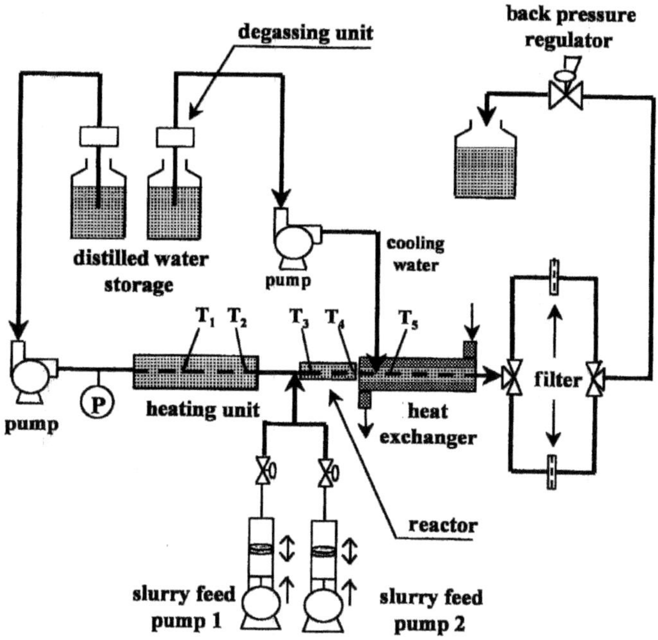

Fig. 6.3.8 Example of experimental setup for hydrolysis reactions [23]. Reprinted from: M. Sasaki, B.M. Kabyemela, R.M. Malaluan, S. Hirose, N. Takeda, T. Adschiri, K. Arai: Cellulose hydrolysis in subcritical and supercritical water, *J. Supercritical Fluids* (Elsevier Science) **13**, 261 (1998)

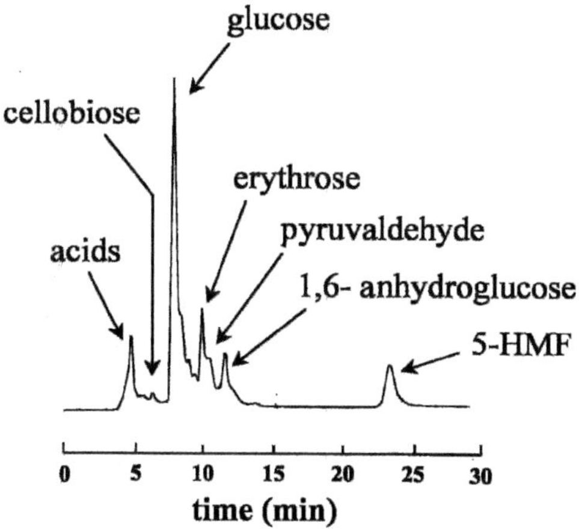

Fig. 6.3.9 HPLC chromatogram of recovered solution (320 °C and 9.9 s) [23]

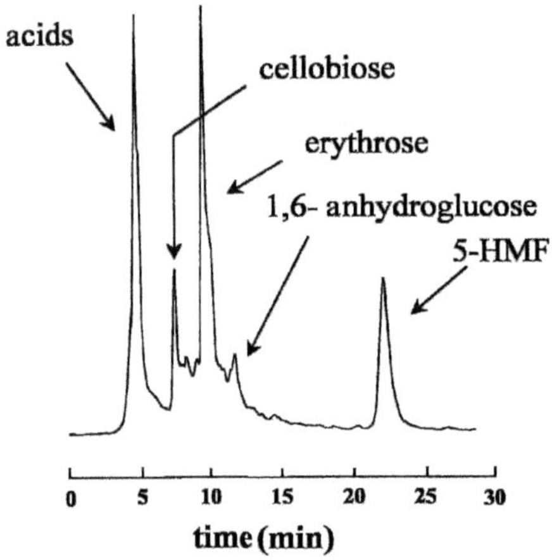

Fig. 6.3.10 HPLC chromatogram of recovered solution (350 °C and 8.8 s) [23]

For explaining the difference of the product distributions among the reaction conditions, kinetic studies were performed. The first-order rate constant, k_v, was evaluated from the conversion of cellulose, X, by the following equation.

$$\frac{\mathrm{d}X}{\mathrm{d}t} = k_v(1 - X). \tag{6.3.6}$$

The evaluated rate constant, k_v (s^{-1}), was plotted against reciprocal temperature in Fig. 6.3.12 [23]. In the range of temperature from 290 to 350 °C, the Arrhenius plot showed a straight line. However, above 350 °C, the reaction rate drastically increased and at 400 °C, a more than one order of magnitude higher reaction rate was obtained than at 350 °C.

The glucose decomposition rate evaluated [26] is also shown in Fig. 6.3.12. At a low-temperature region, glucose conversion is much faster than the hydrolysis rate of cellulose. Thus, even if the hydrolysis product, such as glucose, is formed, its further decomposition rapidly takes place and a high yield of hydrolysis products cannot be obtained. However, at 400 °C, the hydrolysis rate increases and becomes much higher than the glucose decomposition rate. This is the reason why it is possible to obtain high yields of hydrolysis products in SCW.

The reason for the drastic change of cellulose hydrolysis rate was also studied. At lower temperatures, heterogeneous hydrolysis takes place on the cellulose crystal surface. However, it was found that dissolution of cellulose

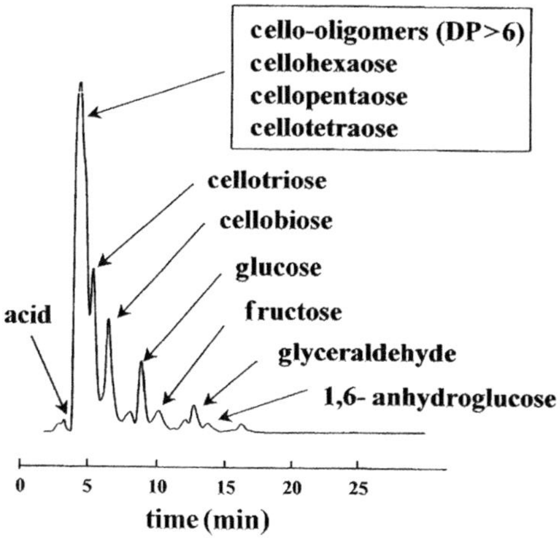

Fig. 6.3.11 HPLC chromatogram of recovered solution (400 °C and 0.05 s) [23]

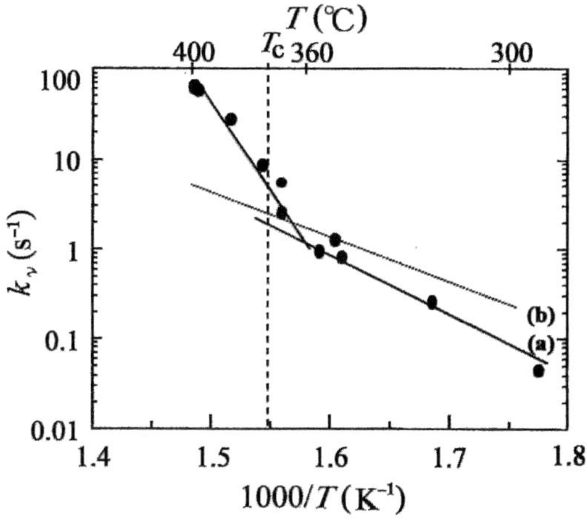

Fig. 6.3.12 Arrhenius plot of decomposition rate constants of cellulose and glucose in subcritical and supercritical water at 25 MPa: (**a**) cellulose [23]; (**b**) glucose [26]

takes place at higher temperatures, which leads to the formation of a homogeneous reaction atmosphere. It was concluded that the change in reaction phase is the reason for the high reaction rate in the supercritical region.

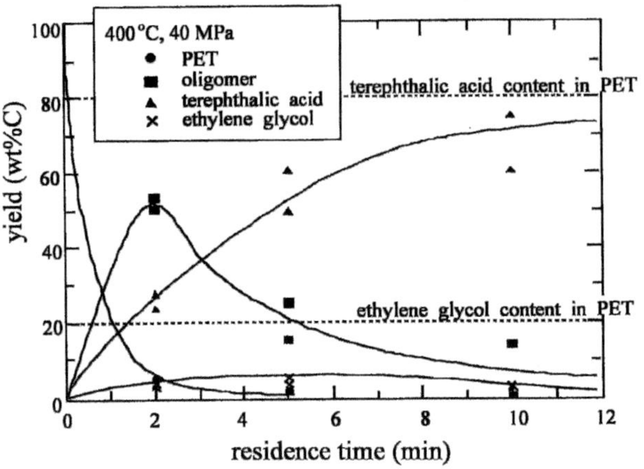

Fig. 6.3.13 Decomposition products of PET in supercritical water (400 °C and 40 MPa) [20]

(d) Chemical Recycling of Waste Polymers

The importance of developing the technology for chemical recycling of waste plastics is widely recognized. Adschiri et al. proposed a process to recover monomers from condensation polymers by noncatalytic hydrolysis in SCW [20].

Results for the decomposition of polyethylene terephthalate (PET) in SCW are shown in Fig. 6.3.13. The reaction primarily produced its monomer, i.e. terephthalic acid and ethylene glycol. The formation of other products such as benzoic acid, benzene, acetic acid and carbon dioxide due to further decomposition or decarboxylation were negligible under the conditions used. Consequently, little or no gaseous products were formed in these reactions. The yield of terephthalic acid reached close to 100% with a purity of greater than 97% under the conditions of 673 K, 40 MPa and a reaction time of 30 min. Increasing reaction pressure is effective for suppressing char formation or carbon dioxide production during decomposition. For comparison, 90 min reaction time was required at 573 K to obtain 90% recovery of telephthalic acid.

This method can also be employed for recovering monomers from heavy distillates. For producing bisphenol-A (BPA) after the dehydration reaction of phenol and acetone, BPA is separated from the other byproducts or residual reactants by distillation. The distillation bottoms have a phenol content of more than 50 wt%, but further distillation is difficult for economical reasons. The bottom liquids contain BPA, phenol, and isomers of BPA and thus by hydrolysis, phenol recovery is possible. A new process for BPA tar hydrolysis in SCW to recover phenol was proposed by Adschiri et al. [27]. First, the decomposition of BPA in SCW was studied at 673 K and in the pressure range from 28 to 38 MPa. Although polymer formation took place in an argon atmosphere, little or no polymerization occurred in SCW. In SCW, BPA was hydrolyzed to produce phenol and 2-(4-hydroxy phenyl)-2-propanol (HPP). HPP was further hydrolyzed to produce phenol and, at the same time, dehydrated to produce 4-isopropenylphenol (IPP), which was stable under the conditions used. With increasing water density, the contribution of HPP hydrolysis became more significant than that of dehydration, and as a result, the yield of phenol increased. The maximum phenol yield was 40 wt% of the raw tar, which corresponded to around 60 mol% recovery of the phenol structures in the original tar. This high phenol yield was obtained even at tar/water ratios as high as 30 wt% which would be required for the process to be economically feasible. Based on these same concepts, Kobe Steel developed a commercial plant of recycling tolylenediamine by the hydrolysis of tolylenediisocyanate for a pharmaceutical company in 1998 [28].

As explained above, hydrolysis without any acid or base catalyst in high-temperature water is a promising technology for the chemical recycling of waste condensation polymers. For the further development of the process and its application to various fields, fundamental information on kinetics and mechanism of the reactions and that of the reaction phase, and the establishment of the unit operation methods of SCW processes are required. This would lead to wider development and application of SCW hydrolysis in industrial processes.

6.4 Organic and Catalytic Reactions in Supercritical Carbon Dioxide

in Supercritical Carbon Dioxide

6.4.1 Introduction

When the chemical reactions in solvents are performed, the solvent molecules surround a solute molecule to form a solvent shell which is caused by various kinds of intermolecular noncovalent forces such as van der Waals forces, charge–charge, charge–dipole, or dipole–dipole interaction, and hydrogen bonding forces. In the liquid phase, solvent molecules are densely packed to form a solvation sphere, where there is continuous rotation or exchange of the solvent shell molecules. On the other hand, in the gas phase, molecules have high mobility and interact without the formation of rigid aggregation of these molecules. The properties of supercritical fluids (SCFs) can vary continuously between those of liquids and gases, even though the inherent molecular structures of the fluids are maintained. In particular, near the critical point, molecules in the supercritical phase are not uniformly distributed in space, but the solvent molecules aggregate around the solute by intermolecular solvent–solute interactions to form clusters, where the aggregated molecules are in dynamic equilibrium with free solvent molecules. The state of the solvation depends strongly on the density of the SCF and differs from that in the liquid solution or gas phase.

SCFs offer many advantages as reaction media. Improved performance or novel behavior of stoichiometric organic reactions, particularly in supercritical carbon dioxide ($scCO_2$) and supercritical water (SCW), have been extensively investigated [1–6]. Replacement of conventional liquid solvents by SCFs can also increase the rate and change the selectivity of homogeneously catalyzed reactions for the following reasons:

(i) high solubility of reactant gases in SCFs [7–13];
(ii) rapid diffusion of solutes into, out of, and within the supercritical phase;
(iii) weakening of the solvation around reacting species in comparison to that in the liquid phase;
(iv) local clustering of solutes or solvents resulting in an appreciable increase in the local concentration of solute or solvent molecules;
(v) reduction of the cage effect in radical reactions;
(vi) large compressibility near the critical point resulting in very large negative activation volumes [14, 15].

Many reports on organic reactions in SCFs have been published, examining the unique solvent effects of these media on reaction performance [5, 6, 16]. Examples of organometallic homogeneous catalysis in SCFs have been increasing since around 1990, but are still rare. This section was written to emphasize the recent development of organic reactions and homogeneous catalysis using transition metal-based complexes in SCF reactions which are homogeneously catalyzed by metal-containing complexes, although some heterogeneously catalyzed reactions will be noted for illustrative purposes.

6.4.2 Recent Advances in Organic Reactions in Supercritical Fluids

The unique physical properties of SCFs offer the possibility of a dramatic change in reaction rates. In addition to these effects of SCFs, one must consider the potentially strong pressure effects on the rate constant which can be understood in terms of transition-state theory. In particular, near the critical point, SCFs are highly compressible so that the density varies from gas-like to liquid-like over three orders of magnitude. These density changes can elongate the average intermolecular distance by a factor of 10 in comparison to that in the ordinary liquid phase. Since the first report of an unusual rate enhancement for the oxidation of ammonia to ammonium nitrate around the critical point over a Pt/C catalyst achieved by Toriumi in 1946 [17], there have been a number of studies to examine the possibility of tuning the reactivity and selectivity or improving the productivity over or around the critical point [3, 4, 18–20].

The pressure tunable rate of the Diels–Alder reaction has been well-investigated [12, 14, 19, 21]. The rate constants of the reaction of 2-methylbuta-1,3-diene and maleic anhydride obtained in the near-critical region are much larger than those under a higher-pressure condition or in liquid solvents as shown in Fig. 6.4.1. A similar behavior was already illustrated in Fig. 6.1.6. These reactions are kinetically controlled and thus significantly influenced by the local statistical structure of solvents around the reactant molecules. A large negative activation volume ($\Delta V^{\neq} = -4\,000$ cm^3/mol) near the critical point can be estimated, while the $\Delta V^{\neq}$ values obtained in high-pressure $scCO_2$ (296 atm) is −55 cm^3/mol, the value of which is comparable to that for the reaction in ethyl acetate, −37 cm^3/mol [15]. A similar rate enhancement in variation of pressure can be observed in unimolecular decomposition of α-chlorobenzyl methyl ether [14]. The rate constant increases by an order of magnitude as the pressure decreases from 59 to 44 atm at around the

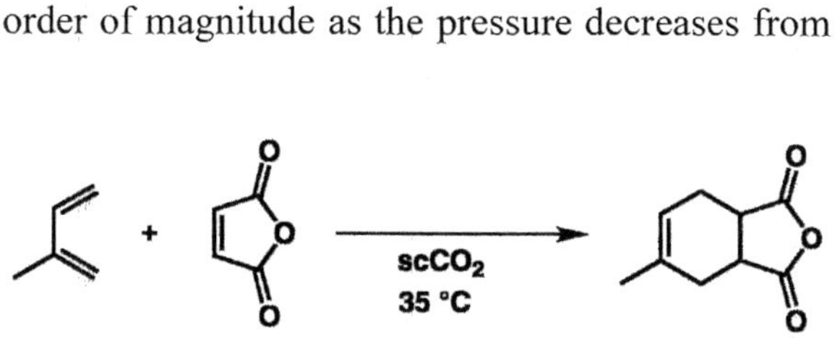

pressure (atm)	$-\Delta V^{\ddagger}$ (cm^3 mol^{-1})
75	4000
200	70

Fig. 6.4.1 Diels–Alder reaction of isoprene and maleic anhydride in $scCO_2$.

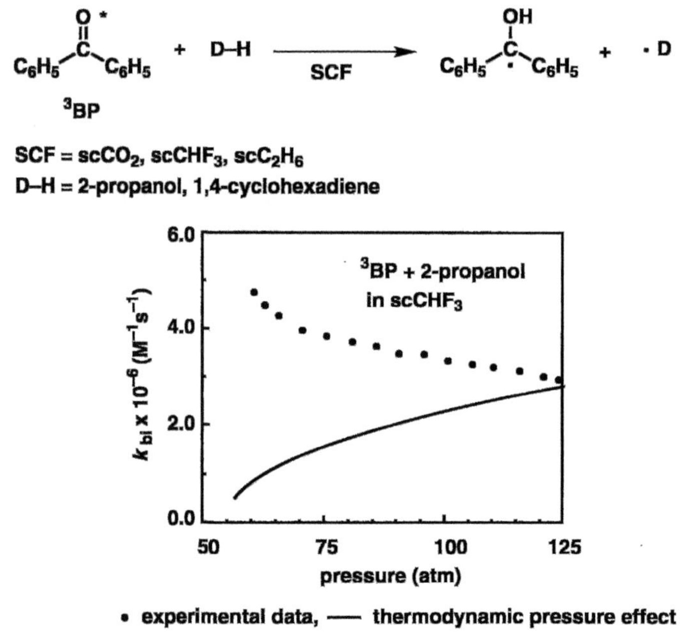

Fig. 6.4.2 Reaction of triplet benzophenone and 2-propanol or 1,4-cyclohexadiene in SCFs

critical point of 1,1-difluoroethane. The activation volume value is as low as −6 000 cm^3/mol near the critical point, but its value is −72 cm^3/mol at a higher-pressure region, $P/P_C = 6.1$, similar to that observed in conventional liquid solvents [14]. These rate enhancements are attributed to the increase in the local density at near-critical conditions.

Solute–solute clustering which results in an increase in local concentration of reactants often facilitates bimolecular reactions at around the near-critical point. The hydrogen atom transfer from 2-propanol or 1,4-cyclohexadiene to triplet-state benzophenone generated by laser flash photolysis in $scCO_2$, scC_2H_6, or $scCHF_3$ proceeds via the same mechanism as that in liquids. The reaction rates obtained in these SCFs are the same order of magnitude as those in liquids, which are approximately three orders of magnitude below diffusion control. As shown in Fig. 6.4.2, the second-order rate constant decreases sharply to about one fourth with an increase in the pressure above the critical pressure [22, 23]. Since the thermodynamic pressure effect on the rate constant suggests an increase in the rate constant with pressure, the observed unusual pressure effect on the reaction rate at a near-critical point is probably due to an increase in the local concentration of hydrogen donors around the triplet benzophenone.

The esterification of phthalic anhydride with methanol to methyl phthalate in $scCO_2$ provides a 25-fold decrease in the bimolecular rate constant when the pressure is increased from 96 to 164 atm [24, 25]. The higher rate constants observed at lower pressures are interpreted as a considerable increase in the local concentration of methanol around anhydride near the critical point. Spectroscopic studies for methanol around the solute in $scCO_2$ show that the local composition increase is of the magnitude that would be required to create the observed discrepancy [25].

The rate of the Michael addition of piperadine to methyl propiolate in $scCHF_3$, a polar fluid, at 37 °C depends on the fluid density [26]. The rate constant in this medium is linearly related to the pressure except near the critical point. The observed pressure dependence on the rate constant is attributed to solvent polarity at higher pressures, as would be consistent with the stabilization of a highly polar transition state. The unusually large rates observed near the critical points are interpreted in terms of solvent–solute clustering in $scCHF_3$.

The selectivity or product distribution in the chemical reactions is also tunable by using SCFs. As shown in Fig. 6.4.3, photodimerization of isophorone gives three major products, an *anti*-configurated head-to-head (HH) dimer and *anti*- and *syn*-configurated head-to-tail (HT) isomers [27]. The product distribution varies as a function of the pressure of the SCFs, the ratio of the HH to total HT dimers being considerably higher in $scCHF_3$ than in $scCO_2$. This change in the selectivity is attributed to the polarity of the fluid significantly affecting the polar transition state as well as the solvent reorganization. The stereoselectivity is influenced largely by solvent reorganization, whereas the regiochemistry is affected by both solvent polarity and solvent reorganization. The dielectric constant of $scCHF_3$ varies from 2.5 to 8.4 with

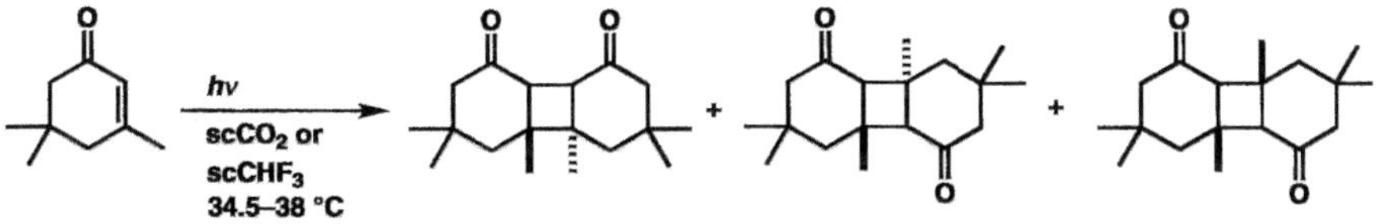

anti-HH I | *anti*-HT II | *syn*-HT III

medium	pressure (atm)	ε	HH:HT (I:II+III)	*anti*:*syn* (II:III)
$scCO_2$	88–280	1.34–1.35	0.09–0.1	2.00–3.95
$scCHF_3$	56–270	2.5–8.4	0.74–1.0	2.56–3.32

Fig. 6.4.3 Photodimerization of isophorone in SCFs: HH = head-to-head dimer, HT = head-to-tail dimer

pressure at 34.5 °C [28], while the dielectric constant of $scCO_2$ does not change significantly over the entire pressure range.

Control of stereoselectivity by changing the density of $scCO_2$ has been studied in the Diels–Alder reaction of methyl acrylate and cyclopentadiene, where *endo*- and *exo*-products are obtained [29] as shown in Fig. 6.4.4. A maximum in the *endo*- to *exo*-product ratio from this reaction can be observed at a density higher (700 kg/m^3) than the critical density (466 kg/m^3). This shift of the maximum from the near-critical point may be attributed to tuning of the average distance between the solute molecules and the transition states as the pressure and density vary. Theoretical studies suggest that this behavior of the product distribution can be predicted on the basis of the van der Waals equation [30]. The solvation energy of the solute in this range can be controlled by a change in the density and this will cause variations in the physical or chemical equilibrium or the rate of chemical reaction.

Cage effects of SCFs on radical reactions have been well investigated compared to those in conventional solvents. Because SCF are less viscous than liquid solvents, a small cage effect might be expected in SCF media, which can directly affect rapid reactions of reactive species in close proximity in a solvent cavity. For example, the initiation efficiency f of 2,2'-azobis(isobutyronitrile) (AIBN) for radical polymerization in $scCO_2$ ($f = 0.83$) is 1.5 times greater than in benzene, suggesting a weaker cage effect [31, 32]. On the other hand, the Norrish type I photolysis of 1-(*p*-methylphenyl)-3-phenyl-2-propanone in $scCO_2$ and scC_2H_6 gives a statistical mixture of the bibenzyls formed by random coupling of the benzyl and *p*-methylbenzyl radicals to 1,2-diphenylethane, 1-(*p*-methylphenyl)-2-phenyl-ethane and 1,2-bis(*p*-methylphenyl)ethane in a 1:2:1 molar ratio indicating that no solvent cage effect is operative in these SCFs even near their critical points [33]. Similarly, photoinitiated free-radical chlorination of cyclohexane in $scCO_2$ and conventional solvents provides a mixture of monochloride generated from the cage-escape reaction, and polychlorides formed by in-cage abstraction of hydrogen atoms from cyclohexyl chloride. The dependence of the molar ratio of monochloride to polychlorides on the inverse of the viscosity is linear over a range of viscosities from conventional solvents to $scCO_2$, indicating no enhanced cage effect even near the critical point of $scCO_2$ [34]. Thus, the product ratio is affected only by solvent viscosity.

The clustering between a solute molecule and solvent molecules causes an enhancement of the cage effect due to an increase in local density of a solvent molecule about a radical rather than the bulk density. The photo-Fries rearrangement of naphthyl acetate shown in Fig. 6.4.5 is known to proceed via a short-lived singlet radical pair with a lifetime of ca. 25 ps to form naphthol by abstraction of hydrogen atoms in the absence of a solvent cage. The photolysis

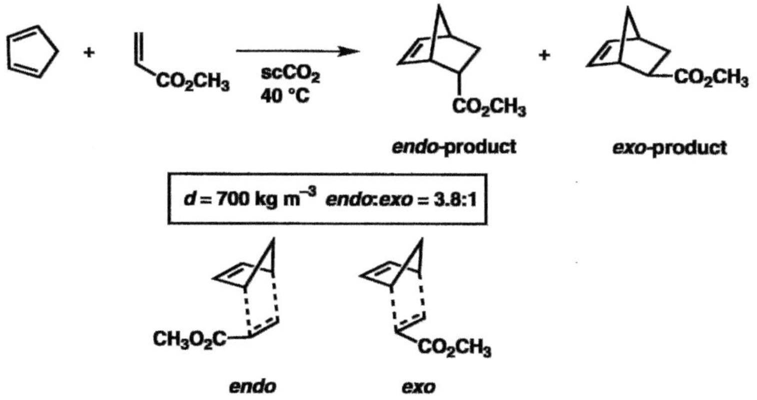

Fig. 6.4.4 Diels–Alder reaction of methyl acrylate and cyclopentadiene in $scCO_2$

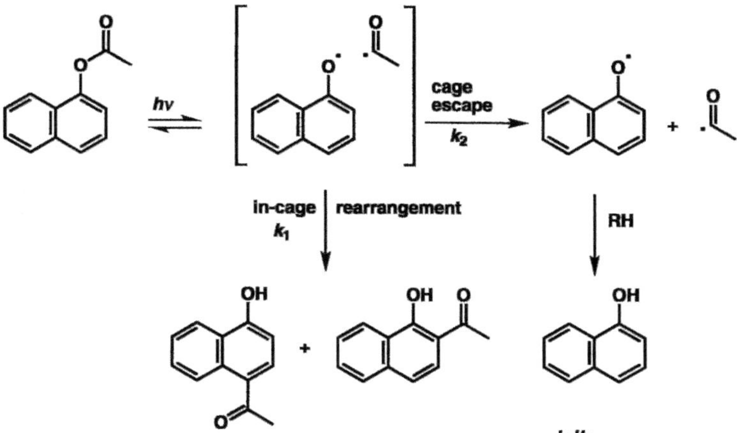

		k_2/k_1		
solvent	temp (°C)	76	80	110 (atm)
liqCO_2	27	4.1	4.1	4.1
scCO_2	47	5.2	13.5	4.0

Fig. 6.4.5 Photo-Fries rearrangement of 1-naphthyl acetate

in $scCO_2$ containing 2-propanol provides the products arising from an in-cage reaction of the geminate radical pair and products from a cage-escape reaction [35]. The ratio of the in-cage product to cage-escape product is invariant with pressure except near the critical point where the in-cage rearrangement product is predominantly formed because of a significant cage effect.

6.4.3 Homogeneous Catalytic Reactions in Supercritical Fluids

The pressure-tunable physical properties of SCFs could uniquely and subtly influence the nature of the molecules and molecular associates. These factors have already been demonstrated for a number of stoichiometric reactions, but their application to homogeneous catalysis is far more complicated. The principles developed with stoichiometric reactions, if applied to the product-determining or turnover-limiting step of the catalytic cycle, should allow one to explain the effect of supercritical conditions on homogeneous catalysis. Furthermore, SCFs offer a great opportunity to replace conventional organic solvents with environmentally benign solvents such as CO_2. They can also allow the deliberate control of the phase behavior, dissolution of the reactants, or precipitation of the products or catalysts with moderate changes in temperature and pressure.

Homogeneous catalysis in SCFs is considered a relatively new field, only gaining wide attention since around 1990 [36]. Although an early example was $AlCl_3$-catalyzed oligomerization of supercritical ethene at 240 °C reported by Ipatiev in 1913 [37], the catalytic reaction in $scCO_2$ has not been the subject of serious study until more recently. Part of the reason for this is the poor solubilizing power of $scCO_2$, which has restricted the choice of catalysts, reagents, and substrates to a narrow range of nonpolar, nonionic, and low molecular weight compounds. The recently developed utilization of surfactants, cosolvents, and other reagents can dramatically increase the solubility of polar or charged species in $scCO_2$ [38].

Transition metal complexes have been known to dissolve in $scCO_2$, but for which quantitative data are not available, including phosphine complexes [39–41], porphyrin complexes [42], metallocenes [43], metal carbonyls [44], and metal diketonates [45, 46]. Polar complexes, charged complexes, and even nonpolar complexes with many aryl-substituted ligands often have insufficient solubility to be used as catalysts. Ligand modifications such as replacement of ligands containing aryl substituents with ligands bearing alkyl groups [40] or perfluoroalkyl groups have been developed to increase the solubility of the transition metal complexes in $scCO_2$ [47–50]. Fig. 6.4.6 shows typical examples of $scCO_2$ soluble phosphine complexes. Introduction of CO_2-philic ponytails such as polyfluoroalkyl, fluoro ether, or silicone groups, often as substituents on aryl rings can also increase the solubility of metal complexes [49, 51, 52]. The longer the fluorinated alkyl chain, the better the solubility [52]. Often, the first two carbons in the ponytail are unfluorinated, so that the electron-withdrawing effect of the group is mitigated [49, 50, 53]. The use of other fluorinated phosphine ligands such as $P(C_6F_5)_3$, $P(C_6H_3\text{-}3,5\text{-}(CF_3)_2)_3$, and $P(C_6H_5)(CH_2CH_2C_6F_{13})_2$ also increases the solubility of the complexes [54,

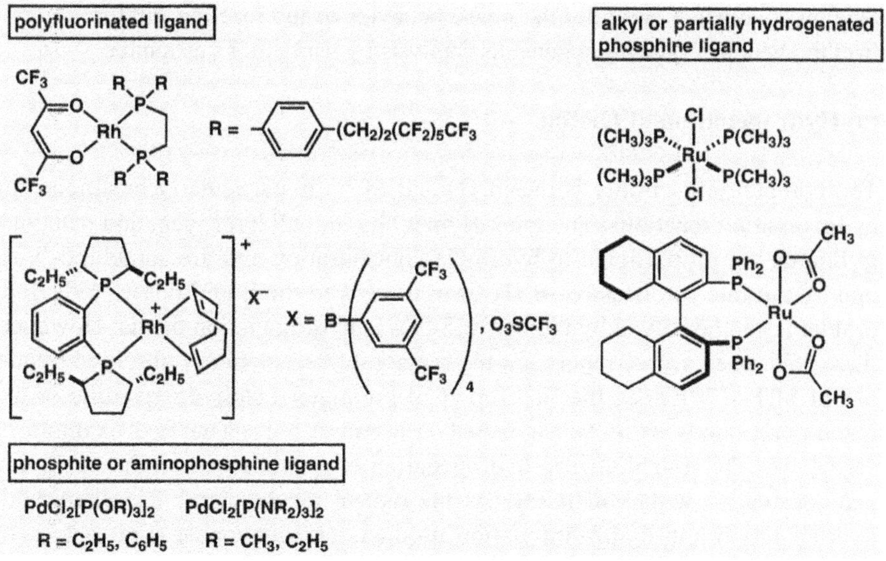

Fig. 6.4.6 $scCO_2$ soluble phosphine complexes

55]. Recently, Ikariya's group confirmed that the use of trialkylphosphites as ligands can increase the solubility of the transition metal complexes [56].

Careful choice of counter-ions can greatly increase the solubility of charged complexes which are generally insoluble in $scCO_2$. The use of $[(3,5\text{-}(CF_3)_2C_6H_3)_4B]^-$ ("BArF$^-$") [57, 58] and $CF_3SO_3^-$ as anions improves the solubility of their cationic rhodium complexes [59]. Addition of cosolvents also improves the solubility of complexes [60].

Surfactants have been developed which can create aqueous reverse-micelles in $scCO_2$ [61, 62], and hence solubilize hydrophilic reagents such as $KMnO_4$ [63, 64]. The surfactants generally have siloxane [65] or fluorinated lipophilic tails to increase their solubility in $scCO_2$ [66], although nonfluorinated surfactants are more appropriate for use in scC_2H_6 [67, 68].

The phase behavior of the reaction mixture is crucial subject to the chemical reactions in SCFs because the outcome of the reaction can be strongly affected by whether a particular catalysis system is single or multiphase. The phase separation often occurs during the progress of a reaction even if all components including the molecular catalyst, starting material, and cosolvent are all soluble in SCFs at the start of the reaction. The values of T_C and P_C of the reaction mixture can change during the reaction and the actual critical parameter of the reaction mixture may be substantially different from that of the pure fluid.

One must consider carefully the phase behavior of the reaction mixture to confirm the benefit of the homogeneous catalysis by using SCFs as media.

(a) Hydrogenation of Olefins

The complete miscibility between H_2 and SCFs is particularly beneficial for hydrogenation reactions. The rates of many but not all hydrogenation reactions in liquids are proportional to hydrogen concentration and are sometimes limited by the rate of diffusion of H_2 from the gas to the liquid phase [69]. Both problems can be solved by the use of SCFs as hydrogenation media. However, there have been scarce reports on the remarkable acceleration observed by the use of SCFs [70] since the first report of homogeneous hydrogenation of organic compounds in SCW published in a patent [71]. The first example in $scCO_2$ was the stoichiometric hydrogenation of an activated olefin, cyclopropene derivative with $MnH(CO)_5$ via the radical mechanism [72], where both hydrogenation and hydroformylation products are obtainable (Fig. 6.4.7). In these cases the selectivity for hydrogenation over hydroformylation has been

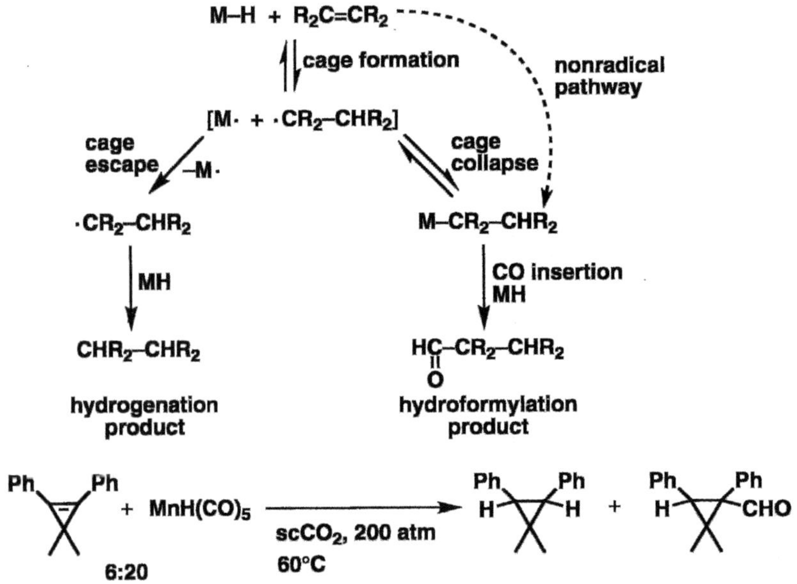

solvent	yield (cis:trans)	
$scCO_2$	66 (6:1)	34 (mostly cis)
hexane	66 (7:1)	34 (7:1)

Fig. 6.4.7 Reaction of 3,3-dimethyl-1,2-diphenylcyclopropene with $MnH(CO)_5$ in $scCO_2$

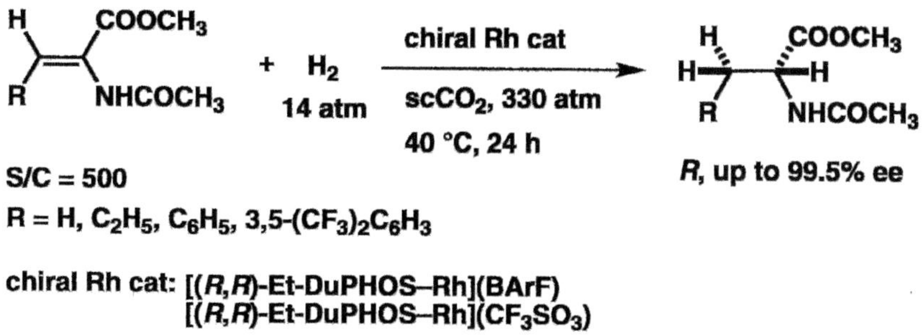

Fig. 6.4.8 Asymmetric hydrogenation of α-enamides catalyzed by chiral Rh complexes in $scCO_2$

used as a measure of the strength of the solvent cage, with stronger cages favoring hydroformylation [73]. The observed selectivity in $scCO_2$ was 66–70% [72], close to the selectivity observed in pentane or hexane, and suggesting a solvent cage strength comparable to that of liquid alkanes. Alternatively, the aldehydes may be formed by nonradical pathways which are independent of solvent viscosity.

A recent report on continuous hydrogenation in $scCO_2$ or scC_3H_8 using heterogeneous group 10 metal catalysts on Deloxan aminopolysiloxane sheds light on the practical application of SCFs. The combination of SCFs and a continuous flow reactor has considerable promise both for laboratory-scale reactions and for the large-scale production of chemicals. This continuous reaction system is applicable to the practical hydrogenation of a wide range of organic compounds including alkenes, ketones, aldehydes, epoxides, phenols, oximes, nitro compounds, and nitriles [74].

(b) Asymmetric Hydrogenation of Olefins

The enantioselectivity of asymmetric hydrogenation of a prochiral olefin often depends strongly on the hydrogen concentration [75]. The miscibility of H_2 in SCFs might lead to better enantioselectivity. Asymmetric hydrogenation of olefins in $scCO_2$ by chiral transition metal catalysts have been reported by Burk and Tumas's and Ikariya and Noyori's groups, independently.

Tumas et al. found that a $scCO_2$ soluble cationic Rh complex with chiral (*R,R*)-Et-DuPHOS is an efficient catalyst for the asymmetric hydrogenation of α-enamides to α-amino acid derivatives in $scCO_2$. The key to success of the reaction was the choice of a counter-ion, either $CF_3SO_3^-$ or $BarF^-$, which causes an increase in the solubility of the catalyst (Fig. 6.4.8) [59]. The reac-

Table 6.4.1 Hydrogenation of *β,β*-disubstituted *α*-enamides catalyzed by [Rh((*R*, *R*)-Et-Duphos)(cod)][$B(C_6H_3\text{-}3,5\text{-}(CF_3)_2)_4$] in $scCO_2$ and other media

product	%ee		
	$scCO_2$	CH_3OH	hexane
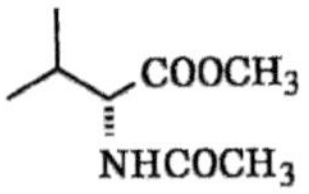	96.8	81.8	76.2
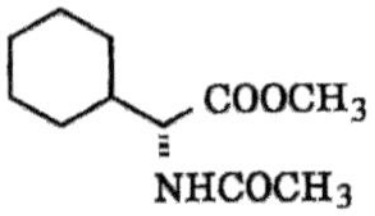	84.7	62.6	69.5

tion proceeds homogeneously under the reaction conditions tested. The enantiomeric excesses (ee's) were fair to excellent and generally comparable to those obtained in conventional solvents, although in the case of a valine derivative the ee values was superior to that obtained in either methanol or hexane (Table 6.4.1) [76].

Ikariya et al. demonstrated that a chiral Ru complex, $Ru(OCOCH_3)_2$ (H_8-binap), is an effective catalyst for asymmetric hydrogenation of *α,β*-unsaturated carboxylic acids such as tiglic acid in $scCO_2$. The ee value of the product obtained in $scCO_2$ (81%) was comparable to that in methanol (82%) and greater than that in hexane (73%), all at 50 °C and 30 atm H_2. The addition of fluorinated alcohol $CF_3(CF_2)_6CH_2OH$ to the reaction in $scCO_2$ improves not only the conversion of the substrate but also the enantioselectivity by up to 89% ee. The alcohol may serve to increase the catalyst solubility or to change the reaction mechanism (Fig. 6.4.9).

Noticeably, the hydrogenation of tiglic acid by $Ru(OCOCH_3)_2(H_8$-binap) in $scCHF_3$ is dependent not only on the H_2 partial pressure but also on the pressure of $scCHF_3$. The ee of the product is higher at 188 atm CHF_3 than it is at 82 atm (H_2 partial pressure constant at 12 atm) (Fig. 6.4.9) [77]. This change in the selectivity correlates well with the change in the dielectric constant of the solvent as observed in conventional liquid solvents [78]. The notable pressure effect in the hydrogenation in methanol on the enantioselectivity indicates the coexistence of a highly stereoselective monohydride mechanism with a less selective polyhydride mechanism [75]. The increase in enantioselectivity with the increase of polarity of the SCF phase is ascribable to the enhanced contribution of the monohydride mechanism. Below that

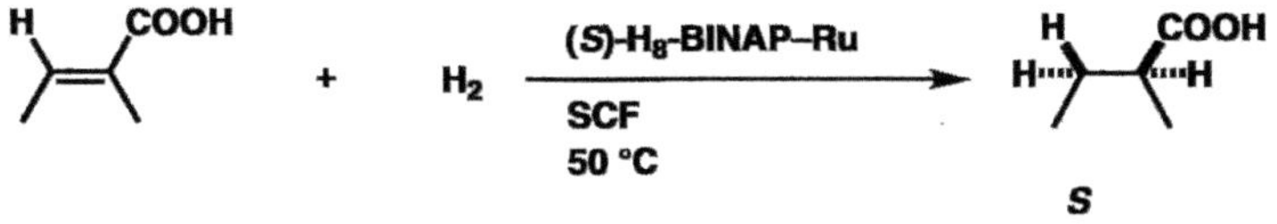

Ru catalyst:

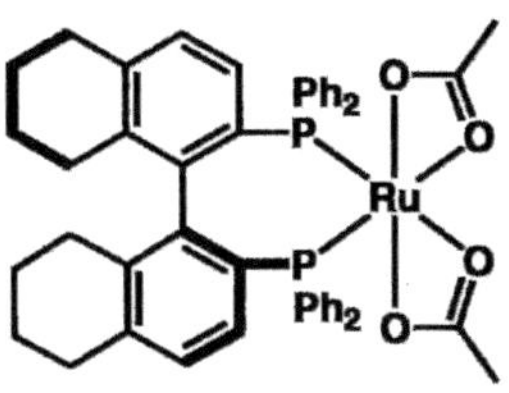

medium (atm)	H_2, atm	yield, %	% ee
CO_2 (170)	33	99	81
CO_2/R_FOH (170)	5	99	89
CHF_3 (188)	12	>99	90
CHF_3 (82)	12	99	83
CH_3OH	12	100	91
hexane	5	91	74

Conditions: Ru cat, 4.7 μmol; S/C = 150–160
$R_FOH = CF_3(CF_2)_6CH_2OH$.

Fig. 6.4.9 Asymmetric hydrogenation of tiglic acid with (*S*)-H_8-BINAP-Ru in $scCO_2$

pressure, a liquid phase formed, in which the reaction occurs resulting in an increase in the ee values of the product to 93% as observed in the polar solvents. Thus, the phase behavior affects significantly the outcome of the asymmetric hydrogenation. A similar pressure tunable enantioselectivity was reported in asymmetric cyclopropanation with chiral Rh complexes in $scCHF_3$ [79].

For comparison, it is worth pointing out that the enantioselectivity of heterogeneously catalyzed reactions are also dependent on H_2 concentration. The enantioselectivity of the hydrogenation of ethyl pyruvate using cinchonidine-modified Pt on alumina in scC_2H_6 was equal to that in toluene and greater than that in ethanol [80]. The observed ee increased with increasing H_2 partial pressure up to 60 atm (74% ee), beyond which no further increase in the ee value was observed. In $scCO_2$, however, the reaction was slower because of the reduction of $scCO_2$ to CO on the surface of the catalyst [81].

(c) Hydrogenation of Carbon Dioxide

The high miscibility between H_2 and $scCO_2$ can allow highly efficient hydrogenation of carbon dioxide which is one of the most profitable reactions for practical CO_2 fixation in which $scCO_2$ is both the reaction medium and reactant [82].

Ikariya et al. established that $scCO_2$ can be efficiently hydrogenated to formic acid by Ru-$P(CH_3)_3$ complexes in the presence of $N(C_2H_5)_3$, a thermodynamic sink, and a trace amount of water, as a promoter. The $P(CH_3)_3$

complexes are the best catalysts for this reaction because of their high solubility in this medium. The conventional $P(C_6H_5)_3$-based catalyst $RuH_2[P(C_6H_5)_3]_4$ is less active, possibly because of its low solubility in $scCO_2$. The catalyst efficiency of 7 200 TON and initial rate or TOF (turnover frequency, TON per hour) of 1 400 are greater than any previously reported for reactions in conventional liquid solvents (Fig. 6.4.10) [10, 83]. Addition of a trace amount of an alcohol as a cosolvent to pure $scCO_2$ results in a significant improvement in the rate of the reaction, TOF exceeding 4 000. A homogeneous reaction system is crucial at the start of the reaction to attain such a high productivity. When the catalysis is conducted in two phases from the start by using a large amount of the amine, water, THF, or CH_3CN as a liquid solvent, the reaction proceeds slowly. Use of methanol or primary and secondary amines in the CO_2 hydrogenation system permits the efficient production of methyl formate or formamide derivatives, respectively [40, 84]. The syntheses of methyl formate and *N,N*-dimethylformamide (DMF) in $scCO_2$ have far greater TON and TOF than those reported for the same reactions in liquid solvents. For example, DMF is produced from dimethylamine and CO_2 with up to 420 000 TON and in 99% selectivity. The high rates and productivity are due to the high concentration of H_2 in $scCO_2$, rapid mass transfer between phases, and high reactivity of the catalyst in the supercritical phase. Baiker and coworkers reported a further improvement of the catalytic activity for the formation of DMF and methyl formate under super- and subcritical conditions by using CO_2-insoluble Ru complexes, $[RuCl_2(dppe)_2]$ (DPPE = 1,2-bis(diphenylphosphino)ethane) in 500 ml reaction vessel. The TOF for DMF formation exceeded 360 000 h^{-1} at the initial 18% conversion for 2 h [85]. It should be noted that the rate of the reaction under supercritical condition de-

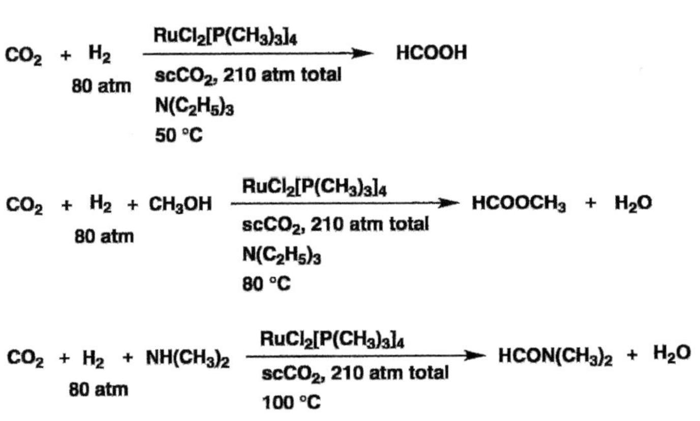

Fig. 6.4.10 Hydrogenation of $scCO_2$ catalyzed by Ru(II) complex

creases gradually in both $P(CH_3)_3$ and dppe catalyst systems probably because of the inhibition effect of water formed during the reaction [86]. Baiker also showed that the Ru complex supported on a silica matrix by means of sol-gel process exhibits a remarkable activity for the synthesis of DMF [87].

(d) Hydroformylation of Olefins

Rathke [9, 88] demonstrated the first homogeneous hydroformylation of olefin in $scCO_2$ catalyzed by $Co_2(CO)_8$ (Fig. 6.4.11). The reaction of propene with $Co_2(CO)_8$, at 80 °C at $P_{H_2} = P_{CO} = 56$ atm occurs at a slightly lower rate in $scCO_2$ ($d = 0.5$ g/ml) than in hydrocarbon solvents such as methylcyclohexane and heptane. The selectivity for the linear aldehyde, 88%, is greater than the value, 83%, obtained in benzene. The linear to branch ratio is slightly influenced by the pressure and temperature [89]. The linear selectivity in products increases from 73 to 81% as the pressure doubles at 88 °C. The major advantage of the using SCFs, the elimination of gas-to-liquid mass transfer problems or the order-of-magnitude lower viscosity of $scCO_2$ compared to typical liquids, has a beneficial line-narrowing effect on quadrupolar nuclei in NMR analysis [90]. High-pressure ^{59}Co NMR studies revealed that the equilibrium constant between $Co_2(CO)_8$ with H_2 and $HCo(CO)_4$ in $scCO_2$ is close to that in heptane (Fig. 6.4.11) [9, 91]. Furthermore, the rate of hydrogen atom transfer from $HCo(CO)_4$ to $Co_2(CO)_8$ is proved to be much higher than the rate of hydroformylation. This suggests that Co carbonyl radicals are possible intermediates in the hydrogen transfer reaction and also in the hydroformylation reaction.

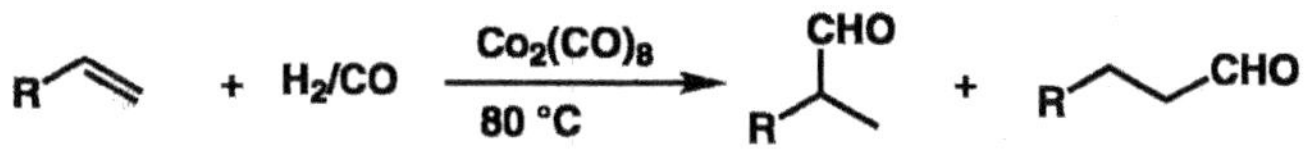

R = CH_3, n-C_6H_{13}

olefin	medium	H_2/CO (atm)	d[aldehyde]/dt ($M\ s^{-1}$)
propene	$scCO_2$, 165 atm	56/56	0.77×10^{-5}
1-octene	c-$C_6H_{11}CH_3$	48/48	1.2×10^{-5}

$$Co_2(CO)_8 + H_2 \underset{80\ °C}{\overset{scCO_2}{\rightleftharpoons}} 2\ HCo(CO)_4$$

$K_{eq} = 0.025$, $K_p = 8.8 \times 10^{-4}$ M atm^{-1}

Fig. 6.4.11 Hydroformylation of olefins catalyzed by $Co_2(CO)_8$ in $scCO_2$

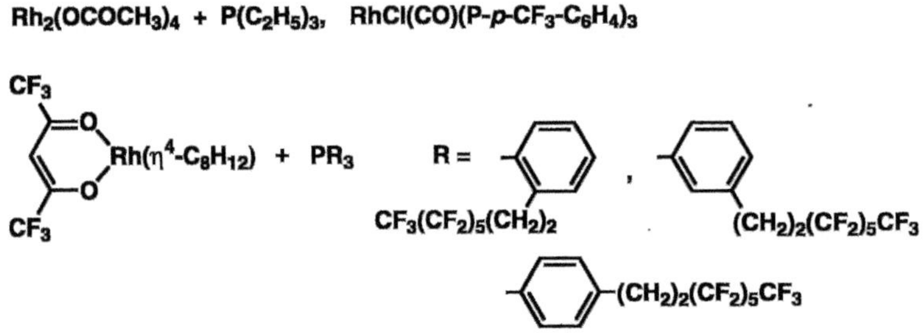

Fig. 6.4.12 $scCO_2$ soluble Rh phosphine complexes for hydroformylation

Recently, several groups examined the hydroformylation of olefins catalyzed by $scCO_2$-soluble Rh complexes (Fig. 6.4.12). The ponytail ligand, polyfluoroalkyl substituted triarylphosphine-rhodium complexes effect the hydroformylation of 1-octene to afford a linear aldehyde both in a good yield and selectivity (linear/branch = 4.6) without the undesired hydrogenation or isomerization products [51, 92]. The trifluoromethyl-substituted arylphosphine complex, *trans*-$RhCl(CO)[P(p\text{-}CF_3C_6H_4)_3]_2$ which is soluble in $scCO_2$ also promotes the hydroformylation of 1-octene to give the corresponding aldehydes with a lower linear to branch ratio of ca. 2.4 [93]. In the reaction of 1-hexene with the Rh-$P(C_2H_5)_3$ complex, by-products, heptanols are obtained in addition to the aldehyde products with a linear/branch molar ratio of 2.4 [94].

(e) Cyclization and Other C–C Bond Formation Reactions

Friedel–Crafts alkylation of mesitylene and anisole with supercritical propene or 2-propanol in $scCO_2$ in the presence of a solid acid catalyst provides selective formation of the mono-alkylated products [95]. Use of a continuous flow reactor prevents catalyst deactivation and minimizes the reactor size compared with a batch reactor. High diffusivity of the SCFs allows the reduction of the mass transport restriction at the surface of the solid catalyst.

The Pauson–Khand reaction, the catalytic cocyclization of alkynes with alkenes and carbon monoxide, proceeds smoothly in $scCO_2$ to give cyclopentenones. As shown in Fig. 6.4.13, the ene-ynes react with carbon monoxide in the presence of 2–5 mol% of $Co_2(CO)_8$ as the catalyst, affording the cyclization products in moderate to good yields. Temperatures of 90–100 °C, well beyond T_C, and CO pressures of 15–30 atm are required [96]. It is noted that the high CO pressure prevented the deactivation of the catalyst, whereas the high pressures of CO tend to decrease the yield obtained in liquids [97].

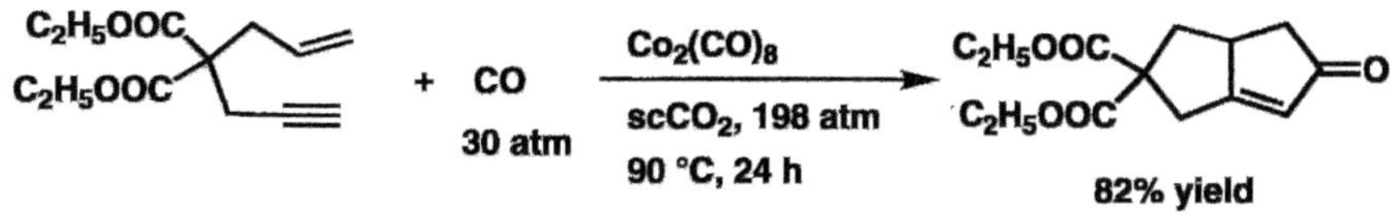

Fig. 6.4.13 Pauson–Khand reaction catalyzed by $Co_2(CO)_8$ in $scCO_2$

However, the solubilities of the ene-ynes in $scCO_2$ and phase behavior during the reaction were not reported.

The coupling of alkynes with CO_2 catalyzed by $Ni(cod)_2/P(CH_3)_3$ proceeds well in $scCO_2$ to give 2-pyrone derivatives with over 90% selectivity and a TON of 3 for 50 h at 40 °C [44, 98–100]. The TON could be improved to 9 by increasing the temperature to 95 °C after about 9 h.

Cyclotrimerization of alkynes to substituted benzene derivatives is homogeneously promoted by $CpCo(CO)_2$ in SCW [101]. Thermal decomposition of the complex is not a problem even at the high temperature of SCW. Selectivity for the 1,2,4-isomer of the product is comparable to that found in organic solvents. In these catalytic conditions some of metallorganic compounds could be identified [102]. The results suggest a mechanism, working in SCW, similar to that in organic solvents.

(f) Olefin Metathesis

Recent results demonstrate that supercritical fluids can be useful reaction media for the metathesis of olefins. The Ru catalyst, $[Ru(H_2O)_6](OTs)_2$ (Ts = *p*-toluenesulfonyl) and isolable metal-carbene complexes shown in Fig. 6.4.14 catalyzed the ring-opening metathesis polymerization (ROMP) of norbornene and the reaction proceeds smoothly in $scCO_2$ to give an off-white spongy-textured polymer [103–105]. The chemical yield and molecular weight of the polymers are comparable to those in conventional solvents. Although the Ru catalyst is not completely soluble, the change in the color of the reaction mixture, from clear colorless at the beginning to orange during the reaction, strongly indicates that the catalyst is sufficiently soluble for the completion of the reaction. Addition of methanol as cosolvent causes an increase in the *trans*-olefin content to 70% from 17% in the polymers obtained in $scCO_2$.

Fürstner and Leitner found that these same carbene complexes effected ring-closing metathesis (RCM) of various dienes to cyclic olefins [70, 103]. Even dienes with functionalized groups such as the NH group can be converted, leading to carbo- and heterocyclic compounds in good yields. The density of the $scCO_2$ significantly affects the reaction pathway. An acyclic diene

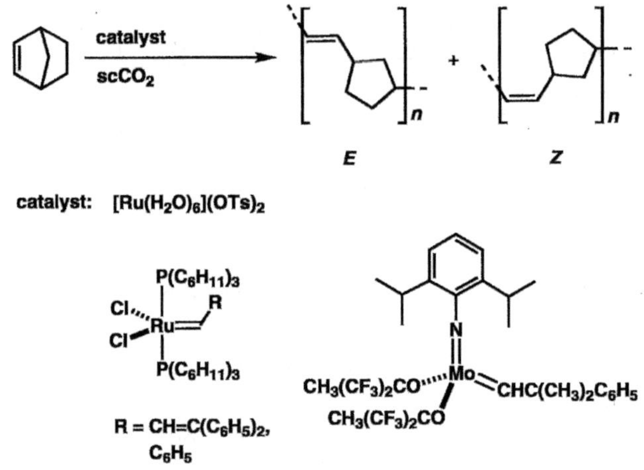

Fig. 6.4.14 Ring-opening metathesis polymerization of norbornene in $scCO_2$

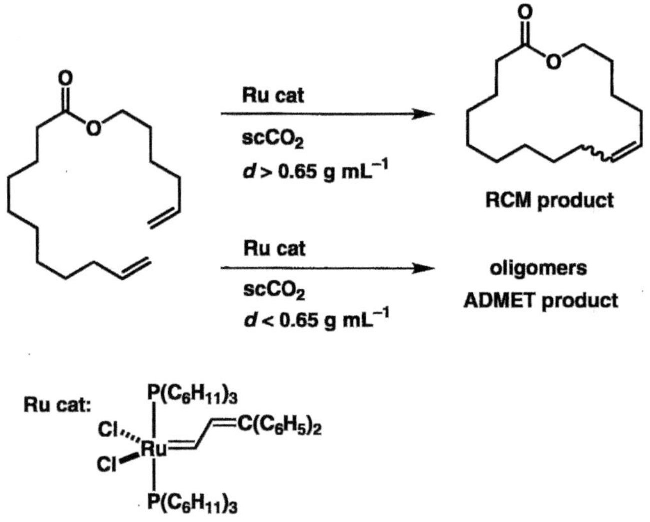

Fig. 6.4.15 Ru-carbene catalyzed olefin metathesis depending on the density of $scCO_2$

leads to 16-membered cyclic compounds with RCM, while its acyclic diene metathesis (ADMET) gives a polymer or oligomer as illustrated in Fig. 6.4.15. At the densities of CO_2, $d > 0.65$ g/ml, the cyclic compound is produced in an excellent yield, while mainly oligomers are obtainable at lower densities. The density effect on the reaction pathway is not fully understood.

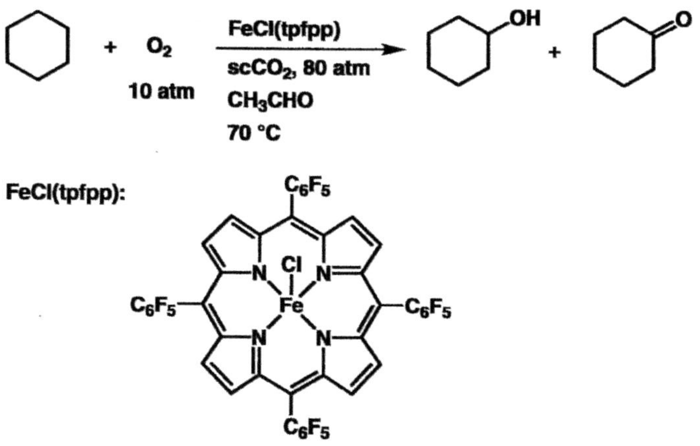

Fig. 6.4.16 Oxidation of cyclohexane catalyzed Fe(II) complex in $scCO_2$

(g) Oxidation Reaction

Apart form the miscibility of O_2 with $scCO_2$, and the nonflammability of CO_2, there have been few homogeneously catalyzed oxidations in SCFs.

The oxidation of cyclohexane with 10 atm O_2 at 70 °C catalyzed by FeCl(tpfpp) (TPFPP= 5,10,15,20-tetrakis(pentafluorophenyl)porphyrin) provided a 1:1 mixture of cyclohexanone and cyclohexanol in $scCO_2$ [106] as illustrated in Fig. 6.4.16, whereas the reaction in organic solvents preferentially afforded cyclohexanone [107]. The greatest yield obtained was 20% after 1 h at 70 °C, based on aldehyde initial concentration. In comparison, a yield of 85% after 24 h at the same temperature in liquid ethyl acetate was obtained [107]; however, the rate of the reaction in $scCO_2$ is comparable to that in a liquid solvent.

Oxidation of alkenes by hydroperoxides in the presence of a soluble catalyst $Mo(CO)_6$ in $scCO_2$ proceeds to give mainly the corresponding epoxides. The oxidation of cyclooctene takes place at 45 °C to form the corresponding epoxide as a sole product [108], but 1,2-diol is selectively obtained by the reaction of cyclohexene at 95 °C (Fig. 6.4.17) [44]. The epoxidation of 2,3-dimethylbutene is also successful by use of cumene hydroperoxide as a oxidant in $scCO_2$ doped with 1,1,2,2-tetrachloroethane as a cosolvent [109]. When $Ti[OCH(CH_3)_2]_4$ was used as a catalyst for the oxidation of cyclooctene in $scCO_2$, the epoxide and a small amount of 1,2-diol were obtained. The oxidation of cyclohexene gave the corresponding epoxide and 1,2-diol with formation of a condensation product 2-*tert*-butoxycyclohexanone.

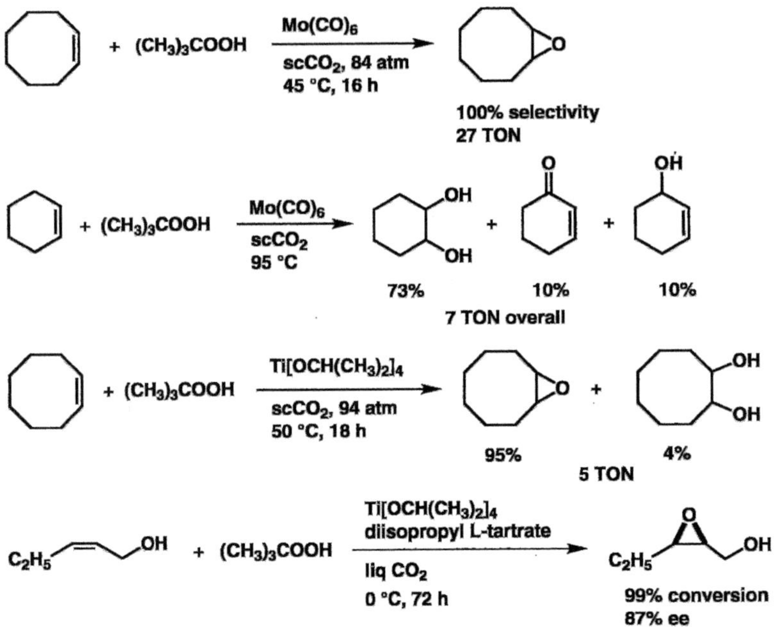

Fig. 6.4.17 Oxidation of olefins by hydroperoxides in $scCO_2$ and liquid CO_2

Allylic and homoallylic alcohols can be oxidized selectively with the oxovanadium complex, $VO[OCH(CH_3)_2]_3^{2-}$ in liquid CO_2 to the corresponding epoxides, whereas $VO(acac)_2$, a conventional epoxidation catalyst in liquid solvents, has no remarkable conversion in this medium probably due to its poor solubility in CO_2. Sharpless epoxidation of allylic alcohols takes place in liquid CO_2. Early results indicate that up to 87% ee is attainable for the epoxidation of (*E*)-hex-2-ene-1-ol by $(CH_3)_3COOH$ with $Ti[OCH(CH_3)_2]_4$ and diisopropyl *L*-tartarate [108].

(h) Palladium-Catalyzed C–C Bond Formation Reactions

Pd catalyzed C–C bond formation is one of the most powerful tools in organic synthetic chemistry. Since a remarkable acceleration in Heck–Mizoroki olefination of aryl halides was observed by performing the reaction under high-pressure conditions (1.0×10^4 atm) [110], several Pd-catalyzed reactions in $scCO_2$ have been tested. As shown in Fig. 6.4.18, the two groups of Holmes [54] and Tumas [55] have reported on Heck–Mizoroki olefination, Suzuki-, Sonogashira-, and Stille-coupling reaction under supercritical conditions, independently. These reactions proceed well in $scCO_2$ but the use of Pd(II) complexes with fluorinated

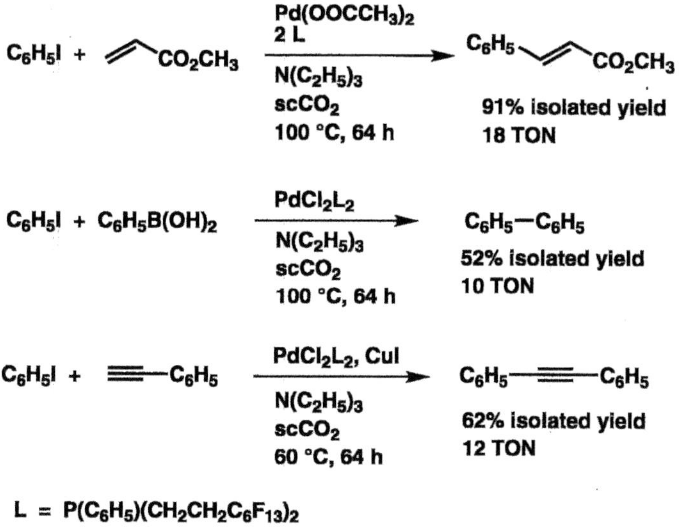

Fig. 6.4.18 Pd(II) catalyzed C–C bond formation in $scCO_2$

phosphine ligands such as $P(C_6H_5)_{3-n}(CH_2CH_2C_6F_{13})_n$ (n = 1 or 2), $P(C_6F_5)_3$, or $P(C_6H_3\text{-}3,5\text{-}(CF_3)_2)_3$ causes no significant improvement in the catalytic activity for the coupling reactions, limited turnover numbers (TON = 10–31) being obtained for the Heck–Mizoroki reaction.

Recently, Ikariya's group found that the incorporation of an alkoxy group onto the phosphine ligand resulted in a significant increase in the solubility of the Pd catalyst precursor in $scCO_2$ [56]. For example, 1×10^{-2} mmol (5.1 mg) of the triethylphosphite-Pd(II) complex, $PdCl_2[P(OC_2H_5)_3]_2$ is dissolved in $scCO_2$ (300 atm, 130 °C) in 50 ml of vessel. Thus, the tris(2,2,2-trifluoroethyl)phosphite complex, $PdCl_2[P(OCH_2CF_3)_3]_2$ efficiently effects the Heck–Mizoroki reaction of PhI and ethyl acrylate to give ethyl cinnamate with a high TON of 83 600 at 130 °C [111]. The success of this strategy is also shown by the rapid intramolecular carbonylation of 2-iodobenzylalcohol catalyzed by the same Pd complex, $PdCl_2[P(OC_2H_5)_3]_2$ in a supercritical mixture of CO_2 and CO (P_{CO_2}, 200 atm and P_{CO}, 10 atm) at 130 °C (Fig. 6.4.19). The rate of the reaction is higher than that attained in solution phase reactions. The advantage of performing the carbonylation in $scCO_2$ is that there is no marked effect of the CO pressure on the TON values obtained in $scCO_2$ because of the higher diffusivity of $scCO_2$ as well as the high solubility of CO in this medium. The yield of the product is independent of the CO pressure over a range of 1–5 atm in $scCO_2$ (200 atm), the reaction with S/C = 5 000 being to near completion.

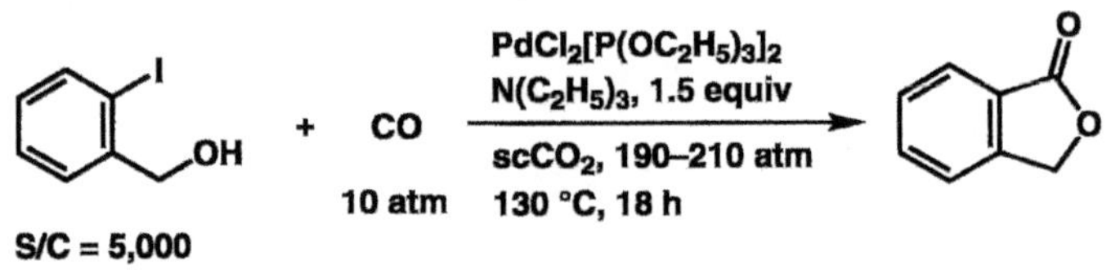

Fig. 6.4.19 Pd(II) catalyzed intramolecular carbonylation in $scCO_2$

6.4.4 Concluding Remarks

SCFs have proved to be unique and very practical media for chemical reactions, providing a third reaction phase besides conventional gas and liquid phases. The reaction performance can be easily tuned by changing only the temperature and pressure. The utility of SCFs is not limited to the chemical reactions described herein. Potential advantages of SCFs as reaction media have been demonstrated in uncatalyzed reactions [1–6, 16, 112, 113], heterogeneously catalyzed reactions [114], enzyme catalyzed reactions [115–117], phase transfer catalysis [118–121] and polymerization [4, 16, 112]. In addition, the following recent topics illustrate the special benefits. Use of SCFs allows the elimination of solvent residues and wastes and facile separation of reactants, catalysts, and products after reaction by selective precipitation. These characteristics offer a great opportunity for rapid synthesis, workup, purification, and analysis of reaction products containing a radionuclide with a short half-life such as ^{11}C ($t_{1/2}$ = 20.3 min). In fact, highly efficient synthesis of ^{11}C-labeled monosubstituted guanidines has been achieved in $scNH_3$ (T_C = 132.6 °C, P_C = 111.3 atm) by reacting aliphatic or aromatic amines and [^{11}C]cyanogen bromide followed by treatment of ammonia. The syntheses are performed within 10 min and with high reproducibility. The ^{11}C-labeled products are used for both *in vivo* positron emission topography (PET) studies and *in vitro* assays in modern bioscience [122].

Fluoropolymers can be produced efficiently in $scCO_2$, as exemplified by the AIBN-initiated homogeneous free-radical polymerization of 1,1-dihydroperfluorooctyl acrylate giving CO_2-soluble fluoropolymers. In $scCO_2$, the reaction occurs with a 1.5 times higher initiation efficiency than in benzene, but with a lower reaction rate due to the weaker cage effect of $scCO_2$ [123]. Inert $scCO_2$ can avoid the environmental problems caused by CFCs. DeSimone has cleverly used soluble fluoropolymers in small amounts as surfactants in the polymerization of methyl methacrylate [124, 125] or acrylamide [126] in $scCO_2$, resulting in greater rates and uniformity of product particle size, respectively.

The industrial outlook for SCFs as reaction media is bright, mainly because of the beneficial environmental effect of dispensing with organic liquid solvents. In particular, SCW is a very attractive medium for waste-clean technology. The application of SCW as a reaction medium has been extensively reviewed [4, 6]. High-pressure operations would be facilitated in large-scale production processes and, especially, continuous flow systems for which SCFs are well suited. Favorable mass transfer between phases or control of phase behavior facilitates the catalytic hydration of olefins. Reaction of 2-butene and water in the presence of an aqueous heteropolyacid as catalyst at 200 °C and 200 atm in scC_4H_8 proceeds effectively to produce 2-butanol [127]. The use of supercritical conditions can overcome the equilibrium problems of this hydration reaction; the product 2-butanol can be effectively separated from the catalyst solution by extraction with scC_4H_8 because of the high solubility of 2-butanol in scC_4H_8. This process is now working on a large scale, 40 000 ton/year, at Idemitsu Petrochemical in Japan.

6.5 Reactions of Enzymes and Molecular Aggregates in Supercritical Fluids

6.5.1 Introduction

Considerable attention has been directed to the highly compressible properties of supercritical fluids (SCFs) in the near-critical region, from the theoretical and industrial application viewpoints [1–5]. Interactions between solute and solvent in SCFs are of fundamental importance because small changes in pressure or temperature near the critical point may alter the reactivity in chemical and biochemical processes [6–9]. Enzymes are well known to be a class of proteins and have a phenomenal ability to accelerate the rates of reactions. Strong effects of pressure on rates of enzymatic reactions in supercritical carbon dioxide ($scCO_2$) have been reported [6, 7]. Furthermore, the solvent strength of SCFs also varies widely with pressure and temperature [10–13], and this sensitivity could prove useful for controlling solute–solvent or solute–solute intermolecular forces in SCF media where molecular aggregates such as micelles are formed. Salts of such large organic ions possessing large cations and anions with long hydrocarbon chains are often highly aggregated in dilute aqueous or organic solution. The resulting structured aggregates, together with counter-ions localized near their periphery by Coulomb forces, are termed micelles. The alternation of chemical reactions by incorporation of reactant molecules into micelles has received considerable investigation; for example, references [14–17], but little work in SCFs [18, 19]. This sensitivity in the solvent strength of SCFs may lead to the control of catalytic actions of micelles formed in SCFs by changing pressure and temperature, and one can expect that the rate observed in SCFs would be much higher than those obtained under normal conditions. The roles of micelles in reactions in SCFs must therefore be determined at the microscopic level.

Biocatalytic reactions in $scCO_2$ have been reported [8, 9]. The initial rate of lipase-catalyzed alcoholysis of methyl methacrylate by 2-ethyl hexanol increases with decreasing pressure [20]. Up to now, the effects of the conformations of active enzymes on reactivity have not been clarified.

l-Terpene esters are stereoselectively synthesized by Candida cylindracea (CCL) lipase from acyl donors and secondary alcohols in nonaqueous solutions, but not from primary alcohols [21]. Ikushima et al. established a CCL-catalyzed ester synthesis from a primary alcohol of (±)-citronellol and oleic acid in $scCO_2$ at 304.1 K. Studies were made to determine the reason for the stereoselectivity in the near-critical region. Certain interactions among the enzyme, reactants, and solvent may possibly occur in the near-critical region. Furthermore, Russell et al. [20] found the pressure dependence on stereoselectivity for the reaction between *N*-acetyl-(*L* or *D*)-phenylalanine ethyl ester and methanol catalyzed by Aspergillus protease in supercritical fluoroform at 323.1 K. Hence one can apply the pressure and temperature of SCFs to effectively regulate not only the stability and activity, but also the specificity in enzymatic reactions in SCFs. To resolve the reasons for the stereoselectivities, structure-function relations and factors that trigger activation should be determined under conditions of controlled catalytic properties of the enzymes.

Aerosol OT(Sodium bis-2-ethylhexylsulfosuccinate), or AOT, creates reverse (inverted) micelles and water-in-oil (w/o) microemulsions in a variety of apolar solvents, solubilizing a large amount of water in the aqueous core of the AOT reverse micelles [22, 23]. The active site of an enzyme is in a polar atmosphere surrounded by a hydrophobic environment, and thus the structure of an AOT reverse micelle is similar to that of the active site. The technological relevance of the AOT reverse microemulsions in apolar media includes the fields of oil recovery, lubrication and detergents, besides catalysis [24]. Reverse micelles or microemulsion phases are stable in continuous phases of supercritical alkanes, such as ethane and propane, though at lower water content than that in usual liquid solvent systems, such as isooctane and hexane [25–27]. These molecular aggregates in SCFs are of scientific and industrial importance because their marked pressure and temperature dependence should facilitate the control of catalytic actions of AOT reverse microemulsions, leading to rate enhancement. Therefore, it is essential to first determine structural characteristics and chemical microenvironments in H_2O/AOT/SCF systems;

direct (*in situ*) spectroscopic measurements will certainly provide detailed information at the molecular level.

The chemical reactions in such AOT microemulsions that have been studied so far include emulsion polymerization [18, 19], precipitation reactions [28], enzyme catalyzed reactions [29], the reduction of nitroxide [30], and so on [31]. Both the interactions between water and AOT molecules and hydration processes of the water-containing AOT aggregates in liquid solvent systems have been extensively studied by NMR and ESR spectroscopy [32, 33], fluorescence decay [34, 35], light [36], X-ray [37] and neutron [38] scattering. In SCF media, spectroscopic studies have also been done on such systems in SCF media, but to a lesser extent [25]. Dynamic light scattering in scC_2H_6 and scC_3H_8 has been investigated to determine the effects of pressure, temperature, and water-to-surfactant mole ratio (W) on AOT reverse micelle size [27, 39]. However, to shed light on the causes of the improved chemical reactivities, no detailed study has been done on either the microstructural properties, including conformational characteristics, or the hydration process of AOT reverse micellar aggregates in SCF media.

To determine the microstructural characteristics of enzymes and/or micelles in SCF media, *in situ* measurements with vibrational spectroscopy, such as infrared and Raman spectroscopy, should be made since these techniques have sufficient sensitivity to detect changes in microenvironments and intermolecular combinations. In this section, (i) the influence of CO_2 clusters on the conformation of the CCL lipase, (ii) involvement of conformational changes of the CCL lipase in reactivity, (iii) promotion effect of the chemical reaction in AOT micelles by controlling water-to-surfactant mole ratio and pressure, and (iv) participation of the structure of the surfactant/water interface in the promotion effect, will be mainly illustrated.

6.5.2 Enzymatic Reactions in Supercritical Carbon Dioxide

A lot of metabolic reactions in nature are catalyzed by enzymes. The specific stereoselectivity of these enzymes makes their use as catalysts desirable. However, many reactions that enzymes catalyze occur in an aqueous medium because many proteins are labile and frequently modified in apolar organic solvents. Hydrophobic substrates can be dissolved in apolar media, and enzymes in such environments that can catalyze reactions on water-insoluble substrates will be advantageous from the industrial point of view. Furthermore, apolar media using organic solvents like haxane can reduce microbial contamination because bacteria cannot grow in such nonaqueous media.

SCFs are an interesting class of nonaqueous media. Enzymatic reactions in SCFs are considered to be valid because of the high solubility of hydrophobic substrates, no solvent contamination, and the low viscosity and high diffusivity of SCFs. These characteristics increase the mass transfer coefficients of solutes, leading to rate enhancement. In addition, SCFs can offer the flexibility of changing physical properties without dramatic changes in chemical structures, and it is convenient to correlate the physical properties of SCFs to the behavior of enzymes. In conventional solvents one solvent interacts with an enzyme, while another one may not [40], and such a successful correlation may be hindered by unavoidable differences in chemical structure among conventional organic solvents.

ScCO_2 is favorable for its availability, cost and low toxicity, and enzymatic reactions in scCO_2 have been carried out in several model systems [7, 8], where the rate enhancement compared with that in organic solvents was ascribed to increases in the solubilities of water or substrates in scCO_2 relative to organic media. The stability of enzymes in scCO_2 has been further examined [8]. Loss of activity was attributed to the pressure effect on enzymes in scCO_2, in which CO_2 sorption in the hydrated enzyme molecules leads to a pH change, and an explosive release of CO_2 after reduction in pressure.

Esterification Catalyzed by Lipases

The industrial importance of organic syntheses catalyzed by lipases emphasizes a comprehensive understanding of their conformation and properties. However, few studies have been performed to elucidate the effects of scCO_2 on both enzymatic specificity and stereoselectivity of lipases. In exploring and controlling the catalytic properties of the lipase in scCO_2, the structure-function and factors that trigger the activation should be clarified at the molecular level. To understand these questions, examples of the influence of CO_2 clusters on the conformation of a lipase, and involvement of the conformational changes in reactivity will be shown below.

Ikushima et al. [41] demonstrated an interesting pressure effect, that the rate of esterification of *n*-valeric acid and citronellol catalyzed by Candida cylindracea (CCL) lipase is maximized at a certain pressure in the near-critical region with increasing pressure. Figure 6.5.1 shows the rate of esterification as a function of pressure at a fixed temperature of 308.1 K. The rate of esterification of *n*-valeric acid and citronellol at 308.1 K significantly increases, reaches a maximum of 282 mmol/h per 1 g of the enzyme at 7.55 MPa near the critical point and then decreases with increasing pressure. The rate of reaction at 24.5 MPa is 101 mmol/h per 1 g of the enzyme. The pressure dependence of the Kamlet–Taft α-value of scCO_2 is very similar to that of the rate of esterification observed, and the electron accepting power of the scCO_2 is of significance for the esterification studied.

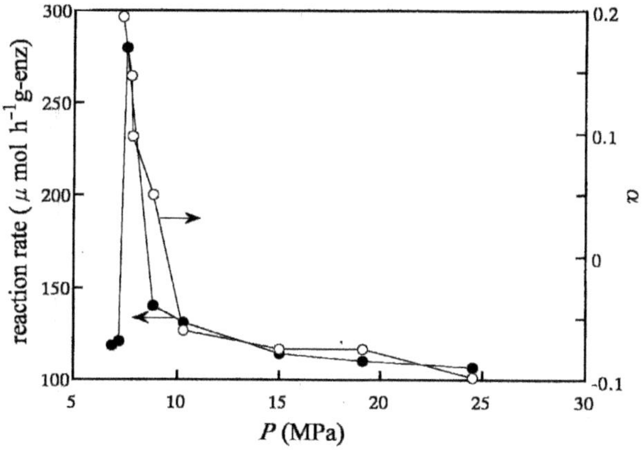

Fig. 6.5.1 Rate of esterification between *n*-valeric acid and citronellol catalyzed by Candida cylindracea (CCL) in $scCO_2$ (●) and α-value of $scCO_2$ (○) as a function of pressure at a temperature of 308.1 K [58]. Reprinted from: Y. Ikushima, Z. Shervani, N. Saito, M. Arai: The promotion of alkaline fading of crystal violet by AOT micelles dispersed in supercritical ethane medium, *J. Colloid Interface Sci.* (Academic Press) **191**, 177 (1997)

In the esterification of oleic acid and citronellol, only oleic acid 3,7-dimethyl-6-octenyl ester was synthesized as a product in all runs [42]. Table 6.5.1 shows the optical purity, i.e., % enantiomeric excess (e.e.) of (S)-(−)-oleic acid 3,7-dimethyl-6-octenyl ester calculated from the specific rotation $[\alpha]_D$ determined as $[\alpha]_D = \alpha/(C \times l)$, where α is the optical rotation observed, C the concentration of the ester, and l the cell length. The optical pu-

Table 6.5.1 Optical purity of the oleic acid 3,7-dimethyl-6-octenyl ester product in the esterification of (±)-citronellol and oleic acid in $scCO_2$

run	pressure (MPa)	temperature (K)	$[\alpha]_D$	optical purity[a] (% e.e.)
1	7.58	308.1	-0.08	4.1
2	8.41	308.1	-0.45	23.4
3	10.15	308.1	-0.04	2.1
4	15.03	308.1	-0.04	2.1
5	19.03	308.1	0.00	0.0
6	7.58	304.1	-0.50	25.7
7	8.41	304.1	-1.90	98.9
8	8.41	313.1	-0.04	2.1

[a] determined by dividing $[\alpha]_D$ observed by $[\alpha]_D$-1.92 for the ester formed from (*S*)-(-)-citronellol and oleic acid at 353 K in the presence of concentrated sulfuric acid.

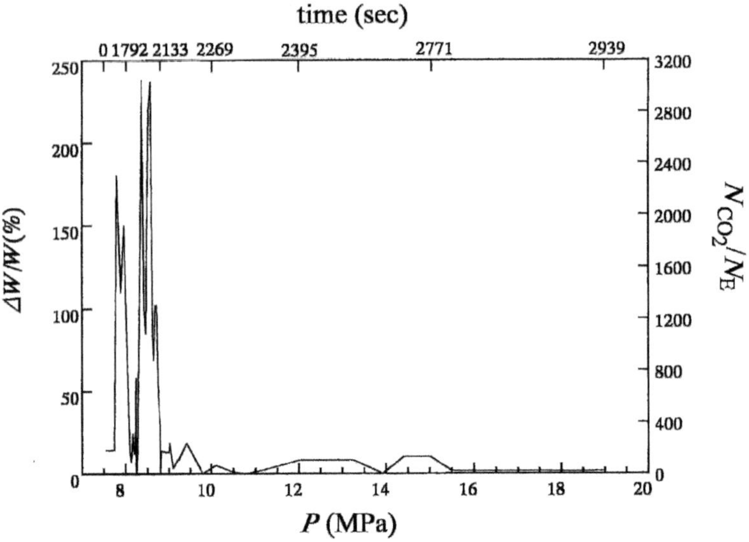

Fig. 6.5.2 Pressure dependence of the change of the weight of CCL film at pressures of 6.86−19.12 MPa at a temperature of 304.1 K. $\Delta W / W$ represents the changes in the weight of the CCL film relative to the initial value

rity of the rotation product, (S)-(−)-oleic acid 3,7-dimethyl-6-octenyl ester, was found to be very sensitive to the reaction pressure; it was at least 98% e.e. at 8 .41 MPa, indicating that (S)-(−)-oleic acid 3,7-dimethyl-6-octenyl ester was stereoselectively formed, whereas it was 26%e.e. at 7.58 MPa and 4.1%e.e. at 10.15 MPa. In water-saturated cyclohexane solution, the stereoselective synthesis of (S)-(−)-oleic acid 3,7-dimethyl-6-octenyl ester failed to occur. Although esterification from 5-phenylvaleric acid and (±)-citronellol by CCL is carried out in water-saturated cyclohexane solution [21], stereoselective esterification does not take place.

The pressure dependence of the change in weight as well as the secondary structure of the CCL at 304.1 K in scCO_2 was measured using FT-IR spectroscopy in conjunction with microgravimetry [43]. Figure 6.5.2 gives the change in weight of the CCL film with increasing pressure. A large increase in the weight was noted at 7.7−8.7 MPa. Carbon dioxide molecules thus considerably aggregate into the enzyme, and the mole ratio of CO_2 to CCL (MW = 56 000) molecules reached a maximum of about 3 000, indicating that strong interactions occurred between CO_2 and CCL at 7.7−8.7 MPa. These interactions were not stable nor were they reversible with pressure.

The amide I bands are intimately related to the molecular geometry and ultimately determine the protein secondary structure [44]. Figure 6.5.3 shows

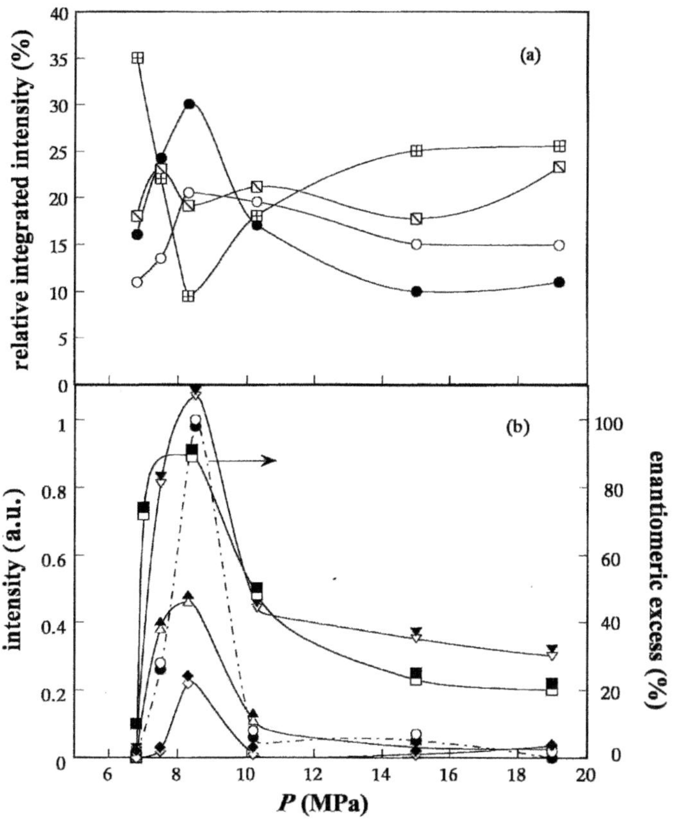

Fig. 6.5.3 (**a**) Pressure dependence of the relative intensities of α-helix at 1657 cm^{-1} (□ with plus), irregular at 1646 cm^{-1} (●), turn at 1666 cm^{-1} (○), and β-sheet-rich at 1635 and 1622 cm^{-1} (□ with slant). X-ray: 31% α-helix, 17% β-sheet. (**b**) Changes of intensities of absorption bands at 1722 (△,▲), 1760 (◇,◆), 2963 (□,■), and 2860 cm^{-1} (▽,▼) during upward (open symbols) and downward (filled symbols) pressure cycles (6.86–19.12–6.86 MPa) and of % e.e. (×) of (S)-(−)-oleic acid 3,7-dimethyl-6-octenyl ester on the esterification from oleic acid and (±)-citronellol

an example of FT-IR results, giving information on the secondary structure of CCL at different pressures in $scCO_2$. The assignment of bands was based on FT-IR data of proteins reported previously [45, 46]. Four interesting facts are implicit in Fig. 6.5.3. First, as shown in Fig. 6.5.3 (a), in a very limited pressure range, between 7.7 and 8.5 MPa, the band intensity at 1646 cm^{-1} due to the appearance of an increase in irregular structures significantly increases, due to a remarkable decrease in that at 1657 cm^{-1} corresponding to the appearance of a decrease in α-helical structures. Second, in Fig. 6.5.3 (b) only in the same near-critical region did the band intensity at 1722, 1760, 2963, and 2860 cm^{-1} suddenly increases, indicating the formation of either diester (R-OCO-COO-R') or amido (R-CONH-R') species. Carbon dioxide molecules and alkyl chains or amino groups of amino acids in α-helix combined directly to produce ester or amido species linked to different chains of amino acids. The band intensity at 1722, 1760, 2963, and 2860 cm^{-1} is sharp and very active, and each species protrudes spatially from the outside of the protein surface, being supported by the increase in turn (1666 cm^{-1}) and irregular structures and by the decrease in α-helical structures shown in Fig. 6.5.3 (a). Third, Fig. 6.5.3 (b) shows that the spectra recorded during the increase and decrease of the pressure agreed very well. This reversible pressure-induced unfolding and refolding of a protein have been studied very little. Fourth, the pressure at which there is optical purity in the CCL-catalyzed esterification was much the same as that for the sudden conformational changes in CCL.

The characteristic structure of CCL has a hydrophobic tunnel occluded by a "lid" and a closed conformation [47]. The "open" form of CCL requires only movement of the lid and shows a greater hydrophobic surface area near the active site. Lipase activity is strongly associated with activation of the helical lid movement [48, 49]. The IR data show that a conformational change of CCL occurred by the swinging of the helical portion of the lid. Since $scCO_2$ strongly interacts with hydrophobic compounds, alkyl chains or amino groups of amino acids that form an α-helix in the lid may be bound together through CO_2 molecules, thus producing a short irregular helix. In other words, the lid moves away from hydrophobic surface, and elongated hydrophobic species bound together through CO_2 protrude from the active site. The lid then opens to clear the entrance of the tunnel, thus allowing substrate molecules to easily approach the active site.

In the present stereoselective esterification, conformational change to the open structure in the near-critical region promote the approach of substrate molecules to the active site. A direct covalent bond is formed between Oγ of the Ser residue and carbon atoms of fatty acid, thus producing an oxyanion hole in a tetrahedral intermediate that is stabilized by interactions with NH groups [50], forming an oxyanion hole as shown in Figure 6.5.4. The resulting hydrophobic face may prevent the approach of (R)-(+)-citronellol to the active site that would result in the tetrahedral intermediate. Interactions between CO_2 and CCL molecules in the near-critical region may help to push the fatty acyl chain into the tunnel, like jamming a sting into a hole, to move a side-chain of residues along the tunnel, followed by accommodation of longer chain length of the intermediate. Thus, the solvent triggers the activation of CCL by causing lid movement and creates a hydrophobic active-site groove; this is the CCL stereoselective mechanism.

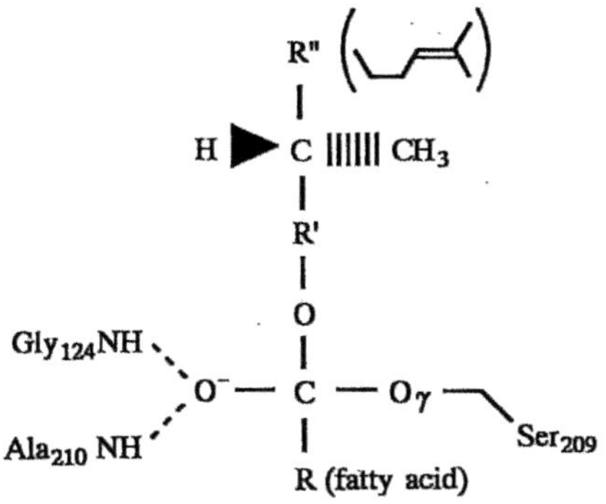

Fig. 6.5.4 Schematic representation of an inferred enzyme-substrate complex. The covalently modified Oγ atom of Ser 209 and oxyanion hole is shown schematically. The R" (C−C−C−C=C<) chain of (R)-(+)-citronellol might be blocked by the hydrophobic species (R_1−OCO−COO−R_2) sticking out from the surface in the formation of the intermediate, and the resulting product from the esterification would be the (S)-(−)-ester alone

6.5.3 Reactions in Microemulsions Formed in Supercritical Fluids

The incorporation of a microemulsion phase into SCFs creates a new class of solvent, and one can expect that this system will be a useful field for reaction processes. The improvement in mass transport properties due to the high diffusivities and the low viscosities of SCFs implies that in reaction processes the diffusion of reactants or recovery of products will be greatly facilitated. The hydrocarbon tail region divides the phase into two parts, the extremely low dielectric constant domain of the continuous SCF phase and the high dielectric constant microdomain of the aqueous core. The surfactant structure, interactions between the surfactant head group counter-ion and aqueous core, and interactions between the hydrocarbon chains and SCF phase affect molecular aggregation or orientation, leading to the appearance of a significant catalytic action contained in the aqueous core. Moreover, the interfacial region has intermediate properties between the aqueous core and the SCF phase, and many of the reactants that possesses slightly polar functional groups can be preferentially located at this region. Thus, the slightly polar reactants tends to meet near the interface region by decreasing the pressure of the SCF phase, and the increased local concentration of reactants can result in rate enhancement. It would also be convenient to recover the surfactant and/or products by adjusting the pressure of the SCF. Bright et al. have succeeded in forming micellar aggregates in $scCO_2$, in which the pH value in the aqueous core is signifi-

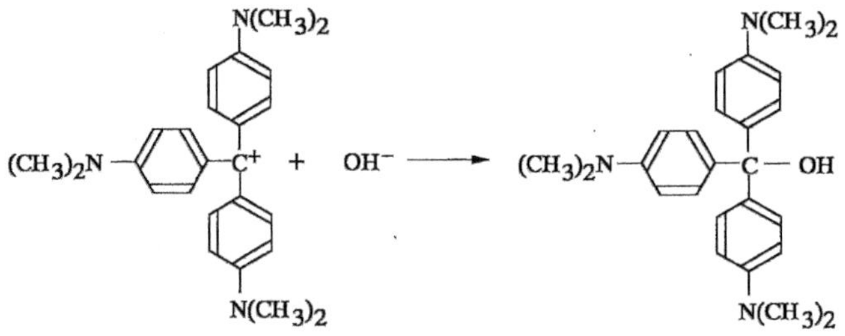

Fig. 6.5.5 Scheme for the alkaline fading of CV

cantly decreased due to the solubilization of CO_2 into the aqueous core [51]. These types of micelles seem to catalyze reactions which should be promoted by acid catalysts.

Preliminary studies of reactions in microemulsions formed in SCFs have been carried out. Matson et al. reported a reaction in which corpuscle particles of $Al(OH)_3$ were produced in a microemulsion in scC_3H_8 [28]. Aqueous $Al(NO_3)_3$ present in the core of AOT micelles reacted with NH_3 dissolved in supercritical propane and produced particles in the core. Beckmann et al. further reported the polymerization of water-soluble acrylamide in alcohol/water/ethane and propane systems under the supercritical states including the near-critical region [52]. The reactivities such as the molecular weight and yield of the polymer depend on the degree of clustering of the micelles and pressure of the SCF phase.

Fading of Crystal Violet

Although reaction processes in microemulsion systems in SCFs will be means of great promise from the scientific and industrial points of view, little work has been reported. Thus, an interesting example of a chemical reaction in AOT micelles formed in supercritical ethane as a continuous phase will be shown below, selecting the alkaline fading of crystal violet (CV) as a first test reaction (see Scheme 1 in Fig. 6.5.5). Figure 6.5.6 (a) shows the pseudo-first-order rate constant, k, as a function of the molar ratio of water to AOT, W, at 35.0 MPa. The rate constant gradually increases from $1 \times 10^{-5}\ s^{-1}$ to $3 \times 10^{-5}\ s^{-1}$ with increasing W up to 5. These values are smaller than $11.7 \times 10^{-4}\ s^{-1}$ observed for the reference liquid phase fading; however, as W further increases, k increases markedly and has a maximum value of about $1 \times 10^{-2}\ s^{-1}$ at $W = 6–7$, being two orders of magnitude larger than the rate constant for aqueous solution fading. This 100-fold or more enhancement in the rate constant is one of the most dra-

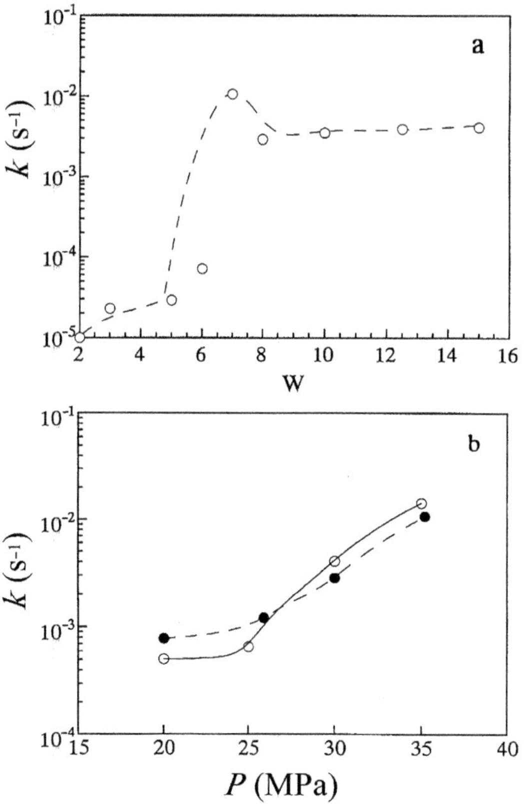

Fig. 6.5.6 The pseudo-first-rate constant, k, as a function of W (**a**) and pressure (**b**) for the fading of CV in H_2O/AOT/ethane w/o microemulsion under supercritical conditions at 306.1 K. (**a**) pressure = 35.0 MPa. (**b**) W = 6.5 (○) and 7.0 (●)

matic increases reported for reactions in SCFs [53, 54]. It is also much larger compared with the rate constant observed by Tamura et al. [55] for the same CV fading in H_2O/AOT/heptane at high pressures under similar conditions. The present system was in the transparent single phase when W was smaller than 9. At large W, however, the system was shown to change into a two-phase state: one was ethane including AOT micelles solubilizing some water and the other was water dissolving some AOT molecules. For the two-phase system at larger W, Figure 6.5.6 (a) shows the rate constant to depend little on W.

Figure 6.5.6 (b) shows the pseudo-first-order rate constant as a function of pressure for $W = 6.5$ and $W = 7$, in which the system is in the single-phase state. The rate constant increases with pressure, ranging from 7×10^{-4} s^{-1} to 1×10^{-2} s^{-1}, which is larger than those for the above-mentioned reactions under the reference liquid phase conditions and in H_2O/AOT/heptane at high pressures [55]. In addition, the observed increase of k with pressure is opposite to that reported by Tamura et al. [55] for the fading in H_2O/AOT/heptane in which k decreases with pressure in the range 0.1–98.1 MPa.

What is responsible for this unique specificity of rapid reaction appearing only in the AOT microemulsion formed in scC_2H_6? It is probable that the conformation of the AOT molecule (rotational isomerism), the local concentration of OH^-, and the micelle size influence the reactivities. In previous work, the characteristics of the interior water molecules and AOT molecules were examined by high-pressure FT-IR spectroscopy [56, 57]. The effect of W seems to depend on the type of interior water molecule and the conformation of AOT molecules. The effect of pressure is suggested to be related to the intermicellar and AOT–ethane interactions [58]. It can be said that chemical reactions in microemulsions in SCFs will be interesting useful processes from practical point of view.

6.5.4 Concluding Remarks

Among the enzymatic reactions in SCFs investigated so far, the use of lipase shows commercial promise; this method is positively applicable to bimolecular synthesis and modification. SCFs are ideal solvents for improving and examining an enzyme's function, and a greater understanding of the characteristics and mechanisms in SCFs can offer many opportunities for the control of enzyme activity and stability. The unique pressure or temperature dependence of SCF physico-chemical properties can influence the solvation state of substrates, products or enzymes, leading to high reaction rates and yields. The superior characteristics of SCFs over conventional liquid solvents may be further manifested in the enzymatic specificity such as regioselectivity, stereoselectivity, and chemoselectivity as well as separation bioprocesses. One can thus expect that the results of future investigation will enable the industrial production of, e.g., pharmaceutical or chiral compounds.

Establishment of a microemulsion system in the SCF phase can provide a means to overcome the difficulties in dissolving polar materials in apolar SCFs. The studies regarding microemulsion systems formed in SCFs have offered interesting details as to the pressure and temperature dependence on the phase behavior, the core properties, and micelle size; however, there has been little work on the catalytic actions. The pressure- and temperature-induced variability of the polarity in the aqueous core can facilitate the selective solubilization or separation of solutes, and the feasibility of performing both organic and inorganic reactions in the SCF microemulsion systems would be demonstrated, in which interesting pressure or temperature effects on reactivities may be revealed. Indeed, by just adjusting pressure and temperature of SCFs without changing the chemical nature of the system, properties of the microemulsion systems may be ultimately improved, and it is likely that a number of reactions will also arise.

6.6 Heterogeneous Catalysis in Supercritical Fluids

6.6.1 Introduction

As a new field related to supercritical fluid (SCF) technology, use of SCF solvents as reaction media for catalytic reactions, in which the SCF either actively participates in the relations or functions solely as the solvent medium for the reactants, catalysts, and products, has attracted increasing attention recently.

By making use of the characteristic features of SCFs, the reaction performance of a lot of catalytic reactions can be remarkably improved. Through exploiting the unique solvent properties of SCFs, it may be possible to enhance reaction rates while maintaining or improving selectivity. Also separating products from reactants or catalyst can be greatly facilitated by the ease with which the solvent power of the SCF solvent can be adjusted.

If the catalytic reaction field is a supercritical fluid, the following merits can be expected:

(i) Gaseous reactants such as hydrogen are able to well mix with SCF at high concentration. The reaction rate, dependent on gas concentration, will be enhanced. Its high diffusion coefficient can enhance the reaction rate in the diffusion-controlled reaction regime.

(ii) Organic compounds become soluble in SCFs to form a homogeneous reaction phase, leading to high reactivity and selectivity.

(iii) Through pressure controlling, separation between catalyst and reactant/product is easily realized. This property shows a strong effect on the reaction equilibrium shift, suppression of subreaction and extraction of the catalyst poison precursor. Pressure change is all that is needed to control the formation of the homogeneous phase or heterogeneous phase, and the transfer between the homogeneous phase and heterogeneous phase.

(iv) The high mass transfer efficiency and effective heat transfer capacity of SCFs make process-control easier and stabilize a lot of catalytic reactions.

Based on the unique properties of SCFs, not only can known chemical reaction performance be obviously improved, but also new reactions become possible in an SCF medium, where most of them need a catalyst. Some typical examples will be introduced below.

6.6.2 Fischer–Tropsch Synthesis in Supercritical Fluids

The Fischer–Tropsch (F–T) synthesis reaction conducted in a solid-catalyzed gas-phase reaction system, which has been commercially operated or will be operated in several countries, is inevitably accompanied by local overheating of the catalyst surface as well as by the production of heavy wax (paraffin higher than C20). Local overheating of the catalyst may lead to the deactivation of the catalyst and also to the increase in methane selectivity. Heavy wax may plug micropores of the catalyst and the catalyst bed itself, resulting in catalyst deactivation as well.

The slurry-phase F–T reaction process, on the other hand, in which a slurry, composed of fine powdery catalyst and mineral oil, is used as a reaction medium, has been developed to overcome defects of the gas-phase process [1]. The diffusion of synthesis gas in the micropores of the catalysts is, however, so slow in the slurry phase that the overall reaction rate is markedly lower than that in the gas-phase reaction [2]. Further, the concentration of solid catalyst particles in the slurry medium is limited to a low level (< 20 wt%), in order to maintain slurry fluidity. Other disadvantages of the slurry reactors are: the accumulation of high molecular weight products in the reactor during operation, and the *in situ* separation of fine catalyst particles from the heavy products.

Fujimoto and Fan developed an F–T synthesis reaction in the supercritical phase, and compared its reaction performance with those in the liquid phase and gas phase [3–5]. The supercritical phase F–T reaction shows unique characteristics such as quick diffusion of reactant gas, effective removal of reaction heat, and *in situ* extraction of high molecular weight hydrocarbons (wax). Bukur et al. conducted the F–T reaction on iron catalysts in supercritical phase propane, and a similar observation was obtained [6, 7].

The supercritical fluid was selected using the following criteria:

(i) Its critical temperature and pressure should be slightly lower than the typical reaction temperature and pressure.
(ii) It should not poison the catalysts and should be stable under the reaction conditions.
(iii) It should have a high affinity for aliphatic hydrocarbons to extract wax from the catalyst surface and reactor.

n-Hexane (C_6H_{14}), whose critical temperature and critical pressure are 233.7 °C and 29.7 atm, respectively, was chosen as the supercritical solvent. The standard reaction conditions were $T = 240$ °C, P (total) = 45 atm, P ($CO + H_2$) = 10 atm, P (balance gas) = 35 atm, CO/H_2 = 1/2, W / F ($CO + H_2$) = 10 g of catalyst h/mol.

(a) Diffusion Behavior of Synthesis Gas

Figure 6.6.1 shows the Arrhenius plots for ruthenium-catalyzed reactions in the three reaction phases, which indicates that the reaction rate fell in the following order: gas phase > supercritical phase > liquid phase.

Apparent activation energies were 23 kcal/mol for the gas-phase reaction, 21 kcal/mol for the supercritical-phase reaction and 17 kcal/mol for the liquid-phase reaction. These results lead to the conclusion that the mass transfer efficiency in the supercritical-phase reaction is higher than that in the liquid-phase reaction and is close to that in the gas-phase reaction.

The lower catalytic activity and the smaller apparent activation energy in the liquid phase reaction should be attributed to the lower rate of synthesis gas diffusion inside the catalyst particles compared to the rate of the surface reaction, because the activation energy of diffusion should be somewhat smaller than the intrinsic reaction activation energy. As a result, the reactant concentrations inside the catalyst particles were so low that the effectiveness factor of the catalyst had dropped.

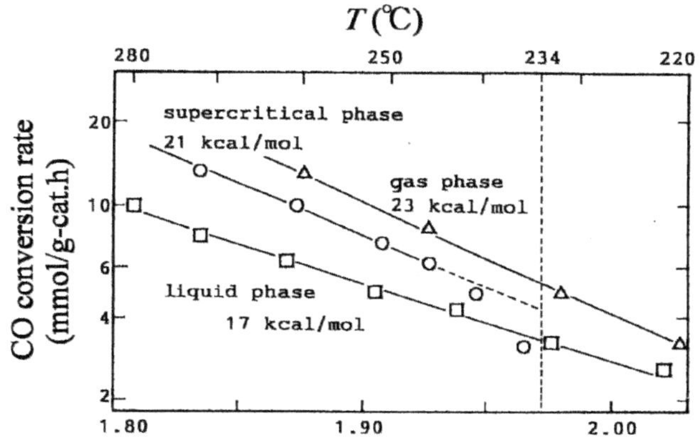

Fig. 6.6.1 Arrhenius activation energy of F-T reaction in various phases. Co-La/SiO_2 catalyst, standard reaction conditions, W / F ($CO + H_2$) = 3 g·cat.h/mol

(b) Diffusion and Reaction of Products

In the F-T synthesis reaction, 1-olefins are produced as the primary products and are subjected to various secondary reactions such as hydrogenation, C–C chain propagation and hydrocracking [8].

Shown in Fig. 6.6.2 is a comparison of the olefin content in the products in each reaction phase on a Ru/Al_2O_3 catalyst. The olefin content in the products decreased in each reaction phase with an increase in the carbon number of the products. The decrease in the olefin content with increasing carbon number should be attributed to the increase in the hydrogenation rate relative to the diffusion rate with increasing carbon number. This is due to the longer residence time on the catalyst surface derived from the slower diffusion rate of these olefins.

Although the mass transfer rate in the supercritical fluid is slightly lower than that in the gas phase, the olefin content in the produced hydrocarbons in

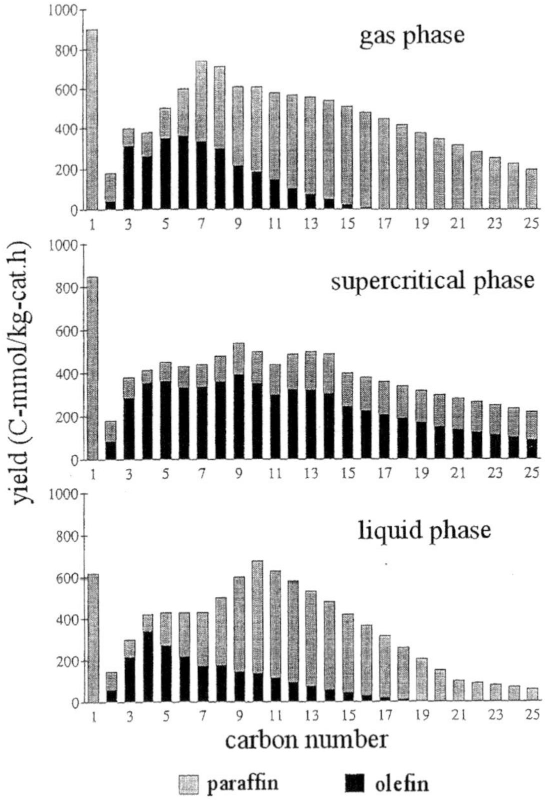

Fig. 6.6.2 Olefin distribution in hydrocarbon products of F-T reaction in various phases. Ru/Al_2O_3 catalyst, standard reaction conditions

supercritical-phase reaction was much higher than that in the gas-phase reaction. This means that the fine balance between the rates of product desorption from the catalyst surface and its transfer from the catalyst pores, is the characteristic feature for the effective overall transfer of the products in the supercritical phase reaction. The olefins, therefore, were extracted and transported by the supercritical fluid out of the catalyst particles quickly before it was readsorbed and hydrogenated. On the other hand, the primary products, especially those of high molecular weight, in the gas-phase reaction, hardly desorbed to the gas phase flow medium and had to be subjected to further hydrogenation, resulting in their low olefin contents. In the liquid-phase reaction, the low olefin content was mainly due to the slow diffusion of these olefins inside the catalyst pores, filled with the liquid reaction medium, and the enhanced possibility of successive hydrogenation, even if these primary olefins desorbed fast enough from catalyst surface.

(c) Wax Production: Addition of Heavy Olefins in the Supercritical Phase

F–T wax, owing to its merits such as high melting point, high hardness value, low viscosity, as well as being nitrogen-, sulfur-, and aromatics-free, has been highly favored. It can be used not only directly in many fine chemical fields such as cosmetics, packing material and adhesive, but also as a feedstock through different degrees of cracking reactions, to produce different-sized hydrocarbon products ranging from kerosene to aviation fuels.

It is extremely difficult to selectively synthesize waxy hydrocarbons through the F–T reaction. The main reason is that the F–T products fall in the Anderson–Schultz–Flory (ASF) distribution [8]. The development of a new type of F–T reaction, free from the ASF constraint on its product selectivity, is of great importance for wax production, as modification of the F–T catalyst alone is not enough to increase wax selectivity significantly.

Here, addition of a small amount of heavy 1-olefin into the supercritical phase F–T reaction can significantly promote chain growth and greatly enhance the selectivity of waxy products. As a matter of interest, this phenomenon does not occur in the gas phase reaction [9, 10].

Shown in Fig. 6.6.3 is the F-T product distribution profile for a 4 mol% (CO base) addition of 1-tetradecene or 1-hexadecene into the supercritical *n*-pentane (T_C : 196.6 °C; P_C : 33.3 atm) fluid. It is clear from the figure that the product distributions in the olefin-added systems are very flat, in marked contrast with that in the supercritical phase F–T reaction without addition of olefin. The selectivities for hydrocarbons lower than C14 were higher in the F–T reaction without addition of olefin. The reverse was true, however, for heavy products with carbon numbers higher than 14; the selectivity to waxy products

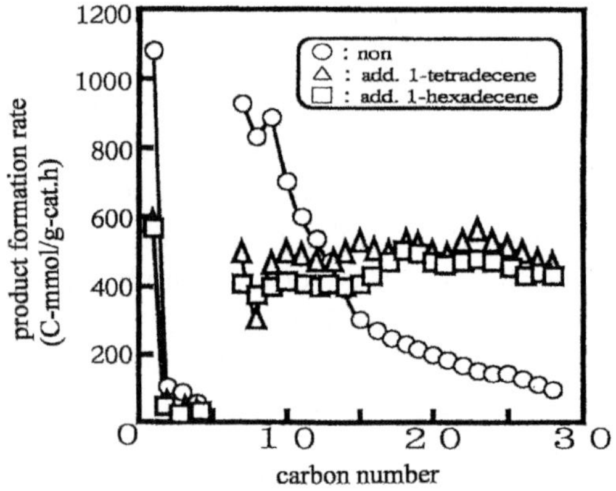

Fig. 6.6.3 Addition of heavy olefins into supercritical phase F–T reactions: anti-ASF distribution Co-La/SiO$_2$ catalyst, T = 220 °C, P (n-pentane) = 35 atm, P (CO + H$_2$) = 10 atm, W/F (CO + H$_2$) = 9 g·cat.h/mol, added olefin amount: 4 % (CO base)

was remarkably enhanced in the olefin-added reactions. Another remarkable phenomenon was the suppression of methane formation in the olefin-added F–T reactions.

Similar behavior was observed for the addition of 1-heptene or 1,7-octadiene to the supercritical phase reaction [9, 10].

6.6.3 Synthesis of Fine Chemicals in Supercritical Fluids

Paliakoff et al. introduced a supercritical phase into the Friedel–Crafts alkylation reaction by using supercritical CO_2 or making propene, one reactant, to a supercritical state [11]. This heterogeneous supercritical phase reaction was conducted continuously and the selectivity was very high, if a solid acid, Deloxan catalyst, was utilized.

Vieville et al. conducted the esterification of oleic acid with methanol in supercritical CO_2 using a cation exchange resin [12]. They reported ester yields of about 50% for reactions at 40 °C and 160 atm, as shown below. They also found that the reaction rate was limited by transportation of the reactants to the external surface of the resin,

$$\mathrm{RCOOH} + \mathrm{CH_3OH} \underset{\text{acidic resin}}{\overset{\text{scCO}_2}{==}} \mathrm{RCOOCH_3} + \mathrm{H_2O}. \tag{6.6.1}$$

(R: $CH_3(CH_2)_7CH{=}CH(CH_2)_7$)

The enantioselective hydrogenation of ethyl pyruvate to (R)-ethyl lactate has been studied by Minder et al. [13] using SCFs. The catalyst was a Pt/Al_2O_3 modified with cinchonidine. In scC_2H_6 the reaction time could be reduced by a factor of 3.5 compared to toluene under similar conditions, without any loss in enantioselectivity. A further advantage of ethane is that the enantioselectivity remained high even at high catalyst/reactant ratio, which indicated application of a continuous fixed-bed reactor for this reaction. But a strong catalyst deactivation was observed in supercritical CO_2, due to the reduction of CO_2 on Pt.

Hybrid catalysts derived from group 8 metal-chloro complexes by cocondensation with $Si(OEt)_4$ via a sol-gel process were highly active for the synthesis of N, N-dimethylformamide from CO_2, H_2 and dimethylamine under supercritical conditions, affording turnover numbers up to 100 800 at 100% selectivity [14]. These catalysts, containing methylphosphine ligands, showed the activity as follows: Ru > Ir > Pt, Pd > Rh. It seemed that the highest activity of silica matrix-stabilized ruthenium complexes was due to the formation of an active hydride intermediate by hydrogenolysis of the Ru-Cl bond,

$$CO_2 + H_2 + Me_2NH = Me_2NC(O)H + H_2O. \tag{6.6.2}$$

From the viewpoint of reaction engineering, some studies provide interesting insights on the role of SCF. The poly-siloxane-supported noble metal catalysts were used for hydrogenation of a wide range of organic functional group by H_2, where $scCO_2$ or scC_3H_8 was introduced to the flow-type reactor [15]. As many reaction condition parameters such as temperature, pressure or feed ratio could be adjusted independently and continuously, the selectivity and yield of a particular product were optimized.

For oxidation reaction, Dooley and Knopf conducted toluene oxidation by air in $scCO_2$ [16]. If a redox catalyst or acid catalyst was used, toluene was converted to benzaldehyde effectively. It was revealed that Al_2O_3-supported CoO catalyst, partly oxidized to Co^{3+}, was the most active and selective catalyst, due to the Co^{3+}/Co^{2+} redox couple. If this reaction was conducted in the liquid phase alternatively, CO_2 formed in large amounts and partial oxidation product selectivity was suppressed.

6.6.4 Kinetics of Heterogeneous Catalysis in Supercritical Fluids

(a) Syngas Transfer in the Fischer–Tropsch Reaction

It is well known that in the gas-phase F-T reaction, after reaching the steady state, the catalyst pores were filled with heavy hydrocarbons in which CO and H_2 must be dissolved and subsequently diffuse to reach the inner surface of the catalyst. The same situation happens in the liquid-phase reaction but the catalyst pores are filled with *n*-hexadecane. In the supercritical-phase F-T reaction, due to the higher solubility of SCF, a large part of these heavy hydrocarbons could be dissolved and removed *in situ* by an SCF whose density and viscosity were much smaller than those of the heavy hydrocarbons. In order to interpret the differences in gas-phase, liquid-phase and supercritical-phase reaction over the F–T catalyst, the syngas diffusion coefficient and its concentration profiles inside catalyst pores were calculated and simulated.

By supposing a spherical catalyst with radius R, the material balances of CO and H_2 within a unit shell with radius r and thickness $\mathrm{d}r$ can be calculated:

$$D_{\mathrm{CO}}\left(\frac{\mathrm{d}^2 C_{\mathrm{CO}}}{\mathrm{d}r^2}+\frac{2\,\mathrm{d}C_{\mathrm{CO}}}{r\,\mathrm{d}r}\right)-V_{\mathrm{CO}}=0. \tag{6.6.3}$$

$$D_{\mathrm{H2}}\left(\frac{\mathrm{d}^2 C_{\mathrm{H2}}}{\mathrm{d}r^2}+\frac{2\,\mathrm{d}C_{\mathrm{H2}}}{r\,\mathrm{d}r}\right)-V_{\mathrm{H2}}=0. \tag{6.6.4}$$

Here r is the radius of the shell; D_{CO}, D_{H2} are the effective diffusion coefficients for H_2 and CO in the reaction medium; V_{CO}, V_{H2} is the reaction rate per unit volume; and C_{CO} and C_{H2} are the concentrations of the reactants.

The boundary conditions are:

$$r=R \qquad C_{\mathrm{H2}}=C_{\mathrm{H2,0}},\ C_{\mathrm{CO}}=C_{\mathrm{CO,0}}. \tag{6.6.5}$$

$$r=0 \quad \frac{\mathrm{d}C_{\mathrm{H2}}}{\mathrm{d}r}=\frac{\mathrm{d}C_{\mathrm{CO}}}{\mathrm{d}r}=0. \tag{6.6.6}$$

It was found that the F–T synthesis proceeded almost exclusively via the reaction:

$$n\mathrm{CO}+m\mathrm{H_2}\rightarrow \mathrm{C}_n\mathrm{H}_{2(m-n)}+n\mathrm{H_2O}. \tag{6.6.7}$$

Further reaction of H_2O with CO, known as the water–gas shift reaction, can be neglected. The reaction rate equation of the F–T reaction can be expressed as follows.

$$v=kP_{\mathrm{H2}}^{m}P_{\mathrm{CO}}^{n}. \tag{6.6.8}$$

Here k is the rate constant and m and n are the exponential dependencies on the H_2 and CO partial pressure, respectively. It is often recognized that the reaction orders of the F–T synthesis reaction with respect to hydrogen and carbon monoxide are one and zero, respectively [17], since the conversion rate depends on the hydrogen pressure rather than on the carbon monoxide pressure. The relationship between V_{H} and V_{CO} follows the stoichiometry considerations:

$$V_{\mathrm{CO}} = kC_{\mathrm{H2}} = (1/2)V_{\mathrm{H2}}, \tag{6.6.9}$$

where k is the reaction constant. From (6.6.3), (6.6.4) and (6.6.9), the profile of the hydrogen concentration along the radius of the catalyst sphere can be expressed as follows:

$$\frac{C_{\mathrm{H2}}}{C_{\mathrm{H2,0}}} = \frac{\sinh(3\phi r/R)}{(r/R)\sinh(3\phi)}, \tag{6.6.10}$$

where ϕ is the Thiele modulus

$$\phi = \frac{R}{3}\left(\frac{k}{D_{\mathrm{H2}}}\right)^{\frac{1}{2}}. \tag{6.6.11}$$

From (6.6.3), (6.6.4) and (6.6.9), the relationship between the hydrogen and carbon monoxide concentration can be obtained.

$$C_{\mathrm{CO}} = \frac{D_{\mathrm{H2}}}{2D_{\mathrm{CO}}}(C_{\mathrm{H2}} - C_{\mathrm{H2,0}}) + C_{\mathrm{CO,0}}, \tag{6.6.12}$$

so the catalyst effectiveness factor η can be derived as:

$$\eta = \frac{1}{\phi}\left(\frac{1}{\tanh(3\phi)} - \frac{1}{3\phi}\right). \tag{6.6.13}$$

According to the study of Debenedetti and Reid [18], the method of Wilke and Chang is the most appropriate to calculate the diffusion coefficients in the SCF systems [19], if the Chapman–Enskog method [20] is used for calculation of the viscosity of the solvent. For liquid-phase and gas-phase reactions, the Stokes–Einstein equation can be utilized to obtain diffusion coefficients of CO and H_2 in the reaction medium [4], where the influence on the solution composition can be corrected via Vignes equation [21].

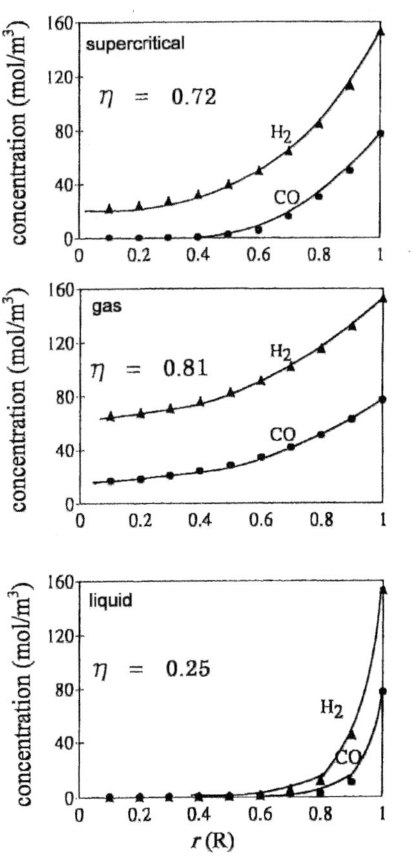

Fig. 6.6.4 Calculated concentration distribution of synthesis gas inside catalyst pellets Co-La/SiO_2 catalyst, D_P = 1 mm, standard reaction conditions

The calculated results and simulated syngas concentration profiles inside the catalyst pores in gas-phase, liquid-phase and supercritical-phase reactions are compared in Fig. 6.6.4.

From the concentration profiles of syngas inside the catalyst pores, it is clear that the diffusivity of syngas in the liquid-phase reactions is much lower than that in the supercritical-phase and gas-phase reactions, which leads to significant profiles inside the catalyst particles, and also to a decrease of the catalyst effectiveness factor. This model was also used to simulate the catalyst pore size effect in supercritical-phase F−T reaction [4].

(b) Relationship Between Catalyst Deactivation and Product Diffusion in Supercritical Fluids

Subramaniam et al. studied the 1-hexene isomerization reaction in the supercritical phase. Establishing a single-pore model to describe the physico-chemical processes including coke formation and *in situ* coke extraction, Baptist-Nguyen and Subramaniam showed that for an increase in pressure along a near-critical isotherm, there exists an optimum pressure, or reaction mixture density, at which catalyst activity is maximized [22]. At lower than optimum densities, the catalyst undergoes deactivation due to a lack of coke extraction, while at higher than optimum densities, the catalyst activity decreases due to pore-diffusion limitation in the liquid-like reaction mixtures.

Subramaniam and McCoy described a kinetic scheme in which coke compound formation from olefin oligomerization and their adsorption–desorption at the catalyst surface were simulated. A model was set up based on continuous mixture kinetics to interpret the molecular weight distributions for the oligomers and aromatic coke compounds which were obtained in the experiments [23, 24]. The model was consistent with observations of the deactivation rates of the deactivated catalyst when coke desorption rates approached coke production rates. This model was also successful in predicting the catalyst activity maintenance at supercritical conditions.

6.6.5 Concluding Remarks

Heterogeneous catalysis happens in multiphase systems. If a new phase, a supercritical phase, is suitably introduced into this system, remarkable change can be expected. The introduction of a supercritical phase to heterogeneous catalyst systems brings about many effects, including effective transfer of reactants and products inside catalyst pores and extraction of high molecular weight materials which do not vaporize under the reaction conditions and have the potential for deactivating the catalyst. A successful supercritical-phase heterogeneous catalytic reaction can be realized, as long as these parameters are carefully designed. With obviously higher activity, catalyst lifetime or selectivity than those in gas-phase reaction or liquid-phase reaction, some supercritical-phase reactions are of industrial potential.

References

6.1 Reactions and Reaction Theory in Supercritical Fluids

[1] G. Brunner, "Gas Extraction. An Introduction to Fundamentals of Supercritical Fluids and the Application to Separation Processes", Springer, New York (1994).
[2] M. McHugh and V. J. Krukonis, "Supercritical Fluid Extraction: Principle and Practice", 1st ed., Butterworths, Boston (1986).
[3] S. Saito, J. Supercritical Fluids, **8**, 177 (1995).
[4] P. E. Savage, S. Gopalan, T. I. Mizan, C. J. Martino and E. E. Brock, AIChE J., **41**, 1723 (1995).
[5] P. E. Savage, Chem. Rev., **99**, 603 (1999).
[6] A. Saul and W. Wagner, J. Phys. Chem. Ref. Data, **18**, 1537 (1989).
[7] M. Uematsu and E. U. Franck, J. Phys. Chem. Ref. Data, **9**, 1291 (1980).
[8] G. M. Schneider, Ber. Bunsenges. Phys. Chem., **76**, 325 (1972).
[9] E. U. Franck, Pure & Appl. Chem., **53**, 1401 (1981).
[10] C. B. Roberts, J. Zhang, J. F. Brennecke and J. E. Chateauneuf, J. Phys. Chem., **97**, 5618 (1993).
[11] K. J. Laidler, "Chemical Kinetics", 3rd ed., Harper & Row, New York, p. 183 (1987).
[12] E. S. Amis and J. F. Hinton, "Solvent Effects on Chemical Phenomena", Vol. 1, Academic Press, New York (1973).
[13] P. Debye and J. McAulay, Physik. Z., **26**, 22 (1925).
[14] Y. Ikushima, N. Saito and M. Arai, J. Phys. Chem., **96**, 2293 (1992).

6.2 Characteristics of Supercritical Fluids as Reaction Solvents

[1] J. M. DeSimone, E. E. Maury, Y. Z. Menceloglu, J. B. McClain, T. J. Romack and J. R. Combes, Science, **265**, 356 (1994).
[2] C. Y. Tsang and W. B. Streett, Chem. Eng. Sci., **36**, 993 (1981).
[3] M. Uematsu and E. U. Franck, J. Phys. Chem. Ref. Data, **9**, 1291 (1980).
[4] W. L. Marshall and E. U. Franck, J. Phys. Chem. Ref. Data, **10**, 295 (1981).
[5] D. S. Bulgarevich, T. Sako, T. Sugeta, K. Otake, Y. Takebayashi and C.Kamizawa, J. Chem. Phys., **111**, 4239 (1999).
[6] N. Asahi and Y. Nakamura, J. Chem. Phys., **109**, 9879 (1998).
[7] T. Ebukuro, A. Takami, Y. Oshima and S. Koda, J. Supercritical Fluids, **15**, 73 (1999).
[8] T. Sako, T. Sugeta, K. Otake, N. Nakazawa and M. Sato, J. Chem. Eng. Japan, **30**, 342 (1997).
[9] T. Sako, T. Sugeta, K. Otake, Y. Takebayashi, C. Kamizawa, M. Tsugumi and M. Hongo, Kobunshi Ronbunshu, **55**, 685 (1998).
[10] T. Sako and M. Suzuki, Chemistry and Chemical Industry, **52**, 288 (1999).

6.3 Oxidation and Hydrolysis Reactions in Supercritical Water

[1] H. R. Holgate, P. A. Webley, J. W. Tester and R. K. Helling, Energy and Fuels, **6**, 586 (1992).
[2] H. R. Holgate and J. W. Tester, Combust. Sci. Tech., **88**, 369 (1993).
[3] P. A. Webley and J. W. Tester, Energy and Fuels, **5**, 411 (1991).
[4] R. R. Steeper, S. F. Rice, I. M. Kennedy and J. D. Aiken, J. Phys. Chem., **100**, 184 (1996).
[5] J. W. Tester, P. A. Webley and H. R. Holgate, Ind. Eng. Chem. Res., **32**, 236 (1993).
[6] E. E. Brock, Y. Oshima, P. E. Savage and J. R. Barker, J. Phys. Chem., **100**, 15834 (1996).
[7] S. F. Rice, T. B. Hunter, A. C. Ryden and R. G. Hanush, Ind. Eng. Chem. Res., **35**, 2161 (1996).
[8] P. E. Savage and S. Gopalan, AIChE J., **41**, 1864 (1995).
[9] M. Krajnc and J. Levec, AIChE J., **42**, 1977 (1996).
[10] M. Koo, W. K. Lee and C. H. Lee, Chem. Eng. Sci., **52**, 1201 (1997).
[11] Y. Oshima, K. Hori, M. Toda, T. Chommanad and S. Koda, J. Supercrit. Fluids, **13**, 241 (1998).
[12] H. R. Holgate and J. W. Tester, J. Phys. Chem., **98**, 810 (1994).
[13] E. E. Brock and P. E. Savage, AIChE J., **41**, 1874 (1995).
[14] S. Gopalan and P. E. Savage, ACS Symposium Series, **608**, Chapter 14 (1995).
[15] L. T. Boock and M. T. Klein, Ind. Eng. Chem. Res., **32**, 2464 (1993).
[16] L. T. Boock and M. T. Klein, Ind. Eng. Chem. Res., **33**, 2554 (1994).
[17] T. Sako, T. Sugeta, K. Otake, M. Sato and M. Tsugumi, J. Chem. Eng. Japan, **30**, 744 (1997).
[18] S. H. Townsend, M. A. Abraham, G. L. Huppert, M. T. Klein and S. C. Paspek, Ind. Eng. Chem. Res., **27**, 143 (1988).
[19] C. L. Harrell, M. T. Klein and T. Adschiri, Advances in Environmental Research, **1**, 373 (1997).
[20] T. Adschiri, Kagaku Kogaku Ronbunshu, **23**, 505 (1997).
[21] T. Adschiri, S. Hirose, R. M. Malaluan and K. Arai, J. Chem. Eng. Jpn., **26**, 676 (1993).
[22] R. M. Malaluan, Ph.D. dissertation, Tohoku University, Sendai, Japan (1995).
[23] M. Sasaki, B. M. Kabyemela, R. M. Malaluan, S. Hirose, N. Takeda, T. Adschiri and K. Arai, J. Supercrit. Fluids, **13**, 261 (1998).
[24] K. Arai and T. Adschiri, Fluid Phase Equilibria, in press.
[25] E. S. Amis and J. F. Hinton, "Solvent Influence on Rates and Mechanisms", in "Solvent Effects on Chemical Phenomena", ed. E. S. Amis and J. F. Hinton, Academic Press, New York (1973).
[26] B. M. Kabyemela, T. Adschiri, R. M. Malaluan and K. Arai, Ind. Eng. Chem. Res., **36**, 1552 (1997).
[27] T. Adschiri and R. Shibata; K. Arai, Sekiyu Gakkaishi, **40(4)**, 291 (1997).
[28] European Chemical News, 1–7 June (1998).

6.4 Organic and Catalytic Reactions in Supercritical Carbon Dioxide

[1] B. Subramaniam and M. A. McHugh, Ind. Eng. Chem. Proc. Des. Dev., **25**, 1 (1986).
[2] J. F. Brennecke, "Supercritical Fluid Engineering Science Fundamentals and Applications", ACS Symposium Series, **514**, ed. E. Kiran and J. F. Brennecke, American Chemical Society, Washington DC, p. 201 (1993).
[3] A. A. Clifford, "Supercritical Fluids Fundamentals for Application of the NATO ASI Series E", ed. E. Kiran and J. M. H. L. Sengers, Kluwer, Amsterdam, p. 449 (1994).
[4] P. E. Savage, S. Gopalan, T. I. Mizan, C. J. Martino and E. E. Brock, AIChE J., **41**, 1723 (1995).
[5] C. A. Eckert, B. L. Knutson and P. G. Debenedetti, Nature, **383**, 313 (1996).
[6] T. Clifford and K. Bartle, Chem. Ind., 449 (1996).
[7] M. Poliakoff, S. M. Howdle and S. G. Kazarian, Angew. Chem., Int. Ed. Engl., **34**, 1275 (1995).
[8] G. M. Kramer and F. Leder, U.S. Patent 3,880,945 (1975).
[9] J. W. Rathke, R. J. Klingler and T. R. Krause, Organometallics, **10**, 1350 (1991).
[10] P. G. Jessop, T. Ikariya and R. Noyori, Nature, **368**, 231 (1994).
[11] S. M. Howdle, P. Grebenik, R. N. Perutz and M. Poliakoff, J. Chem. Soc., Chem. Commun., 1517 (1989).
[12] S. M. Howdle, M. A. Healy and M. Poliakoff, J. Am. Chem. Soc., **112**, 4804 (1990).
[13] S. M. Howdle and M. Poliakoff, J. Chem. Soc., Chem. Commun., 1099 (1989).
[14] K. P. Johnston and C. Haynes, AIChE J., **33**, 2017 (1987).
[15] S. Kim and K. P. Johnston, "Supercritical Fluids, Chemical and Engineering, Principles and Applications", ed. T. G. Squires and M. E. Paulaitis, ACS Symposium Series, **329**, American Chemical Society, Washington DC, p. 42 (1987).
[16] G. Kaupp, Angew. Chem., Int. Ed. Engl., **33**, 1452 (1994).
[17] T. Toriumi, J. Sakai, T. Kawakami, D. Osawa and M. Azuma, J. Soc. Chem. Ind. Japan, **49**, 1 (1946).
[18] J. F. Brennecke, "Supercritical Fluid Engineering Science", ed. E. Kiran and J. F. Brennecke, ACS Symposium Series, **514**, American Chemical Society, Washington DC, p. 201 (1993).
[19] B. L. Knutson, A. K. Dillow, C. L. Liotta and C. A. Eckert, "Innovations in Supercritical Fluids, Science and Technology", ed. K. W. Hutchenson and N. R. Foster, ACS Symposium Series, **608**, American Chemical Society, Washington DC, p. 166 (1995).
[20] R. D. Weinstein, A. D. Renslo, R. L. Danheiser, J. G. Harris and J. W. Tester, J. Phys. Chem., **100**, 12337 (1996).
[21] M. E. Paulaitis and G. C. Alexander, Pure Appl. Chem., **59**, 61 (1987).

[22] C. B. Roberts, J. Zhang, J. E. Chateauneuf and J. F. Brennecke, J. Am. Chem. Soc., **114**, 8455 (1992).
[23] C. B. Roberts, J. F. Brennecke and J. E. Chateauneuf, AIChE J., **41**, 1306 (1995).
[24] J. B. Ellington and J. F. Brennecke, J. Chem. Soc., Chem. Commun., 1094 (1993).
[25] J. B. Ellington, K. M. Park and J. F. Brennecke, Ind. Eng. Chem. Res., **33**, 965 (1994).
[26] T. A. Rhodes, K. O'Shea, G. Bennett, K. P. Johnston and M. A. Fox, J. Phys. Chem., **99**, 9903 (1995).
[27] B. J. Hrnjez, A. J. Mehta, M. A. Fox and K. P. Johnston, J. Am. Chem. Soc., **111**, 2662 (1989).
[28] S. J. Kamat, E. J. Beckman and A. J. Russell, J. Am. Chem. Soc., **115**, 8845 (1993).
[29] S. Kim and K. P. Johnson, Chem. Eng. Commun., **63**, 49 (1988).
[30] A. A. Clifford, K. Pople, W. J. Gaskill, K. D. Bartle and C. M. Rayner, Faraday Trans., **94**, 1451 (1998).
[31] C. Reichardt, Chem. Rev., **94**, 2319 (1994).
[32] Z. Guan, J. R. Combes, Y. Z. Menceloglu and J. M. DeSimone, Macromolecules, **26**, 2663 (1993).
[33] K. E. O'Shea, J. R. Combes, M. A. Fox and K. P. Johnston, Photochemistry and Photobiology, **54**, 571 (1991).
[34] J. M. Tanko, N. K. Suleman and B. Fletcher, J. Am. Chem. Soc., **118**, 11958 (1996).
[35] D. Andrew, B. T. Des Islet, A. Margaritis and A. C. Weedon, J. Am. Chem. Soc., **117**, 6132 (1995).
[36] P. G. Jessop, T. Ikariya and R. Noyori, Chem. Rev., **99**, 475 (1999).
[37] V. Ipatiev and O. Rutala, Ber., **46**, 1748 (1913).
[38] J. A. Darr and M. Poliakoff, Chem. Rev., **99**, 495 (1999).
[39] D. R. Palo and C. Erkey, J. Chem. Eng. Data, **43**, 47 (1998).
[40] P. G. Jessop, Y. Hsiao, T. Ikariya and R. Noyori, J. Am. Chem. Soc., **118**, 344 (1996).
[41] M. J. Carrott, B. E. Waller, N. G. Smart and C. M. Wai, Chem. Commun., 373 (1998).
[42] M. A-Khorassani and L. T. Taylor, J. Chlomatogr. Sci., **27**, 329 (1989).
[43] M. A-Khorassani, J. W. Hellgeth and L. T. Taylor, Anal. Chem., **59**, 2077 (1987).
[44] U. Kreher, S. Schebesta and D. Walther, Anorg. Alleg. Chem., **624**, 602 (1998).
[45] N. Saito, Y. Ikushima and T. Goto, Bull. Chem. Soc. Jpn., **63**, 1532 (1990).
[46] F. Bickmann, B. Wenclawiak and Z. Fresenius, Anal. Chem., **320**, 261 (1985).
[47] A. F. Lagante, B. N. Hansen, T. J. Bruno and R. E. Sievers, Inorg. Chem., **34**, 5781 (1995).
[48] Y. H. Lin, N. G. Smart, and C. M. Wai, Trends Anal. Chem., **14**, 123 (1995).
[49] A. V. Yazdi and E. J. Beckman, Ind. Eng. Chem. Res., **36**, 2368 (1997).

[50] R. P. Hughes and H. A. Trujillo, Organometallics, **15**, 286 (1996).
[51] S. Kainz, D. Koch, W. Baumann and W. Leitner, Angew. Chem., Int. Ed. Engl., **36**, 1628 (1997).
[52] N. G. Smart, T. Carleson, T. Kast, A. A. Clifford, M. D. Burford and C. M. Wai, Talanta, **44**, 137, (1997).
[53] D. Koch, W. Leitner, J. Am. Chem. Soc., **120**, 13398 (1998).
[54] M. A. Carroll and A. B. Holmes, Chem. Commun., 1395 (1998).
[55] D. K. Morita, D. R. Pesiri, S. A. David, W. H. Glaze and W. Tumas, Chem. Commun., 1397 (1998).
[56] Y. Kayaki, Y. Noguchi, S. Iwasa, T. Ikariya and R. Noyori, Chem. Commun., 1235 (1999).
[57] H. Nishida, N. Takada and M. Yoshimura, Bull. Chem. Soc. Jpn., **57**, 2600 (1984).
[58] M. Brookhart, B. Grant and A. F. Volpe Jr., Organometallics, **11**, 3920 (1992).
[59] M. J. Burk, S. Feng, M. F. Gross and W. Tumas, J. Am. Chem. Soc., **117**, 8277 (1995).
[60] C. M. Cowey, K. D. Bartle, M. D. Burford, A. A. Clifford, S. Zhu, N. G. Smart and N. D. Tinker, J. Chem. Eng. Data, **40**, 1217 (1995).
[61] G. J. McFann, K. P. Johnston and S. M. Howdle, AIChE J., **40**, 543 (1994).
[62] K. A. Bartscherer, M. Minier and H. Renon, Fluid Phase Equilibria, **107**, 93 (1995).
[63] M. J. Clarke, K. L. Harrison, K. P. Johnston and S. M. Howdle, J. Am. Chem. Soc., **119**, 6399 (1997).
[64] K. P. Johnston, K. L. Harrison, M. J. Clarke, S. M. Howdle, M. P. Heitz, F. V. Bright, C. Carlier and T. W. Randolph, Science, **271**, 624 (1996).
[65] A. V. Yazdi, C. Lepilleur, E. J. Singley, W. Liu, F. A. Adamsky, R. M. Enick and E. J. Beckman, Fluid Phase Equilibria, **117**, 297 (1996).
[66] K. A. Consani and R. D. Smith, J. Supercrit. Fluids, **3**, 51 (1990).
[67] Y. Ikushima, N. Saito and M. Arai, J. Colloid Interface Sci., **186**, 254 (1997).
[68] Y. Ikushima, Z. Shervani, N. Saito and M. Arai, J. Colloid Interface Sci., **191**, 177 (1997).
[69] Y. Sun, C. LeBlond, J. Wang and D. G. Blackmond, J. Am. Chem. Soc., **117**, 12647 (1995).
[70] S. Kainz, D. Koch and W. Leitner, "Selective Reactions of Metal Activated Molecules", ed. H. Werner and W. Schreier, Vieweg, Wiesbaden, p. 151 (1998).
[71] H. Coenen, R. Hagen and E. Kriegel, US Patent 4,485,003 (1984).
[72] P. G. Jessop, T. Ikariya and R. Noyori, Organometallics, **14**, 1510 (1995).
[73] Y. Matsui and M. Orchin, J. Organomet. Chem., **244**, 369 (1983).
[74] M. G. Hitzler and M. Poliakoff, Chem. Commun., 1667 (1997).
[75] R. Noyori, "Asymmetric Catalysis in Organic Synthesis", John Wiley and Sons, New York (1994).

[76] M. J. Burk, S. Feng, M. F. Gross and W. Tumas, J. Am. Chem. Soc., **117**, 8277 (1995).
[77] T. Ikariya and R. Noyori, "Transition Metal Catalyzed Reactions", ed. S.-I. Murahashi and S. G. Davies, Blackwell Science, Oxford, p. 1 (1999).
[78] T. Uemura, X. Zhang, K. Matsumura, N. Sayo, H. Kumobayashi, T. Ohta, K. Nozaki and H. Takaya, J. Org. Chem., **61**, 5510 (1996).
[79] D. C. Wynne and P. G. Jessop, Angew. Chem., Int. Ed. Engl., **38**, 1143 (1999).
[80] B. Minder, T. Mallat, K. H. Pickel, K. Steiner and A. Baiker, Catal. Lett., **34**, 1 (1995).
[81] B. Minder, T. Mallat and A. Baiker, "High Pressure Chemical Engineering: Proceedings of the 3rd International Symposium on High Pressure Chemical Engineering", ed. P. R. von Rohr and C. Trepp, Elsevier, Amsterdam, p. 139 (1996).
[82] P. G. Jessop, T. Ikariya and R. Noyori, Chem. Rev., **95**, 259 (1995).
[83] P. G. Jessop, Y. Hsiao, T. Ikariya and R. Noyori, J. Am. Chem. Soc., **116**, 8851 (1994).
[84] P. G. Jessop, Y. Hsiao, T. Ikariya and R. Noyori, J. Chem. Soc., Chem. Commun., 707 (1995).
[85] O. Kröcher, R. A. Köppel and A. Baiker, Chem. Commun., 453 (1997).
[86] T. Suzuki, Y. Kayaki and T. Ikariya, in preparation.
[87] O. Kröcher, R. A. Köppel and A. Baiker, Chem. Commun., 1497 (1996).
[88] J. W. Rathke, R. J. Klingler and T. R. Krause, Organometallics, **11**, 585 (1992).
[89] Y. Guo and A. Akgerman, Ind. Eng. Chem. Res., **36**, 4581 (1997).
[90] S. Gaemers and C. J. Elsevier, Chem. Soc. Rev., **28**, 135 (1999).
[91] R. J. Klingler and J. W. Rathke, J. Am. Chem. Soc., **116**, 4772 (1994).
[92] D. Koch and W. Leitner, J. Am. Chem. Soc., **120**, 13398 (1998).
[93] D. R. Palo and C. Erkey, Ind. Eng. Chem. Res., **37**, 4203 (1998).
[94] I. Bach and D. J. Cole-Hamilton, Chem. Commun., 1463 (1998).
[95] M.G. Hitzler, F. R. Smail, S. K. Ross and M. Poliakoff, Chem. Commun., 359 (1998).
[96] N. Jeong, S. H. Hwang, Y. W. Lee and J. S. Lim, J. Am. Chem. Soc., **119**, 10549 (1997).
[97] N. Jeong, S. H. Hwang, Y. Lee and Y. K. Chung, J. Am. Chem. Soc., **116**, 3159 (1994).
[98] E. Dinjus and R. Fornika, "Applied Homogeneous Catalysis with Organometallic Compounds"; ed. B. Cornils and W. A. Herrmann, VCH, Weinheim, **2**, p. 1048 (1996).
[99] M. T. Reetz, W. Könen and T. Strack, Chimia, **47**, 493 (1993).
[100] J. A. Banister, P. D. Lee and M. Poliakoff, Organometallics, **14**, 3876 (1995).
[101] K. S. Jerome and E. J. Parsons, Organometallics, **12**, 2991 (1993).
[102] H. Borwieck, O. Walter, E. Dinjus and J. Rebizant, J. Organomet. Chem., **570**, 121 (1998).

[103] A. Fürstner, D. Koch, K. Langemann, W. Leitner and C. Six, Angew. Chem., Int. Ed. Engl., **36**, 2466 (1997).
[104] C. D. Mistele, J. M. DeSimone and H. H. Thorp, J. Macromol. Sci., Pure Appl. Chem., **A33**, 953 (1996).
[105] J. G. Hamilton, J. J. Rooney, J. M. DeSimone and C. D. Mistele, Macromolecules, **31**, 4387 (1998).
[106] X.-W. Wu, Y. Oshima and S. Koda, Chem. Lett., 1045 (1997).
[107] S.-I. Murahashi, T. Naota and N. Komiya, Tetrahedron Lett., **36**, 8059 (1995).
[108] D. R. Pesiri, D. K. Morita, W. Glaze and W. Tumas, Chem. Commun., 1015 (1998).
[109] P. G. Jessop, Top. Catal., **5**, 95 (1998).
[110] K. Voigt, U. Schick, F. E. Meyer and A. de Meijere, Synlett., 189 (1994).
[111] Y. Kayaki, Y. Noguchi and T. Ikariya, in preparation.
[112] M. Buback, Angew. Chem., Int. Ed. Engl., **30**, 641 (1991).
[113] J. M. Tanko and J. F. Blackert, Science, **263**, 203 (1994).
[114] K. Yokota and K. Fujimoto, Ind. Eng. Chem. Res., **30**, 95 (1991).
[115] O. Aaltonen and M. Rantakylä, Chemtech, **21**, 240 (1991).
[116] A. J. Russell, E. J. Beckman and A. K. Chaudhary, Chemtech, **24**, 33 (1994).
[117] A. K. Chaudhary, S. V. Kamat, E. J. Beckman, D. Nurok, R. M. Kleyle, P. Hajdu and A. J. Russell, J. Am. Chem. Soc., **118**, 12891 (1996).
[118] A. K. Dillow, S. L. J. Yun, D. Suleiman, D. L. Boatright, C. L. Liotta and C. A. Eckert, Ind. Eng. Chem. Res., **35**, 1801 (1996).
[119] V. S. Bhise, US Patent 4,400,559 (1983).
[120] D. A. Morgenstern, R. M. LeLacheur, D. K. Morita, S. L. Borkowsky, S. Feng, G. H. Brown, L. Luan, M. F. Gross, M. J. Burk and W. Tumas, "Green Chemistry: Designing Chemistry for the Environment", ed. P. T. Anastas and P. T. Williamson, ACS Symposium Series, **626**, American Chemical Society, Washington DC, p. 132 (1996).
[121] K. Chandler, C. W. Culp, D. R. Lamb, C. L. Liotta and C. A. Eckert, Ind. Eng. Chem. Res., **37**, 3252 (1998).
[122] G. B. Jacobson, G. Westerberg, K. E. Markides and B. Långström, J. Am. Chem. Soc., **118**, 6868 (1996).
[123] J. M. DeSimone, Z. Guan and C. S. Elsbernd, Science, **257**, 945 (1992).
[124] J. M. DeSimone, E. E. Maury, Y. Z. Menceloglu, J. B. McClain, T. J. Romack and J. R. Combes, Science, **265**, 356 (1994).
[125] K. A. Shaffer, T. A. Jones, D. A. Canelas and J. M. DeSimone, Macromolecules, **29**, 2704 (1996).
[126] F. A. Adamsky and E. J. Beckman, Macromolecules, **27**, 312 (1994).
[127] R. Fukuzato, Proceedings of the 3rd International Symposium on Supercritical Fluid, ed. M. A. McHugh, John Hopkins University, Baltimore, p. 196 (1991).

6.5 Reactions of Enzymes and Molecular Aggregates in Supercritical Fluids

[1] B. Subramanian and M. A. McHugh, Ind. Eng. Chem. Process Des. Dev., **25**, 1 (1986).
[2] D. M. Pfund, L. L. Lee and H. D. Cochran, Fluid Phase Equilibria, **39**, 161 (1988).
[3] O. Kajomoto, M. Futakami, T. Kobayashi and K. Yamasaki, J. Phys. Chem., **92**, 1347 (1988).
[4] R. Wu, L. L. Lee and H. D. Cochran, Ind. Eng. Chem. Res., **29**, 977 (1990).
[5] S. Kim and K. P. Johnston, AIChE J., **33**, 1603 (1987).
[6] J. C. Erickson, P. Schyns and C. L. Cooney, AIChE J., **36**, 299 (1990).
[7] T. W. Randolph, H. W. Blanch and J. M. Prausnitz, AIChE J., **34**, 1354 (1988).
[8] K. Nakamura, TIBTECH, **8**, 288 (1990).
[9] O. Altonen and M. Rantakyla, CHEMTECH, 240 (1991).
[10] Y. Ikushima, N. Saito and M. Arai, J. Phys. Chem., **96**, 2293 (1992).
[11] Y. Ikushima, N. Saito, M. Arai and K. Arai, Bull. Chem. Soc. Jpn., **64**, 2224 (1991).
[12] N.A. Collins, P. G. Debenedetti and S. Sundaresan, AIChE J., **34**, 1211 (1988).
[13] A. N. Maitra and H. F. Eicke, J. Phys. Chem., **85**, 2687 (1981).
[14] J. K. Thomas, Chem. Rev., **80**, 283 (1980).
[15] D. G. Whitten, J. C. Russell and R. H. Schmehl, Tetrahedron, **38**, 2455 (1982).
[16] J. H. Fendler, J. Chem. Educ., **60**, 872 (1983).
[17] V. Ramamurthy, Tetrahedron, **42**, 5753 (1986).
[18] E. J. Beckmann and R. D. Smith, J. Phys. Chem., **94**, 3729 (1990).
[19] F. A. Adamsky and E. J. Beckmann, Macromolecules, **27**, 312 (1994).
[20] S. Kamat, J. Barrera, E. J. Beckmann and A.J. Russell, Biotech. Bioeng., **40**, 158 (1992).
[21] K. Sonomoto and A. Tanaka, Ann. N.Y. Acad. Sci., **542**, 235 (1988).
[22] P. I. Luisi and B. Straub, "Reverse Micelles", Plenum Press, New York (1984).
[23] J. H. Fendler and E. J. Fendler, "Catalysis in Micellar and Micromolecular Systems", Acadamic Press, London (1975).
[24] J. L. Fulton and R. D. Smith, J. Phys. Chem., **92**, 2903 (1988).
[25] R. D. Smith and J. L. Fulton, J. Phys. Chem., **94**, 781 (1990).
[26] J. P. Blitz, J. L. Fulton and R. D. Smith, J. Phys. Chem., **92**, 2707 (1988).
[27] S. Kim and K. P. Johnston, Ind. Eng, Chem . Res., **26**, 1206 (1987).
[28] D. W. Matson, J. L. Fulton and R. D. Smith, Mater. Lett., **6**, 31 (1987).
[29] T. W. Randolph, H.W. Blanch and J.M. Prausnitz, AIChE J., **34**, 1354 (1988).
[30] K. P. Johnston, K. L. Harrison, M. J. Clarke, S. M. Howdle, M. P. Heitz, F. V. Bright, C. Carlier and T.W. Randolph, Science, **271**, 624 (1996).

[31] M. Boutonnet, C. Anderson and R. Carsson, Acta. Chem. Scand. A, **34**, 639 (1980).
[32] A. Llor and P. Rigny, J. Am. Chem. Soc., **108**, 7533 (1986).
[33] J. Lang, A. Jada and A. Mallians, J. Phys. Chem., **92**, 1946 (1988).
[34] M. Belletete and G. Durocher, J. Colloid Interface Sci. **134**, 289 (1990).
[35] M. Zulauf and H. F. Eicke, J. Phys. Chem., **83**, 480 (1979).
[36] S.-H. Chen, Physica B, **137**, 183 (1986).
[37] M. Kotlarchyk, J. S. Huang and S.-H. Chen, J. Phys. Chem., **89**, 4382 (1985).
[38] R. D. Smith, J. P. Biltz and J. L. Fulton, “Supercritical Fluid Science and Technology”, ACS Symposium Series, **406**, American Chemical Society (1989).
[39] M. J. Politi, O. Brandt and J. H. Fendler, J. Phys. Chem., **89**, 2345 (1985).
[40] Z. Yang, D. Zacherl and A. Russell, J. Am. Chem. Soc., **115**, 12251 (1993).
[41] Y. Ikushima, N. Saito and M. Arai, J. Chem. Eng. Jpn., **29**, 551 (1996).
[42] Y. Ikushima, N. Saito, T. Yokoyama, K. Hatakeda, S. Ito, M. Arai and H.W. Blanch, Chem. Lett., 109 (1993).
[43] Y. Ikushima, N. Saito and M. Arai, J. Phys. Chem., **99**, 8941 (1995).
[44] S. Krimm and J. Bandekar, Adv. Protein Chem., **38**, 181 (1986).
[45] L. Siquano, C.-Y. Huang, F. McClelland and D. J. Graves, Anal. Biochem., **216**, 67 (1994).
[46] C. L. Wilder, A. D. Friedrich, R. O. Potts, G. O. Daumy and M. L. Francoeur, Biochemistry, **31**, 27 (1992).
[47] P. Grochulski, Y. Li, J.D. Schrag and M. Cygler, Protein Sci., **3**, 82 (1994).
[48] J. D. Schrag, Y. Li, S. Wu and M. Clyger, Nature, **351**, 761 (1991).
[49] H. van Tilbeutgh, M.-P. Egloff, C. Martinez, N. Rugani, R. Verger and C. Cambillau, Nature, **362**, 814 (1993).
[50] C. Chapus and M. Semeriva, Biochemistry, **15**, 4988 (1976).
[51] E.D. Niemeyer and F.V. Bright, J. Phys. Chem. B **102**, 1474 (1998).
[52] E. J. Beckmann, J. L. Fulton, D. W. Matson and R. D. Smith, “Supercritical Fluid Science and Technology”, ACS Symposium Series, **406**, American Chemical Society (1989).
[53] P. G. Jessop, T. Ikariya and R. Noyori, Science, **269**, 1065 (1995).
[54] J. B. Ellington and J. F. Brennecke, J. Chem. Soc. Chem. Commun., 1094 (1993).
[55] K. Tamura and N. Nii, J. Phys. Chem., **93**, 4825 (1989).
[56] Y. Ikushima, N. Saito and M. Arai, J. Colloid Interface Sci., **186**, 254 (1997).
[57] Y. Ikushima, Adv. Colloid Interface Sci., **71–72**, 259 (1997).
[58] Y. Ikushima, Z. Shervani, N. Saito and M. Arai, J. Colloid Interface Sci., **191**, 177 (1997).

6.6 Heterogeneous Catalysis in Supercritical Fluids

[1] H. Koelbel and M. Ralek, Catal. Rev. Sci. Eng., **21**, 225 (1980).
[2] D. Stern, A. T. Bell and H. Heinemann, Chem. Eng. Sci., **38**, 597 (1983).
[3] K. Yokota and K. Fujimoto, Ind. Eng. Chem. Res., **30**, 95 (1991).
[4] L. Fan, K. Yokota and K. Fujimoto, AIChE J, **38**, 1639 (1992).
[5] L. Fan, K. Yokota and K. Fujimoto, Topics in Catal., **2**, 267 (1995).
[6] X. Lang, A. Akgerman and D. B. Bukur, Ind. Eng. Chem. Res., **34**, 72 (1995).
[7] D. B. Bukur, X. Lang, A. Akgerman and Z. Feng, Ind. Eng. Chem. Res., **36**, 2580 (1997).
[8] R. B. Anderson, "The Fischer-Tropsch Synthesis", Academic Press, Orlando, FL (1984).
[9] K. Fujimoto, L. Fan and K. Yoshii, Topics in Catal., **2**, 259 (1995).
[10] K. Fujimoto, K. Ohdan and K. Yoshii, Jpn. Kokai Tokkyo Koho 1992-181293 (1992), 1993-138534 (1993) and 1993-138535 (1993).
[11] M. Hitzler, F. Smail, S. Ross and M. Paliakoff, Chem. Comm. 359 (1998).
[12] C. Vieville, Z. Mouloungui and A. Gaset, Ind. Eng. Chem. Res., **32**, 2065 (1993).
[13] B. Minder, T. Mallat, K. Pickel, K. Steiner and A. Baiker, Catal. Letters, **34**, 1 (1995).
[14] O. Krocher, R. Koppel and A. Baiker, Chem. Comm., 1497 (1996).
[15] M. Hitzler and M. Paliakoff, Chem. Comm. 1667 (1997).
[16] K. Dooley and F. Knopf, Ind. Eng. Chem. Res., **26**, 1910 (1987).
[17] M. F. M. Post, A. C. Hoog, J. K. Minderhoud and S. T. Sie, AIChE J., **35**, 1107 (1989).
[18] P. G. Debenedetti and R. C. Reid, AIChE J., **32**, 2034 (1986).
[19] C. F. Wilke and P. Chang, AIChE J., **1**, 264 (1955).
[20] Japan Society of Chemical Engineers, "Handbook of Chemical Engineering", 5th ed. Maruzen, Tokyo (1988).
[21] A. Vignes, Ind. Eng. Chem. Fundam., **5**, 189 (1966).
[22] S. Baptist-Nguyen and B. Subramaniam, AIChE J., **38**, 1027 (1992).
[23] B. Subramaniam and B. J. McCoy, Ind. Eng. Chem. Res., **33**, 504 (1994).
[24] B. J. McCoy and B. Subramaniam, AIChE J., **40**, 131 (1994).

Appendices

Appendix-A : Critical and basic constants for selected fluids

substance		molar mass M [g/mol]	critical temp. T_C [K]	critical pres. P_C [MPa]	critical dens. ρ_C [kg/m^3]	acentric factor ω	dipole moment μ [debye]
methane	CH_4	16.043	190.55	4.60	162	0.011	0.0
ethylene	C_2H_4	28.054	282.4	5.08	218	0.089	0.0
fluoroform	CHF_3	70.014	299.3	4.83	525	0.260	1.6
carbon dioxide	CO_2	44.010	304.18	7.38	468	0.239	0.0
ethane	C_2H_6	30.069	305.3	4.87	205	0.099	0.0
propane	C_3H_8	44.096	369.8	4.25	217	0.153	0.0
ammonia	NH_3	17.031	405.41	11.35	240	0.250	1.5
methanol	CH_3OH	32.042	512.6	8.10	272	0.556	1.7
ethanol	C_2H_5OH	46.069	516.2	6.38	276	0.644	1.7
benzene	C_6H_6	78.113	562.1	4.90	302	0.212	0.0
toluene	C_7H_8	92.140	591.8	4.11	292	0.263	0.4
water	H_2O	18.015	647.096	22.064	322	0.344	1.8
naphthalene	$C_{10}H_8$	128.173	748.4	4.11	314	0.302	0.0

Appendix-B : Comparison of the range of solvent properties between gas, liquid and supercritical fluid (SCF) [M. L. Lee and K. E. Markides, *Science*, 235, 1342 (1987)]

state		gas (1atm, 25 °C)	SCF (T_C, P_C)	SCF (T_C, $4P_C$)	liquid (1atm, 25 °C)
density	ρ (kg/m^3)	0.6 ~ 2	200 ~ 500	400 ~ 900	600 ~ 1,600
viscosity coefficient	$\eta \times 10^5$ (kg/m·s)	1 ~ 3	1 ~ 3	3 ~ 9	20 ~ 300
diffusion coefficient	$D \times 10^8$ (m/s)	1000 ~ 4000	5 ~ 40	1 ~ 10	0.02 ~ 0.2
thermal conductivity	$\lambda \times 10^3$ (W/ m·K)	4 ~ 30	20 ~ 80	40 ~ 150	80 ~ 250

Appendix-C : Conversion table of units of pressure

	Pa	MPa	bar	atm	psi
1 Pa	1	10^{-6}	10^{-5}	9.86923×10^{-6}	1.45038×10^{-4}
1 MPa	10^{6}	1	10	9.86923	145.038
1 bar	10^{5}	0.1	1	0.986923	14.5038
1 atm	101325	0.101325	1.01325	1	14.6960
1 psi	6894.76	6.89476×10^{-3}	6.89476×10^{-2}	6.80460×10^{-2}	1

Subject Index

Printing (Computer to Film): Saladruck Berlin
Binding: Stürtz AG, Würzburg

GPSR Compliance

The European Union's (EU) General Product Safety Regulation (GPSR) is a set of rules that requires consumer products to be safe and our obligations to ensure this.

If you have any concerns about our products, you can contact us on ProductSafety@springernature.com

In case Publisher is established outside the EU, the EU authorized representative is:

Springer Nature Customer Service Center GmbH
Europaplatz 3
69115 Heidelberg, Germany

Batch number: 10755284

Printed by Printforce, the Netherlands